PROCESSING AND USE OF ORGANIC SLUDGE AND LIQUID AGRICULTURAL WASTES

Proceedings of the Fourth International Symposium held in Rome,
8–11 October 1985 and jointly organized by
– National Research Council of Italy
  Water Research Institute
and
– Commission of the European Communities
  Directorate-General for Science, Research and Development
  Environment Research Proramme

Commission of the European Communities

PROCESSING AND USE OF ORGANIC SLUDGE AND LIQUID AGRICULTURAL WASTES

Proceedings of the Fourth International Symposium held in Rome, Italy, 8–11 October 1985

Edited by

P. L'HERMITE

Commission of the European Communities,
Directorate-General Science, Research and Development, Brussels, Belgium

D. REIDEL PUBLISHING COMPANY

A MEMBER OF THE KLUWER ACADEMIC PUBLISHERS GROUP

DORDRECHT / BOSTON / LANCASTER / TOKYO

Library of Congress Cataloging in Publication Data

Processing and use of organic sludge and liquid agricultural wastes.

    (EUR 10361)
    At head of title: Commission of the European Communities.
    "Symposium . . . jointly organized by National Research Council of Italy, Water
Research Institute, and Commission of the European Communities, Directorate-General for
Science, Research, and Development, Environment Research Programme"—Prelim. p.
    Includes index.
    1. Sewage sludge—Congresses. 2. Agricultural wastes—Congresses. 3. Organic
wastes—Congresses. I. L'Hermite, P. (Pierre), 1936‑ . II. Commission of the Euro-
pean Communities. Environment Research Programme. III. Istituto di ricerca sulle acque
(Italy) IV. Series.
TD767.P75 1986   628.3'6   86–20210

ISBN 978-94-010-8613-4   ISBN 978-94-009-4756-6 (eBook)
DOI 10.1007/978-94-009-4756-6

Publication arrangements by
Commission of the European Communities
Directorate-General Information Market and Innovation, Luxembourg

EUR 10361
© 1986 ECSC, EEC, EAEC, Brussels and Luxembourg

LEGAL NOTICE
Neither the Commission of the European Communities nor any person acting on behalf of the
Commission is responsible for the use which might be made of the following information.

Published by D. Reidel Publishing Company
P.O. Box 17, 3300 AA Dordrecht, Holland

Sold and distributed in the U.S.A. and Canada
by Kluwer Academic Publishers,
101 Philip Drive, Assinippi Park, Norwell, MA 02061, U.S.A.

In all other countries, sold and distributed
by Kluwer Academic Publishers Group,
P.O. Box 322, 3300 AH Dordrecht, Holland

PREFACE

Disposal of organic sludge and liquid agricultural wastes is a universal problem. Their production cannot be halted and as steps are taken to maintain or improve the quality of rivers and lakes it grows in quantity. The Commission's early awareness of the need for action to prepare for substantial growth in the Community's sludge disposal problem led to the setting up of the COST 68 project to coordinate and guide European research and development work with particular emphasis on recycling sludge to agricultural land. Two years ago the field of research activities was extended to liquid agricultural wastes. This Symposium is the latest opportunity to provide a comprehensive review of the results of the project, to define current trends in practice and to establish by discussion the priorities for research over the next few years.

The development of instrumentation and of analytical techniques during the period has extended our knowledge of the organic and inorganic constituents of sewage sludge and agricultural wastes and enabled us more readily to identify and measure the risks to which our general environment may be exposed when disposing of it. This evolution of understanding is a continuing process and an essential guide to the modification of disposal practices to achieve safer and more efficient operations.

However, it is important to take a broad view of the application of research findings in the light of the considerable contrast in conditions in different parts of the world. Even in the European Community there are wide variations in climate, agricultural practices, population density and industrial capacity. Conclusions from research may have application in some areas but not in others, and practices proved over time to be satisfactory in some regions may not be of general application because of these variations.

The agricultural disposal of sludge and agricultural wastes to land can supply nutrients to only a small part of the productive land and although of great value to farmers it has no major impact on national agricultural operations.

It is the aim in disposing of sludge to protect public health and provide a good environment at realistic cost by utilising best practice based on sound scientific knowledge. Operations should be carried out under controlled conditions with provision for adequate monitoring by sampling and analysis and one finds that the authorities responsible for waste water treatment and sewage sludge treatment are well fitted to provide the scientific and technical organisation needed for supervisory control and to give farmers adequate information on the quality of sludge being spread on their land. It is the correlation of this practical experience with new information from research work which can build up systems of guidance.

The original objectives of the Symposium, jointly organised by the Commission of the European Communities and the Water Research Institute of the National Research Council of Italy, were to give a comprehensive view on completed research in Europe in general, and more specifically on the results obtained from COST 681 "Treatment and Use of Organic Sludge and Liquid Agricultural Wastes" in the last two years.

P. L'HERMITE

October 1985

C O N T E N T S

Preface   v

SESSION 1 : PROCESSING AND COSTS

Overview of sewage sludge treatment processes : An outlook
    U. MOLLER, Ruhr-Universität Bochum, Federal Republic of
    Germany   2

Treatment processes for farm wastes
    V.C. NIELSEN, Farm Waste Unit, A.D.A.S., Ministry of
    Agriculture, Fisheries and Food, United Kingdom   25

Stabilisation of sewage sludges and liquid animal manures by
mesophilic anaerobic digestion - an overview
    A.M. BRUCE, Water Research Centre, Stevenage, Herts, United
    Kingdom   39

"State of the Art" on sludge composting
    L.E. DUVOORT-VAN ENGERS, National Institute of Public
    Health and Environmental Hygiene (RIVM-LAE), The
    Netherlands and S. COPPOLA, Istituto di microbiologia
    agraria e statione di microbiologia industriale, Portici,
    Italy   59

Optimization of processing, managing and disposing of sewage
sludge   76
Part I : Treatment processes models
    L. SPINOSA, V. LOTITO, Istituto di Ricerca sulle Acque,
    Bari, Italy   77
Part II : Disposal options
    A.C. DI PINTO and G. MININNI, Istituto di Ricerca sulle
    Acque, Roma, Italy   86

Alternative uses of sewage sludge
    R.C. FROST, WRc Processes, Stevenage, Hertfordshire, United
    Kingdom and H.W. CAMPBELL, Wastewater Technology Centre,
    Environment Canada, Burlington, Ontario, Canada   94

SESSION 2 : CHARACTERIZATION OF SLUDGE

Sampling techniques for sludge, soil and plants
    A. GOMEZ, I.N.R.A., Station d'Agronomie, Centre de
    Recherches de Bordeaux, France; R. LESCHBER, Institut für
    Wasser-, Boden-, und Lufthygiene des Bundesgesund-
    heitsamtes, Berlin, Federal Republic of Germany and
    F. COLIN, Institut de Recherches Hydrologiques (I.R.H.),
    Nancy, France   112

Abundance and analysis of PCBs in sewage sludges
    J. TARRADELLAS, Institut du Génie de l'Environnement, Ecole
    Polytechnique Fédérale de Lausanne, Switzerland; H. MUNTAU,
    Commission of the European Communities, Joint Research
    Centre Ispra, Italy and H. BECK, Bundesgesundheitsamt,
    Berlin, Federal Republic of Germany   124

Chemical methods for the biological characterization of metal in
sludge and soil
    H. HÄNI and S. GUPTA, Swiss Federal Research Station for
    Agricultural Chemistry and Hygiene of Environment,
    Liebefeld-Berne, Switzerland   157

Characterization of the physical nature of sewage sludge with
particular regard to its suitability as landfill
    U. LOLL, Abwasser - Abfall - Aquatechnik, Darmstadt,
    Federal Republic of Germany   168

Microbiological specifications of disinfected sewage sludge
    D. STRAUCH, Institute of Animal Medicine and Animal
    Hygiene, University of Hohenheim, Stuttgart, Federal
    Republic of Germany and M. DE BERTOLDI, Institute of
    Agricultural Microbiology, University of Pisa, Study Center
    for Soil Microbiology, C.N.R., Pisa, Italy   178

Inactivation of parasitic ova during disinfection and
stabilisation of sludge
    E.B. PIKE and E.G. CARRINGTON, Water Research Centre,
    Medmenham, Marlow, United Kingdom   198

Epidemiological studies related to the use of sewage sludge in
agriculture
    A.H. HAVELAAR, National Institute of Public Health and
    Environmental Hygiene, Bilthoven, The Netherlands and
    J.C. BLOCK, Centre des Sciences de l'Environnement,
    Université de Metz, France   210

Transport of viruses from sludge application sites
    P.H. JØRGENSEN and E. LUND, Department of Veterinary
    Virology and Immunology, Royal Veterinary and Agricultural
    University of Copenhagen, Frederiksberg, Denmark   215

Health risks of microbes and chemicals in sewage sludge applied
to land - recommendations to the world health organisation
    P.J. MATTHEWS, European Water Pollution Control Association
    (EWPCA) and Anglian Water Huntingdon, United Kingdom;
    E. LUND, Royal Veterinary and Agricultural University,
    Copenhagen, Denmark and R. LESCHBER, Institute of Water,
    Soil and Air Hygiene, Federal Health Office, Berlin,
    Federal Republic of Germany   225

SESSION 3 : MANAGEMENT CONSIDERATIONS IN SLUDGE AND EFFLUENT DISPOSAL

Recommendations on olfactometric measurements
    M. HANGARTNER, Institut für Hygiene und Arbeitsphysiologie,
    Zürich, Switzerland; J. HARTUNG, Institut für Tierhygiene,
    Hannover, Federal Republic of Germany and J.H. VOORBURG,
    Government Agricultural Wastewater Service, Arnhem, The
    Netherlands   236

Manure nutrient composition : rapid methods of assessment
    H. TUNNEY, The Agricultural Institute, Johnstown Castle
    Research Centre, Wexford, Ireland   243

Efficiency of utilisation of nitrogen in sludges and slurries
    J.H. WILLIAMS, Ministry of Agriculture, Fisheries and Food,
    Wolverhampton, West Midlands, United Kingdom, and
    J.E. HALL, Water Research Centre, Medmenham, Bucks, United
    Kingdom   258

Effect of a long term sludge disposal on the soil organic matter
characteristics
    M. LINERES, C. JUSTE, J. TAUZIN and A. GOMEZ, I.N.R.A.,
    Station d'Agronomie, Centre de Recherches de Bordeaux,
    France   290

Contamination problems in relation to land use
    G.A. FLEMING, Plant Nutrition and Biochemistry Department,
    An Foras Taluntais, Johnstown Castle Research Centre,
    Wexford, Ireland, and R.D. DAVIS, Environmental Impact
    Group, Water Research Centre, Medmenham, Buckinghamshire,
    United Kingdom   304

Long-term effects of contaminants
    D.R. SAUERBECK and P. STYPEREK, Institute of Plant
    Nutrition and Soil Science, Federal Research Center of
    Agriculture Braunschweig-Völkenrode, Federal Republic of
    Germany   318

Effect on a long term sludge disposal on cadmium and nickel
toxicity to a continuous maize crop
    C. JUSTE and P. SOLDA, Station d'Agronomie, I.N.R.A., Pont
    de la Maye, France   336

Trace metal regulations for sludge utilization in agriculture; a
critical review
    J.C. TJELL, Department of Environmental Engineering,
    Technical University of Denmark, Lyngby, Denmark   348

POSTER SESSION

Leaching resulting from land application of sewage sludge
    M. MELKAS and M. MELANEN, Water Research Institute,
National Board of Waters, Helsinki, Finland; A. JAAKKOLA,
Department of Agricultural Chemistry, University of
Helsinki, Finland and M. AHTIAINEN, North Karelia Water
District Office, Joensuu, Finland   366

Germination tests for the determination of sludges agricultural
value
    L. BARIDEAU, Groupe Valorisation des Boues, Faculté des
Sciences Agronomiques, Gembloux, Belgique   374

Utilization of sewage sludge as fertilizer in energy plantations
on peatland
    T. BRAMRYD, Depatment of Plant Ecology, University of Lund,
Ecology Building, Lund, Sweden   381

The use of composted agricultural waste as peat substitute in
horticulture
    Y. CHEN, Dept. of Soil and Water Sciences and Y. HADAR,
Dept. of Plant Pathology and Microbiology, The Hebrew
University of Jerusalem, Israel   389

Labscale simulation of isman-cotton batch phase anaerobic
digestion for prediction of plant design and performance
    P. AMMANN, Ecole Polytechnique Fédérale de Lausanne Génie
biologique, Lausanne, Switzerland; A. COTTON, A. Cotton
S.A., Vandoeuvres-Genève, Switzerland and N. MAIRE, ACEPSA,
Oulens, Switzerland   393

Solid phase anaerobic digestion of farm wastes mixed with sewage
sludge
    A. COTTON, Biogaz-Bureau d'études techniques et
réalisations, A. Cotton S.A., Vandoeuvres, Switzerland   402

Aerobic stabilization in the solid state of partially dewatered
sewage sludge
    S. COPPOLA and F. VILLANI, Istituto di Microbiologia
agraria e Stazione di Microbiologia industriale, Portici,
Italy and F. ROMANO, Istituto di Meccanica agraria,
Università degli Studi di Napoli, Portici, Italy   407

Environmental impact assessment of agricultural use of sewage
sludge
    E. DE FRAJA FRANGIPANE, R. VISMARA and V. MEZZANOTTE,
Istituto di Ingegneria Sanitaria del Politecnico di Milano   414

Analytical characterization of biological sewage sludges
especially about some organic compounds
    P.L. GENEVINI, Istituto di Chimica Agraria, Università di
Milano, Italy, and M.C. NEGRI, Ecodeco s.p.a., Giussago,
Pavia, Italy   427

Sludge composting and utilization in West Switzerland
    P. REGAMEY and G. HUBERT, Institut de Génie Rural,
    E.P.F.L., Lausanne, Switzerland   431

Productivity and quality of cereal crops grown on sludge-treated
soils
    M. CONSIGLIO and R. BARBERIS, Istituto per le Piante da
    Legno e l'Ambiente, Torino, Italy; G. PICCONE and
    G. DE LUCA, Istituto di Chimica Agraria, Università di
    Torino, Italy and A. TROMBETTA, Assessorato Ambiente ed
    Energia, Regione Piemonte, Torino, Italy   436

Biological evaluation of sludge phytotoxicity
    P. NAPPI, R. BARBERIS, M. CONSIGLIO and R. JODICE, Istituto
    per le Piante da Legno e l'Ambiente, Torino, Italy and
    A. TROMBETTA, Assessorato all' Ambiente ed Energia, Regione
    Piemonte, Torino, Italy   441

The fate and behaviour of selected heavy metals during pyrolysis
of sewage sludge
    R.C. KISTLER, F. WIDMER and P.H. BRUNNER, Swiss Federal
    Institute of Technology, Zürich, Switzerland   448

Latest test results in sludge dewatering with the CHP-filter
press
    U. LOLL, Abwasser-Abfall-Aquatechnik, Darmstadt, Federal
    Republic of Germany   452

A pot experience with a high level copper sewage sludge
    A. MANRIQUE, I. ARROYO, A.M. SANZ, J.M. GARCIA DE BUSTOS
    and A.M. NEBREDA, Seccion de Analisis Ambiental, Servicio
    de Investigaciones Agrarias, Junta de Castilla y Leon,
    Spain   460

Chemical properties of sewage sludges produced in the Piedmontese
area (Italy)
    G. PICCONE, C. SAPETTI, B. BIASIOL, G. DE LUCA and
    F. AJMONE MARSAN, Istituto di Chimica agraria, University
    of Turin, Italy and A. TROMBETTA, Assessorato Ambiente ed
    Energia, Regione Piemonte, Italy   465

Autothermic sludge incineration
    P. OCKIER, W.Z.K., Coastal Water Authority, Ostend, Belgium   470

Modification of heavy metals solubility in soil treated with
sewage sludge
    G. PETRUZZELLI, G. GUIDI and L. LUBRANO, Institute of Soil
    Chemistry C.N.R., Pisa, Italy   478

Methods of treating pig slurry to increase the volumes which can
be used on crops
    S. PICCININI, L. CORTELLINI and G. BONAZZI, Centro Ricerche
    Produzioni Animali, Reggio Emilia, Italy   485

Chemical composition of sludges from sewage treatment systems in
the Emilia Romagna region
    N. ROSSI, R. RASTELLI and P.L. GRAZIANO, Istituto Chimica
    Agraria, Università Bologna, Italy and N. DE MARTIN, Ente
    Regionale Sviluppo Agricolo per l'Emilia Romagna, Bologna,
    Italy   491

Application of $CaCl_2$-extraction for assessment of cadmium and
zinc mobility in a wastewater-polluted soil
    C. SALT, Institute for Ecology, Botanical Section,
    Technical University, Berlin, Federal Republic of Germany
    and A. KLOKE, Federal Biological Research Centre for
    Agriculture and Forestry, Berlin, Federal Republic of
    Germany   499

Anaerobic contact digestion of biochemical sludge (from
simultaneous precipitation) - results of a semi full-scale study
    U. TANTTU, Plancenter Ltd, Finland   505

The alternative "earthworm" in the organic wastes recycle
    U. TOMATI, A. GRAPPELLI and E. GALLI, Institute of
    Radiobiochemistry and Plant Ecophysiology, National
    Research Council, Monterotondo Scalo, Italy   510

Slurry meter instructions
    H. TUNNEY, The Agricultural Institute, Wexford, Ireland   517

Composting of sewage sludge containing polyelectrolytes
    J.J. VAN DEN BERG, Grontmij n.v., De Bilt , The Netherlands   518

An alternative application for sewage sludge : black earth
    J.J. VAN DEN BERG, Grontmij n.v., De Bilt, the Netherlands   523

Toxic organic substances in sewage sludges : case study of
transfer between soil and plant
    M.A. WEGMANN, R. CH. DANIEL, H. HÄNI, A. IANNONE, Swiss
    federal research station for agricultural chemistry and
    hygiene of environment, Liebefeld-Bern, Switzerland   532

Varietal tolerance in cereals to metal contamination in a sewage
treated soil
    J.H. WILLIAMS, K.A. SMITH and J.R. JONES, Soil Science
    Department, MAFF, Wolverhampton, West Midlands, United
    Kingdom   537

Effect of sample storage on the extraction of metals from raw,
activated and digested sludges
    J.V. TOWNER, J.A. CAMPBELL and R.D. DAVIS, Environmental
    Impact Group, Water Research Centre, Medmenham,
    Buckinghamshire, United Kingdom   543

DISCUSSION   548

SESSION REPORTS

Processing and costs
    A.M. BRUCE, Water Research Centre, Stevenage, United
    Kingdom   552

Characterization of sludge
    A.H. HAVELAAR, National Institute of Public Health,
    Bilthoven, The Netherlands and E.B. PIKE and R.D. DAVIS,
    Water Research Centre, Environmental Directorate,
    Medmenham, United Kingdom   556

Management considerations in sludge and effluent disposal
    J.E. HALL and R.D. DAVIS, Water Research Centre,
    Environment Directorate, Medmenham, United Kingdom   559

CONCLUSIONS AND RECOMMENDATIONS   563

LIST OF PARTICIPANTS   567

INDEX OF AUTHORS   575

SESSION 1 : PROCESSING AND COSTS

Overview of sewage sludge treatment processes : An outlook

Treatment processes for farm wastes

Stabilisation of sewage sludges and liquid animal manures by mesophilic anaerobic digestion - an overview

"State of the Art" on sludge composting

Optimization of processing, managing and disposing of sewage sludge

Alternative uses of sewage sludge

OVERVIEW OF SEWAGE SLUDGE TREATMENT PROCESSES : AN OUTLOOK

Professor Ulrich Möller
Ruhr-Universität Bochum

Summary

English abstract of Professor Möller's paper "General review of treatment processes for sewage sludge; looking forward into the future" (abstract prepared by editors).
The paper gives an over-view of sludge disposal routes and of stabilization, dewatering and disinfection techniques. It is predicted that future legislation will necessitate a closer and more critical examination of disposal routes and lead to the introduction of increasingly sophisticated techniques of treatment. As an example, disposal to landfills would require dewatered sludges of a minimum solids content. Sea-disposal of sludge would be unacceptable and recycling on land or disposal by tipping would be the only options in the future. Following the practice of the Federal Republic of Germany, future legislation would specify sludge disinfection (at least a reduction of pathogens e.g. salmonella).

1. Treatment, including pretreatment, as a prerequisite of safe disposal

Purification of wastewater (sewage) means first and foremost the removal or elimination of polluting matter [1] (Figure 1).
Safe disposal of the residues, i.e. disposal avoiding harm to the environment, is thus an integral part of sewage treatment and water pollution control. In other words, final disposal of the residues must be accomplished in such a way that they can have no prejudicial effects on the environment.
Of the residues from sewage treatment, sludge presents the greatest problem. The draft of the new German standard DIN 4045, Waste Water Engineering Terms, defines sewage sludge as the water-containing solid matter excluding any screenings, sievings and grit which can be separated from the sewage [2].
As early as 1951, Dr. Karl Imhoff stated, in the 14th edition of his "Taschenbuch der Stadtentwässerung" (Municipal Sewerage Handbook) [3], that "it is essential that sewage treatment should result in the complete disposal of the sludge. A plant which fails to accomplish this is worthless... the sludge can be finally disposed only by application to agricultural land, dumping in the sea or on a landfill...". Presently, the term "controlled tip(ping)" is preferred.
In Germany and in certain other countries, dumping of sludge at sea has been prohibited for some time now, but in others, it is still disposed of in this manner. Around 24 % of sewage sludge in the United Kingdom was disposed of by ocean dumping in 1981 [4] and there are even plans to increase the volume substantially.
According to reports from the United Kingdom, this method has not caused any harm to the environment [4], but I personally feel that it is no longer acceptable since sludge discharged to the sea also contains

potentially toxic elements which are dispersed in a manner that is not amenable to long-term control.

Thus, sewage sludge can be disposed of safely only by :
(i)   utilizing it in some way, in particular by returning it to the natural cycle, or
(ii) withdrawing it from the cycle by means of dumping, i.e. controlled tipping [1], (Figure 2).

However, owing to its harmful properties and their potentially undesirable and deleterious effects, sewage sludge cannot be disposed of in the raw state. It must first undergo some suitable form of pretreatment, the aim being to render it fit for disposal, i.e. to alter and improve its characteristics through various processes so that the ultimate means of disposal selected has no detrimental effects on the environment. Given this objective, the type of treatment and its extent and result are geared to the requirements of the disposal method.

The properties and composition of raw sewage sludge can have undesirable and harmful effects. These characteristics include :
(i)   a high proportion of digestible organic matter,
(ii) a high water content,
(iii) the presence of a critically high load of pollutants,
(iv) the presence of pathogens.
(See Table 1).

This state of affairs will not change significantly and the undesirable or harmful effects of sewage sludge will be subject to much closer and more critical scrutiny in the future. It is therefore to be expected that treatment and the techniques employed will have to comply with much more stringent requirements, especially as regards quality of the sludge and reliability of the techniques used. Treatment processes in general will have to be more sophisticated. This will result in the application of structured processes tailored very much to the requirements of the particular method of safe disposal selected, and it goes without saying that treatment techniques and processes will have to be optimized both technically and economically.

2. Basic steps

The process of altering and improving the properties and composition of sewage sludge comprises the following basic steps :
(i)   stabilization
(ii) separation of the sludge liquor, and
(iii) disinfection of the sludge.

Stabilization is a crucial step, both where the treated sludge is to be spread on farmland or where tipping is the method of disposal chosen [5]. As a rule, tipping should not take place until the liquid content of the sludge has been reduced to a satisfactory level.

Where the sludge is to be used in agriculture, forestry or horticulture, disinfection is often essential and the concentration of potentially toxic substances must in any case be suitably low.

These basic steps must form part of any sludge treatment and disposal system or process. The choice of steps and their chronological order are variable, as illustrated in Figure 2.

Future treatment and disposal techniques will also have to take account of the prerequisites enumerated above.

3. Stabilization
The organic elements of the raw sludge resulting from treatment

processes are not stabilized. Their constituents, which under natural environmental conditions are degraded as a rule by the action of microorganisms, undergo an uncontrolled - but only partial - anaerobic decomposition process, known as acid digestion. The critical products of this digestion are low-molecular organic acids such as butyric acid and acetic acid, which can be extremely malodorous.

Inadequately stabilized sludge can also give rise to other aesthetically or hygienically undesirable conditions.

So one of the key objectives of stabilization of sewage sludge or its critical constituents by means of controlled processes is to prevent the uncontrolled anaerobic partial decomposition of the organic substances.

The purpose of controlled stabilization is to reduce the organic matter content susceptible to putrefactive bacteria and/or alter its characteristics so that the sludge can no longer be a nuisance to the environment. The degradation and transformation processes associated with sludge stabilization generally involve conversion of the constituents from an unstable, generally high-molecular and energy-rich form into a stable, generally low-molecular state with a low energy content. Transformation of the constituents into inert substances (mineralization) is the outcome of these processes.

Sludge can thus be stabilized by means of biological (aerobic or anaerobic), physical and/or chemical and, in particular, thermal decomposition and transformation processes (Table 2).

By contrast, stabilization by the addition of lime involves temporary inhibition of the degradation of the organic substances by putrefactive bacteria. This is achieved by raising the pH to a value above 10.

The primary objective of the stabilization process is to stabilize the substrate, with a substantial reduction in any constituents liable to produce odours. Secondary objectives include :
(i)   a reduction in the volume of sludge (solids),
(ii)   enhancement of the dewaterability of the sludge,
(iii) reduction of pathogens,
(iv)   production of biogas (only under anaerobic conditions), and
(v)   creation of storage capacity.

The objectives of the process, and thus the degree of stabilization, should be geared to the destination of the sludge, be it further treatment, utilization or disposal.

The key objectives of the stabilization process (see Table 3) are thus :
(i)   application to agricultural land (in liquid form),
(ii)   application to agricultural land in dewatered state,
(iii) storage of liquid sludge in sludge lagoons,
(iv)   tipping in dewatered state,
(v)   tipping of residues after incineration.

Table 4 shows the relevant stabilization parameters, together with the corresponding degrees of stabilization observed by various authors. The values for loss on ignition are in correspondence [8] with the original proposal [7] , while the other parameters correspond to the definitions given by the authors cited [9], [10], [11], [12]. Experience shows that the range of values for acetic acid and fatty acids content are achieved with most types of sludge after a shorter period of stabilization than for loss on ignition or the BOD/COD ratio. This is particularly true in the case of "full stabilization" (i.e. the maximum degree of stabilization technically attainable, or the "practical digestion limit") under anaerobic conditions. (See also [2]).

Compared with the other parameters, "loss on ignition " has the advantage of being more straightforward and less demanding in terms of analytical effort. The results achieved are often not reliable enough.

In practice, stabilization of the constituents of sewage sludge is effected primarily by biological techniques (Figure 3).

4. Separation of the sludge liquor

With increasing concentration of the solid matter, or reduction in the water content, not only the volume of the sludge but also its properties are altered (Table 5/Figure 4). From the point of view of volume, concentration of the solids beyond 30 %, i.e. reducing the water content to under 70 %, is of little practical use, but when it comes to further treatment and disposal of the sludge it could be important to withdraw enough liquor to eliminate the sticky properties of the sludge in particular. Table 5 indicates the physical characteristics of municipal sewage sludges with varying solids/water content and an insignificant industrial effluent component.

The solids/water content of sewage sludge is not really a good yardstick for determining its suitability for tipping. However, in the absence of a technique for measuring and assessing the suitability of sewage sludge for tipping, the approach in Germany has been to use water content as a criterion for assessing whether the sludge can be tipped together with normal refuse 14 . So a new method has now been developed in Germany to determine the suitability of sewage sludge for tipping 15 , 16 .

The reason for the high water content of sewage sludge, and thus the technical problems involved in separating the water from the solid matter, is its considerable capacity to bind readily, both physically and chemically, to the solid matter. The extent of this waterbinding capacity is a function of the constituents of the sludge.

The organic constituents of sludge bind very readily to water, but the colloidal and gel-like constituents present in secondary sludges from activated sludge plants and hydroxide sludges in particular bind more readily still, thus contributing substantially to the considerable waterbinding capacity of such sludges.

This also explains why different types of sludges with quite different composition and binding capacity are produced by the processes used at each individual sewage treatment stage to eliminate the various constituents.

It is generally acknowledged that organic matter accounts for between 65 and 70 % of the solids present in raw sewage sludges of domestic origin containing an insignificant amount of industrial effluents. If the sludge does contain a substantial amount of organic industrial effluents the organic component of the solid matter can rise appreciably, but this component may drop significantly in raw sludges with a largely inorganic pollution load from industrial sources.

Although the breakdown of the organic constituents of wastewater or sewage sludge during biological treatment in particular or during stabilization results in a reduction in such constituents, and particularly in the proportion of colloidal and gel-like components, this does not affect the different binding capacities of the different types of sludge produced at each stage of treatment.

On the contrary, these differences continue to play a crucial role in all the subsequent phases of sludge treatment and disposal, and in particular in all the processes and steps involved in the separation of the

sludge liquor, namely :
(i)   conditioning; the appropriate dosing depends on the type and amount of the conditioning agent,
(ii)   thickening and dewatering; the specific initial solids content and the specific process result - the latter chiefly involving the throughput attainable, the separation effect achieved and the final solids content are decisive variables,
(iii) drying, as a separate stage or in combination with incineration; the initial solids content of the feed is a decisive variable.

In addition, the capacity of sludge to bind to water - expressed, for example, in terms of its thickening capacity and dewaterability - can even be reduced as a result of stabilization. This is one of the undesirable side-effects typical of "cold" aerobic stabilization. Moreover, sludge which has already been partially digested does not lend itself at all well to conditioning or dewatering (Figure 5).

As a consequence of these associations, it is more necessary than ever before to come up with the best possible combination of quality requirements for wastewaters discharged to the sewerage network, the choice, combination and form of treatment processes, and the specific steps involved in the requisite sludge treatment and disposal.

5. Disinfection of sewage sludge

The purpose of disinfection is to render the sludge safe from a health point of view [18], [19], [20] whether or not the sludge liquor has been separated, and independent of the basic step of stabilization.

Prior disinfection is essential (see Table 6) if sewage sludge is to be used in agriculture where :
(a) the sludge is to be spread on arable land, in the growing season [20],
(b) the sludge is applied to grassland and fodder crops 21 from 1 January 1987 : application permitted throughout the year; until 31 December 1986 (transitional arrangement) : where sludge is applied during the growing season, from the beginning of the year until end of utilization or completion of the harvest.

It is worth noting here that maize does not count as a fodder crop [22], [23].

The rules contained in guide-line (Merkblatt) M7 [20] on the application of sewage sludge to arable land remain in force, whereas those rules on the application of sludge to grassland and land given over to fodder crops will expire after a transitional period on 31 December 1986. The latter will be replaced as of 1 January 1987 by the Sewage Sludge Regulations, which lay down a more stringent requirement. The new requirement will then have direct legal effect [24].

The purpose of sewage sludge disinfection is a preventive one, the idea being to break the possibly existing chain of infection (sewage - sludge - soil - plant - animals - humans) with salmonella and other pathogens.

Disinfection will also help to eliminate parasites where sludge is used to fertilize grassland and to interrupt the host and intermediate host cycle [25].

5.1. Sewage sludge disinfection : definitions

The Sewage Sludge Regulations of 25 June 1982 contain the following definition (chapter 2 (2)) :

"A safe sewage sludge is one which has been subjected to chemical or thermal conditioning, thermal drying, heating, composting, chemical stabilization or any other process in such a way that pathogens are killed ...".

This definition enumerates the processes which can be used to disinfect sewage sludge. Guideline M7 [20] gives a more precise description of these processes in its section 5.3 to 5.6 below :

"Conditioned sludge
Chemical or thermal conditioning should above all enhance the dewatering potential of the sludge.

Sludge which has been chemically conditioned by the addition of metallic salts, alkaline earths or polymers, etc. may be used in agriculture provided that such use is not ruled out by the presence of the chemicals added. Sludge thus treated, like stabilized sludge, has to undergo further treatment before it can be considered hygienically safe.

If the pH value of the sludge is raised to 10 by the addition of alkaline earths and the sludge is then disinfected after a certain reaction period, the range of application of the sludge may be extended so that it corresponds to that of thermally conditioned sludge.

Thermally conditioned sludge subjected to temperatures and reaction times corresponding at least to those applied in heating wet sludge poses no health problems and, in common with the types of sludge described below, can also be used for gardening and vegetable growing, or during the growing period on arable and grassland.

Thermally dried sludge
Thermal drying not only removes water but can alos kill pathogens. Thermally dried sludge poses no health problem if the drying process has the same effect as the heating of wet sludge.

Heated wet sludge (pasteurized sludge)
Sludge subjected to temperatures of over 65° C for at least 30 minutes may be considered as hygienically safe and can therefore be used for various applications in agriculture.

Composted sludge
Composted sludge may be used in agriculture provided it has been subjected to the requisite composting conditions at temperatures of at least 65° C."

The following rules apply to stabilized sludge :

"Stabilized sludge
Liquid, dewatered or naturally dried stabilized sludge treated anaerobically (fully digested in digestion tanks or chambers up to the technical digestion limit) or aerobically (in aeration tanks), and thus rendered inoffensive as a result of the reduction in its putrescibility, has to undergo further treatment before it can be considered as safe."

Guideline M7 therefore not only enumerates the processes that can be used to disinfect sewage sludge but also indicates the appropriate mechanisms for achieving that end.

These German definitions can be regarded as indirect definitions since, unlike the Swiss definition, they are not based directly on the findings of an examination of the sewage sludge that was to be labelled as safe before disposal. Article 1 of the Swiss Sludge Regulations [26] defines sewage sludge as follows :

"Sewage sludge is considered as hygienically safe, if it contains no more than 100 enterobacteriaceae per gram and is free of helminth eggs when delivered by the sewage works owner."

5.2. Established sewage sludge disinfection processes

Three different mechanisms (see also Table 1) can be used to disinfect sewage sludge :

a) Adequate heat treatment, by means of :
   (i)   an external heat source in :
          – pasteurization,
          – thermal drying,
          – thermal conditioning or similar treatment, or
   (ii)   Spontaneous heating using various composting methods, or
   (iii) Development of heat by adding unslaked lime to the sludge,

b) Adequate radiation treatment

c) Adequate adjustment of the pH value, by adding calcium hydroxide, eg during conditioning, or quicklime.

Established and proved disinfection techniques include :

a) those involving an external source of heat :
   (i)   pasteurization, [27], [28], [29]
   (ii)   thermal conditioning,
   (iii) thermal drying,

b) those involving spontaneous heating (see also Table 1) :
   (i)   aerobic-thermophilic sludge stabilization [30], [31], [32],
   (ii)   bio-reactors [33], [34], [35],
   (iii) composting jointly with refuse [36],
   (iv)   addition of quicklime [37], [38];

c) irradiation techniques :
   (i)   gamma irradiation [39], [40], [41]

d) adjustment of the pH value
   (i)   addition of hydrated lime, eg during conditioning [38],
   (ii)   addition of quicklime [37], [38].

6. Summary

Owing to its harmful properties and their potentially undesirable and deleterious effects, sewage sludge cannot be disposed of in the raw state. It must first undergo some suitable form of pretreatment, the aim being to render the sludge fit for disposal, i.e. to alter and improve its characteristics through various processes so that the ultimate means of disposal selected has no detrimental effects on the environment. Given the objective, the type of treatment and its extent and results are geared to the requirements of the disposal method.
   Treatment comprises the following basic steps :
   (i)   stabilization,
   (ii)   separation of the sludge liquor, and
   (iii) disinfection of the sludge.
Each of these steps involves a variety of processes.

This state of affairs will not change significantly and the undesirable or harmful effects of sewage sludge will be subject to much closer and more critical scrutiny in the future. It is therefore to be expected that treatment or preliminary treatment and the techniques employed will have to comply with much more stringent requirements than in the past, especially as regards quality of the sludge and reliability of the techniques used.

In general, more sophisticated techniques will have to be applied. This will result in the application of structured processes tailored very much to the requirements of the particular method of safe disposal selected, and it goes without saying that treatment techniques and processes will have to be further optimized, among other things, with an eye to their cost.

LITERATURE

[1] MOLLER, U :
"Schlammbehandlung und -beseitigung" VDI-Zeitschrift "Umwelt", H. 1, 1983, S. 37-64;

[2] MOLLER, U. und LESCHBER, R. :
"Viele neue Begriffe zum Thema "Schlamm" in der neuen DIN 4045 "Abwassertechnik, Begriffe"", veröffentlicht in der Zeitschrift "Korrespondenz Abwasser" 31, 1984, S. 971 - 975;

[3] IMHOFF, K. :
"Taschenbuch der Stadtentwässerung", 14. Auflage, 1951, R. Oldenbourg Verlag, München-Wien;

[4] FISH, H. :
"Sea Disposal of Sludge : The United Kingdom Experience"; "Water Science and Technology", Vol. 15, Capetown pp 77 - 87, 1983;

[5] MOLLER, U. :
"Schlammuntersuchung und -charakterisierung als Grundlage der Schlammbehandlung und - beseitigung - Wechselbeziehungen zwischen Schlammeigenschaften und Schlammbehandlungs-verfahren sowie der Beseitigung der Schlämme", Band 11 der Technisch-Wissenschaftlichen Schriftenreihe der ATV "Aus Wissenschaft und Praxis", 1984, S. 49 - 130;

[6] MOLLER, U. :
"Podiumsgespräche "Das Klärschlammproblem" Klärschlamm und Müll - Frischschlamm oder Faulschlamm - Schlammstabilisierung", Berichte der Abwassertechnischen Vereinigung, Band 18, "Die Frankfurter Tagung 1965", 1967, S. 37 - 51;

[7] ROEDIGER, H. :   "Die anaerobe alkalische Schlamm-
faulung", gwf-Schriftenreihe "Wasser -
Abwasser", Oldenbourg-Verlag, München,
Band 1, 1967;

[8] LOLL, U. und   "Prozeßziele der Klärschlamm-
      MOLLER, U. :   stabilisierung und deren Zusammenhang
mit Dimensionierung-, Betriebs- und
Kontrollparametern", veröffentlicht in
der Zeitschrift "Korrespondenz
Abwasser " $\underline{31}$, 1984, S. 940 - 945;

[9] NIEMITZ, W. :   "Beschaffenheit des häuslichen und
industriellen Klärschlammes", in ATV
"Lehr- und Handbuch der Abwassertechnik", Band III, Verlag W. Ernst u.
Sohn, Berlin, 2. Auflage, 1978, S. 16;

[10] DICHTL, N. :   "Die Stabilisation von Klärschlämmen
unter besonderer Berücksichtigung
einer zweistufigen aeroben/anaeroben
Prozeßführung", veröffentlicht in Band
5 der Schriftenreihe Siedlungswasserwirtschaft Bochum des Vereins
zur Förderung des Lehrstuhls für
Wasserwirtschaft und Umwelttechnik II
(Siedlungswasserwirtschaft) an der
Ruhr-Universität Bochum e.V., 1984;

[11] BAUMGART, H. CHR. :   "Aerobe und anaerobe Stabilisierung",
ATV-Fortbildungskurs für Wassergüte-
wirtschaft, Abwasser- und Abfall-
technik, Teil B/2, 1979, V 1-1/20;

[12] LOLL, U. :   "Stabilisierung hochkonzentrierter organischer Abwässer und Abwasser-
schlämme durch aerobthermophile
Abbauprozesse", Zeitschrift "Das Gas-
und Wasserfach" $\underline{115}$, 1974, S. 191 -
198;

[13] RIEGLER, G. :   "Eine Verfahrensgegenüberstellung von
Varianten zur Klärschlammstabili-
sierung", Schriftenreihe WAR des
Institutes für Wasserversorgung,
Abwasserbeseitigung und Raumplanung
der Technischen Hochschule Darmstadt,
Band 7, 1981;

[14] LAGA :   Merkblatt "Die geordnete Ablagerung
von Abfällen (Deponie-Merkblatt, Stand
1.9.79");

[15] MOLLER, U.,   "Neudefinition der Deponierfähigkeit
      OTTE-WITTE, R.,   von Abwasserschlämmen - BMFT-Projekt",
      KASSNER, W.,   Referat auf dem 4. Internationalen

LOLL, U. und GAY, CH.G. :	Recycling-Congress vom 30.10. – 1.11.1984 in Berlin, veröffentlicht im Berichtsband dieses Kongresses "Recycling International", S. 446 – 451;
[16] MOLLER, U., KASSNER, W., KOHLHOFF, D. LOLL, U. und OTTE-WITTE, R. :	"Neudefinition der Deponierfähigkeit von Abwasserschlämmen", veröffentlicht in der Zeitschrift "Korrespondenz Abwasser" 31, 1984, S. 928 – 933;
[17] LOLL, U. :	"Optimale Gasproduktion bei der Ausfaulung organischer Substanzen" Schriftenreihe "Gewässerschutz – Wasser – Abwasser", Band 45, 1981 , S. 315 – 335;
[18] MOLLER, U. :	"Forderungen an die Technik aus der heutigen Sicht des Gewässerschutzes für den Bereich der Schlamm- und Abfallbeseitigung, Notwendigkeit und Möglichkeiten für eine geordnete Schlamm – und Abfallbeseitigung – Grenzen des Dispositions-spielraumes und verbleibende Freiheitsgrade für das notwendige Handeln – weitere Konsequenzen", Berichtsheft der ATV Nr. 25 "Die Kasseler Tagung 1971", 1972, S. 181 – 206;
[19] REPLOH, H. :	"Klärschlammverwertung im Landbau", Zeitschrift "Kommunalwirtschaft", H. 9, S. 378 – 382;
[20] ZENTRALSTELLE FUR ABFALLBESEITIGUNG (ZfA) beim Bundesgesundheitsamt gemeinsam mit der LAWA :	Merkblatt M 7 "Behandlung und Beseitigung von Klärschlämmen unter besonderer Berücksichtigung ihrer seuchenhygienisch unbedenklichen Ver- wertung im Landbau", Handbuch "Müll- und Abfallbeseitigung", Herausgeber : Kumpf, W., Maas, K. und Straub, H. bzw. Hösel, G., Straub H. und Schenkel, W., 1964/85 Kennzahl 6854, S. 1 , – 11, Erich-Schmidt-Verlag Berlin-Bielefeld;
[21] BUNDESMINISTER DES INNERN :	Verordnung über das Aufbringen von Klärschlamm – Klärschlammverordnung (AbfKlärV)", Bundesgesetzblatt 1982, Teil 1, S. 734 – 739;
[22] DEUTSCHE LANDWIRT- SCHAFTSGESELLSCHAFT (DLG) :	DLG-Merkblatt Nr. 209; "Klärschlammanwendung in der Landwirt- schaft – Was ist zu beachten ?";

[23] MINISTER FUR ERNÄHRUNG,   "Vorläufige Verwaltungsvorschrift zum
LANDWIRTSCHAFT UND   Vollzug der Klärschlammverordnung
FORSTEN DES LANDES   (AbfKlärV) vom 25. Juni 1982",
NORDRHEIN-WESTFALEN :   Ministerialblatt für das Land
    Nordrhein-Westfalen Nr. 22, vom 30.
    März 1983, S. 337 - 340;

[24] KREFT, H. :   "Die neue Klärschlammverordnung",
    Zeitschrift "Korrespondenz Abwasser"
    29, 1982, S. 246 - 250;

[25] BUNDESMINISTER DES   Begründung zur "Verordnung über das
INNERN :   Aufbringen von Klärschlamm - Klär-
    schlamm verordnung (AbfKlärV)",
    Begründung zu § 4 Abs.3,
    Bundesratsdrucksache 56/82;

[26] EIDGENOSSISCHES   Richtlinie vom 19. Juli 1982 zur
DEPARTMENT DES INNERN,   "Abnahmeuntersuchung von Hygieni-
Bundesamt für Umwelt-   sierungsanlagen für Klärschlamm";
schutz, Bern :

[27] KUGEL, G. :   "Geordnete Klärschlammbeseitigung im
    Niersgebiet aus neuerer Sicht",
    Zeitschrift "Kommunalwirtschaft",
    1972, Heft 9, S. 382 - 388;

[28] KUGEL, G. :   "Pasteurisierung von Roh- und
    Faulschlamm" in "Behandlung und
    Beseitigung von Abwasserschlämmen",
    Band 6 der Schriftenreihe
    "Gewässerschutz - Wasser - Abwasser",
    Aachen, 1971, S. 311 - 324;

Pollutants affecting disposal	Adverse effects	Points of treatment or elimination prior to discharge to sewerage system	Processes used or measures of avoidance
Organic matter that can be degraded by microorganisms	Formation of odours	Sewage works	Stabilization : biological : aerobic (including composting) anaerobic (digestion) thermal oxidation : incineration
High water content	Problems associated with suitability for tipping	Sewage works	Separating of water by means of a) Thickening : by flotation, or mechanically b) Dewatering : mechanically, or naturally (including conditioning) c) Drying
Pathogens	Possible restrictions on use in agriculture for health reasons	Sewage works	Disinfection by means of : Heat treatment External heat source (e.g. pasteurization) Spontaneous heating Irradiation Adjustment of pH value
Pollutants	Potential for pollutants to enter the food chain through soil treated with sludge	Prior to treatment	Elimination prior to discharge of the contaminated waste water into the public sewerage system

Table 1 : Constituents of sewage sludge adversely affecting disposal, possible harmful effects, and treatment processes used

Type of process/action	Conditions	Phase	Process Characteristics	Process	Application	Practical significance
biological	aerobic	liquid	without efficient self heating	extended aeration e.g. with B_{TS} = 0,05	especially for small treatment plants	essential to considerable
		liquid	with (efficient) self heating	aerobic-thermophilic sludge stabilization (liquid composting)	small to medium size treatment plants	up to now, low; obviously positive development
		de-watered	with (efficient) self heating	composting (e.g. in so called "bio-reactors")	small to medium size treatment plants	up to now, low
	anaerobic	liquid	without external heating	Emscher tanks	small treatment plants	considerable
				open unheated digestion tank	formerly : small to middle sized treatment plants	no longer applicable
			with external heating	independently heated digestion tank	medium to large size treatment plants	high to very high
chemical	aerobic	liquid	with external energy input	wet oxidation	very large treatment plants	very low
		dried	with/without external energy input	incineration/combustion	if possible, only large treatment plants	limited because very expensive
chemical, prevention of a temporary transformation of substances by drastic rise of pH	(aerobic)	de-watered	with addition of reactants	so-called lime stabilization		disputed; as a solution in the long run refused by professionals /6/

Table 2 : Review of the methods/processes of sludge stabilization

Objective of Stabilization	Requisite degree of stabilization
Application in agriculture in) liquid state))	Fully stabilized
Application in agriculture in) dewatered state)))	Where the sewage sludge is desinfected : also incompletely or even unstabilized
storage of liquid sludge in lagoons	Fully stabilized
Tipping in dewatered state	(i.e. up to the technical limit)
Tipping after incineration	Unstabilized prior to incineration

Table 3 : Degree of stabilization required for different objectives of the stabilization process

Biological sewage sludge stabilization	Aerobic and anaerobic treatment	Anaerobic treatment			Aerobic treatment, liquid	
Parameter	Loss on ignition as % of dry matter	Acetic acid content mg/l	Acetic acid content mg/l	Fatty acids content mg/kg	BOD_5/COD ratio	BOD_5/COD ratio
Degree of stabilization	LOLL and MÖLLER /8/ after ROEDIGER /7/	according to NIEMITZ /9/	according to DICHTL /10/	according to BAUMGART /11/	according to LOLL /12/	according to DICHTL /10/
Raw sludge (unstabilized)	90 % and more *)	1800 – 3600	no data	>1000	> 0,25	no data
Partially stabilized	(50) – 60 % *)	1000 – 2500 (well digested) /6/	no data	–	ca. 0,25 – 0,18	no data
Stabilized (depending on objective of treatment and initial value)	45 – 55 %	100 – 1000	no data	300 – 1000	ca. 0,18 – 0,15	no data
Fully stabilized	$\leq$ 45 % **)	< 100	$\leq$ 200 ***)	$\leq$ 300	$\leq$ 0,15	$\leq$ 0,10 ***)

*) Depending on initial value
**) Applies only to sludge of standard composition and stabilization capacity
***) Including sufficient safety margin to take account of imponderables in practice

Table 4 : Range of values for various stabilization parameters in respect of sludges predominantly of municipal origin

Dry matter content	Moisture content	Characteristics
< 15 – 20 %	> 85 – 80 %	Fluid to mushy, capable of being pumped
– 20 – 30 %	– 80 – 70 %	Generally suitable for shovelling, still plastic, sticky, thixotropic
> 35 – 40 %	> 65 – 60 %	Crumbly, no longer sticky, leachable only to a limited extent
> 60 – 65 %	< 40 – 35 %	Capable of being spread, consolidated
> 85 – 90 %	< 15 – 10 %	Powdery

Table 5 : Characteristics of sewage sludge types with different moisture content.

Table 6 : Rules for the safe application of sewage sludge in agriculture as laid down in the Sewage Sludge Regulation and Notice No M7 [21] [20].

		I	II		III		IV	V	VI	VII
		Unutilized land	Arable land		Grassland and fodder crops d)		Vineyards	Fruit growing	Forestry	Vegetable growing, parks and gardens
		a)	F b)	G	F	G	F			e)
5.1.1 Raw sludge	nhr	-	-	-	-	-	-	-	-	-
5.1.2 Sludge from small sewage works and cesspools	nhr	+	+	-c)	-	-	-	-	+)3	-
5.2.1 Liquid — anaerobically or	nhr									
5.2.2 Dewatered — aerobically stabilized sludge	nhr	+	+	-c)	-)1 +)2	-	+	-	+)3	-
5.2.3 Naturally dried	nhr									
5.3.1.1 Chemically conditioned sludge (without adequate disinfectant effect)	nhr									
5.3.1.2 Chemically conditioned sludge (with adequate disinfectant effect)										
5.3.2 Thermally conditioned sludge										
5.4 Thermally dried sludge		+	+		+		+	-	+)3	-
5.5 Heated liquid sludge (untreated, stabilized)										
5.6 Composted sludge										
5.7 Chemically stabilized										

nhr = no health risk

F = fallow season (see also footnote b)
G = growing season

+ = spreading permitted
− = may not be spread

a) This category includes arable land not under crops, marginal land, ploughed-over grassland, newly planted forest land, waste land and reclamation land.

b) The fallow period includes the winter months and the period in which the land is free of plants and fruit grown for human or animal consumption. Outside the winter months sludge should be spread in a way that is not detrimental to other farm land.

c) During the growing season sludge posing no health risk may be spread on harvested arable land provided it is worked immediately into the soil and provided that no vegetables or fodder crops are sown immediately after spreading. Where grain cereals are concerned, this period may be extended until just before shoot emergence, and where grain maize is concerned until the plants reach a height of 25 - 30 cm.

d) Typically, clover and fodder catch crops. Maize does not come into the latter category within the meaning of Article 4 (3) (1) of the Sewage Sludge Regulations.

e) Potatoes, which are root crops, do not count as vegetables.

)1 After 31.12.1986

)2 Only until 31.12.1986 (transitional arrangement); until then only from completion of use / harvest until the end of the year.

)3 Only with special consent.

)4 Spreading of raw sludge is not covered by the "Sewage Sludge Regulations". It comes under the regulations on waste disposal.

)5 This category includes sludge subject to stabilization in multi-chamber facilities or plant with similar effect.

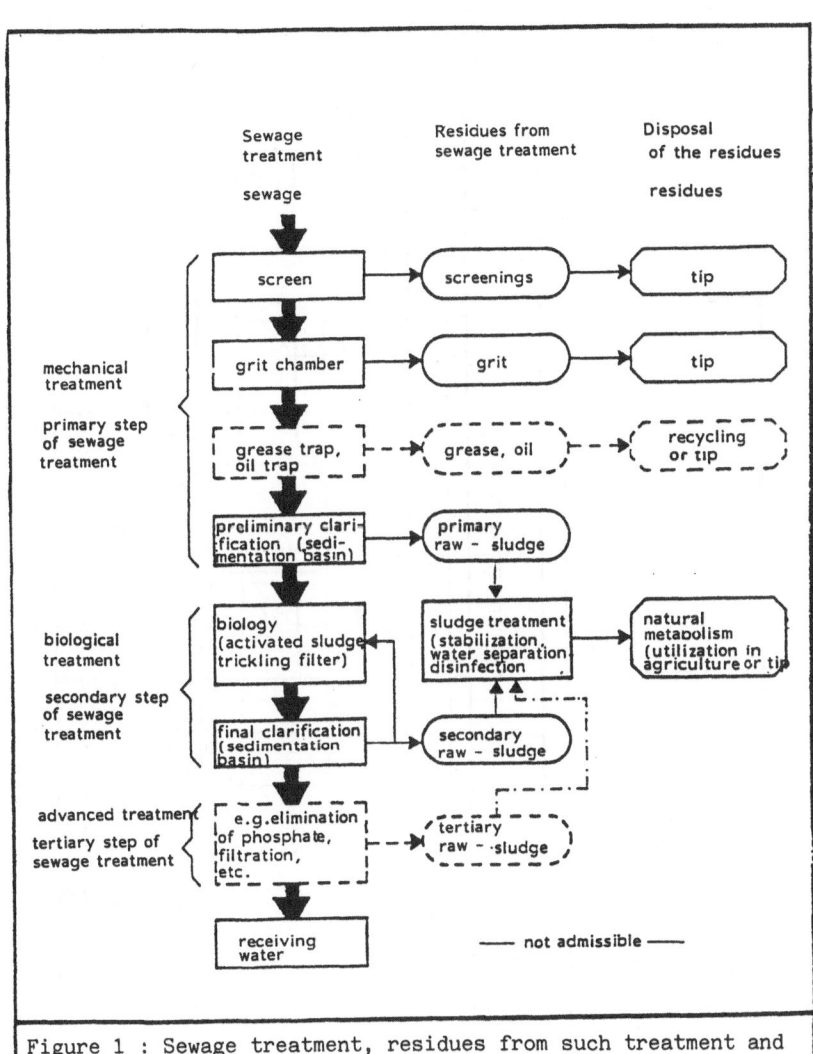

Figure 1 : Sewage treatment, residues from such treatment and
their disposal in a municipal sewage works

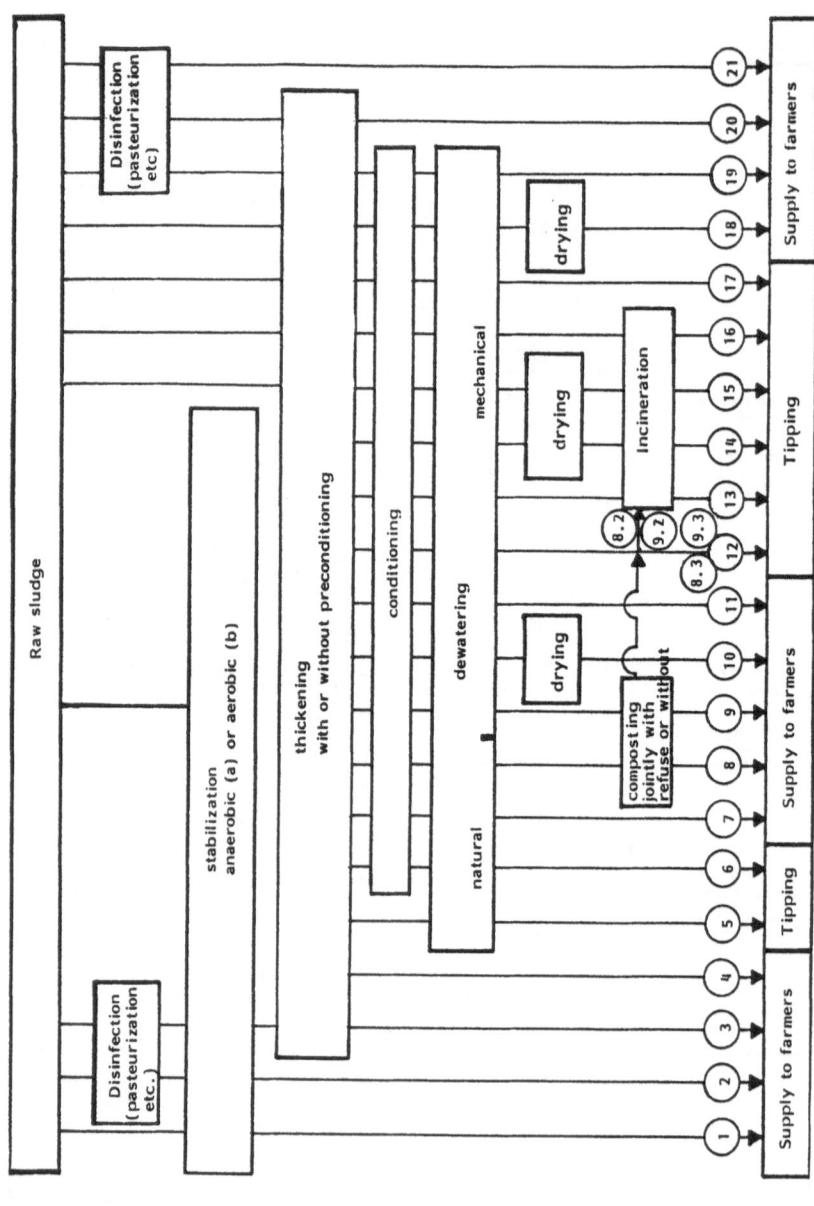

Figure 2 : Sludge treatment, utilization and disposal

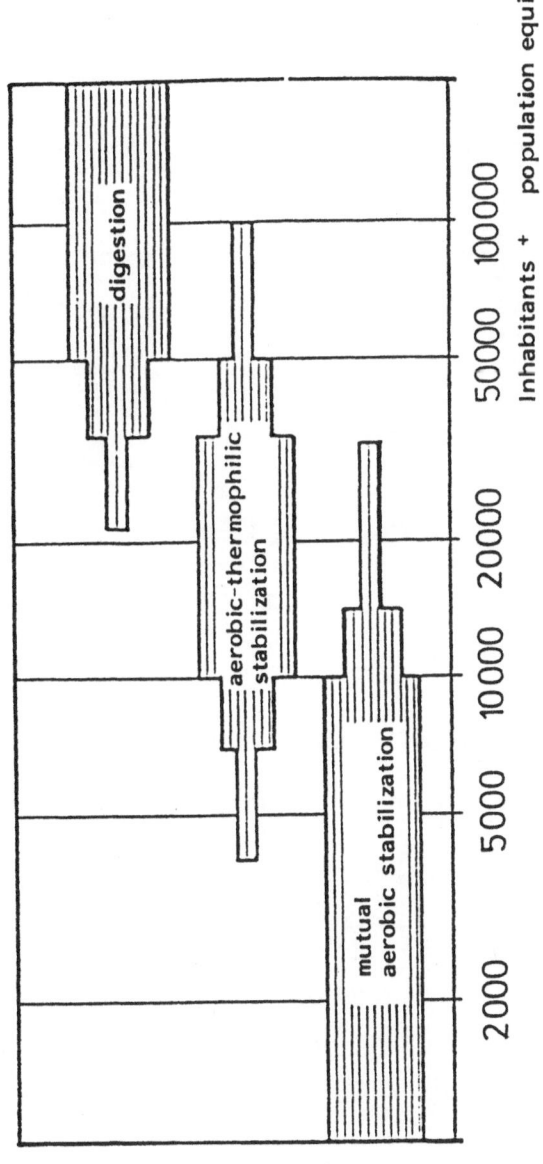

Figure 3 : Recommended ranges of use for the different variants of biological stabilization of sewage sludge with reference to the capacity of sewage works (according to RIEGLER /13/)

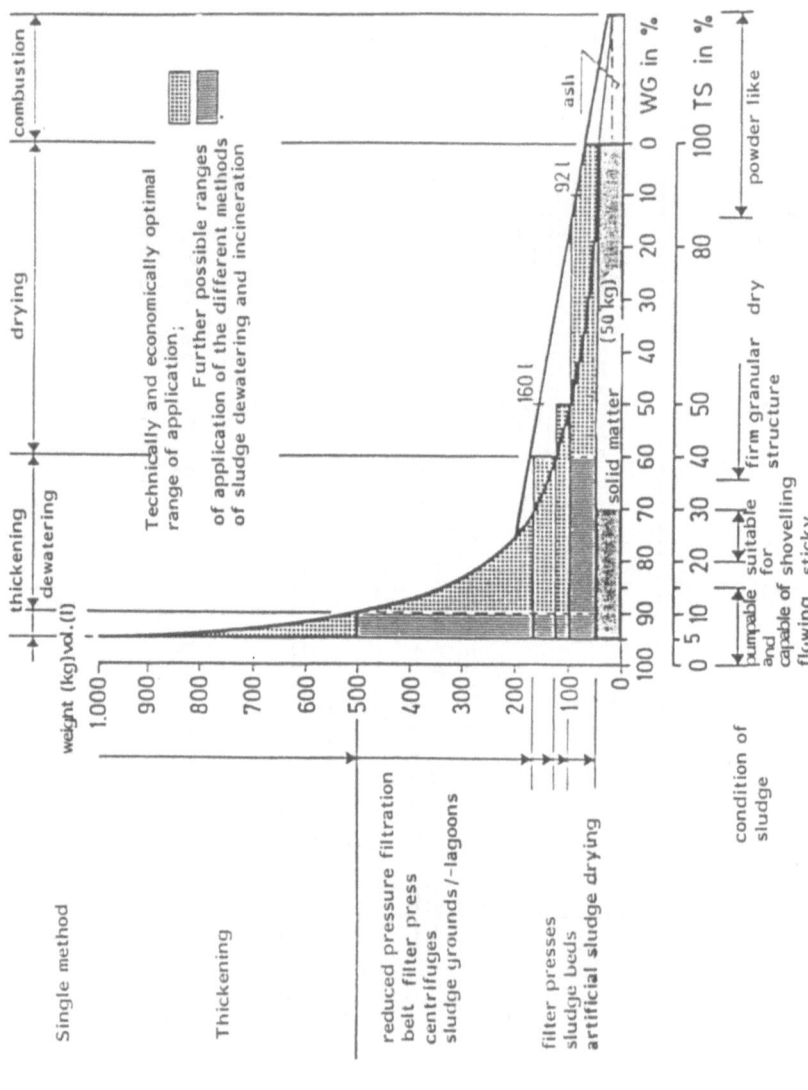

Figure 4 : Technically and economically optimal volume reduction through water separation and incineration

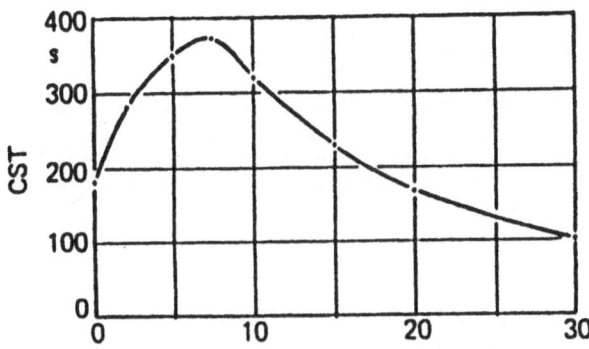

Figure 5 : Dewatering potential (CST value) of municipal sewage sludge as a
function of sludge digestion period.
According to LOLL (17)

DISCUSSION

G. MININNI

Are there laws in the Federal Republic of Germany to control the
treatment and disposal of sewage sludge?

R.L. LESCHBER

Regulations, with the force of law, for the control of the treatment and
disposal of sewage sludge were introduced into the Federal Republic of
Germany in 1982 by the Federal Ministry of the Interior. These
Regulations cover many aspects such as sludge and soil quality,
application rates and, very importantly, hygiene aspects.

In the Federal Republic, we have tried to reduce environmental pollution
by following 3 main principles:-

1. Concern for the environment -

    this has led to practices designed to reduce the pollution potential
    of the sludge, eg. heavy metal content, in the first place. One way
    of achieving this is by a stricter control of the discharge of
    industrial effluents containing heavy metals to the public sewerage
    system.

2. Communication -

    a number of Working Parties of experts from both research and
    operational organisations have been established to help diseminate
    the "state of the art".

3. "Polluter pays" -

    this principle has been shown to work well but can only operate after
    the nature and source of the pollution has been identified.

A further way of removing heavy metals from sewage sludge is through
bacteriological leaching. Battelle Research Laboratories are examining
the process at the present time. It appears, however, to be so expensive
that it only could be used as a final resort.

TREATMENT PROCESSES FOR FARM WASTES

V C NIELSEN
Head of the Farm Waste Unit, A.D.A.S.
Ministry of Agriculture, Fisheries and Food

Summary

This paper reviews the current state of biological treatment methods for liquid manures to control environmental pollution and to achieve better utilisation of the fertiliser value.

Improved performance of both aerobic and anaerobic treatment systems can be achieved by better evaluation of the farm problem. A method of carrying out a farm survey is described.

The parameters which control aerobic treatment of farm slurries have been defined under laboratory and in small pilot plants. Farm scale designs based on these parameters have yet to be evaluated. The design and operation of aerators in farm systems need further development to improve performance in terms of KgO2/kWh.

Laboratory-scale experiments have established the parameters which control anaerobic digestion but application to farm scale digesters remains a problem.

Efficient utilisation of the biogas produced has yet to be achieved. Odour control is an important factor in the economics of anaerobic treatment, better evaluation of its effectiveness needs to be demonstrated.

1. INTRODUCTION

This paper reviews the research and development and farm scale operation of biological treatment systems for liquid manures, for the control of environmental pollution and improved utilisation as a fertiliser. The increase in environmental pollution arising from intensive livestock husbandry practices has led to public concern and this has pressurised government departments into finding solutions. As a result there has been a general tightening of planning controls for livestock farms and control of pollution legislation. The need to find solutions has encouraged industry and research centres to develop novel and improved systems of handling and storing farm wastes and for a range of treatment systems which will enable farmers to comply with the standards required.

The aim of treatment must be to enable intensive livestock farming to continue while operating within the constraints of pollution legislation and at a cost which can be sustained by the enterprise. One approach is to develop systems which intensify the natural microbial processes of decomposition which take place in the soil, (1). Treatment must produce a stabilised product which can be returned to the land and utilised as a source of plant nutrients without causing pollution of the environment.

Biological treatment systems for liquid farm manures must operate within the following constraints:
1. The plant must be able to treat the daily input of livestock manure throughout the year.
2. The output should comply with the regulations set by enforcement authorities.

3. The installation, running and maintenance costs must be realistic and be within the economics of livestock production.
4. Systems must be relatively easy to operate by farm staff and have a low labour requirement.
5. Systems must be robust (eg an operational life of at least ten years) and be capable of being repaired and maintained quickly and easily.

Treatment systems designed to stabilise farmyard manures (faeces, urine and bedding materials) or slurries (faeces, urine plus some additional water from pen cleaning etc), should be assessed on their ability to:-
1. Control odours during storage and following application to land.
2. Stabilise soluble organic matter to prevent water pollution following land spreading.
3. Control the spread of livestock diseases, (bacteria, viruses and helminths), and plant pathogens.
4. Improve the utilisation of the major plant nutrients, (N P & K).
5. Control N content of treated slurry when required.
6. Utilise where possible surplus heat, energy and other valuable by-products to help offset running costs.

2. BIOLOGICAL TREATMENT PROCESSES IN USE
Treatment processes which are available for the treatment of liquid manures, (faeces, urine, some additional water = slurry) are:
1. Aeration
2. Anaerobic digestion

Biological treatment systems depend on the conditions in the slurry. The availability of oxygen has a fundamental effect on the microbial processes within the treatment system. An unrestricted supply of oxygen will result in the development of aerobic reactions. In the absence of oxygen the reaction will result in anaerobic fermentation.

The total solids content of the manure will affect the method of treatment. Wastes of high total solid contents (over 12 per cent) such as farmyard manure, poultry manure and solids from the mechanical separation of slurries are most effectively treated by composting. Total solid contents of less than 12 per cent are liquid in character and are most effectively treated by aeration or anaerobic processes.

The control of the volume and dilution of the substances from which the mixed microflora obtain the essential components for continued metabolism is often overlooked in the design and operation of farm scale plants. Volume and dilution must be controlled and be uniform to provide meaningful parameters on which to base the size of the treatment tank and to achieve the desired mean residence time for treatment. Wide variations in the daily volume caused by the uncontrolled entry of rainwater and washing water into the collection tanks will alter the total solids content and volume and lead to failure of the system. Alternative arrangements for dealing with exceptional conditions must be incorporated into the design of the system.

Temperature affects microbial activity, increasing it will speed up activity. The microbial species will change depending on their optimal growth temperature. Extremes will inhibit activity. In temperate climates the ambient temperature will not affect aerobic treatment. Low winter temperatures will affect the operation of anaerobic digesters and thought should be given to preheating slurries before treatment to avoid temperature shock.

The presence of toxic metals, disinfectants or medicants for the

livestock inhibit treatment. It is essential to the success of the
system that these substances are separated and kept out or the manure containing them does not enter the treatment system.

2.1 Farm surveys

Any farm with a pollution problem should be surveyed. This survey
should be carried out as part of the investigations to determine the extent and type of pollution and to obtain essential background information
on the livestock husbandry practices on the unit, this should include:-
1.   The cause, extent and type of pollution.
2.   The layout and design of livestock buildings, stores and waste
handling facilities.
3.   The situation of the buildings in relation to natural drainage,
wind direction and proximity to private houses.
4.   The age range and numbers of stock at various stages of growth.
The diet, method of feeding and availability of drinking water.
5.   The characteristics and volume of manure to be treated. It is
usual to carry out analysis for: total solids, (TS), volatile solids
(VS), volatile fatty acids, (VFA), chemical oxygen demand, (COD,
biochemical oxygen demand, (BOD), total nitrogen, (TN), ammonium
nitrogen, (NH_4-N), nitrate nitrogen, (NO_3-N), phosphorus, (P),
potassium, (K), copper, (Cu), zinc (Z), and for pH.
Monitoring should establish when possible variations in volume are likely
to occur, daily, weekly or monthly and at what point in the system.
Variations in climate likely to affect treatment, prolonged dry spells,
heavy rainfall or snow, duration of severe frosts. This information
will influence system design.

3.   AEROBIC TREATMENT

The aim is to maintain livestock manures in an aerobic state prior to
application to agricultural land. It is used primarily to control odours
during the storage and spreading of slurry, (2, 3 and 4). Other advantages are that the treated slurry contains less soluble organic matter than
fresh or anaerobically stored slurry and is less likely to lead to organic
pollution of water courses, (5, 6 and 7). Manipulation of the dissolved
oxygen concentration during treatment and mean residence time can be used
to remove up to 70 per cent of the NH_4-N, (8 and 9). Treatment can
destroy harmful pathogenic bacteria, (10), and aerobic-thermophilic
treatment is an effective treatment for many pathogenic viruses, (11).
Systems can be designed and operated to recover the exothermic heat of the
microbial reactions to produce surplus low grade heat which can be used on
the livestock unit, (12 and 13).

The parameters which control continuous aerobic treatment have been
defined for pig slurry (8) and for cattle and poultry manures, (14).
These parameters are as follows:-

3.1 Oxygen demands

When oxygen is present, micro-organisms break down organic chemicals
to simpler substances or carbon dioxide and oxidise nitrogen containing
chemicals to nitrate. The nature and speed of these processes depends on
the amount of oxygen available, the temperature and the time. Equations
for carbon (15) and nitrogen (10 and 15) have been given and combinations
of them define the total oxygen demand of the system (16 and 18).

3.2. Oxygen Availability

If minimal oxygen is present, odorous substances present can be oxidised, removing the immediate smell but leaving chemicals which can reconvert to odorous substances in anaerobic conditions. As oxygen input is increased (or more time is allowed) these substances, too, are broken down, these processes use up oxygen sufficiently quickly to keep the oxygen dissolved in the liquid at very low (almost undetectable) levels. Finally, when oxygen is more freely available, more extensive oxidation of organic chemicals occurs.

3.3 Choice of aeration rates and treatment time

At a given temperature (determined by microbial activity in the slurry as well as by ambient temperature) the length of time and the rate of aeration can be varied to alter the odour offensiveness of the treated slurry and the potential to regenerate odours during subsequent storage before spreading. To achieve an odour offensive rating of 2 (faintly offensive, 18 and 19), will require a minimum treatment of 2 days. A treatment of 5 days will prevent odour regeneration for several months (10 and 15).

Similarly the level of dissolved oxygen to be maintained varies depending on the objectives/aims of treatment. The options are to:

1. spread slurry immediately after treatment,
2. store treated slurry for several months before spreading,
3. combine odour control with a reduction in water pollution potential,
4. control the residual nitrogen content,
5. for odour control with heat recovery.

A computer program to select and solve the most appropriate set of equations has been produced by Evans et al, (8), for any of these five options. The results of these equations suggests the retention times and oxygen demands and hence aerator sizes need to achieve the treatment options.

Laboratory scale work has demonstrated that similar effects to those described by the equations, (8), can be achieved using aerobic treatment with no detectable dissolved oxygen by controlling the Redox potential of the mixed liquor to Omv Eh and producing an acceptable odour free, stable slurry, (16). Other work has shown successful odour control maintaining the Redox potential of the mixed liquor at levels as low as -50mV (Eh), (19).

3.4 On-farm treatment systems

Farm scale aeration systems are in use in many countries to control odours, water pollution and to reduce the nitrogen content of the slurry. Heat is recovered on some sites and the methods used have often been improvised by the farmer, (20 and 21).

The most common design of treatment system is the semi continuous flow process, that is the daily slurry load is pumped often in small batches on a regular basis into the treatment tank. Pretreatment in the form of mechanical separation to remove coarse solids from the liquid fraction may be used to improve pumping, avoid blockages and reduce the energy demand when mixing the contents of the treatment tank. Separator designs in use have been evaluated by Pain et al, (22). Low total solid slurries of less than 3 per cent are normally treated without separation.

The whole or separated liquid fraction of the slurry is either pumped or flows by gravity into the treatment tank. The size and configuration of the tank invariably depends on the size and shape of tanks already used for storage purposes on the farm or when new tanks are obtained, the size is determined by the mean treatment time required. There are no published design data available for slurry treatment tanks in relation to the type and size of aerator used.

There are five basic types of aerators available to treat farm slurries, these are: mechanical surface, mechanical sub-surface, combined compressed air with either diffusers or mechanical mixing, and pumped liquid, (venturi or jet type). The performance criteria which need to be determined for efficient aeration have been reviewed by Cumby, (23). The most important of these is the ability of the aerator to provide adequate dissolved oxygen to the entire volume of the treatment tank. The efficiency of the aerator is the main criterion of its performance. This is usually expressed as the ratio of the mass of oxygen transferred to the energy consumed: kg O_2/kWh. Transfer efficiencies of less than 1 kg O_2/kWh are considered poor and inefficient, good transfer efficiencies range between 1 and 2 kg O_2/kWh, a few experimental machines have achieved more than 2 kg O_2/kWh, (24). The parameters which control oxygen transfer include the T.S. content of the slurry, air/slurry contact time, the area of the air/slurry interface per unit volume of air. These factors can have a significant effect on the operational performance of an aerator. A secondary requirement of an aerator is that it must completely mix the contents of the aeration tank. The concept of energy density has been used to assess the mixing ability of mechanical aerators. 30 W/m^3 has been estimated to be necessary to achieve complete mixing, (depending upon tank size and shape), however 3 W/m^3 is usually the level found in farm aeration systems, (25). Using energy density seems an indirect way of providing design data. It would be better if manufacturers used oxygen inputs.

Electricity charges make up between 20 and 50 per cent of the annual running costs of aeration systems, (26), depending on the design of the system. It has been reported that the electricity required to obtain odour control per pig place a year ranges between 52.5 kWh and 55 kWh and the total running costs per pig produced range between £1.30 and £1.40; using 20-100% of the profits from pig production, (23 and 27). It is therefore important to the profitability of the pig farm that the oxygen transfer efficiency and the aeration system design take advantage of high oxygen transfer aerators operating in correctly designed aeration tanks. There is scope for a significant reduction in the costs of operation.

At present the treated slurry is either discharged continuously or by displacement by the addition of fresh untreated slurry. It may be stored for up to one week or several months where soil conditions are unsuitable, before being spread. If stored for any length of time secondary aeration is often carried out in the store to prevent the regeneration of odours. Very few aerobic treatment systems have been designed and operated according to the parameters described by Evans and Baines (8).

3.5 Conclusions on aeration systems

Regimes for odour control have been defined in the laboratory and at pilot scale in terms of retention time, level of aeration, (dissolved oxygen/redox potential) for pig slurry. Similar regimes have been suggested for cattle and poultry manures but not tested comprehensively. The application of these regimes to commercially available aerators on farms is yet to be tested.

Farm scale systems in use can achieve odour control and produce usable surplus heat. However aerator performance in terms of aerator efficiency kg O2/kWh is often not known. Aerators are often sold using the concept of energy density, as the measure of performance with no reference to oxygen demand. The calculation is based on the size of the available treatment vessel after applying an energy density factor, $(W/m3)$, (28).

There are no design parameters for the correct configuration of the treatment tank in relation to the type of aerator used. Therefore complete mixing of the tank contents to achieve a uniform dissolved oxygen content and prevent settlement is often less efficient than it could be.

These factors and others mentioned earlier indicate that better designed systems could achieve substantial reductions in the costs of operating these systems.

4. ANAEROBIC TREATMENT

Anaerobic digestion of livestock slurries is primarily carried out to obtain energy in the form of biogas, (a mixture of methane and carbon dioxide with traces of hydrogen sulphide and ammonia). Biogas can be used as:
1. fuel for heating, (buildings or water), 2. as a fuel for engines to power generators to produce electricity and hot water, 3. as a fuel for engines to drive vehicles, (29). The digestion of livestock slurries has other advantages:
1. The digested slurry is less odorous than untreated slurry, but odour is not completely eliminated by the process, (30, 31 and 32).
2. The digested slurry is stabilised and can be stored for long periods at varying temperatures without the regeneration of odozous substances, (30).
3. The digested slurry contains less organic matter than untreated slurry and is less likely to cause organic pollution of water courses, when applied to land as a fertiliser, (37).
4. Treatment reduces the number of pathogenic organisms which can infect livestock, (33 and 34).
5. Treatment reduces the number of plant pathogens in the slurry, (35).

4.1 The biomethanation process

It is an anaerobic process in which a series of communities of microorganisms operate simultaneously to reduce the organic oompounds present in livestock slurries to biogas, biomass and a stabilised sludge. The sludge has similar concentrations of N, P and K to the untreated slurry, but the nitrogen content will contain higher concentrations of $NH4-N$ to organic N than in the original slurry.

The process takes place in an enclosed tank called a digester. The microbial reactions which occur in the digester have been described by Pffeffer, (36) and Hobson et al (37). The reactions are complex and rely on the correct balance being maintained to ensure the operation of a continuous process. The following parameters affect the operation of the process.
1. pH, digesters operating on farm slurries require a pH of between 7 and 7.2.
2. Temperature. There are two temperature ranges over which these complex reactions take place, $25-45^{\circ}C$ mesophilic range and $55-70^{\circ}C$ thermophilic range. Most farm operated digesters in temperate climates are operated in the mesophilic range at $30-35^{\circ}C$. This

temperature range has been selected as most economic one in terms of biogas produced per unit mass of substrate and energy required to maintain the reaction.

3. Substrate. Farm digesters normally use a mixture of faeces, urine and some washing/cleaning water. The total solids content of neat faeces and urine for cattle and pigs is usually between 10-12 per cent depending on the diet, and 23-25 per cent for battery laying hens. In most farm situations dilution occurs, dairy cattle slurries usually range between 6-10 per cent, pig slurry from large units, between 2-5 per cent. Poultry manure remains high around 20 per cent due to the method of collection. Hawkes (38) has shown that slurries with a T.S. content less than 4 per cent will not attain net energy production.

Inhibition by dissolved chemicals in the slurry may occur. Hydrogen sulphide gas in solution at concentrations greater than 100 mg/l and ammonia at more than 3,000 mg/l will affect digester performance. (37).

The volatile solid (VS) content of the slurry is an indication of the substrate concentration. It can range between 40-90 per cent of the TS content, typical farm waste values are 70-75 per cent of the TS, (37). The efficiency of the digestion process is often quoted as the m^3 biogas/kg V.S; typical values for the different species are dairy cows 0.22 m^3/kg V.S., pigs 0.38 m^3/kg V.S., poultry 0.49 m^3/kg V.S. (3). Hobson et al, (37), found with laboratory scale digesters the following results, dairy cattle 0.22 m^3/kg V.S., pig 0.43 m^3/kg V.S. and poultry 0.66 m^3/kg V.S. An alternative measurement used is m^3 biogas/kg TS, typical figures for: dairy cattle 0.21 m^3/kg TS, pigs 0.30 m^3/kg TS, and poultry 0.38 m^3/kg TS. (37).

4. Retention time. This is the calculated residence time of the slurry in the digester. It is a compromise between diminishing gas yield per unit of substrate and the unit cost of the digester per m^3. The optimum return per unit cost is at about 70-80 per cent of the theoretical biogas yield per kg of substrate. Retention times generally quoted: dairy cattle 20-30 days (37 and 38), pigs 10-15 days, (32 and 37), poultry, 20 days (37 and 38).

5. Volumetric loading rate, kg VS (or TS) per m^3 digester volume per day. It is controlled by the VS and TS of the input slurry and the retention time. On many farm systems, there is great difficulty in maintaining the TS content of the input slurry. Hobson et al (37) link the gas yield m^3/kg TS to an optimum TS content of the input slurry, for dairy cattle 7 per cent at a 20 day retention time, and for pig and poultry slurry 6 per cent at retention times of 10 and 20 days. The optimum volumetric loading rate for dairy cattle will be 3.5 kg TS per m^3 digester volume per day.

6. Biogas composition. The composition of biogas is affected by the species and diet fed. The methane content varies between 55 and 75 volume per cent. Its calorific value varies from 20 to 26 MJ/m^3. Cattle slurries generally produce gas with the lower value, poultry and pig slurries achieve the higher value. Hydrogen sulphide gas is present in the gas, the content varies between 0.5 and 1.3 per cent, (40). At these concentrations the gas is toxic to livestock and man and highly corrosive to metal in the presence of water vapour, (41).

4.2 Design of farm digester systems

Most agricultural systems are designed to treat slurries with a TS content of between 2 to 10 per cent and are therefore liquid handling systems. 68 per cent of the European systems surveyed, (29) were semi continuous flow, completely mixed digesters, single stage, 16 per cent plug flow digesters and the rest of experimental designs.

Semi-continuous flow, completely mixed digesters consist of a collection tank into which slurry is either piped or scraped from build-ings. The tank is usually designed to hold one day's supply, and is built below ground in either concrete or steel. The tank may be stirred or mixed mechanically to maintain an even supply of solids in suspension to the digester. Most digesters are supplied by pumping, usually for a predetermined time each hour. The discharge point into the digester can be at the base, top or half-way down, there are no design criteria.

The digester tank shape, height and width vary according to the manufacturer, prefabricated tanks are generally modifications of existing slurry or feed storage tanks. The tanks are usually made of concrete, steel glass enamelled steel sheets or polymeric materials. Digesters are usually erected above ground, set on a concrete base.

The tank walls and roof are often insulated, either internally or externally by synthetic polymeric materials; polyurethane or polystyrene, glass wool or rock wool. Design parameters to obtain maximum insulation at $0^{o}C$ have been described by Hawkes, (38). It has been shown that internal insulation may damage prefabricated enamelled steel tanks due to distortion by solar heat.

Heating digesters can be achieved by direct heating using internal heat exchangers or by re-circulation through external heat exchangers. There are a large variety of systems in use, at present, (29), these have never been evaluated to determine the most efficient system in terms of heat exchange and ease of operation, (fouling and blockages to both the exchangers and the digester).

Mixing the digester contents is necessary to maintain even tempera-ture gradients through the digester, to maintain an even supply of sub-strate and prevent settlement and the formation of a surface crust. There are a variety of mixing systems, (29), ranging from pumping, gas mixing to mechanical stirring. An evaluation of the efficiency of mixing, energy requirements and ease of maintenance needs to be made.

Discharge of the treated effluent from the digester can be by gravity flow through a gas trap or pumping. The position of the draw off point within the tank varies with the design of system. Few digesters are designed to remove heavy settled sludges, which develop due to the ingress of small particles of grit, cement dust and the percipitation of inert minerals during digestion. Most systems have to be shut down and cleaned out manually.

Gas storage is needed to balance supply and demand and thus contin-uity of supply. The storage containers are kept at a constant pressure to maintain the gas supply at the correct volume for utilisation. The gas store may be used as a reservoir to maintain supplies over periods of peak demand or during breakdown or maintenance.

Farm gas stores usually contain enough gas for 15 minutes to 24 hours supply. The stores are generally of two types, variable pressure and fixed volume or variable volume and fixed pressure. The latter are most often used on farms. Types most commonly found are floating gasholders made of steel and plastic membrane balloons, (29). Volumes stored range from 20 m^3 to over 830 m^3, (41). Farm gas stores design and the construction materials, need to be evaluated and engineering and design standards set, for practical working systems and safety.

Biogas purification and scrubbing systems to remove water vapour, hydrogen sulphide and carbon dioxide gases are rarely used on farms. The systems employed are simple and rarely completely effective. There is a need for scientific design criteria.

Energy saving systems to retain heat from treated slurry and to pre-heat incoming untreated slurry are being developed. Preheating systems are used in some countries such as the hot well designs used in Danish digesters. There is a need for an evaluation of present systems and to establish design parameters.

Pre-treatment of cattle slurry by mechanical separation to remove coarse solids of low digestibility has been examined at laboratory scale by Hawkes et al, (42). The results indicate that positive net energy yields can be obtained throughout the range of 10-20 day retention time, at temperature ranges from 25-35°C, with the optimum at 30°, and 20 day RT. Farm scale development by Pain et al, (43), compared anaerobic digestion of whole cow slurry 7 per cent TS at 20d RT with separated slurry at 4 per cent TS at 20 and 15d RT. Gas yields for whole slurry were $0.204 \, m^3/kg$ TS at 20d RT with calorific value of 21.3 MJ/m^3 and for separated slurry 0.279 m^3 and 0.251 m3/kg TS for 20d and 15d retention times with a calorific value of 21.7 MJ/m3. Mechanical separation has advantages in reducing costs by shortening retention time and hence reducing digester capacity.

4.3 Utilisation of biogas and treated slurry on farms

Efficient biogas utilisation throughout the year is essential for the successful economic performance of a digester on a farm. Matching supply to demand has been found to be one of the most difficult operations within the decision making process on the feasibility of anaerobic treatment for a particular farm. It is essential that the site survey described, be carried out to determine the volume and suitability of the substrate/ slurry. It is equally essential that an appraisal of the farms energy requirement is carried out, detailing the farm energy costs for all operations including domestic and industrial. It is then possible to determine which operations can be supplied by the proposed system, (44).

Most systems are designed to produce electricity. Converted petrol or diesel engines are commonly used with generators or purpose made combined heat and power units, (40). Engine cooling water is frequently used to provide heat to digester, buildings and for domestic use. Smaller digesters more frequently produce biogas for space heating livestock buildings and for domestic use. Surplus gas is often vented through lack of storage or occasionally surplus electricity is sold to the national grid. In Italy 50 per cent of biogas plants generate electricity using cogeneration combined heat and power units, 35 per cent produce hot water for space heating, some is used for cheese making. 10 per cent of plants have no use for the biogas, (45).

Utilisation of treated slurry as a fertiliser has been compared to untreated slurry in a number of studies. Comparison of barley and beet yields indicated that there was no significant difference (46), similar results were obtained for grassland, (47).

4.4 Economic performance

An appraisal of performance of biogas plants in the UK, (26) indicated that plants could only be economically viable provided that a

monetary value for odour control was included in the assessment.   There
are two important aspects of production and utilisation which must be
satisfied, these are the efficiency of gas production in terms of output
per unit capital cost and the utilisation efficiency, gas production must
match demand.   A study of the economics of biogas production in the
Netherlands, (27) came to similar conclusions.   Farm appraisals indicate
that plants could not produce and utilise biogas efficiently enough to
operate at a profit.   Including odour pollution control as saving in
other plant can justify the installation of biogas systems.

4.5 Conclusions
      The installation of anaerobic digesters on livestock farms has
advantages as an on farm supply of energy and as a treatment system to
control odours.   The laboratory and pilot scale digestion of livestock
slurries has been carried out successfully to produce good gas yields
and effective odour control.   The translation of this work into farm
scale plants requires further research and development work into the
following:
1.   better quality and control of the feedstock
2.   reducing the energy demands of the digester
3.   fundamental improvements to digester design
4.   improving gas utilisation by producing a cleaner and better quality
      gas
5.   improving gas storage
6.   matching demand and supply
7.   improving energy conversion into alternative methods of utilisation.

5.   RECOMMENDATIONS FOR FURTHER WORK
      Biological treatment systems for liquid manures have been the subject
of much research over many years.   Operational parameters have been
determined for both aerobic and anaerobic treatment systems in the laboratory.   The translation of these operations into farm scale systems has
shown up weaknesses in the design and operation of the plant, (detailed
comments are made in the paper).
      For new buildings and manure handling systems, treatment plants must
be designed to integrate with the total system.
      Further research and development is required into the engineering and
technical design aspects of collection, pre-treatment, treatment, storage
and application equipment to enable a consistently satisfactory standard
of treatment to be maintained together with the optimum utilisation of the
products of treatment.

6.   REFERENCES
      (1)   Baines, S., Evans, M. R., Hissett, R. and Hepherd, R. Q., (1973).
            The principles of treatment of animal slurries.   The Agricul-
            tural Engineer 28, 2, 72-76.
      (2)   Loehr, R. C., (1972) Animal waste management - problems and
            guidelines for solutions.   J. Environ Quality 1 71-78.
      (3)   Owens, J. D., Evans, M. R., Thacker, F. E., Hissett, R and
            Baines, S., (1973).   Water Res 7 1745-1766.
      (4)   Vetter, R., and Ruprich, W. (1980) Investigations into the
            treatment of animal excrements by aeration to reduce smell, aid
            disinfection and reduce volume.   Effluents from Livestock,
            Applied Science Publishers 580-607.
      (5)   Evans, M. R., Hissett, R., Ellam, D. F. and Baines, S. (1975).

Aerobic treatment of piggery waste prior to land treatment. Managing Livestock Wastes, Proced of 3rd Internat Symp. ASAE. 556-559.

(6) McAllister, J. S. V., (1976) Agriculture and Water Quality, MAFF Bulletin 32 H.M.S.O. 418-431.

(7) Nielsen, V.C., (1977). An evaluation of the effects of aerobically treated cow slurry compared with untreated cow slurry when applied to grassland. Completion of M phil thesis, Reading University.

(8) Evans, M. R. and Baines, S. 1985. Aeration and Odour control by heterotrophic and autotrophic micro-organisms. FAO/EEC Workshop on odours, MIAE, Silsoe, UK to be published by the E.C.

(9) Bonazzi, G. Copelli, M. and de Angelis, S. (1985), Aeration of pig slurry to control odours and to reduce nitrogen levels. FAO/EEC Workshop on odours, N.I.A.E. Silsoe, UK to be published by the EC.

(10) Evans, M. R., Svoboda, J. F., and Baines, S. (1985). Heat from aerobic treatment of liquid animal wastes. Composting of agricultural and other wastes. Editor J.K.R. Gasser. Elsevira, Applied Science Publishers.

(11) Bohm, H. O. (1983) The effect of aerobic-thermophilic treatment on liquid manure pits containing different viruses in Hygienic Problems of Animal Manures. Editor D. Strauch. Inst of animal medicine, University of Hohenheim, Stuttgart. 183-198.

(12) Tjernshaugen, O. and Gjervan, J. O. (1980), The slurry pit as a source of heat. Offprint 204. The Inst of Bldg, Tech, National Norwegian College of Agriculture. AD-NLH.

(13) Tjernshaugen, O. and Gjervan, J. O. (1983), Report 187, Inst of Bldg. Tech. National Norwegian College of Agriculture.

(14) Evans, M. R. Hissett, R. Smith, M. P. W. Thacker, F. E. and Williams, A. G. (1980). Aerobic treatment of beef cattle and poultry waste compared with treatment of piggery waste. Agricultural Wastes 2 93-101.

(15) Evans, M. R. Deans, E. A., Hissett, R. Smith, M. P. U., Svoboda J. F. and Thacker, F. E. (1983). The effect of temperature and residence time on aerobic treatment of piggery slurry - - degredation of carbonaceous compounds. Agric wastes 5. 25-36.

(16) Williams, A. G. Shaw, M. Selviah, C. M. and Cumby, T. R. (1985). Oxygen requirements for controlling odours from pig slurry by aeration. FAO/EEC Workshop on odours, N.I.A.E., Silsoe, UK to be published by the EC.

(17) Evans, M. R., Svoboda, J. F. and Baines, S. (1982). Heat from aerobic treatment of slurry in completely mixed reactors. J. Agric. Engng Res 27. 45-50.

(18) Williams, A. G. (1984), Indicators of piggery slurry odour offensiveness Agric wastes 10. 15-36.

(19) Ginnivan, M. J. (1983), The effect of aeration rates on odours and solids of pig slurry. Agric wastes 7. 197-207.

(20) Williams, A. G., Schofield, C. P. (1984). Overseas visit to Norway and Sweden. Ref VRO 171 N.I.A.E. Silsoe, Bedford UK.

(21) Hughes, D. (1983), Pig house heating from aerobic wastes using a heat pump. Paper 122. British Society of Animal Production winter meeting.

(22) Pain, B. F., Hepherd, R. Q., and Pittman, R. J. (1978). Factors

affecting the performance of four slurry separating machines.
J. agric Engng. Res. 23. 231-242.

(23) Cumby, T. R. (1985). A review of slurry aeration 3. Performance of aerators. Submitted for publication J. agric. Engng
Res.

(24) Sneath, R. W. (1978). The performance of a plunging jet
aerator and aerobic treatment of pig slurry. Wat Pollut
Control 77. 408-420.

(25) Cumby, T. R. (1985), A review of slurry aeration 1. Factors
affecting oxygen transfer. Submitted for publication J. agric
Engng. Res.

(26) An economic assessment of anaerobic digestion. A discussion
document of an ADAS/BABA working party. (1982.) Min of Agric.
Fish. and Food.

(27) Voermans, J. A. M. (1985). Farm experiments of anaerobic
digestion to control odours from slurry. FAO/EEC Workshop on
odours, N.I.A.E. Silsoe, UK to be published by EC.

(28) Hume, I. H., (1985) Pers Comm. MAFF. 79/81 Basingstoke Road,
Reading.

(29) Demuynck, M. Nyns, E. J. and Palz, W. (1984) Biogas plants in
Europe. Solar Energy R & D in the European Community Series E
6. D. Reidel publishing company.

(30) Welsh, F. W. Schulte, D. D. Kroeker, E. J. and Lapp. H. M.
(1977) The effect of anaerobic digestion upon swine manure
odours. Canadian agric Engng 19.2. 122-126.

(31) Klarenbeek, J. V. (1982). Odour measurements in Dutch agricul.
Research report 82-2 Instit agric Engng, Wageningen, NL.

(32) Velsen, A. F. M. van (1981). Anaerobic digestion of piggery
waste. Thesis L H Agricultural University, Wageningen, NL.

(33) Lund, E (1983), Inactivation of viruses under anaerobic or
aerobic stabilisation in liquid manure and in sludge from
sewage plants. Hygienic Problems of Animal Manures Ed
D Strauch, Institute of Animal Medicine, University of Hokenheim,
Stuttgart, FRD 199-124.

(34) Demuynck, M, Nyns, E. J. and Naveau, H. P. (1984.) A review of
the effects of anaerobic digestion on odour and on disease
survival. Composting Agricultural and other wastes. Ed. Gasser,
J. K. R. Elsevier Applied Science publishers.

(35) Turner, J. Stafford, D. A. Hughes, D. E. and Clarkson, J (1983).
The reduction of three plant pathogens, (Fusarium, Corybacterium
and Globodera) in anaerobic digesters. Agric Wastes 6.1. 1-11.

(36) Pfeffer, J. T. (1979). Anaerobic Digestion Processes, Anaerobic
Digestion. Ed by D. A. Stafford, B. I. Wheatley and D. E.
Hughes Applied Science Publishers 15-33.

(37) Hobson, P. N. Bousfield, S. and Summers, R. (1981) Methane
Production from Agricultural and Domestic Wastes. Applied
Science Publishers 10-48.

(38) Hawkes, D. L. (1979) Factors affecting net energy production
from mesophilic anaerobic digestion. Anaerobic Digestion
Ed by D. A. Stafford, B. I. Wheatley and D. E. Hughes, Applied
Science Publishers 131-149.

(39) Hayes, T. D., Jewell, W. J., Dell'Orto, S. Fanfoni, K. J.
Leuschner, A. P. and Sherman, D. F. (1979). Anaerobic
digestion of cattle manure Anaerobic Digestion. Ed D. A.
Stafford, B. I. Wheatley and D. E. Hughes, Applied Science
Publishers 255-285.

(40) Friman, R. (1984). Monitoring anaerobic digesters on farms. J. agric Engng Res 29 357-365.

(41) Noren O (1975). Noxious Gases and Odours. JTI-rapport 1 Swedish Institute of Agricultural Engineering, Ultuna, Uppsala.

(42) Meszaros, G. and Matyas, L. (1985). Large scale biogas plants in Hungary. FAO/EEC Workshop on odours, NIAE, Silsoe, UK to be published by EC.

(43) Hawkes, F. R. Rosser, B. L. Hawkes, D. L., and Statham, M. (1984) Mesophilic anaerobic digestion of cattle slurry after passage through a mechanical separator. Factors affecting gas yield. Agric Wastes 10. 4. 241-256.

(44) Pain, B. F., West, R., Oliver, B., and Hawkes, D. L., (1984). Mesophilic anaerobic digestion of dairy cow slurry on a farm scale: First comparisons between digestion before and after solids separation. J. agric Engng Res 29. 249-256.

(45) Hardiman, N. (1984) Rationale for change in biogas utilisation on Bore Place Farm. Report of a symposium on using biogas. RURAL report 3. Bore Place, Chiddingstone, Edenbridge, Kent.

(46) Tilche, A. (1984), Biogas utilisation in Italy. RURAL Report 3, Bore Place Farm, Chiddingstone, Edenbridge, Kent.

(47) Dam Kofoed A. and Sondergard Klausen, P. (1983) Fertilizer effect of fermented slurry from biogas production and non fermented slurry. No. 1641. Statens Planteavlsforsog, Askov, 6600 Vejen, DK.

(48) Suess, A. and Wurzinger, A. (1985). The effect of anaerobic digestion on nutrient value of farm manure. Bavarian State Instit for Soil and Plant Cultivation, Munich, FRD.

DISCUSSION

J.H. VOORBURG

You mentioned reduction of the nitrate concentration during treatment. How is this achieved? For example, through N_2 or NH_3 loss after conversion from the nitrate via bacterial action?

V.C. NIELSEN

The reduction is due to ammonia loss caused by stripping. Ammonia salts can be oxidised to nitrates and so it is possible to achieve both nitrification and denitrification in the same system. In all the reduction can be as high as 70% of the original concentration.

J.H. VOORBURG

Have you considered the use of enzyme additive to speed the treatment of farm slurries?

V.C. NIELSEN

Dr S. Baines and Dr M.R. Evans in Scotland have looked at this possibility and have shown that a very high concentration was required to change the character of the sludge. The bacterial content of bacterial additives was too dilute to have any observable effect. Furthermore, the mutated strains were unable to compete with indigenous bacteria; I have not noticed any effect when these compounds have been used in slurries.

C. JUSTE

Are the sludges produced by the anaerobic or aerobic treatment of farm slurries really useful for agriculture?

V.C. NIELSEN

Treatment either by an anaerobic or an aerobic process removes soluble organic matter which is used by the micro-flora and converted into biomass. Therefore the original organic matter becomes less mobile in the soil and the risk of pollution by run-off is reduced.

STABILISATION OF SEWAGE SLUDGES AND LIQUID ANIMAL
MANURES BY MESOPHILIC ANAEROBIC DIGESTION - AN OVERVIEW

A. M. BRUCE
Water Research Centre
Stevenage, Herts, UK

Summary

Mesophilic anaerobic digestion is the longest-established and most important method of stabilisation of sewage sludges in all the countries participating in COST 681. About half the sludge produced in the EEC (3.3 million tonnes DS per year) is subjected to this process in over 1700 sludge digestion plants. An increase in this number of plants is expected. The application of anaerobic digestion to treatment of animal slurries is a much more recent development and is still on a small scale with about 300 digesters only in Europe treating around 80,000 tonnes DS per year. Farm-scale digestion can be successful but is usually uneconomic for the farmer unless subsidised. Digester gas represents a very small energy source in national or community terms but is important locally especially at sewage works.
In regard to design, the most important developments are in digester shape, in the use of prefabricated steel instead of concrete for tanks, in the use of various forms of gas mixing and in the growing use of pre-heating (for pasteurisation) by aerobic thermophilic digestion and other methods. Mesophilic anaerobic digestion itself is very effective in reducing pathogens but it is recognised that operation of the digestion process as a 'completely-mixed' system can reduce removal efficiency. Ways of overcoming this disadvantage are discussed.
The most important problems still occurring with sewage sludge digesters are operational ones - the feeding of thin sludge, overloading, occasional chemical inhibition and sludge foaming. With farm digesters, the problems relate also to design and engineering aspects and ways of improving these and the economics of farm digestion will need to be faced before it is widely accepted.

1. INTRODUCTION

The stabilisation of putrescible organic wastes by anaerobic digestion represents one of the longest-established and most successful applications of 'biotechnology' in the field of pollution control. It is now over 60 years since the earliest heated and enclosed digestion tanks with gas collection facilities were installed at sewage works in Europe for separate stabilisation of raw sewage sludge. Since then, anaerobic digestion plants have demonstrated their value at many hundreds of sewage works both in Europe and other areas of the world and the number of such

plants is still increasing. There have, of course, been operational
problems in some cases, resulting mainly from poor process control,
overloading or chemical inhibition. As a consequence, anaerobic
digestion has occasionally waned in popularity in some areas but the main
weight of operational experience has shown the process to be robust and
reliable, and effective for its main purpose of odour reduction.

Anaerobic digestion has also found successful application for
treatment of many other types of concentrated organic wastes and, in the
last 10 years especially, there has been particular interest in the use
of the process for stabilisation of pig, cattle and poultry slurries on
individual farms or groups of farms. The main incentive for installation
of farm digesters has been to provide a 'low cost' energy source from the
digester gas but odour reduction, BOD removal, and improvements in slurry
handling properties have also been seen as important benefits of the
process from the environmental aspect. In most countries, much more
animal slurry than sewage sludge is produced so the potential, at least,
for farm-scale digestion is very large.

The scope of the COST 681 activity covers research and development
on the treatment and use of both sewage sludges and liquid animal
manures. There are some broad similarities between the two types of
waste and so far as anaerobic digestion is concerned there is a common
research interest because the process is essentially the same whether
applied to the farm situation or to the sewage works. The main
difference is in the scale of operation and in the financial resources
available to instal and operate a digestion plant - both being generally
much larger in the public 'sewage works sector' than in the largely
private 'farm sector'. There are, however, also some important
differences between the characteristics of sewage sludges and animal
slurries which means there must be some difference in approach to the
design and operation of digestion plant for the two situations. But in
the main, many of the research developments in anaerobic digestion have
applications in both sectors.

The object of this paper is to broadly review the current situation
in regard to anaerobic digestion as applied to sewage sludges and to
animal slurries, to consider recent developments and likely future trends
and to compare the research needs in both areas.

2. CURRENT USAGE OF ANAEROBIC DIGESTION

A recent survey (1) has provided information about the current level
of use of heated anaerobic digestion for stabilising sewage sludges in
most of the countries participating in the COST 681 activity (Table 1).
It is seen that, among the 9 EEC countries (Luxembourg not included)
about 1800 sewage works currently employ anaerobic digestion plants. In
total, these plants treat an estimated 3.3 million tonnes dry sludge
solids (80 million wet tonnes) per year. This quantity represents
sludge from a population of some 110 million people and is over 50 per
cent of the total mass of sewage sludge currently produced in the 9 EEC
countries included in the survey. Earlier surveys (3) had shown that
anaerobic digestion is by far the most widely used method of sewage
sludge stabilisation in all countries.

Most anaerobic digestion plants are located at medium and
large-sized works serving populations ranging from 50,000 to over 2
million. The latest survey shows that there are many hundreds of works
without digesters but these are mainly of very small size (serving less

Table 1. Comparison of use of mesophilic anaerobic digestion for stabilisation of sewage sludges at sewage works and of liquid animal manures on farms in various countries

Country	SEWAGE WORKS					FARMS			
	Annual quantities of sewage sludge (a)		Total number of works (b)	Number of works with anaerobic digestion plants (a)	Total digester capacity (103 m3)	Annual quantities of liquid animal manure		Number of farms with anaerobic digesters (full-scale) (c)	Total digester capacity (103 m3) (c)
	Produced (103 tonnes dry solids)	Anaerobically digested (103 tonnes dry solids)				Produced (103 tonnes dry solids)	Anaerobically digested (103 tonnes dry solids)		
BELGIUM	30	19	100	44	23	*	3.7	18	3.4
DENMARK	130	65	6400	140	79	*	4.9	22	4.5
FRANCE	840†	~400†	>4000	200†	487	*	4.1	29	3.7
FR GERMANY	2500	1625	8000	800	1979	*	10.1	61	9.2
GREECE	10	10	*	1	12	*	0.8	3	0.7
IRELAND	20†	2†	160	1†	3	*	0.04	2	0.03
ITALY	800	350	1500	100	426	*	50.0	55	45.7
NETHERLANDS	258	98	590	220	119	*	6.0	20	5.4
UNITED KINGDOM	1400	740	7800	280	900	*	3.3	11	3.0
EEC (9) TOTAL	5988	3309	28550	1786	4028	>50,000	82.9	221	75.6
CANADA	500	315	1544	133	383	*	*	*	*
FINLAND	130	35	550	13	42	*	*	*	*
SWITZERLAND	150	120	800	650	145	*	14.3	94	13.1

* No data available

(a) Data based on information provided by COST 681 Working Party No. 1 Members (1985) with the exception of † which are estimates from other sources

(b) Data from Ref. 2

(c) Data from Ref. 3

than 5000 population equivalent) and generally not suitable for heated anaerobic digestion. Some of these works employ cold (unheated) anaerobic digestion in lagoons or tanks or use other forms of stabilisation, for example aerobic digestion or lime stabilisation. The scope for a significant increase in the numbers of heated anaerobic digestion plants in the future is still, however, quite large. This applies particularly to existing works of medium size (20,000-50,000 population) and to new larger works currently under constructon or planned. For example, in the UK where just over 50 per cent of all sewage sludge is anaerobically digested, it is estimated that there are about a hundred sewage works serving populations of over 50,000 which still have no digestion facilities. It is evident that future investment in new digestion plant could be considerable particularly if, as expected, environmental pressures on sludge disposal authorities continue to increase. Further research and development to improve the overall economics of anaerobic digestion for sewage sludges is therefore strongly indicated.

The position in regard to the use on farms of heated anaerobic digestion plant is rather different. Results of an EEC-sponsored survey (4) carried out in 1983 are included in Table 1. These show that, two years ago, there were about 220 farms only in the EEC countries with full-scale operational digesters treating liquid slurry from pigs, cattle or poultry. The total increases to 300 plants if Switzerland is included. Apart from the 90 or so digesters in Switzerland, most farm digesters are located in Italy and FR Germany where there are some financial incentives to treat slurries by this means. It is thought that the number of new farm digesters has not increased significantly in the past two years, so the data in Table 1 is still reasonably accurate. The actual quantity of animal slurry being anaerobically digested is estimated at around 80 thousand tonnes dry solids per year. This is only about 2.5 per cent of the amount of sewage sludge being digested and an even smaller proportion of the total mass of 'available' animal slurry produced each year in the EEC countries (estimated at around 50 million tonnes dry solids).

The reason for the relatively low usage of anaerobic digesters on farms is quite clear. Except where the reduction in polluting load from a slurry has an economic value (e.g. in parts of Italy) or where state subsidies are provided for the investment costs of a farm digester (e.g. in FR Germany), the installation and operating costs render the farm digester uneconomic for most farmers. Even with a 'do-it-yourself' approach to construction and with the use of surplus digester gas to provide energy on the farm, the 'pay-back' periods are still usually over 5 years and may be as high as 10 years (4). Operational problems and high maintenance requirements have also given farm digesters a bad reputation in some areas so that there has been little interest in installing them.

The future growth of farm-scale anaerobic digestion is therefore rather uncertain. It will depend to some extent on whether or not environmental constraints on slurry disposal become more severe and on the availability or otherwise of official subsidies to encourage farm-waste treatment. The future popularity of farm digesters will also depend very much on the availability of low-cost and reliable plant and equipment.

3. DESIGN TRENDS

The broad suitability of some of the early designs of anaerobic digester used for sewage sludges is evidenced by the fact that many are still in service after 50-60 years operation. None the less, there have been important design and operational improvements in recent years which have helped to reduce construction costs, improve the efficiency and reliability of the digestion process and reduce maintenance requirements.

3.1 Materials of construction

The traditional use of reinforced concrete for constructing the shell of digestion tanks for sewage works has been maintained in most countries and indeed it is still considered essential for tanks of over about 2000 m^3 capacity. Capacities of individual digesters on large works can range up to 12,000 m^3. Steel tanks have been used to a limited extent for many years but a more dramatic departure from the 'concrete tradition' has occurred in the last 6 years in the UK where the use of prefabricated glass-coated steel tanks (based on farm silo technology) has now become about the most common form of construction for new digesters. Tank volumes of prefabricated digesters may range up to 1000 m^3 which means that for very large installations concrete tanks are still used. The alternative is a multiplicity of smaller prefabricated steel tanks which can prove more costly. A feature of the prefabricated digester is the very effective thermal insulation which can be provided resulting in much reduced thermal losses through the digester wall and an overall reduction in heat requirements especially in the winter.

The reduction in total capital costs by using prefabricated steel digesters instead of concrete tanks is considerable. Costs data for the UK have shown that a halving of capital costs is possible (5) over a range of plant sizes. This saving would not necessarily justify the use of steel tanks if the lifetime of the tanks was likely to be very short or if maintenance costs were high. But the probability is that a properly installed and protected glass-coated steel tank will have a lifetime similar to that of a concrete one. So far, however, the UK appears to be the only country which has adopted to any significant extent the use of prefabricated tanks as a more economic alternative to the conventional concrete tank at sewage works.

In regard to farm digesters, the EEC survey (4) showed that both concrete and steel digesters are in use in all the countries surveyed and in roughly equal numbers. The use of polymeric materials such as glass reinforced plastics (GRP) for tank construction is also known (6) but mainly for the small sizes of digester (i.e. 100 m^3 working volume or less). The use of GRP for sewage sludge digesters has not been encountered.

3.2 Digester shape

The traditional 'low-form' digestion tank (Diameter > Height) employed in the UK and North America for many years (Fig. 1a) is now considered to be less conducive to efficient mixing than tanks with a high aspect-ratio (Height > Diameter). Although the 'low form' tanks often perform quite well they tend to be difficult to mix and to deposit grit. The taller form of digestion tank (Aspect ratios > 1) used traditionally at sewage works in mainland Europe (Fig. 1b) is seen in its most extreme form in the 'egg-shaped' digesters used for some very large plants in FR Germany and a few other countries (Fig. 1c). This shape is

Anaerobic Digesters - Tank Shapes

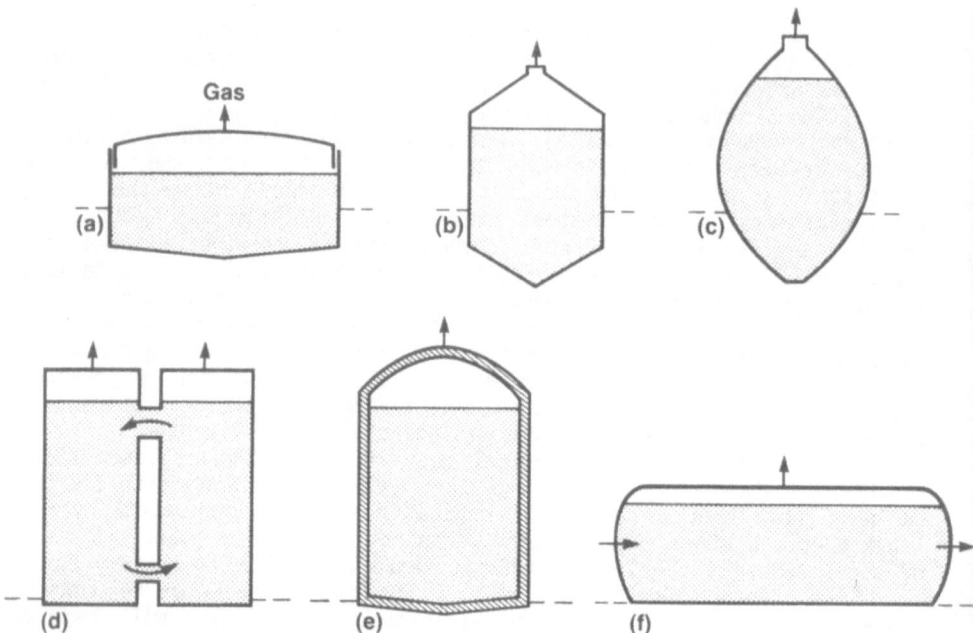

(a) 'Low-form' digester traditional in UK and N. America (sewage sludge)

(b) Traditional form of continental european digester (sewage sludge)

(c) Egg-shape used for very large continental digesters (sewage sludge)

(d) 'High-form' tower digesters with interconnections (sewage sludge)

(e) Pre-fabricated steel type (sewage sludge/animal slurries)

(f) Plug-flow tube digester (animal slurries)

Fig. 1

said to be economical to construct and to be efficient (7) but clearly it is only appropriate for very large plants. Tanks with very high aspect ratios of around 1.5 have also been employed in FR Germany (8) and in the UK (9). In the case of the German plant, pairs of tall tanks are hydraulically inter-connected at top and bottom to give a circular mixing pattern (Fig. 1d). The prefabricated steel digestion tanks now in use tend to have aspect ratios of about 1-1.2 (Fig. 1e).

The most radical departure from the conventional tank shape is seen in some farm digesters which are operated on the 'plug-flow' principle (Fig. 1f). The tanks are usually constructed in GRP are essentially horizontal and tubular in shape with semi-spherical end walls. This form of construction is claimed to reduce costs (6).

3.3 Mixing systems

All anaerobic digesters designed for sewage sludges are operated essentially as completely-stirred tank reactors (CSTRs) and only in the farm digesters mentioned earlier is there an approach to 'plug flow' conditions. In the strict sense, there is no absolute need for mixing in order for the anaerobic digestion process to function. But the mixing system of a digester does serve several important purposes - i.e. uniform dispersal of raw sludge feed throughout the tank, prevention of short circuiting, maintenance of uniform temperature conditions, reduction in grit deposition and scum accumulation, and prevention of stratification.

The trend over the years has been away from mechanical mixing systems (Figs 2a, 2b, 2c) and towards gas mixing (Figs 2d, 2e, 2f). Although gas mixing is, in some cases, less efficient in terms of energy requirements it reduces maintenance problems and can be very effective. It is claimed that unconfined gas mixing (Fig. 2e) is more effective than the confined type but both are used successfully in practice. In most cases, the actual energy requirements for mixing are quite small (5-20 w/m^3 digester volume) and it is usually more important to select an effective system with low maintenance costs than a system requiring minimum energy. Casey (10) has recently reviewed the whole subject of digester mixing and has described a novel method of 'see-saw' mixing which may have some advantages over conventional systems but it is still subject to full-scale evaluation.

3.4 Heating systems

Virtually all heated anaerobic digesters at sewage works in Europe and, as far as is known, all farm digesters are designed to operate in the mesophilic temperature range (32-37°C). The numbers of thermophilic anaerobic digesters (operating in the range 45-55°C) are very small though they are reported (11) to work satisfactorily in Canada and the USA. The use of unheated (cold) anaerobic digesters or lagoons is outside the scope of this paper though they are important locally in some areas (12).

Most digesters are heated by hot water boiler or by waste heat from a gas engine with a water/sludge heat exchange system either internal or external to the digester (Fig. 3a). Some small digesters provide supplementary (or emergency) heating by electrical immersion heating (3a). The use of direct injection of live steam (Fig. 3b) has been successfully revived recently in the UK with some success and claimed low costs (13) and it is also used at a few plants in mainland Europe. A

Anaerobic Digesters - Mixing Systems

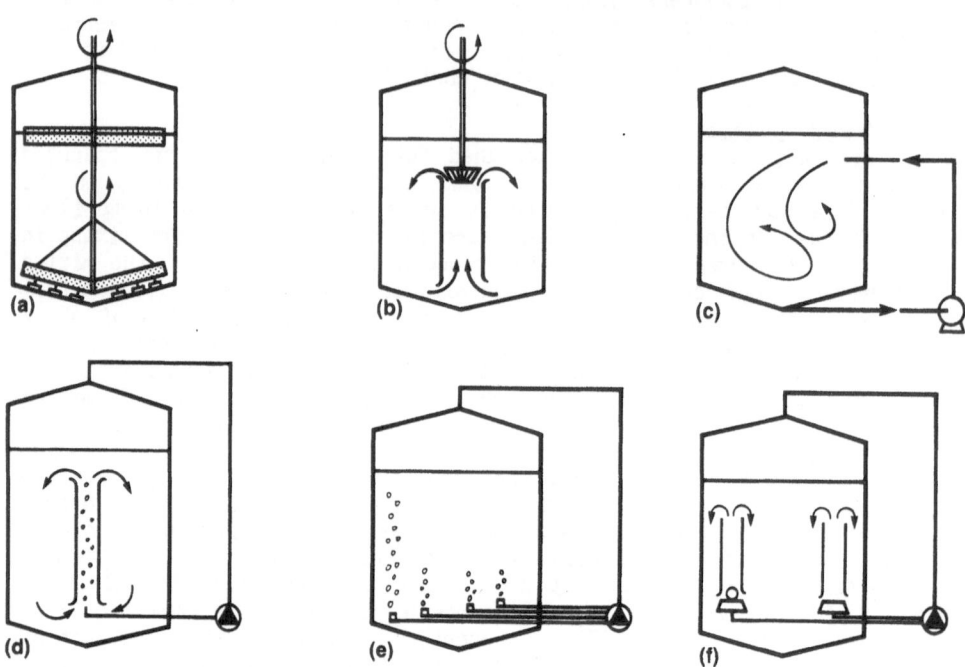

(a) Conventional european mechanical mixing with bottom scraping

(b) Rotating cone with updraft tube

(c) Pumped recirculation

(d) Confined gas lift system

(e) Unconfined gas lift

(f) Bubble-gun/"Burper" type of gas mixing

Fig.2

Anaerobic Digesters - Heating Systems

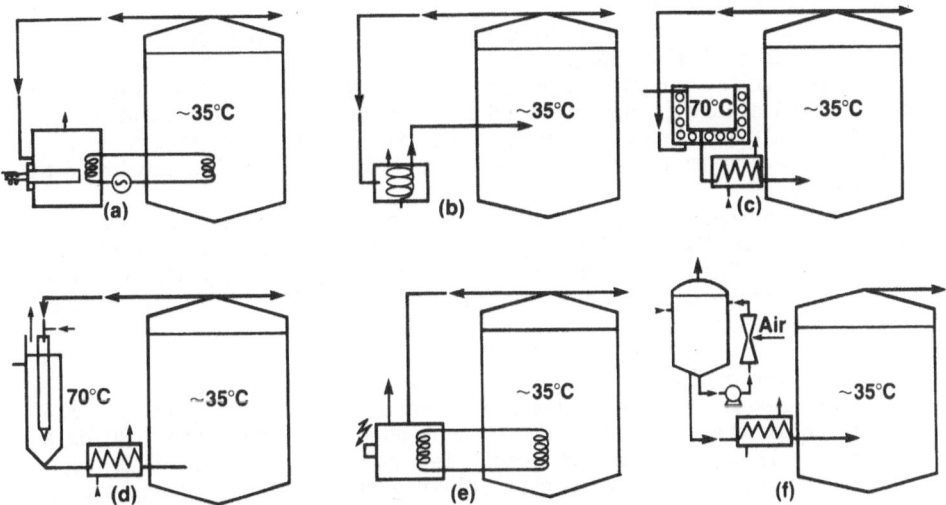

(a) Hot water boiler and sludge/water heat exchange (internal or external). Standby electric immersion heater

(b) Steam boiler and injection of live steam

(c) Pre-pasteurisation (to 70°C) by hot water boiler followed by sludge cooling

(d) Pre-pasteurisation (to 70°C) by submerged combustion followed by sludge cooling

(e) Heat recovery from gas engine (combined heat and power)

(f) Pre-heating by thermophilic aerobic digestion followed by cooling

Fig. 3

major advantage of steam is the direct heat transfer to the sludge and elimination of the need for heat-exchange equipment. The need to deionize the feed water and the necessary safety equipment required for steam boilers represent only minor disadvantages.

Where pre-pasteurisation is required, by heating the sludge to 70°C for a minimum period of 30 minutes, it is still most common to use hot water boilers with heat exchangers (Fig. 3c) followed by sludge cooling before pumping it into the digester. In FR Germany (14) and more recently in the UK (15), submerged combustion has been used successfully for pre-pasteurisation (Fig. 3d) and has the advantage of high heat-transfer efficiency (80-90%). Again, some cooling of the sludge is normally required before it enters the digester but no further heating of the sludge is necessary.

Another successful approach to pre-pasteurisation, and one which is becoming increasingly popular, is to subject the sludge to a short period of thermophilic aerobic digestion (1-2 days at 60-70°C) (16). The pasteurised sludge is again cooled before anaerobic digestion at 35°C.

The only practicable system of heating for farm digesters is by hot water boiler or by using the waste heat from a gas engine (combined heat and power).

In practical terms, the most common problem on farms is insufficient gas to provide the heating requirements for the digester especially in the winter. The root cause of this problem is usually the feeding of slurry of inadequate solids content. A minimum of 6% dry solids is recommended (17).

3.5 Retention period

For sewage sludge digestion plants, there has been a gradual reduction in the accepted design retention period from around 30 days to closer to 15 days. In most countries now, a period of 15-20 days is considered quite adequate to give full digestion but the recent survey (1) showed that many digestion plants still operate at 25 days retention and more. Apart from reduced capital costs, the more intensive digestion achieved with somewhat shorter retention times has advantages in terms of self-mixing through more vigorous gas evolution (18).

For farm digesters, the EEC survey showed that the mean detention times used for digesters treating cattle slurry is about 29 days while a shorter average retention time of 19 days is normal for pig slurries. Some wide extremes around these mean values were observed.

3.6 Feed solids and feed regime

One of the long-standing common problems with both sewage sludge digestion plants and farm digesters is the feeding of sludge or slurry of low solids content (i.e. the feed material is too dilute). This gives rise to various problems but more particularly to that of inadequate gas yield to provide for the heating requirements of the digestion plant in the colder weather.

For sewage sludges, a minimum sludge solids content of about 3.5 per cent is necessary for a positive heat balance in cold weather conditions. The survey (1) showed that the typical range of feed solids in the various countries was 2-5 per cent DS. The introduction of pre-thickening plant is a gradual trend and with such plants it is quite feasible to produce feed sludges of 6-8 per cent DS. The advantage in terms of reduced digester volume requirements are considerable quite

apart from the much more favourable heat balance.   Satisfactory
digestion of feed sludges up to 9 per cent DS has been obtained (18).
    It has been recommended (17) that farm digesters should not be
operated with feed sludge solids content of less than 6 per cent though
this is difficult to obtain in many cases. Very thick slurries (10-11
per cent DS) have proved difficult to heat and mix (16) but the problem
can be overcome by special design features.
    The feeding regime of a digester is important.   Traditionally,
digesters were fed once or twice per day only but now it is much more
common to feed either several times per day or continuously. Continuous
feed is 'ideal' in the sense of maintaining steady loading conditions and
uniform gas production but it may have unfavourable effects in terms of
sludge minimum residence time characteristics and this is important for
pathogen removal (see Section 6). One way to minimise these problems is
to operate on a 'pump-out - pump-in' regime so that the sludge is
retained for a minimum period in the digester rather than be allowed to
continuously overflow.

4. ENERGY ASPECTS
    Anaerobic digestion is unique among waste treatment processes in
being, potentially at least, a net energy producer in the form of surplus
methane gas.   Much attention has been given to the energy aspects of
anaerobic digestion in recent years and 'energy-production' still stands
as one of the most important reasons why digesters are installed on
farms.   The use of anaerobic digestion at sewage works is not based
primarily on the energy-yielding advantages of the process but, none the
less, the exploitation of the energy from the gas is important in most
cases.
    The typical yields of digester gas from sludges and slurries are
well established (Table 2).

Table 2. Normal gas yields from sewage sludges and farm slurries

Type of waste	Typical gas yield per tonne volatile matter fed to the digester (m^3)
Sewage sludge	400-450
Pig slurry	450
Cattle slurry	200-400

    Taking these data and assuming that digester gas (65% methane: 35%
carbon dioxide) has a net calorific value of 22 MJ/m^3, it is possible to
calculate the total energy represented by the gas produced from digestion
of sewage sludge in the EEC countries and that from digestion of animal
slurries. It is seen (Table 3) that the total primary energy 'produced'
by anaerobic digestion in the form of digester gas is extremely small in
relation to the total primary fuel requirements of the 10 countries.
    However, although the energy in digester gas is not very important
in national or Community terms, it can be very important at the local

level. Thus in the UK, the energy in sludge digester gas represents about 30 per cent of the total energy requirements of the UK water industry. At the more local level of the sewage works, it has been appreciated for years that the generation of mechanical or electrical power using gas or dual-fuel engines can satisfy most or all of the energy requirements of the works (Table 4). In the case of works employing biological filters for secondary treatment a surplus of energy is available and cases are known now in the UK where surplus electricity is sold to the electricity supply utility thus yielding an income. This practice still seems to be rare in most of the other countries (1) although it is common almost everywhere to use digester gas to generate electricity for use on the works itself. The overall economics of power generation at sewage works is still a matter for study but it is thought that in many cases it can be an attractive proposition.

Table 3. The estimated total energy represented in digester gas produced annually by sewage sludge digesters and by farm digesters in the EEC in comparison with the total annual primary fuel requirements of the 10 EEC countries

	10 EEC countries	
	Annual energy produced or required	
	10^6 GJ	MTOE*
Total primary fuel requirement	41,000	915
Total energy in digester gas currently produced from digestion of sewage sludge and animal slurries	22	0.5
Maximum possible from sewage sludge	60	1.4
Maximum possible from all sewage sludge and animal slurries	700	15.6

* Million tonnes of oil equivalent

The efficient exploitation of digester gas as a source of energy is crucial to the economics of farm digesters. In many cases, however, too high a proportion of gas produced has to be used to heat the digester itself (3). The use of combined heat and power units (gas engines) is common and in some cases sufficient spare electricity may be produced to allow sale to the electricity supply utility. But often it is difficult to match the energy requirements of the farm with the available gas supply. There are also problems with the maintenance of gas engines especially where the gas is rich in hydrogen sulphide. It is considered most economic to use the gas for heating if practicable (16). The use of gas for lighting and cooking is less common.

Table 4. Energy balance for sewage treatment and digester gas use
(Typical Values)

	Type of Works	
	Biological Filtration	Activated Sludge
	(kWh/Head)	
Annual Energy Consumed		
Sewage Treatment	2.4	17.0
Sludge Treatment	2.0	2.0
Total	4.4	19.0
Annual Energy Produced		
Digester Gas Primary Energy	48.6	53.8
Electrical Energy*	12.1	13.4

* Assumes 25% overall conversion efficiency by gas engines and generators

Loll (19) has described the formation of a 'biogas community' in FR Germany. Five digestion plants in one village are interconnected by a common gas pipeline with a single large biogas storage tank. The system is experimental but is claimed to have several advantages including a balanced supply for users at all times. Apart from the gas producers themselves, other users may also buy gas from the system. The overall economics of this sort of scheme are, however, not yet evaluated.

5. ECONOMIC ASPECTS

The costs involved in the installation and operation of a sludge stabilisation plant are important whether the plant is for sewage sludge or for animal slurry. The most economic form of stabilisation process will normally be selected provided it meets the required process performance specification. As has been indicated earlier, anaerobic digestion is by far the most widely used process for stabilisation of sewage sludges and for farm slurries but there are other potential process options which might be selected in some cases. These are composting, lime addition, and aerobic treatment, and, in particular thermophilic aerobic digestion (TAD). The evidence suggests that the latter process is the most serious practical "competitor" to anaerobic digestion particularly at sewage works serving populations in the range 5000-50,000 (20). Thermophilic aerobic digestion has become fairly well established now in FR Germany where there are over 20 plants in operation (21) and there has been development work in the UK and elsewhere.

Recent comparisons of the capital and operating costs of anaerobic digestion plants and of thermophilic aerobic digestion plants for sewage works serving populations ranging between 5000 and 50,000 are shown in Table 5. The costings apply to UK conditions but the broad comparisons are thought to be valid for other countries as well. It is seen that, as expected, anaerobic digestion plants are much higher in capital costs than the corresponding TAD plants but the latter method incurs much higher operating costs.

Table 5. Comparison of costs of anaerobic digestion and thermophilic aerobic digestion plants (1985 prices)

Population served (thousands)	Sludge volume (m^3/d)	Mesophilic anaerobic digestion (1) Costs			Thermophilic aerobic digestion (2) Costs		
		Capital £k	Operating £k/a	NPC (3) £k	Capital £k	Operating £k/a	NPC (3) £k
5	9.15	71	3.0	109	48	6.6	131
10	18.3	98	5.3	164	56	10.2	183
20	36.6	123	9.2	237	74	17.3	289
50	91.5	294	19.4	536	118	38.3	595

Assumptions
For (1) Retention time 15 days; energy requirements 0.15 kWh/m^3 sludge treated. No value given to digester gas
   (2) Retention time 8 days; energy requirements 20 kWh/m^3 sludge treated
        Prefabricated glass-coated steel tanks with covers used for both types of plant with external insulation
   (3) Net present costs calculated on basis of 5 per cent discount rate over a period of 20 years

        If an economic assessment of the costs of the two processes is made by calculating the Net Present Costs (NPC) based on a 20-year life and 5 per cent discount rate, a direct comparison can be made. It is seen that, for the assumptions given, the NPC of anaerobic digestion plants is consistently less than that for TAD plants over the range of populations considered. However, the accuracy of the cost estimates is not such as to rule out use of TAD as an option and actual cases would need to be individually assessed. In making the comparison, no value has been given to the surplus digester gas that might be available from the anaerobic digestion plants and this could markedly improve the economics of anaerobic digestion. On the other hand, the advantages of thermophilic conditions (45-55°C) in regard to pathogen destruction gives the TAD process some technical advantage over mesophilic anaerobic digestion. In some cases this could represent a real cost benefit (e.g. where disinfection was a requirement) since it dispenses with the need for separate pre-pasteurisation.
        The economic assessment given in Table 5 could apply generally also to the farm situation if the same type of plant and equipment as used at sewage works plants were required. Usually, the farmer is looking for a cheaper approach, possibly on a 'do it yourself basis'. Even then, it has been shown that, at best, pay-back periods of 5 years are required (4).

6. PATHOGEN INACTIVATION

The effect of different types of stabilisation process on the pathogens present in sewage sludges has been the subject of much recent research. This topic wil be discussed in detail in another session of this symposium and it is only appropriate to deal briefly here with the situation regarding mesophilic anaerobic digestion. The pathogens of main concern are Salmonella and the eggs of the beef tapeworm Taenia saginata. It has been reported that anaerobic digestion for 20 days at 35°C has the following effects (22).

Salmonella   -   90% removal (median)
T. saginata eggs - Infectivity destroyed

Therefore, conventional anaerobic digestion has a powerful destructive effect on the pathogens of most concern. Further storage of digested sludge, and/or appropriate disposal methods will minimise any risk of transmission of disease even where there is not a complete removal of pathogens. However, the above levels of inactivation assume that all the sludge resides for 15-20 days in the digester. Since anaerobic digesters are operated as 'completely mixed' reactors there is a certainty that a proportion of sludge will have a residence period less than the nominal mean period. From the point of view of pathogen inactivation, batch operation or plug-flow operation would be ideal but this is not really feasible for sewage sludge digestion. As indicated earlier, some farm digesters have, however, been designed to operate in the plug-flow mode.

Operation in two or more digestion stages provides some improvement in sludge retention characteristics but is undesirable in terms of other process requirements and costs. It is considered that the best solution is to operate the digester on a 'Pump-out: Pump-in' system by which a volume of sludge (equivalent to 1 or 2 days flow) is pumped from the digester and then an equivalent volume of raw sludge is admitted. This guarantees that all raw sludge is retained for a minimum of 1 or 2 days. If such modes of operation are not considered adequate, the only means to ensure complete removal of pathogens is to provide a pre-pasteurisation stage before digestion. This practice is growing in some countries.

7. BIOCHEMISTRY

The successful use of anaerobic digestion for waste treatment came well ahead of a detailed knowledge of the basic biochemistry and microbiology of the process. But such knowledge, in the long term, may be crucial to any significant further advances in process operation or control. The anaerobic digestion process is in fact, very complex chemically and very dependent on the maintenance of a fine balance between at least 4 different groups of bacteria. A detailed exposition of the microbiology has been given by Marty (23) and Mosey (24) has now demonstrated the key (and hitherto unsuspected) role played by hydrogen gas in the maintenance of process equilibrium. Figure 4 gives a very simplified version of Mosey's model. Any build-up in concentration of hydrogen gas can result in a blockage in the conversion of propionic and butyric acids to acetic acid leading to a reduction in pH and eventually a 'stuck' digester. Detection of a rise in the concentration of hydrogen in the gas being produced by a digester could thus give an 'early

METABOLIC PATHWAYS IN ANAEROBIC DIGESTION *

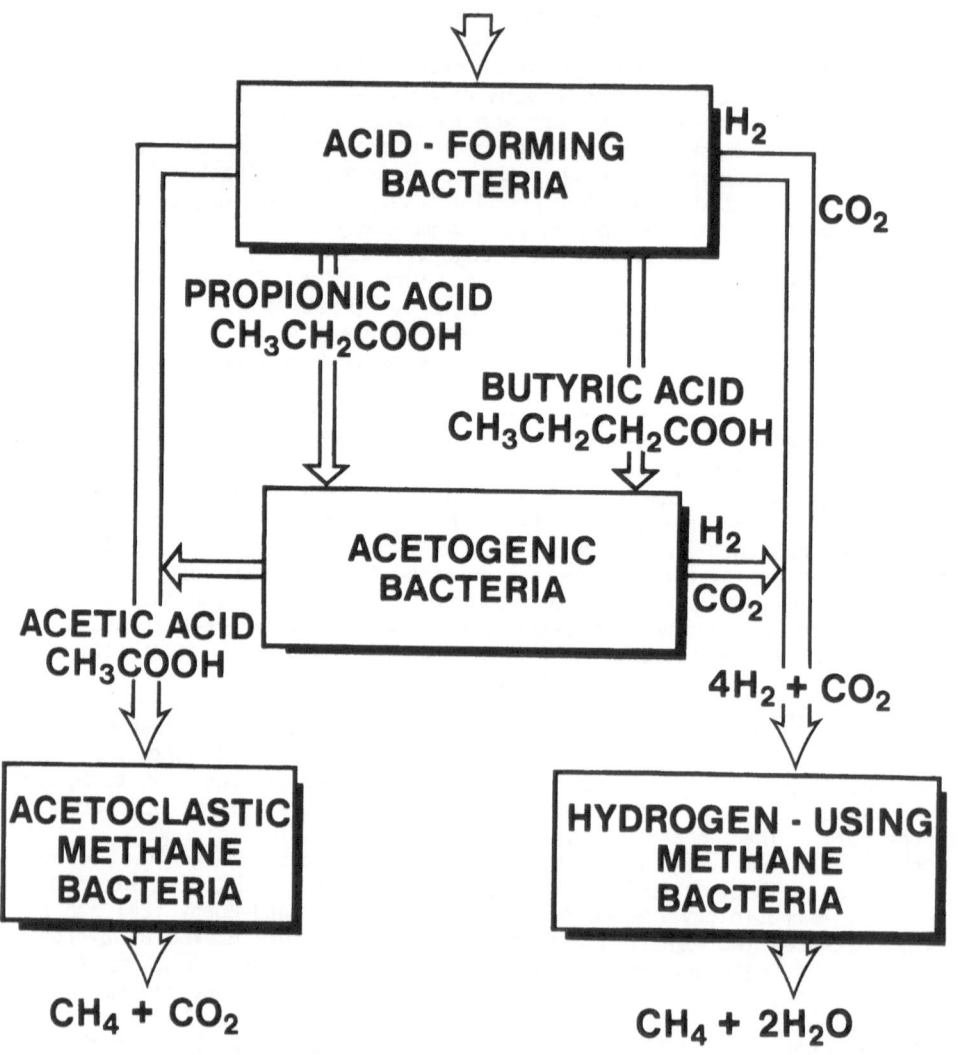

BIODEGRADABLE ORGANIC MATTER (CARBOHYDRATES, FATS, PROTEINS)

ACID - FORMING BACTERIA

H_2

CO_2

PROPIONIC ACID CH_3CH_2COOH

BUTYRIC ACID $CH_3CH_2CH_2COOH$

ACETOGENIC BACTERIA

H_2

CO_2

ACETIC ACID CH_3COOH

$4H_2 + CO_2$

ACETOCLASTIC METHANE BACTERIA

HYDROGEN - USING METHANE BACTERIA

$CH_4 + CO_2$

$CH_4 + 2H_2O$

*After Mosey (Ref.24)   Fig.4

warning' of impending process failure. The availability of an instrument to measure low concentrations of hydrogen gas has allowed this idea to be demonstrated and the objective now is to develop an on-line hydrogen detector to provide for continuous monitoring of the process.

8. PROBLEMS AND RESEARCH NEEDS
Following the relatively long period of technical development of anaerobic digestion plants for sewage sludge stabilisation, the situation has now been reached when the major design and engineering problems have been resolved although improvements and capital costs reductions will continue to be sought. There is still room for debate about the best method of digester mixing or even the necessary degree of mixing that is actually required for satisfactory operation (10). Similarly, the most effective method of heating, or the optimum tank aspect ratio, are still matters for discussion and for possible resolution in due course by definitive experiment.

The main problems that still present themselves regularly are related to digester operation and are as follows:

SEWAGE SLUDGES
Thin feed sludges
The advantages of pre-thickening sludge to above 6 per cent (and preferably to 8 per cent) DS have been well demonstrated, yet the recent survey (1) showed that in most countries feed sludges of 3-4% only are still the norm. This means, essentially, the inefficient use of digester capacity and quite often it leads also to problems of inadequate digester temperature in winter conditions and of stratification of tank contents leading to process failure.

Process inhibition by toxic chemicals
The problem of inhibition is now generally much reduced as a result of a better understanding of the type of chemicals responsible and of improved trade waste control. But process failure through inhibition by, for example, chlorinated hydrocarbons, or by heavy metals (25) still occurs in some areas. The development of 'early warning' monitoring systems to detect and assess toxicity effects is a perceived need (25).

Digester foaming
The phenomenon of sludge foaming, which can give rise to severe operational difficulties, is unpredictable and often unexplained. Further studies to elicit the cause, or causes, and to identify appropriate cures are indicated.

Farm slurries
In regard to farm digesters, the EEC survey identified several clear engineering problems which need to be resolved before anaerobic digestion would become economically viable on farms. There are (a) gas leakages from digesters (b) failure of feed pumps (c) ineffective mixing systems (d) unreliable electricity generating systems and (e) pipe blockages. In addition to these the widespread problem of poor quality feeds (too dilute and too many coarse solids) has been identified by others (17).

CONCLUSIONS

Mesophilic anaerobic digestion will remain as a very important process for stabilisation of sewage sludge with growing numbers of digesters installed in Europe. A variety of design trends have been discernible in recent years and improvements will continue. Some operational problems still need to be tackled. The future for farm-scale anaerobic digestion is less certain and will depend on development of very reliable low-cost systems though financial incentives will probably be determining factors in the rate of growth.

In both cases, there is still scope for research and development to reduce costs and improve efficiency.

REFERENCES

(1) COMMISSION OF THE EUROPEAN COMMUNITIES. COST 681 Working Party No. 1 Survey of Use of Anaerobic Digestion for Sewage Sludges 1985 (unpublished).

(2) HAVELAAR, A.H. and BRUCE, A.M. (1983). Disinfection of Sewage Sludge - an Enquiry among Member Countries of the EEC and other Countries. In Disinfection of Sewage Sludge: Technical, Economic and Microbiological Aspects (Eds. A.M. Bruce, A.H. Havelaar and P. l'Hermite). D. Reidel Publishing Co., Dordrecht, Holland, pp. 245-252.

(3) BRUCE,A.M. AND LOLL, U. (1981). A Review of Methods for Stabilising Sewage Sludges. In Characterisation Treatment and Use of Sewage Sludge (Eds. P. l'Hermite, and H. Ott). D. Reidel Publishing Co., Dordrecht, Holland, pp. 125-139.

(4) DEMUYNCK, M., NYNS, E.J. and PALZ, W. (1984). Biogas Plants in Europe. D. Reidel Publishing Co., Dordrecht, Holland, 1984.

(5) WOLINSKI, W.K. (1984). A Cost Comparison of Prefabricated and Conventional Digesters. In 'Sewage Sludge Stabilisation and Disinfection' (A.M. Bruce, Ed.) Ellis Horwood Ltd., Chichester, UK, pp. 488-498.

(6) CHESSHIRE, M.J. (1985). A Comparison of the Design and Operational Requirements for the Anaerobic Digestion of Animal Slurries and of Sewage Sludge. In 'Anaerobic Digestion of Sewage Sludge and Organic Agricultural Wastes' (Eds. A.M. Bruce, A. Katsiris, P.J. Newman). Elsevier Applied Science Publishers, London (in the press).

(7) HILLS, R.E. and MALLNOWSKI, J.T. (1985). Constructing the Egg: Concrete or Steel? Water Engineering and Management, 132, No. 4, 27-30.

(8) 'NIERSVERBAND 83'. Annual Report Published by the Niersverband, Viersen, FR Germany 1984.

(9) THOMPSON, J.L. and MICHAELSON, A.P. (1984). Design Aspects of the New Anaerobic Digesters at Bury. In Sewage Sludge Stabilisation and Disinfection (A.M. Bruce Ed). Ellis Horwood Ltd, Chichester, UK, pp. 92-106.

(10) CASEY, T.J. (1985). Requirements and Methods for Mixing in Anaerobic Digesters. In 'Anaerobic Digestion of Sewage Sludge and Organic Agricultural Wastes' (Eds. A.M. Bruce, A. Katsiris, P.J. Newman). Elsevier Applied Science Publishers, London (in the press).

(11) BOYKO, B. AND SMART, J. (1977). Full Scale Studies on the Thermophilic Anaerobic Digestion Process. Environment Canada. Research Report No. 59, 90 pp.

(12) BRUCE, A.M. AND FISHER, W.J. (1984). Sludge Stabilisation - Methods and Measurement. In Sewage Sludge Stabilisation and Disinfection (A.M. Bruce Ed). Ellis Horwood Ltd, Chichester, UK, pp. 23-47.

(13) BRADE, C.E., NOONE, G.P., WHYLEY, J. (1984). Progress in Anaerobic Digestion - Heating, Cooling and Thickening. In Sewage Sludge Stabilisation and Disinfection (A.M. Bruce Ed). Ellis Horwood Ltd, Chichester, UK, pp. 158-173.

(14) BRAUN, H.J., KUGEL, G., and ZINGLER, E. (1983). Simultaneous Pasteurisation and Digestion of Sludge (SPD Process). In Disinfection of Sewage Sludge: Technical, Economic and Microbiological Aspects (Eds A.M. Bruce, A.H. Havelaar and P. l'Hermite). D. Reidel Publishing Co., Dordrecht, Holland, pp. 77-93.

(15) KIDSON, R.J., MORRIS, D.L., RAY, D.L. and WAITE, W.M. (1983). Submerged Combustion for Pasteurisation of Sludge before Digestion. In Disinfection of Sewage Sludge: Technical, Economic and Microbiological Aspects (Eds A.M. Bruce, A.H. Havelaar and P. l'Hermite). D. Reidel Publishing Co., Dordrecht, Holland, pp. 94-113.

(16) ZWIEFELHOFFER, H.P. (1983). Experience with Aerobic Thermophilic Disinfection of Sewage Sludge. In Disinfection of Sewage Sludge: Technical, Economic and Microbiological Aspects (Eds A.M. Bruce, A.H. Havelaar and P. l'Hermite). D. Reidel Publishing Co., Dordrecht, Holland, pp. 3-36.

(17) FRIMAN, R.M. (1985). Anaerobic Digestion on Farms in the UK. In 'Anaerobic Digestion of Sewage Sludge and Organic Agricultural Wastes' (Eds. A.M. Bruce, A. Katsiris, P.J. Newman). Elsevier Applied Science Publishers London (in the press).

(18) BRUCE, A.M. (1981). New approaches to Anaerobic Digestion. Journal Institution of Water Engineers and Scientists, 35(3), 231-251.

(19) LOLL, U. (1985). Biogas Plants for Animal Slurries in the Federal Republic of Germany. In Anaerobic Digestion of Sewage Sludge and Organic Agricultural Wastes (Eds. A.M. Bruce, A. Katsiris, P.J. Newman). Elsevier Applied Science Publishers London (in the press).

(20) WOLF, P. Aerobic Thermophilic Stabilisation of Sludge Versus Anaerobic Digestion and Other Kinds of Sludge Treatment at Middle-Sized Plants with respect to Power Conservation and Economy.

(21) FUCHS, H., FUCHS, L., and FUCHS, W. (1980). The Exothermic Aerobic Thermophilic Sludge Stabilisation Process. Umwelt, 3, 312.

(22) PIKE, E.B. and DAVIS, R.D. Stabilisation and Disinfection - Their Relevance to Agricultural Utilisation of Sludge. In Sewage Sludge Stabilisation and Disinfection (A.M. Bruce Ed). Ellis Horwood Ltd, Chichester, UK, pp. 61-91.

(23) MARTY, B. (1985). Microbiology of Anaerobic Digestion (1985). In 'Anaerobic Digestion of Sewage Sludge and Organic Agricultural Wastes' (Eds. A.M. Bruce, A. Katsiris, P.J. Newman). Elsevier Applied Science Publishers London (in the press).

(24) MOSEY, F.E. (1983). Mathematical Modelling of the Anaerobic Digestion Process: Regulatory Mechanisms for the Formation of Short-Chain Volatile Acids from Glucose. Water Science and Technology, 15, 209-232.

(25) KOUZELI-KATSIRI, A. and KARTSONAS, N. (1985). Inhibition of Anaerobic Digestion by Heavy Metals. In 'Anaerobic Digestion of Sewage Sludge and Organic Agricultural Wastes' (Eds A.M. Bruce, A. Katsiris, P.J. Newman). Elsevier Applied Science Publishers London (in the press).

DISCUSSION

G. MININNI

You have mentioned anaerobic and aerobic treatment of sewage sludge.
Have you considered other processes such as 'dual digestion' developed by
Union Carbide?

A.M. BRUCE

Yes we have looked at the dual-digestion process as well as many others.
For example, there is one which is similar to the Union Carbide process
except that it uses air instead of pure oxygen. This has been used in
Switzerland and has been shown to improve the consolidation of sludge.
However, as with all new processes it is important to examine closely the
economics.

P.J. MATTHEWS

You have referred to heated anaerobic digestion. But have you considered
cold digestion? This has been shown to be more effective for killing
Taenia saginata ova than had previously been thought.

A.M. BRUCE

I do refer briefly to cold digestion in the paper. I am not certain that
it has an application to large works where odour problems and land-area
requirements will probably make it unsuitable. Accelerated cold
digestion probably is the most interesting new development for small
works ie. less than about 5000 population-equivalents. Several such
plants are currently being installed in the Yorkshire Water Authority
area of the UK.

There are many other processes available at the moment, for example, one
involving reed beds. But as always it is vital to examine closely the
economics of these processes, not just their technical features.

R.L. LESCHBER

I would like to mention one method which is being developed in Berlin to
save costs. This involves a short anaerobic treatment of less than 4
hours and the partially stabilised sludge is then spread on agricultural
land. It seems to be promising and could be used for communities up to
20000 population equivalents.

A.M. BRUCE

I am aware of a high-rate enzymatic sludge stabilisation process
developed in East Germany and assume that this is what Dr Leschber is
referring to. We have not yet completed our evaluation of this process.

"STATE OF THE ART" ON SLUDGE COMPOSTING

L.E. Duvoort-van Engers
National Institute of Public Health and Environmental
Hygiëne (RIVM-LAE)
Antonie van Leeuwenhoeklaan 9, Postbus 1,
3720 BA Bilthoven, The Netherlands

S. Coppola
Istituto di microbiologia agraria e statione
di microbiologia industriale
80055 Portici (Italy)

Summary

Biological decomposition and stabilization of sewage sludge under conditions which allow development of thermophilic temperatures (composting) can effect considerable drying. This reduces costs of handling
and increases the attractiveness of compost for reuse or disposal.
Microbiological aspects, composting systems and product quality are
described.
A review is given on the "state of the art" of composting of sewage
sludge in Austria, Canada, Denmark, Finland, France, Germany, Ireland,
Italy, The Netherlands, Norway, Sweden, Switzerland and the United
Kingdom.

1. INTRODUCTION
     There is no universally accepted definition of composting. In this pa-
per, composting is defined as the biological decomposition and stabilization
of organic substrates under conditions which allow development of thermophilic temperatures as a result of biologically produced heat, with a final
product sufficiently stable for storage and application on land without adverse environmental effects (Haug, 1980).
     The objectives of composting have traditionally been to biologically
convert putrescible organics to a stabilized form and to destroy organisms
pathogenic to humans. Composting is also capable of destroying plant diseases, weed seeds, insects and insect eggs. Odour potential from use of compost is greatly reduced because organics that remain after proper composting
are relatively stable with low rates of decomposition. Composting can also
effect considerable drying, which is of particular value with wet substrates
as sewage sludges. Decomposition of substrate organics together with drying
during composting can reduce the cost of subsequent handling and increase
the attractiveness of compost for reuse or disposal.
     Compost can be disposed of in a sanitary and usually convenient manner.
If the product is reused, it can accomplish a number of additional purposes
including:
- to improve growth of crops
- to reclaim and reuse certain valuable nutrients

- to serve as a source of organic matter for maintaining or building supplies of soil humus, necessary for proper soil structure and moisture-
holding capacity.

    Generally, composting has been applied to relatively dry materials.
Municipal refuse, agricultural residues, dried animal manures and forest
wastes are examples of substrates commonly used as composting materials.
Dewatered sludge still has a water content of about 75%. The high moisture
content, lack of porosity, tendency to compact and the need to dry dewatered sludge during composting make sludge composting somewhat unique and
often difficult.
    To increase the porosity and to reduce the moisture content a bulking
agent can be added.

2. MICROBIOLOGICAL ASPECTS
    Composting is a process based on microbial activity. It depends on a
sequentially changing microbial ecology involving many types of micro-organisms. At first a large variety of bacteria, always naturally occurring in
all organic wastes, starts the transformation. Changes of trophic conditions
and of factors such as temperature, moisture, pH and aeration progressively
cause changes of microbial populations and their activities.
    An unique and general model can only be approximate since great ecolo-
gical differences may depend on raw material quality (chemical composition,
bulk density, moisture, pH) and technology (closed or open system, aeration
system, static or mixing treatment). According to such a simple scheme an
early progressive increase of biomass and activity of mesophilic micro-organisms characterizes the first stage of the process. The accumulation of microbial waste heat is at first stimulating the development of mesophilic
organisms but turns then inhibitory at temperatures within the $40^{\circ}C$ to $50^{\circ}C$
range. Species variations depending on the exact critical temperature occur
during this second transitional period, involving thermotolerant bacteria.
Thermophilic and spore-forming bacteria as well as thermophilic actinomycetes are the agents of the transformation when temperature exceeds $55^{\circ}C$,
while non-thermotolerant microflora dies off. Then, with decreases of temperature and the lack of easily degradable substances, an important growth
of fungi and mesophilic actinomycetes activity takes place until complete
compost ripening.
    There is not detailed knowledge about microbial species responsible
for every stage of the process, available information certainly depending
on microbiological isolation methods used.
    The Civil Sanitary Engineering Department of the Michigan State Uni-
versity (1955) reported that as the composting progresses there was a shift
from a varied population belonging to genera Pseudomonas, Achromobacter,
Flavobacterium, Micrococcus, and Bacillus, to one dominated by sporeformers
of the genus Bacillus. Niese (1959) identified Bacillus subtilis and Bacillus stearothermophilus. Among spore-forming bacteria belonging to genus
Clostridium, Henssen (1957) identified Cl.thermocellum. Isolation and identification of thermoduric and thermophilic non-sporulating bacteria from
composting materials have never been performed.
    Nakasaki et al. (1983), applying the death rate curve method with UV-
radiation to differentiate between vegetative cells and spores of bacteria,
showed that more than 60% of mesophiles remained in the vegetative form at
$60^{\circ}C$ during thermophilic composting of activated sludge. Concerning Actinomycetes, Lacey (1973) found out that species of Streptomyces were dominant
with mild heating, Thermoactinomyces, Micropolyspora and Thermomonospora
prevailed in hotter piles. Streptomyces thermomonospora curvata,

Thermoactinomyces thalpophilus, Thermoactinomyces glaucus and Pseudonocardia thermophila had been signalized by Henssen (1957). Coppola et al. (1983) isolated 27 species of Streptomyces, 3 of Nocardia, 2 of Streptoverticillium, 1 of Microbispora, 1 of Thermoactinomyces, 2 of Thermomonospora and 1 of Micropolyspora during composting of sewage sludge in mixture with wood chips.

Fungal identity as presented by several authors studying composting of different organic waste materials was reviewed by Finstein and Morris (1975). De Bertoldi et al. (1983) isolated thermophilic and mesophilic species during composting of the organic fraction of urban solid wastes (60%) mixed with sewage sludge (40%). Absidia ramosa, Allescheria terrestris, Mucor pusillus, Chaetomium thermophilum, Thalaromyces thermophilis, Aspergillus fumigatus, Humicola insolens, Humicola lanuginosa, Lenzites sp., Penicillium duponti, Scytalidium thermophilum, Sporotrichum thermophile, Thermoascus aurantiacus and Micelia sterilia were identified among thermophilics. Mesophilics included 6 species of Ascomycotina, 8 species of Basidiomycotina and 29 species belonging to 23 genera of Deuteromycotina.

Microbial activities during composting are as complex as microbial populations. They must assure the achievement of the following main objectives: i) a maximized decomposition rate of organic material; ii) abundant production of substances able to make compost a valid soil conditioner; iii) sanitization of the material.

Finstein et al. (1983) have emphasized the correlation between the decomposition rate and microbial species diversity of the composting ecosystem, as a condition to attain the transformation of a wide range of organic compounds. Since community diversity, as measured by authors, is markedly lower at $60-65°C$ than at $55-61°C$, the temperature ascent must be restrained to less than $60°C$. Thus authors have developed a rational strategy for countering the tendency of the composting ecosystem to self-limit via excessive accumulation of metabolically generated heat. The exceeding heat is removed from the composting mass through ventilation, a practical means also favouring evaporization of water and oxygen supply. This strategy, compared to a conventional approach yielded about four fold more waste treatment in half the processing time (Finstein and Miller, 1985). However, composting is not only a mineralization process. CO_2, H_2O and heat arise from the aerobic metabolism of the easily degradable substances. More complex compounds are humified. New high molecular weight compounds are synthesized.

Solid-state conditions within composting are widely favourable to these processes. Coppola et al. (1983a) measured interesting cation exchange capacities, humic acids and polysaccharides contents in sewage sludges composted in mixture with wood chips or with inert bulking agents. Morel et al. (1985) have found in mature compost relatively high quantities of polysaccharides extractable in acid, likely corresponding with newly synthesized microbial components rather than those fractions which are not yet decomposed.

Moreover composting is widely considered as a disinfection process. It is generally recognized that the temperature effect alone is sufficient to kill pathogenic organisms, eggs and larvae of flies as well as weed seeds. Wiley and Westerberg (1969) showed clearly that indicator pathogens, added to composting primary sewage sludge in enormous number, were rapidly killed. The indicator organisms, Salmonella newport, polio-virus type 1, Ascaris lumbricoides ova, and Candida albicans were all completely killed. Poliovirus was the most sensitive, Candida albicans the most resistant. Microbial antagonism is an additional sanitizing mechanism. Epstein and Wilson (1975) described inoculating Salmonella enteritidis serotype Montevideo into sterile composted sewage sludge (60% solids). Initial counts were measured at 10^3 bacteria per g. After 2 days of incubation, salmonellae counts were found to be greater than 10^9 bacteria per g.

Salmonella regrowth in bagged composted sludge was found by Russ and Yankoo (1981) to occur at 20-40°C, require a moisture content of about or more than 20% and a C/N ratio of 15:1. Population peaks occurred around 5 days under laboratory conditions, followed by subsequent die-off. Authors concluded that only appropriate curing and storage of compost can allow to achieve the necessary levels of safety. Coppola et al. (1983a) put into evidence that by favouring growth of actinomycetes through aeration and drying during composting of sewage sludge in mixture with wood chips, a drastic and permanent reduction of enteric bacteria counts was achieved. Such compost was also unable to support regrowth of Salmonella enteritides. Well controlled composting processes have always shown to provide sanitized products, as reported also by Strauch (1983). Indeed a permanent hygienization can only be warranted if an intensive transformation of chemical, physical and biological characteristics of the medium is achieved, when sludge becomes unsuitable to pathogens growth. Experiments have shown that within such conditions compost exhibits microbial contents very different in comparison with the raw material, especially with reference to Enterobacteriaceae and other gram-negative bacteria (De Bertoldi, Coppola and Spinosa, 1983). So the microbiological quality of compost reflects the composting procedure.

3. COMPOSTING SYSTEMS

Composting systems can be divided into so-called closed and open systems, schematic the systems are as follows:

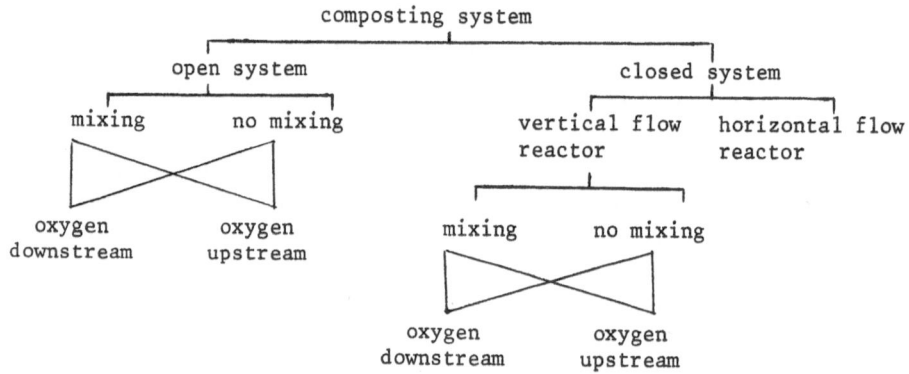

The closed systems have advantages in humid climates. They are, however, expensive both in investment and in exploitation. The output is more or less continuous in a limited number of days.

The open systems are simple and the scale is variable. The investment is relatively low, the exploitation costs are depending on the season. A longer composting period is necessary compared to closed systems.

Open system, mixing (windrow)

The open system, with mixing is a relatively simple system of composting. The sewage sludge is set up in rows, called windrows. Periodically the sludge is turned with a frequency that varies from once a day to once a week. The aeration and oxygen supply takes place naturally during the turning of the material.

The advantage of the windrow composting is the simple way in which dry compost or another bulking agent can be added to the sludge, so that the composting process starts quickly.

Open system, no mixing (aerated static pile)

The open system without mixing often is called after the place where it is developed namely Beltsville in the United States of America. The sludge with bulking agent again is set up in rows. In contrast with the windrows system, however, the material is not turned during the process but forced aeration is applied to maintain aerobic conditions.

Closed system, horizontal flow reactor (DANO)

One of the most important closed systems with a horizontal flow reactor is the DANO-system. The drum is rotating, so that the content is moving continuously. By the intensive mixing, by which moisture- and oxygen supply can be controlled closely, the composting process starts quickly. In a few days the material is that stable that post fermentation (ripening) in the open air can take place, without odour problems.

Close system, vertical flow reactor, mixing (Schnorr)

A vertical flow reactor with mixing is the so-called Schnorr-biocell system. The reactor consists of a vertical tower with 8-10 floors, one above the other. Each floor contains a hydraulically operated valve which allows material to be discharged to the next floor. Oxygen is introduced by forced aeration. Retention time is about three days on each floor. At the lowest floor the material is that stable that no further ripening is needed any more.

Closed system, vertical flow reactor, no mixing (Kneer, BAV)

A vertical flow reactor with no interior floors or other mechanisms is the Kneer-system, recent versions known as BAV-system. There is no mixing. Feed is introduced at the top and flows downward as product is removed by a mechanical scraper from the bottom. Oxygen is supplied by forced aeration from the bottom. Retention time is 10-30 days. Ripening in the open air can take place without odour problems.

Closed system, vertical flow reactor, no mixing (Triga)

Another vertical flow reactor without mixing is the Triga-system, a reactor which is divided into four separate vertical compartments. Oxygen is supplied at the top of the reactor. A screw extractor removes the product from the bottom. The retention time is about 25 days. Ripening takes place in the open air for 2-4 months.

There are, of course, several more firms than the above mentioned building composting reactors; the systems mostly used for sludge composting are described.

4. PRODUCT QUALITY

The quality of the produced compost depends on the quality of the sewage sludge, the bulking agent and the composting system.

Comparing data of sewage sludge and the sludge compost (EPA, 1978) indicates that only one third of the original nitrogen can be found in the compost. This loss is mainly due to the easy leaching of nitrogen and NH_4-evaporation. Phosphate, potassium and magnesium do not leach that quickly, so the decrease of these elements is less.

Almost all heavy metals, both from the sewage sludge and bulking agent, can be found in the compost. As by the composting process a part of the organic matter is converted, the concentration of heavy metals in the dry matter fraction of the compost will increase.

It is impossible to give analyses of an "average of compost". The ratio bulking agent - sewage sludge, and the composition of both differs from place to place.

In general it can be said that in all the countries reviewed there are guidelines (advices or acts) for the use of sewage sludge in agriculture. More and more there are rules that protect the soil for overloading with heavy materials.

If the guidelines for compost quality and use are fixed, on a national or an international level, there may be an increase in popularity of composting in most countries. A review of guidelines for soil protection (maximum permissible concentrations) is beyond the scope of this paper.

5. "STATE OF THE ART" ON SLUDGE COMPOSTING

Austria

In Austria 2,4 million m^3 of sewage sludge is produced yearly with a dry matter content of 5% ($\pm$ 120,000 ton dm). About 2 (volume)% (48,000 m^3) is composted.

Of this quantity only 3,000 m^3 is composted with sawdust in a Kneer bioreactor. After a composting period of 14 days in the reactor, the product ripens in 40 or more days.

The other 45,000 m^3 of sewage sludge is composted together with urban waste in 7 composting plants. Both the DANO-system and the aerated static pile are used.

About 1/3 of the compost is used in agriculture, the rest of it for covering of landfill sites and recultivation of embarkments.

The quantity of sewage sludge composted will probably decrease the next few years because of the guidelines for heavy metals. However, there is a great need for organic matter.

Canada

Sewage sludge composting is done only at a very few locations in Canada. Less than 1% of the yearly production of 290,000 ton dm is composted. In Windsor, Ontario, the windrow method is used; straw is used as bulking agent. The product is sold as a horticultural potting medium and soil conditioner. The same method with wood chips as bulking agent is used in British Columbia (Comox-strathcona Regional District).

The periods of composting and ripening are 21 and 60 days respectively. The compost has been used by the local Parks Department as soil conditioner and mulch, however, the intention is to sell much of it in the future. It is not likely that composting of sewage sludge in Canada will increase very much in future.

Denmark

In Denmark there are two plants with combined sludge and solid urban waste composting.

In Frederikssund the DANO-system is used (24 h) with additional Windrow composting (4 turnings in a 3 month period). They process annually 4,000 ton sludge (18% dm), 20,000 ton sludge (3% dm) mixed with 25,000 ton domestic waste.

The reject output from sieving is 15,000 ton/year and is disposed at a controlled sanitary landfill. The remaining compost output of 6,000 ton/year is partly used on municipal greens (2,000 t/y) while the rest (4,000 t/y) is used as top layer on sanitary landfills.

In Fövling, Jutland, the municipal plant process annually 4,000 ton dewatered sludge (12% dm) $\sim$ 480 ton dm and 6,000 ton sludge of 4% dm $\sim$ 240 ton dm together with 24,000 ton hammermilled domestic waste. The process is locally designed and looks like the aerated static pile method.

The total amount of raw sludge composted in Denmark is approximately 2,000 ton dm/y, corresponding to $\pm$ 1% of the total sludge production.

Finland

The development of composting in Finland is shown in table 1. The number of sewage plants using composting indicates the increasing popularity of composting. In this table the number of inhabitants connected to treatment plants is shown too.

Table 1. Development of composting in Finland

Connected population (x 10^3/1%)		Number of works No/%	
1977	1982	1977	1982
sludge composting 96/3	423/13	13/2	46/8
sewage works 2820/100	3320/100	546/100	563/100

Total population	
4743	4824

Total sludge production m³/a
  930,000   1,130,000
  (100,000 ton)*   130,000 (ton)*   * data not reliable

The composting occurs mainly at small sewage treatment plants with the windrow system. The composting agent is mainly bark and peat. Big cities also use leaves and cut gras etc. and sometimes even sand to control the density of the material.

Big cities, using intensive turning may have a ripening time of 2-3 months. Small units sometimes have a ripening time of more than 1 year in order to include a freezing period.

The justification is – freezing reduces the moisture content and increases the porosity
– freezing and a long retention give more safety margin to pathogen die-off.

The compost is mainly used as topsoil in parks, "green areas", road slopes etc.. The increasing popularity of composting in Finland can be explained as follows:
- low population density: suitable places causing particularly no odour nuisances can be found in the countryside;
- for fairly small plants simple composting provides the only feasible and reliable stabilization method;
- in the northern parts of the country, little or no crops are grown. Fields are used for grazing, and the sludge cannot be used in agriculture. Simultaneously there is a lack of topsoil.

France

In France the total sewage sludge production is nearly 800,000 ton dm/year. About 10,000 ton sludge is composted (thus 1.25%). At this moment there are 5 composting plants: Royan, Nantes (2), Soisson and Blois. Planned composting plants are: Surgères, Besançon and Rodez.

In Royan the Triga process is used, in Soissons and Nantes the BAV process and in Blois the windrow system. The compost is used in bags for garden centers, vineyards and horticulture.

Germany

The annual sludge production in Germany is about 2,4 million ton dm. Approximately 2% (48,000 ton dm) is composted with or without urban waste. Composting of sludge with no other bulking agents than straw, sawdust etc. takes place at 42 locations. Only at 6 locations an open composting system is used. The other ones are mainly vertical flow reactors without mixing. The connected population is 1,1 million in total.

In 1990 about 50-60 sewage sludge composting plants are expected with a connection of 1.2-1.5 million i.e.. The government stimulates recycling of wastes and by this composting of sewage sludge. Of course, a market for the product is necessary and this must be developed further. At the moment the compost is used as topsoil on landfill sites, on public greens and gardens.

Table 2. Sludge composting in Germany (1983)

place	i.e.	system
1. Deidesheim	13,000	closed
2. Denkendorf	15,000	closed
3. Dettenhausen	8,000	closed
4. Gaggenau	121,000	closed
5. Liebenzell	16,000	closed
6. Rastatt	120,000	closed
7. Mittleres Wuhachtal	14,000	closed
8. Althausen	4,200	open
9. Pleidelsheim		open
10. Brilon-Messinghausen	10,000	closed
11. Delbruck	30,000	closed
12. Eddersheim	15,000	closed
13. Ferndorftal	110,000	closed
14. Freigericht	30,000	closed
15. Herbrechtingen	25,000	closed
16. Hochheim	25,000	closed
17. Kandel	10,000	closed
18. Kronberg	30,000	closed
19. Lambsheim	20,000	closed
20. Lollar	25,000	closed
21. Lubeck	15,000	closed
22. Nufringen	15,000	closed
23. Raisdorf	10,000	closed
24. Rosbach	20,000	closed
25. Schofflenztal	25,000	closed
26. Schlüchtern	25,000	closed
27. Steinenbronn	15,000	closed
28. Vlotho	20,000	closed
29. Waldenbuch	15,000	closed
30. Weilheim/Teck	25,000	closed
31. Wilnsdorf-Weisstal	15,000	closed
32. Winterberg	20,000	closed
33. Laupheim	50,000	closed
34. Horn Bad Meinberg	25,000	closed

place	i.e.	system
35. Bickenbach	50,000	closed
36. Oberwald	20,000	closed
37. Neunkirchen/Saar		closed
38. Unteres Kochertal		open
39. Mühlacker		open
40. Oberer Kraichbach	50,000	closed
41. Bad Vilbel		open
42. Edertal		open

Ireland

Any composting that is done in Ireland is on straw, poultry and horse manure and is aimed at mushroom production. This, however, is outside the scope of this review.

Composting of sewage sludge does not take place at all. The total sludge production in Ireland is about 20,000 ton dm/year.

Italy

In Italy the total annual sludge production is approximately 200,000 ton dm. There is only one sludge composting plant; this plant treats 10-20 cubic meter of dewatered sludge a day (700 ton/year) through the BAV-system (mixing with saw dust, 2 weeks bioreactor, 6-8 weeks ripening). The compost is used in plant nurseries and public greens. A second plant is going to start using the same system, with a load of 1500 ton per year.

In Italy 5% of the urban solid waste is composted. An unknown percentage is mixed with sewage sludge in some plants. In total no more than 0,5% of the annual sludge production is composted.

The Netherlands

At the moment composting of sewage sludge takes place at three locations in The Netherlands, two of them on a rather small scale. One company in the western part of Holland treats the sludge from several Water Boards to make compost and black earth on a very big scale. The method used for composting is the aerated static pile. Wood chips are used as bulking agent.

At all locations "open" systems are used, at the biggest plant the composting takes place under a roof. This is to exclude the rainfall and to control odour problems, because all the ventilation air can be lead through compost filters.

The compost is particularly used on public greens.

If the guidelines for compost quality are fixed the quantity of sludge composted will probably increase to 5 percent of the total sludge production of 260,000 ton dm/year.

Norway

The total amount of sewage sludge produced in Norway is approximately 80,000 ton dm, of which 5-7.5% is composted yearly.

1,200 ton sludge is composted by the BAV-method, with sawdust as bulking agent. The retention time is 10-12 days. Ripening takes place in the open air.

In the Oslo area 4,000 ton sludge is composted by the Beltsville method, without using a bulking agent.

During the last ten years 30 plants for combined composting of domestic waste and sewage sludge are built. Most of them, however, are closed; only 4 are still in full activity, but there are problems with the destination of the compost, because of the high lead content.

The sludge compost is of a good quality and the total amount is used on green areas.

Sweden

The annual sludge production in Sweden is approximately 250,000 ton dm. Of which $\pm$ 50,000 ton is composted together with urban waste (20%), mostly in open systems.

The two biorotors constructed are not in use at the moment.

Only in a few cases sewage sludge is composted with bark or sawdust as bulking agent.

The compost is used as landfill cover or in green areas and gardens.

The quantity of sludge composted will probably decrease because of economic reasons.

Switzerland

In Switzerland only one plant is composting sludge, with an amount of 200 ton dm per year. This corresponds to approximately 0.1% of the annual sludge production of 150,000 ton dm. A closed vertical flow reactor is used, with sawdust as bulking agent. Retention time is between 10-15 days. Because of the relatively high content of heavy metals and nutrients in the composted sludge, the application, mainly in vineculture, is strongly restricted to 15 ton per hectare each 15-20 years.

United Kingdom

Currently in the U.K. composting is not widely practiced. Only 1% of all sludge disposed to land is being composted. The total annual sludge production is 1.2 million ton dm.

Composting is not used more extensively mainly because land area and labour requirements for the process are greater than for the preferred method of stabilisation (mesophilic digestion).

About 80% of the sludge composted is lime stabilised, dewatered sludge from plate presses. It is composted in open windrows, which are turned about once a month. Since the piles are only turned infrequently temperatures seldom increase above 40-50°C. After a period of a few months, the organic matter in the sludge is reduced from an initial value of 65% to about 40%.

The composted material is then used on arable land, for land reclamation and for fertiliser manufacture.

Pilot scale trials are currently being carried out by the Water Research Centre to investigate the benefits and costs of composting dewatered sewage sludge using the aerated pile process.

Summarized the "state of the art" of composting of sewage sludge is shown in table 3.

6. CONCLUSIONS

In most of the reviewed countries composting of sewage sludge is still a minor treatment method. The reason for this is probably that the use of liquid sludge in agriculture (mesophilic digested) in general does not meet many difficulties.

Composting can be achieved in open or in closed systems. Most of the closed systems are found in Germany, Austria and France.

In case of open systems mainly the windrow method is used.

Composting may increase in future if the guidelines for the product are fixed, on national or international level, and can be met.

REFERENCES

(1) CIVIL SANITARY ENGINEERING DEPARTMENT, MICHIGAN STATE UNIVERSITY (1955).
     Preliminary report on a study of the composting of garbage and other
     organic solid wastes. Ann. Arbor, Michigan, USA.
(2) COPPOLA, S., DUMONTET, S. and MARINO, P. (1983a). Composting raw sewage
     sludge in mixture with organic or inert bulking agents. International
     Conference on "Composting of solid wastes and slurries", Department of
     Civil Engineering, University of Leeds (UK), 28-30 September.
(3) COPPOLA, S., FERRANTI, E., DUMONTET, S., PARENTE, E., BASILE, G., LUNA,
     M., CUOCOLO, L., DURANTI, A., SANTINI, A., COMEGNA, V., CIOLLARO, G.,
     ROMANO, F., TUCCI, G., CATTARU, G., CARAFA, A., SCALA, A., D'ERRICO,
     F.P. and MICIELI DE BIASE, L. (1983b). Studio dei procedimenti di
     trasformazione dei fanghi risultanti dalla depurazione biologica della
     acque reflue in vista del loro impiego agricolo. III. Compostaggio di
     fanghi grezzi in miscela con altri materiali organic soli di scarto.
     L'Agricolt. Italiana, 5/6, 39-78.
(4) DE BERTOLDI, M., COPPOLA, S. and SPINOSA, L. (1983). Health implica-
     tions in sewage sludge composting. In: Disinfection of sewage sludge:
     technical, economic and microbiological aspects (A.M. Bruce, A.H.
     Havelaar and P. L'Hermite Eds.), D. Reidel Publ. Co., Dordrecht, pp.
     165-178.
(5) DE BERTOLDI, M., VALLINI, G., and PERA, A. (1983). The Biology of
     composting: A review. Waste Manag. & Res., 1, 157-176.
(6) EPA (1978). Sludge treatment and disposal.
(7) EPSTEIN, E. and WILSON, G.B. (1975). Composting raw sludge. In: Munici-
     pal sludge management and disposal. Information transfer, Inc., Rock-
     vill, Madison, USA.
(8) FINSTEIN, M.S. and MILLER, F.C. (1985) Principles of composting loading
     to maximization of decomposition rate, odour control, and costs effec-
     tiviness. In: Composting of agricultural and other wastes (J.K.R.
     Gasser Ed.). Elsevier Appl. Sci. Publ., London, pp. 13-26.
(9) FINSTEIN, M.S., MILLER, F.C., STROM, P.F., MAC GREGOR, S.T. and
     PSARIANOS, K.M. (1983). Composting ecosystem management for waste
     treatment. Biotechnology, 1, 347-353.
(10) FINSTEIN, M.S. and MORRIS, M.L. (1975). Microbiology of municipal
     solid waste composting. Adv. Appl. Microbiol., 19, 113-151.
(11) HAUG, R.T. (1980). Compost engineering; principles and practice. Ann.
     Arbor Science Publishers Inc. Ann. Arbor, Mich.
(12) HENSSEN, A. (1957). Über die Bedeutung der thermophilen Mikro-
     organismen für die Zersetzung des Stallmistes. Ark. f. Mikrobiol.,
     27, 63-81.
(13) LACEY, J. (1973). In: Actinomycetales: Characteristics and practical
     importance (G. Sykes and F.A. Skimmers Eds.), Academic Press, New
     York, pp. 231-251.
(14) MOREL, J.L., COLIN, F., GERMON, J.C., GODIN, P. and JUSTE, C. (1985).
     Methods for the evaluation of the maturity of municipal refuse com-
     post. In: Composting of agricultural and other wastes (J.K.R. Gasser
     Ed.), Elsevier Appl. Sci. Publ., London, pp. 56-72.
(15) NAKASAKI, K., SASAKI, M., SHODA, M. and KUBOTA, H. (1983). Change of
     microbial numbers during thermophilic composting of activated sludge
     with reference to CO_2 evolution rate. In: Biological reclamation and
     land utilization of urban wastes (F. Zucconi, M. De Bertoldi and S.
     Coppola Eds.). La Buona Stampa, Ercolano, Italy, pp. 561-569.

(16) NIESE, G. (1959). Mikrobiologische untersuchungen zur frage der Selbst-
erhitzung organischer Stoffe. Arch. f. Mikrobiol., 34, 285-318.
(17) RUSS, C.F. and YANKOO, W.A. (1981). Factors affecting Salmonellae re-
population in composted sludge. Appl. Environ. Microbiol., 41, 597-602.
(18) STRAUCH, D. (1983). Composting of solid urban wastes and land utilization of compost - A health risk? In: Biological reclamation and land
utilization of urban wastes (F. Zucconi, M. De Bertoldi and S. Coppola
Eds.), La Buona Stampa, Ercolano, Italy, pp. 343-378.
(19) WILEY, J.S. and WESTERBERG, S.C. (1969). Survival of human pathogens
in composted sewage. Appl. Microbiol., 18, 994-1001.

Table 1: development of composting in Finland

	Connected population (x 10^3/%)		number of works No/%	
	1977	1982	1977	1982
sludge composting	96/3	423/13	13/2	46/8
sewage works	2820/100	3320/100	546/100	563/100
Total population	4743	4824		
Total sludge production m^3/a	930.000 (100.000 ton)*	1.130.000 130.000 (ton)*		

*data not reliable.

Table 2. Sludge composting in Germany (1983)

place	i.e.	system	place	i.e.	system
1. Deidesheim	13.000	closed	22. Nufringen	15.000	closed
2. Denkendorf	15.000	"	23. Raisdorf	10.000	"
3. Dettenhausen	8.000	"	24. Rosbach	20.000	"
4. Gaggenau	121.000	"	25. Schofflenztal	25.000	"
5. Liebenzell	16.000	"	26. Schlüchtern	25.000	"
6. Rastatt	120.000	"	27. Steinenbronn	15.000	"
7. Mittleres Wuhachtal	14.000	"	28. Vlotho	20.000	"
8. Althausen	4.200	open	29. Waldenbuch	15.000	"
9. Pleidelsheim		"	30. Weilheim/Teck	25.000	"
10. Brilon-Messinghausen	10.000	closed	31. Wilnsdorf-Weisstal	15.000	"
11. Delbruck	30.000	"	32. Winterberg	20.000	"
12. Eddersheim	15.000	"	33. Laupheim	50.000	"
13. Ferndorftal	110.000	"	34. Horn Bad Meinberg	25.000	"
14. Frei gericht	30.000	"	35. Bickenbach	50.000	"
15. Herbrechtingen	25.000	"	36. Oberwald	20.000	"
16. Hochheim	25.000	"	37. Neunkirchen/Saar		"
17. Kandel	10.000	"	38. Unteres Kochertal		open
18. Kronberg	30.000	"	39. Mühlacker		"
19. Lambsheim	20.000	"	40. Oberer Kraichbach	50.000	closed
20. Lollar	25.000	"	41. Bad Vilbel		open
21. Lubeck	15.000	"	42. Edertal		open

Table 3   The 'state of the art" of composting of sewage sludge

Country	Total Sludge production ton dm	percentage composted %	composting systems	number of plants	destination	remarks
Austria	120.000	0,1	closed	1	agriculture, top soil on landfills land reclamation	decrease (heavy metals)
Canada	290.000	1,9	closed/open	7 together with domestic waste	soil conditioner, horticultural potting medium	
Denmark	200.000	< 1	closed/open	2 together with domestic waste	top soil on landfills, public green	
Finland	±130.000	8[1]	open	46	public green	1) related to the number of sewage treatment plants. increase
France	800.000	1,25	1open/4closed	5	horticultural potting medium, vineyards	
Germany	2.400.000	2	6open/36closed	42	topsoil on landfills, public green	increase
Ireland	20.000	0	--	--	--	--
Italy	200.000	< 0,5	closed	1	horticultural potting medium, public green	
The Netherlands	260.000	1 – 5	open	3	public green	increase?
Norway	80.000	5 – 7,5	open/closed	2 (4 together with domestic waste)	public green	(destination problems by heavy metals)
Sweden	250.000	20	open	10 together with domestic waste	top soil on landfills, public green	
Switzerland	150.000	0,1	closed	1	vineyards	
United Kingdom	1.200.000	1	open		agriculture, landreclamation	

COMPOSTING SYSTEMS

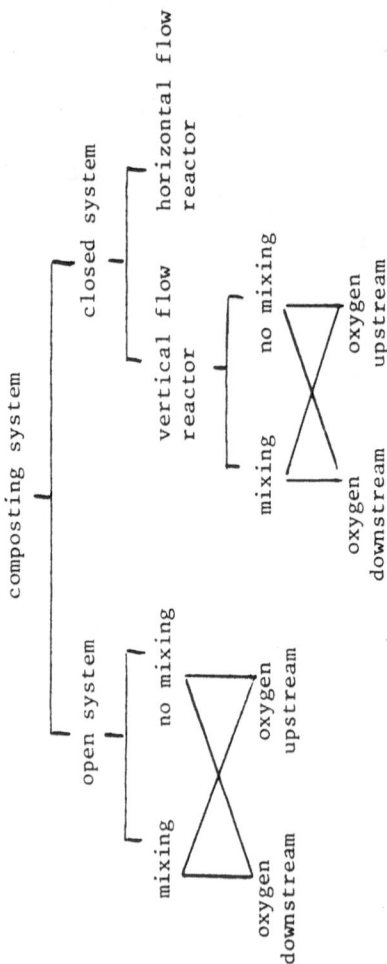

DISCUSSION

G. MININNI

Some countries covered by your survey use open systems whilst others use closed systems. Why is there this difference and does it affect the quality of the composted material?

L.E. DUVOORT-VAN ENGERS

They are both good systems and produce similar composted material. The choice between the two systems really depends on local conditions, for example, if the climate is wet, it is preferable to use a closed system. Furthermore, closed systems, in general, require less land area.

H. KUNZE

What is the "state of the art" view of the economics of composting? For example, how does the price of the composted material compare with cost of production?

M. DE BERTOLDI

A paper dealing with this aspect of composting will be presented next April at the International Symposium on Compost - Production, Quality and use. The Symposium will be held in Udine, Italy and will be jointly organised by the European Commission, the National Research Council of Italy and the International Society of Horticultural Science.

J.E. HALL

I would like to point out that since your survey has been carried out a further composting plant has been commissioned in the UK and now there are two operational plants.

OPTIMIZATION OF PROCESSING , MANAGING AND DISPOSING OF SEWAGE SLUDGE

Istituto di Ricerca sulle Acque, C.N.R.

Summary

Optimization of systems for sludge management requires preliminary
investigations on the disposal options as a function of sludge chara
cteristics, acceptability limits and environmental conditions. Once
the possible disposal alternatives are identified, the processes,
which allow the sludge with the desidered characteristics to be obtained, can be selected. Optimization criteria based solely on cost analysis, are however negatively affected by local conditions and by design and operating parameters, which generally do not take into account the interactions among the treatment processes. If models of
operations, describing how the input sludge characteristics are modified by the process operating variables, are known, then rational optimization criteria can be developed.
Sludge disposal options are practically restricted to spreading it on
the topsoil or disposing of it underground. The practice of dumping,
which is still carried out in some countries, is subjected to very
stringent limits because of the risks it entails for the environment;
incineration and the subsequent disposal of the ashes and composting,
with relevant use of the product obtained, are valid alternatives to
the disposal of more or less dewatered sludge. In some cases it may
be economical to combine disposal of sludges and urban solid waste
(USW) or fractions thereof, either by means of incineration or composting processes or by discharge in landfill. In these cases there is
a possible additional advantage owing to the fact that each component
has a beneficial effect in that: a) USW helps to reduce the auxiliary
fuel requirements in sludge incineration; b) USW acts as a supporting
agent and source of carbon for the sludge composting, which provides
nutrients lacking in the USW; c) USW give the mass greater consistency in the landfill area,thus enhancing its handlability by the vehicles used for packing and covering various layers.
The considerable difficulties encountered in optimizing the treatment-
- disposal system are due to the numerous factors that can affect the
choices to be made and to problems involving the a priori determination of sludge characteristics during the various stages of the process and the definition of the capital and operating costs of each
single operation. The present paper has been subdivided into two parts: Part I deals with the models of the most commonly used treatment
operations and Part II with the main disposal alternatives.

OPTIMIZATION OF PROCESSING , MANAGING AND DISPOSING OF SEWAGE SLUDGE
Part I: TREATMENT PROCESSES MODELS

L. Spinosa and V. Lotito
Istituto di Ricerca sulle Acque-Via F. De Blasio 5 70123 BARI

In this part the following treatment processes are considered and discussed: gravity thickening, aerobic digestion, anaerobic digestion, conditioning and mechanical dewatering.

1. GRAVITY THICKENING

The widely most used mathematical expression correlating settling velocity v to solids concentration c is that utilized by Dick and Young (1):

$$v = a \cdot c^{-n}$$

where a and n are empirically determined constants, which characterize the sedimentation properties of a certain sludge.
The values of these constants range from 0.9×10^{-5} m/h to 2.3×10^{-5} for a and from 1.1 to 2.7 for n.

According to the analysis developed by Hasit ed al.(2), the underflow suspended solids concentration (C_u) can be expressed as a function of the influent flow rate (Q_i) , the influent suspended solids concentration (c_i) and the area of the thickener (A) by means of the following equation:

$$C_u = \left[\frac{a(n-1)(n/n-1)^n \, A}{Q_i c_i} \right]^{1/(n-1)}$$

This equation is valid for thickener operating at the limiting flux conditions and with effluent solids concentration equal to zero. It allows the thickener surface to be calculated as a function of desired underflow sludge concentration.

2. AEROBIC DIGESTION

The residence time in the reactor can be considered the primary variable involved in the aerobic digestion. It affects both sludge dewaterability and volatile solids destruction.

According to Randall et al. (3), the value of Specific Resistance to Filtration is reduced by 10 to 50% in 1-5 days, but then increases in the following 11-15 days to values higher than initial ones. Typical curves for different kinds of sludge are shown in Fig. 1.

In a case studied by El-Gohary and Saleh (4), the Specific Resistance increased by about 280% in 90 h in samples without added nutrients, while it decreased by 40% in 111 h with added nutrients.

However Dick and Hasit (5), proposed neglecting the effect of residence time on dewaterability, because it was not appreciable, data were scarce and not consistent and probably not generally applicable.

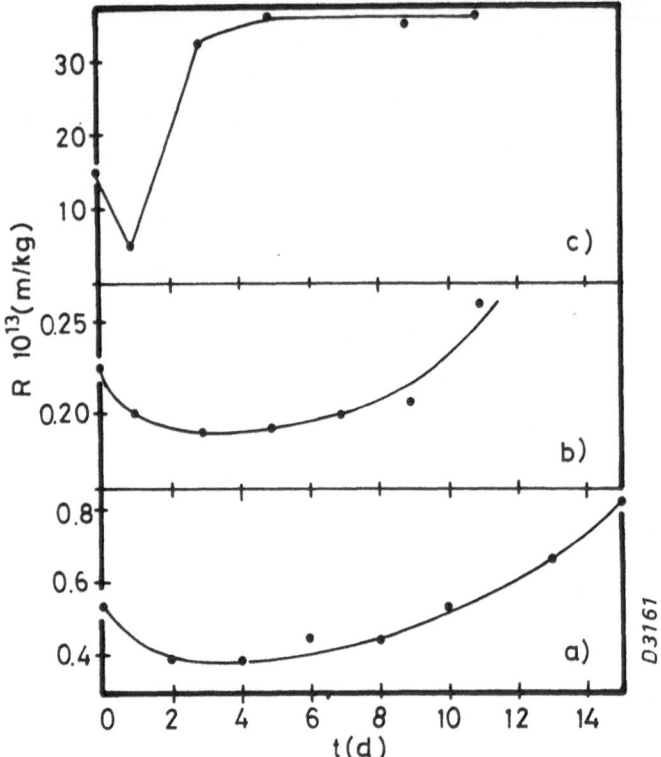

Fig. 1 - Variation of Specific Resistence to
Filtration vs. aerobic digestion time
for different sludges (3)

Volatile solids destruction by aerobic digestion is primarily depen-
dant on retention time in the reactor and temperature. Data deriving from
pilot and full-scale tests of several types of municipal wastewater sludges are shown in fig. 2 (6).
Adams et al. (7), developed the following equation correlating the
volatile solids concentration of digested sludge VS_e (mg/l) as a function
of stabilisation time t(d), once the volatile solids concentration of feed
sludge VS_i (mg/l), the decay rate constant K_b (d^{-1}) and the non degradable
fraction of volatile solids VS_n (mg/l) are known:

$$VS_e = (VS_i + K_b \cdot t \cdot VS_n)/(1+K_b)$$

3. ANAEROBIC DIGESTION

According to Dick and Hasit (5), sewage sludge dewaterability is reduced by anaerobic conditions; a poorer quality supernatant is also produced. However with increasing residence time, the extent of this deterioration diminishes and the mass of dry solids disharged from the digester de-

creases. Generally applicable equations correlating retention time to dewaterability are not available but the cost of sludge conditioning and
dewatering resulted about doublewhen deterioration of dewaterability by
anaerobic digestion was considered.

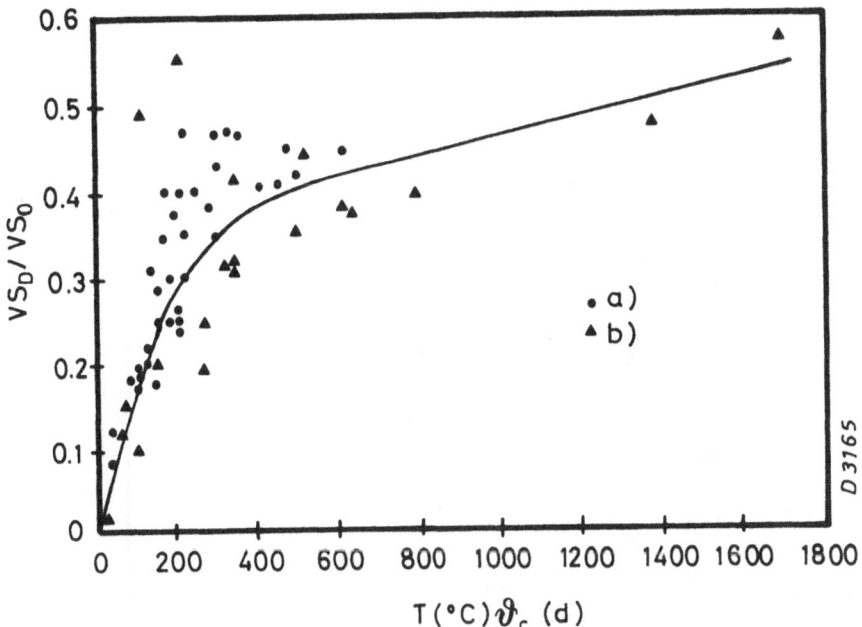

Fig. 2 - Volatile solids reduction (VS$_d$/VS$_0$) vs.temperature
(T) and sludge age (ϑ_c) (6)
a) Pilot tests
b) Full scale tests

The following equation allows the volatile solids abatement r_{vs}
(weight ratio) to be calculated (8):

$$r_{vs} = 0.06\ t/(1+0.06 \cdot t)$$

Since 0.75-1.10 m^3 of biogas (consisting of 65-75% of CH$_4$) are produced
per kg of VSS destroyed, therefore the energy production can be evaluated.

4. CONDITIONING

Among the sludge variables affected by chemical conditioners, the
Specific Resistance to Filtration is that most commonly used for evaluating
and modelling the dewatering operation. It is related to pressure by:

$$R_p = R_0\ (P/Po)^s$$

where s is the coefficient of compressibility.
As regards inorganic chemicals, the following equation was found to

give the best fits of experimental results (5):

$$\frac{R_{oc}}{R_{ou}} = (F+1)^{-4.05} \cdot e^{\,0.47 \cdot F} \cdot (L+1)^{-0.25}$$

where:

R_{oc} = R_o for conditioned sludge
R_{ou} = R_o for unconditioned sludge
F = $FeCl_3$ dosage (%)
L = CaO   dosage (%)

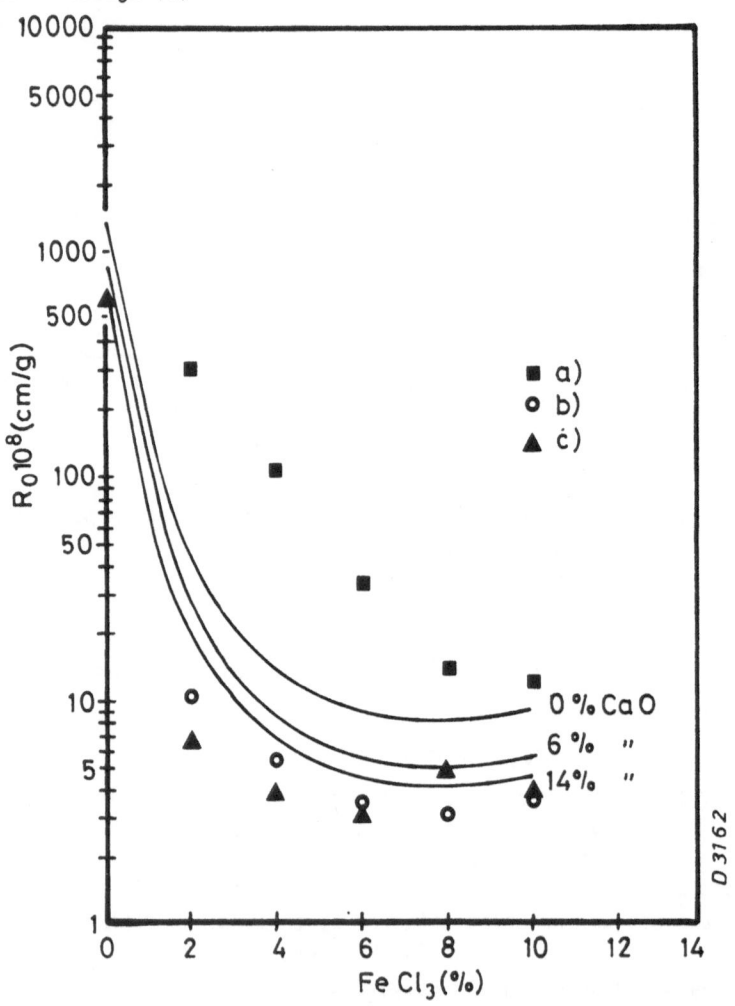

Fig. 3 - Comparison of conditioning model predictions
with observed data   (5)
a) Measured values at 0% CaO
b) Measured values at 6% CaO
c) Measured values at 14% CaO

Figures 3 and 4 show a comparison of this model with experimental results. It is evident that overdosage has a small effect in further improving dewaterability which can sometimes, could be deteriorated.

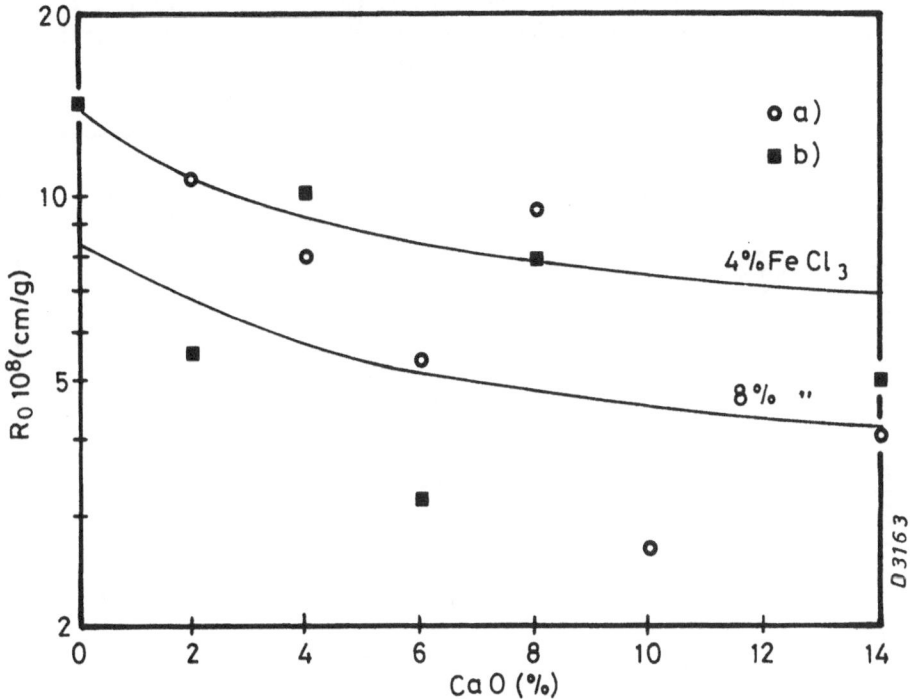

Fig. 4 - Comparison of conditioning model predictions with observed data (5)
a) Measured values at 4% FeCl₃
b) Measured values at 8% FeCl₃

Previous experiments showed also that conditioning does not alter the coefficient of compressibility s (9).

Tests carried out at Water Research Institute with different kinds of sewage sludge and cationic polyelectrolytes, allowed to evaluate the influence of polymer dosage on specific resistance (Fig. 5). The following correlation, valid for a dosage ξ (kg/t) ranging from 2 to 9, was developed:

$$R_c = 20.155 \cdot \xi^{-1.803}$$

These experiments confermed that overdosage could be detrimental.

5. MECHANICAL DEWATERING

Filter-press, belt-press and centrifuge are the most widely utilised machines for sewage sludge dewatering.

The main operating variables affecting dewatering by filter-press are pressure and filtration time. Tests carried out by Mininni et al. (10) led to the development of a model, allowing the coefficients a and b of the equation expressing the filtrate flux vs. time ($\emptyset = a.t^b$) to be correlated with pressure, Specific Resistance to filtration and sludge initial concentration. Once a and b are known, and values of chamber volume, filtration surface and dry solids density fixed, it is possible to estimate the cake solids concentration and the machine yield.

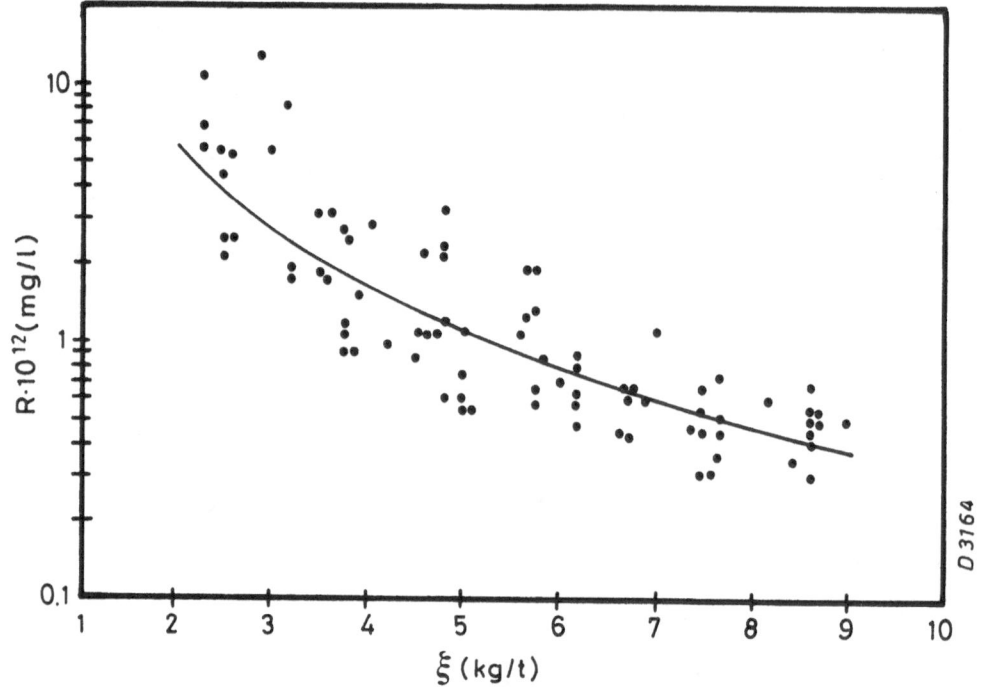

Fig. 5 - Influence of polymer dosage on Specific Resistence R

As far as dewatering by belt-press is concerned, pilot tests carried out by Spinosa et al. (11) allowed the following equation to be developed:

$$C_k = a_0 v_t^{a_1} \cdot F_0^{a_2} \cdot C_0^{a_3}$$

where C_k is the cake solids concentration (%), v_t the belt speed (m/h), F_0 the input sludge flow rate (m³/h) and C_0 the feed concentration (kg/m³). The value of coefficient a_0 and exponents a_1, a_2 and a_3 obtained processing pilot tests results are reported in Tab. I.

The operating variables, which a centrifuge depends on, are bowl speed, liquid ring height, bowl–conveyor differential speed and input sludge flow rate. No widely accepted indications on the effects of such variables are to be found in literature and it is commonly known that for some of them opposite results may also ensue due to the values of other variables. Moreover, a parameter for assessing sludge centrifugability has not yet

been defined because it has not been possible to reproduce on a laboratory-scale the actual conditions occurring in a full-scale machine.

Table I - Statistical correlations of belt-press tests:

$$C_k = a_0 \cdot v_t^{a1} \cdot F_0^{a2} \cdot C_0^{a3} \quad (*)$$

Notes	a_0	a_1	a_2	a_3	Corr. coeff.
24 **tests** for sludge A	76.5	-0.261	0.269	0.024	0.922
24 tests sl. A	83.7	-0.264	0.272	-	0.920
23 tests sl. B	22.3	-0.179	0.074	0.299	0.787
14 tests sl. B with $F_0 < 2$ m^3/h	17.7	-0.199	-0.039	0.401	0.909
8 tests sl. B with $F_0 > 2$ m^3/h	21.8	-0.065	-0.245	0.196	0.669
24 tests sl. A + 14 sludge B	51.5	-0.238	0.165	0.118	0.828

sludge A: activated slaughterhouse (C_0 = 12-43 kg/m^3)
sludge B: digested primary + activated (C_0 = 16-31 kg/m^3)
(*) C_k dewatered sludge concentraction (%)
    V_t belt speed (m/h)
    F_0 input sludge flow rate (m^3/h)
    C_0 feed sludge concentration (kg/m^3)

Statistical correlations obtained by pilot-scale tests are reported
in Table II (12), thus allowing the cake solids concentration to be estimated as a function of machine operating variables.

6. CONCLUSIONS
Any optimization procedure of sewage sludge management systems requires the knowledge of performace models for each treatment operation. Howe
wer, these models are in many cases not available, while in the others
they generally do not apply to all situations because based on experimental results obtained in particular conditions.
Another problem arises from the difficulty to define the values to
assign to coefficients appearing in the models, when either the characteristics of sludges produced in similar conditions or the sludge itself are
not available.
It follows that efforts must be directed towards the knowledge of the
sludge properties variations during its treatment and disposal and the modelling of all these operations.

Tab. II - Statistical correlations of centrifuge tests:
$$y = ax_1^{a_1}\ x_2^{a_2} \ldots x_3^{a_3}$$

| | Sludge type | Coeffic. a | Exponents of x_1 = ... | | | | |
			$x_1 = t_b$ (s)	h_{lr} (mm)	C_{su} (kg/m³)	$\Delta\omega$ (rpm)	$\emptyset_s$ (kg/h)
Dewatered sludge concent.	1	73.75	0.215	0.273	-	-	-
	2	81.79	0.203	0.093	-	-	-
	3	61.25	0.244	0.263	-	-	-
	4	1.22	0.123	-0.056	0.717	-	-
	5	11.97	0.092	-0.272	0.084	-	-
	6	0.0122	0.230	0.399	1.745	-	-
Solid recovery	4	93.95	-	-	-	0.313	-0.946
	5	397.30	-	-	-	0.399	-1.594
	6	96.78	-	-	-	0.311	-0.493

t_b sludge resid. time on the beach
h_{lr} liquid ring height
C_{su} conc. after 2 h settling 1 l cylinder
$\Delta\omega$ differential speed
$\emptyset_s$ solids input flow rate/liquid ring volume (solids flux)
1: 4.8% Aerob. Dig. Activated
2: 3.2% Aerob. Dig. Activated with dephospatation by $Al_2(SO_4)_3$
3: 4.5% Aerob. Dig. Activated whit dephospatation by $FeSO_4$
4: 0.7% Aerob. Dig. Activated using pure O_2
5: 1.6% Aerob. Dig. Activated using pure O_2
6: 2.1% Aerob. Dig. Activated using air

REFERENCES
(1) DICK, R.I. and YOUNG, K.W. (1972). "Analysis of thickening performance of final settling tanks", Proc. 27th. Ind. W. Conf. Purdue University, May 2-4, 33-54.
(2) HASIT, Y., SIMMONS, D.L. and DICK, R.I. (1983). "Modeling of wastewater and sludge management systems", Mathematical Modelling, Vol. 4, 427-438 (Pergamon Press).
(3) RANDALL, C.W., PARKER, D.G. and RIVERA-CORDERO, A. (1973). "Optimal procedures for the processing of waste activated sludge", Bulletin 61, Virginia Water Resources Research Center, Virginia Polytechnic Institute and State University Blacksburg, 86 pp.
(4) EL-GOHARY, F.A. and SALEH, M. (1976). "Conditioning improves sludge filterability", Water and Sewage Works, 123, March, 72-76.

(5) DICK, R.I. and HASIT, Y. (1981). "Integration of processes for waste-water residuals management", Project Report, National Science Foundation, 170 pp.

(6) U.S. EPA (1979). "Process design manual for sludge treatment and disposal", Report 625/1-79-011.

(7) ADAMS, C.E. Jr. ECKENFELDER, W.W. Jr., STEIN, R.M. (1974). "Modifications to aerobic digester design" Water Research, 8 (4), 213.

(8) KOS, P., MEIER, P.M., JOYCE, J.M. (1974). "Economic analysis of the processing and disposal of refuse sludges" EPA 670/274.

(9) DICK, R.I., SIMMONS, D.L., BALL, R.O., PERLIN, K. and O'HARA M.W. (1978). "Process selection for optimal management of Regional Waste-water residuals", National Science Foundation Report, 484 pp.

(10) MININNI, G., SPINOSA, L. and MISITI, A. (1984). "Evaluation of filter press performance for sludge dewatering", Journal WPCF, Vol. 56, 4, 331-336.

(11) SPINOSA, L., MININNI, G., BARILE, G. and LORE', F. (1984). "Study of belt-press operation for sludge dewatering" Proc. of the Conference on "Solids/liquids separation practice and the influence of new techniques", Dept. of Chemical Engineering, University of Leeds, April 2-5, 85-96.

(12) SPINOSA, L., MININNI, G. and MISITI, A. (1984). "Sludge mechanical dewatering", Ingegneria Sanitaria, Anno XXXII, n°4, lug.-ag., 34-42

OPTIMIZATION OF PROCESSING , MANAGING AND DISPOSING OF SEWAGE SLUDGE
PART II: DISPOSAL OPTIONS

A.C. DI PINTO and G. MININNI
Istituto di Ricerca sulle Acque-Via Reno, 1 00198 ROMA

In this Part the main sludge disposal systems are discussed, putting in evidence the relevant characteristics, problems involved and costs with reference to Italian situation.

1. AGRICULTURAL UTILIZATION

Sludge spreading on farmland allow the recovery of organic content and nutrients contained in the sludge; the main problems encountered in this field are due to the presence of metals and pathogenic micro-organisms in excessive concentrations. As far as nutrients are concerned, the limiting factor in sludge application is phosphorus; in fact, the requirement of this nutrient for most crops is considerably lower than that of nitrogen (Table I). The nitrogen and phosphorus requirements are in the ranges 89 - 500 and 22 - 183, whit average values of 210 and 70 kg/ha, respectively. Considering that the mean nitrogen and phosphorus concentrations in a sludge (expressed as N_T and P_2O_5), are respectively about 4% and 2.5% with respect to dry solids (2), it is possible to evaluate the quantity of sludge theoretically needed to satisfy the nitrogen and phosphorus requirements of the crops. As an example, some data are presented in Tab. II. These data confirm that phosphorus is the limiting factor in sludge application and that in some cases (alfa-alfa) the sludge dosages needed to provide the required amount of nitrogen is more than three times that necessary to satisfy phosphorus requirements.

The quantity of usable sludge is, however, limited by heavy metals rather than by phosphorus, when metals concentrations in sewage sludge exceed typical values. For example, in the case of cadmium, at the average concentrations of 16 mg/kg (3), the quantity of disposable sludge,obtained by respecting the proposed EEC Directive, is higher than that required for phosphorus balance.

The cost of distribution of wet sludges can vary considerably according to the method used. If the sludge can be used in an irrigation network or other suitable existing on-site installations, the costs are of the order of 300 Lit/m^3. However, if road tankers and other equipment have to be used, the costs are much higher (of the order of 2000 Lit/m^3) and vary according to the peculiar management conditions (road network, practicability of equipment on the ground, etc). The costs of dewatered sludge spreading are of the order of 5500 Lit/t if ordinary manure spreaders can be used. It is not economical to use the centrifugal fertilizer spreaders commonly used for mineral fertilizers.

Tab. I - Annual nitrogen and phosphorus utilization by selected crops

CROP	NUTRIENTS REQUIREMENT kg/ha	
	N	P_2O_5
Barley	194	72
Corn (grain)	267	111
Rise	122	67
Soybeans	372	72
Sugar beets	306	94
Wheat	194	89
Broccoli	89	33
Cobbage	256	72
Celery	311	183
Potato (Irish)	278	128
Tomatoes	278	89
Apples	111	50
Grapes	117	50
Oranges	133	44
Peaches	106	44
Alfa alfa	500	89
Mean	210	70

Tab. II - Sludge dosages (t/ha) satisfying nitrogen and phosphorus requirements

CROP	NITROGEN BALANCING	PHOSPHORUS BALANCING
Corn	6.67	4.44
Weath	4.85	3.56
Celery	7.77	7.32
Alfa-alfa	12.5	3.56
Average for vaious crops	5.27	2.8

2. LANDFILL

In order to evaluate volume and surface of a sludge landfill site, con centration, mixing with solid wastes or other organic material and the type of landfill must be taken into account. However, it should be pointed out that the sludge is generally disposed in landfill in conjunction with urban solid wastes because of the advantages deriving from optimal volume utilization. Sludge can in fact be used to fill the voids within refuse, especially at comparatively low concentrations. The annual volume of landfill requested for sludge disposal is given by:

$$V = \frac{F_S}{X} \cdot 365 + (P \cdot 10^{-3}-L-E) \cdot A$$

were:

F_S	daily dry solids production	(t)
X	sludge solids concentration	(weight fraction)
P	mean annual rainfall	(mm)
L	annual leachate quantity	(m^3/m^2)
E	annual evaporation	(m^3/m^2)
A	landfill surface	(m^2)

Bulking agent must be taken into account in the computation of landfill volume.

P,L and E are characteristics of each site and of the operation modalities of landfill. When landfill is lined both on the ground and on the surface, P and E equal zero and the quantity of leachate tends to diminish with time. The most economical system, for leachate treatment, appears to be the recycling on the top of landfill. In this may the anaerobic conditions are enhanced and thus also the refuse stabilization process. Moreover the leachate volume gradually becomes reduced due to evaporation.

In Italy, sludge is disposed of in landfill in conjunction with urban solid wastes. The operationg costs are 20,000-50,000 Lit/t of wet sludge.

3. INCINERATION

The cost of incineration and subsequent disposal of the ash is largely dependent on the preliminary treatment.

Incineration costs using a multiple hearth furnace in plant serving 500,000 inhabitants for currently used treatment lines are shown in Tab.III. The data shown in the table are based on the assumptions made in Mininni et al. (4). Amortization has been calculated over a 15 years period at an interest rate of 10%. Furthermore, it has been hypothesized that, on the lines including anaerobic digestion, biogas could be used to cover incineration energy requirements. This would explain the fact that unlike line D, fuel costs in line C are zero, although the sludge characteristics are almost identical. The most representative components of the overall operating costs are amortization (up to 55%) and fuel consumption (up to 53%) for the not selfsustaining lines. The other cost items (personnel and maintenance) are comparable.

The quantity of dry solids treated varies considerably according to the treatment line involved, and reaches its peak in lines without digestion and with filter-press dewatering because of the considerable quantities of inorganic chemicals involved. This means that specific cost trends

of incineration, expressed in units of dry solids, do not coincide with
that of the specific costs per inhabitant served.

If post-combustion at 1,000 - 1,100°C of the exhaust gas is required,
auxiliary fuel consumption rises dramatically; if no energy can be recove
red, the costs become prohibitive.

Tab. III - Incineration costs (without ashes transport and disposal) (4)

Line (*)	A	B	C	D
Dry sludge to be treated (t SS/d)	22	33	24	56
% SS	18.7	36	27	27
% VS	61	47	61	60
Costs (Lit x 10^6/year)				
Fuel	325	0	0	1,230
Energy	75	62	60	110
Personnel	75	75	75	75
Maintenance	100	88	85	124
Amortization	435	385	375	550
Total	1,010	610	595	2,089
Specific cost (Lit/kg SS)	126	51	68	102
Specific cost (Lit/hd · year)	2,000	1,200	1,200	4,200

(*) Line composition
   A: Pre-thickening, anaerobic digestion, post-thickening, centrifugation,
       incineration
   B: Pre-thickening, anaerobic digestion, post-thickening, filter-pressing,
       incineration
   C: Pre-thickening, anaerobic digestion, post-thickening, belt-pressing,
       incineration
   D: Thickening, filter-pressing, incineration

4. TRANSPORT AND STORAGE

Wet or dry sludges are almost never disposed of on-site, so that it
is necessary to provide for their transport; even in the case of incineration, which is generally carried out at the sludge production site, the
ashes still have to be transported. Furthermore, since the sludge disposal
cycles do not always coincide with treatment cycles, it becomes necessary
to make adequate stockpiling provisions. While stockpiling cost are generally
not easy to evaluate, since they depend on a large number of local circumstances, transport costs can be determined fairly accurately once the distance from the disposal point to the production site is known. Especially
when medium-long distances are involved, these costs account for a large
proportion of overall treatment costs.

Storage volumes are very high in the case of agricultural utilization as spreading cannot be carried out all year round. In the U.S.A. these volumes vary from a minimum accumulation time of 30 days, in the case of hot dry regions, to 200 days for cold damp climates. The most common storage systems used are stockpiles and lagoons in the case of dry sludges and closed or open tanks or digesters for wet sludges. Generally speaking, the most economical storage system is lagooning. Covered or tank storage is preferable in the vicinity of residential areas.

Liquid or dewatered sludge can be transported by pipe, truck, barge or railway. In this paper, only transport by truck will be taken into consideration. In fact transport by barge or railway is limited to particular situations, when suitable watercourses or railway stations are in the vicinity of the wastewater treatment plant.

Pipe transport can be used economically for liquid sludges (maximum concentration 8%) (5), only for large quantities. In any case, as for rail trasport, it requires high-cost permanent installations which lack the operating flexibility offered by truck transport.

Costs of sludge transport by truck were evaluated with particular regard to Italian situation (6). The following data, inclusive of amortization, were derived considering the utilization of large trucks:

liquid sludge   8 km (one way)   4,400 - 5,700 Lit/m^3
                          30 km ("   ")   10,500 - 12,500   "
                        120 km ("   ")   28,000 - 29,700   "

dewatered sludge   8 km (one way)   10,000 - 21,300   Lit/t
                          30 km ("   ")   20,000 - 28,000   "
                        120 km ("   ")   31,200 - 37,000   "

6. FINAL CONSIDERATIONS

Several evaluations referring to the cost of different disposal alteratives examined for different distances between the place of origin of the sludge and that of its disposal are shown in Tab. IV.

The data in the table show that, with the assumptions on which the analysis is based, liquid sludge disposal is not economical owing to the high transport costs involved. It can become economical if the disposal site is located in the immediate vicinity of the sludge production site. In such a case it could be economical to use pipes, which are cost-effective over short distances.

In the case of dewatered sludge, if long-distance transport is ruled out, the most economical alternative is sludge disposal on farmland. However, it must be borne in mind that no account was taken in the evaluation of either the costs of stockpiling or the savings in fertilizer obtained by using the sludges.

Of the two remaining alternatives, disposal of dewatered sludges in landfill and incineration, the latter is found to be the more economical in most cases. Only when the costs of landfill lie at the lower end of the cost range shown in the table, this solution is more economical than incineration.

Tab. IV – Analysis of disposal costs

Sludge type	Distance one way (km)	Disposal	Sludge transport cost (Lit/m³ or t wet sludge)	Sludge disposal cost (Lit/m³ or t wet sludge)	Ash transport and disposal cost (Lit/t sludge dry solids)	Total cost (Lit/t dry solids)
Liquid (1)	8	Agriculture	4,400– 5,700	2,000 (3)	–	160,000–192,500
"	30	"	10,500–12,500	2,000 (3)	–	312,500–362,500
"	120	"	28,000–29,700	2,000 (3)	–	750,000–792,500
Dewatered (2)	8	Agriculture	10,000–21,300	5,000 (4)	–	50,000– 87,700
"	30	"	20,000–28,000	5,000 (4)	–	83,300–110,000
"	120	"	31,200–37,000	5,000 (4)	–	120,700–140,000
Dewatered (2)	8	Landfill	10,000–21,300	20,000–50,000	–	100,000–237,700
"	30	"	20,000–28,000	20,000–50,000	–	133,300–260,000
"	120	"	31,200–37,000	20,000–50,000	–	170,700–290,000
Dewatered	8	Incineration	–	18,000–28,000	15,000–35,600	70,000–160,000
"	30	"	–	18,000–28,000	20,000–39,000	76,500–164,000
"	120	"	–	18,000–28,000	25,600–43,500	83,800–168,000

1) Concentration 4%; 2) Concentration 30%; 3) Mean cost with field tankers; 4) Mean cost with manure spreaders

It must not be overlooked, however, that the cost of the biogas produced in the digestion phase and used as an auxiliary fuel in incineration has not been included in the cost analysis.

Since transport costs weigh more heavily for disposal in landfill than for incineration, the former alternative can be found more economical than incineration over distances shorter than those considered.

If spreading on farmland is not possible, disposal in landifill becomes necessary for smaller quantities as incineration would not be feasible.

It is hardly necessary to point out that these considerations are based solely on technical/economical evaluations referring to the disposal and that the inclusion in the analysis of the costs relative to the treatment stage could lead to different results.

REFERENCES

(1) U.S. EPA (1981). "Evaluation of sludge management systems" PB 81-108805.
(2) METCALF and EDDY (1979). "Wastewater Enginnering: treatment, disposal, reuse" Second Edition (London McGraw-Hill).
(3) DAVIS, R.D. and LEWIS, W.M. (1978). "Utilization of sewage sludge on farmland: the gaps in our knowledge" Conference on "Utilization of sewage sludge on land" Oxford 9-14 April 1978 W.R.C. Paper n. 32.
(4) MININNI, G., DI PINTO, A.C., MENDICELLI, A. and SANTORI, M. (1985). "Analisi tecnico-economica del trattamento dei fanghi derivanti dalla depurazione delle acque di scarico urbane" Istituto di Ricerca sulle Acque - Quaderno n. 70.
(5) U.S. EPA (1979). "Process design manual for sludge treatment and disposal" EPA 625/1-79-011.
(6) DI PINTO, A.C., MININNI, G., PASSINO, R., and SANTORI, M. (1982). "Lo smaltimento dei fanghi di depurazione" X Convegno Nazionale Ambiente e Risorse - Bressanone 6-11 Settembre 1982.

DISCUSSION

F. COLIN

Is there one procedure that you would recommend for the optimisation of
the sludge treatment and disposal system?

L. SPINOSA

At the moment, we are not able to recommend one particular procedure. We
need to know much more about the operational and cost parameters before
we will be able to do that. As I mentioned in my paper we only know
about a few cases and much more work is required.

P. BALMER

What are the main differences between the optimisation criteria discussed
in your paper and those given in the literature?

L. SPINOSA

In the past, optimisation of a sludge treatment system was attempted by
considering each step of the process in isolation and optimising that.
This paid no attention to the interaction between the various processes.
What we had tried to do is to consider the system as a whole.

ALTERNATIVE USES OF SEWAGE SLUDGE

R.C. FROST
WRc Processes, Stevenage, Hertfordshire, England

H.W. CAMPBELL
Wastewater Technology Centre, Environment Canada,
Burlington, Ontario, Canada

Summary

Increasing pressure on established routes for the treatment and disposal of sewage sludge is causing sludge disposal authorities to look for alternative outlets. In this context the beneficial use of sludge is attractive in principle since it offers the prospect of limiting any increase in expenditure. The chief potential value of sludge lies in its protein and fat content and a survey of known and potential uses reflects this fact. To illustrate some of the principal features of potential sludge disposal schemes that involve using sludge, five cases are presented in more detail. These are: fat and protein extraction; precious metal recovery; use as building material; electricity generation; and fuel production through low temperature pyrolysis.

1. INTRODUCTION

Sewage sludge is viewed in a number of ways by different sections within a community and by different communities. So that in practice the methods of treatment and disposal adopted locally are determined by a compromise of various and in some ways conflicting views. Essentially, these are:-

- sludge presents - at least potentially - an environmental threat and nuisance
- sludge is an unwholesome, undesirable but inevitable by-product of sewage purification, and as such should be disposed of as cheaply as possible
- sludge is a potentially valuable resource which should, in principle, be exploited to the maximum and not wasted

Any sludge disposal philosophy reflects a balance of these considerations but the balance is of a dynamic nature and, as a consequence of this, the 'best' treatment and disposal routes will tend to change with time.

In recent years public concern over metals, organic and inorganic chemicals, parasitic organisms, and more general pollution caused by sludge has not been significantly assuaged by the scientific evidence. And as a side-effect of the consumerist movement the general public is no longer - if it ever was - prepared to simply put up with nuisances such as that caused by odour. Coupled with these factors the trend is for modern societies to have an ever reducing tolerance towards risk-taking on environmental and safety issues. A net effect is that increasing public and political pressure is being focused on sludge disposal.

Much of this pressure has been directed at the practices of sludge disposal to agricultural land and to sea. Since these routes can, given the right circumstances, often be the cheapest options open to a sludge disposal authority, the increased costs resulting from the imposition of further restrictions can be very significant. Because of the potential need for changes in practice - changes that will cost more money - and their desire to minimise the net increase in expenditure, sludge disposal Authorities are increasingly looking at different and further ways of exploiting the resource value of sludge.

There are a number of ways in which sludge can or could be profitably used, and some of these currently constitute standard practice. These will be mentioned in passing, but the paper will address in detail only those uses which are not so common or are at the development and evaluation stage.

2. SEWAGE SLUDGE AS A RESOURCE

Sewage sludge can be considered the archetypal waste material, the waste not just of a single process but of multitudinous processes and sources: comprising not only human faeces and sanitary-based material, but also other domestic wastes and the wastes from innumerable trades. In consequence, whilst crude protein and fat generally comprise the bulk of raw sludge solids, there are many other constituents as well and the concentration of any given constituent can cover a wide range. Some representative concentrations are listed in Table 1 together with a representive calorific value.

It is fairly clear that, in general, the principal value of sludge as a resource lies in its fat and protein content, and that there exist a number of possible strategies for tapping this resource. These range from using whole sludge as an animal foodstuff, through application to land as a fertiliser, direct extraction to produce feedstock for the chemical industry, down to conversion into fuel products. Locally high concentrations of other constituents of high value may also make extraction of specifics worthwhile in some cases. An example of this, described later, is the possible extraction of gold and other precious metals from some incinerator ashes.

Table 2 contains a few details on a number of known and potential uses of sewage sludge, and five of these are described more fully in following sections of the paper. These uses have been selected on the grounds of example and potential significance at large sewage treatment works. A significant omission from Table 2 is any information on costs. Generally this issue has been poorly addressed by the public literature, and makes an informed assessment and ranking of the alternative uses a very difficult job indeed. This is exacerbated by local factors such as the scale of operation, market outlets and the availability of government grants for capital expenditure, and by the variability of commodity prices.

3. EXTRACTION OF FAT AND PROTEIN

The most thorough investigation of this option for using sewage sludge was carried out in the United Kingdom in the late 1970's by Thames Water Authority and Unilever Ltd[2]. A process was developed for extracting both fat and protein from co-settled primary and activated sludge at Beckton where the sewage from 2.2×10^6 people is treated. The production of fat was chosen for optimisation because of the potential

Table 1. Approximate resource content of co-settled primary
and activated sludge

Constituent	Representative concentration (kg/tonne dry solids)	Nominal market value of constituent (£/kg)	Nominal gross value of resource (£/tonne dry solids)
crude protein	320	0.2 (low grade feedstuff)	64
fat	150	0.2	30
total nitrogen	52	0.25 (as fertiliser)	13
total phosphorus	16	0.6 (as fertiliser)	9.6
potassium	3.5	0.15 (as fertiliser)	0.5
mineral oil	31	0.17	5.3
fibre	120	0.01	1.2
starch	9	-	-
sugar	4	-	-
vitamin B_{12}	0.0025	4200	10
cadmium*	0.016	1.2	0.02
copper*	0.69	1	0.7
lead*	0.57	0.3	0.2
nickel*	0.13	4	0.5
zinc*	1.4	0.6	0.8
chromium*	0.78	4.1	3.2
mercury*	0.003	8.5	0.03
tin*	0.098	9.5	0.9
silver*	0.017	150	2.6
gold*	0.001	8470	8.5
Energy content (MJ/tonne dry solids)	18000	0.0020 to 0.0042 (£/MJ as primary fuel: coal, gas, oil)	36-76

* concentration covers a wide range

Table 2. A summary of current and potential uses of sewage sludge

Use	
Fertiliser (N,P,K)	Application of liquid or cake sludge to grassland, arable land and forested land
Humus/soil* supplement	Application of liquid, cake and composted sludge to land, including the restoration of derelict land.
Animal/fish* foodstuff	Culture of earthworms and aquatic detritivores to produce compost and animal foodstuff. Use of whole sludge or protein extract as a food supplement. Variable results in trials. Metals a problem. Biggest problem is public acceptance.
Chemical feedstock	Extraction of fat, principally, for the manufacture of paints and detergent, grease. See Section 3.
Building material	Principally as a fuel supplement for brick and cement production. See Section 5.
Metals recovery	Not economic unless high concentrations of high value metals such as gold. See Section 4.
Combined heat and power	Use of digester gas for this purpose is well established especially at the larger works. Generation of electricity from the waste heat of incineration less so. See Section 6.
Primary fuel production	Pyrolytic, liquefaction and gasification processes - see Section 7. Combustion of dried sludge in cement kilns and other solid fuel combustion plant.

* For more detail see Reference (1).

for non-edible uses such as the manufacture of soap, and less effort was put into the production of protein. Essentially the process for fat production consisted of dewatering by belt press, extraction with hexane, crystallisation at low temperatures, saponification, removal of mineral oil by solvent extraction, and precipitation of fatty acids using mineral acid. The product was similar in chain length to that of beef tallow, and in degree of unsaturation to that of palm oil, see Fig. 1.

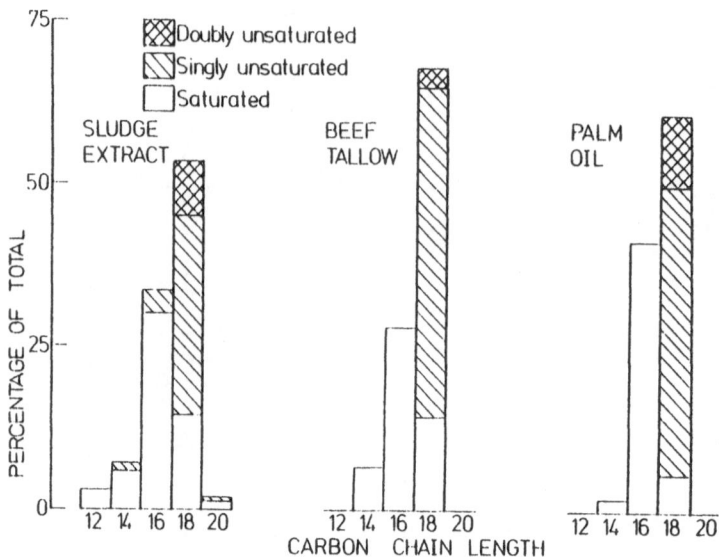

Fig. 1. Comparison of the fat extract from raw sludge with
         beef tallow and palm oil.

Protein recovery was accomplished by treating the residue from the fat extraction with alkali followed by acid. The product contained about 50% true protein and further work would have been needed on optimisation, particularly as regards the removal of heavy metals. Residues, one containing organic and mineral solids and another containing heavy metals, would have remained for disposal.

Financial analysis showed that the viability of the process was critically dependent on the market price of tallow. Many other variables entered into the financial equation including the cost of losing some of the gas production at the digesters. The process was not considered financially viable in the early 1980's because of the future projected price of alternative feedstocks open to Unilever Ltd, and hence was not implemented.

4. PRECIOUS METAL RECOVERY

The metallic content of sewage sludge has many origins. Effluents from industry generally contribute the major load but domestic sources and the load introduced by urban run-off are also significant, and the concentrations of total metals and individual metals span a very wide range. An amount of work has been undertaken to examine the recovery of heavy metals such as cadmium, chromium etc but the costs of extraction have generally been found to rule out this option. A general consensus exists therefore that the heavy metals in sludge constitute a contaminant rather than a resource, and this view is likely to be reinforced by changes in industrial practice and structure, still stricter trade effluent control and further reduction or abolition of lead in petrol.

However, the great variability in concentration suggests that there might be some circumstances in which the recovery of high value constituents might be economically and financially worthwhile. A classic example of this is the situation concerning gold and silver. Mumma et al[3] presented concentration data for 31 sludges from cities across the USA, the results for gold being shown in histogram form in Fig. 2.

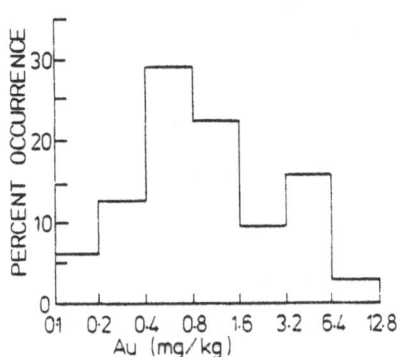

Fig 2. Variation in the gold content of a number of sludges in the USA. (After Reference[3])

The higher concentrations are of the same order as found in North American ores, and since the higher values were found in sludges produced in the more western regions it was speculated that the variation in gold concentration was a reflection of regional geology.

In parallel studies Environment Canada in 1984 funded a project to determine the concentration of gold, silver and platinum group elements in the sludge or ash from fourteen Canadian sewage treatment plants, and to carry out a preliminary economic evaluation of precious metals extraction from the more promising sludges or ashes[4].

Levels of the platinum group elements were sufficiently low as to be of no economic interest. By contrast all sewage sludges and ashes were highly concentrated in gold and silver, in comparison with the abundance of these elements in the earth's crust. Gold concentrations ranged from 0.29 to 1.55 g/t for sludges and from 1.3 to 4.75 g/t for incinerator ashes. The corresponding figures for silver were 23 to 95 g/t for sludges and 60 to 328 g/t for ashes. Out of the fourteen plants surveyed there was only one (Asbridges Bay, Toronto) where the combination of metal concentration and plant capacity were sufficiently large to suggest that recovery would be economical. The sludge at this plant contained 1.55 and 95 g/t of gold and silver respectively. Because gold and silver are both conservative elements they become concentrated when the sludge is incinerated. This results in concentrations in the ash of 4.75 g/t gold and 328 g/t silver. Based on 1984 prices (Cdn) of $400/ounce for gold and $12/ounce for silver, the gross metal value of the incinerator ash is $203/t. This compares to an average gross metal value for Canadian and US gold and silver ores of $108/t.

Sewage ash is a complex material which is not amenable to conventional processes for gold/silver extraction. A preliminary screening suggested that segregation roasting/ flotation concentration process should be applicable. Using this as the basis for the conceptual design, it was estimated that the capital cost for an 80 t/d plant would be approximately $4 million with an annual operating cost of $1.5 million. Based on projected metal prices (Cdn) of $500/ounce for gold and $12/ounce for silver, this would provide a payback period of 1.7 years. Experimental data is required to validate the projected metal recoveries and cost estimates. Although the number of plants where this technology would be applicable is obviously small, it could be quite profitable for those which do qualify.

5. SLUDGE AS A BUILDING MATERIAL

An innovative method for combining the use and disposal of sludge is its use as a component of building materials. The organic, mineral and water content of sludge all have potential value, and significant attempts have been made to exploit this in the production of bricks and cement. Some work has also been undertaken on the manufacture of lightweight aggregate from incinerated sludge ash[5,6]; the use of liquid anaerobically digested sludge as a mix water for concrete[7]; and the compression and heating of sludge with other materials to produce a wall-board[8].

A recent study conducted at the University of Maryland evaluated the use of wastewater sludges as a substitute for other organic materials such as sawdust in the production of light-weight building bricks[9]. The project included production of bricks at bench-scale, a full-scale production run of 35 000 bricks and a further full-scale run of 400 000 bricks.

During the bench-scale experiments, sludges at a nominal 25% solids concentration were added to the clay mixture at ratios of up to 50% by volume. After firing, the sludge amended bricks were subjected to various tests in order to compare their properties with those of conventional bricks. Density of the bricks decreased as the volume of sludge increased; a 50% addition of sludge reduced the density by about 20%. Bricks produced with 40% or less sludge, met the compressive strength standards as defined by ASTM. The addition of sludge increased water adsorption which potentially improves mortar bonding, mortar curing characteristics and freeze-thaw resistance.

The bricks produced at full-scale were significantly improved over those at bench-scale e.g. compressive strength of the full-scale run was approximately double that of the bench-scale bricks. In general, the quality of sludge amended bricks were considered to be equivalent to that of the manufacturer's normal production. There were no discernible differences in the appearance, texture or smell of the amended bricks. Although the type of sludge did affect the colour of the brick e.g. iron conditioned sludge resulted in red bricks, this was not perceived to represent a problem.

Calculations indicated that in the United States, 15% of the annual municipal sludge production could be incorporated in brick production. Allowing for the inevitable disparity between where sludge is produced and where bricks are manufactured, this application could still represent a realistic market for a significant fraction of the sludge produced. Potential benefits could be realised by both the brick and the sludge industries. With respect to brick-making, the addition of sludge results in a less dense product with increased water adsorption properties, reduces water consumption because the water in the sludge provides for the necessary fluidity and may reduce energy costs in the brick kiln due to the inherent calorific value of the sludge. Most importantly, it is not unrealistic to expect that the brick manufacturer would receive a subsidy from the sludge generator because he is in fact accomplishing the necessary task of sludge disposal. With respect to sludge disposal, the extremely high temperature (1000 to 1200ºC) achieved in the brick kiln should oxidise all organic matter and destroy all pathogenic organisms. Although the sludges used in the study did not contain high levels of heavy metals, it is probable that heavy metals in sludge would be dispersed throughout the ceramic matrix in a relatively immobile form. Finally, because hot kiln exhaust gases are routinely captured and recycled to the initial drying stage, the provision of adequate air pollution control is relatively simple.

Though Bonomo et al[10] examined the use of industrial sludges, there is no known European work on the use of sewage sludge for brick manufacture. However a proposal for its use in cement production has been reported[11]. Anaerobically digested sludge from works serving a population equivalent of 150 000 would be dried to 70-75% dry solids content using a thin-layer contact dryer with integral vapour compression. The dried sludge would be hauled by road vehicle to a bunker at the cement works, from which it would be metered onto a belt weigher and blown into the lower section of the heat recovery system of the cement kiln, see Fig. 3.

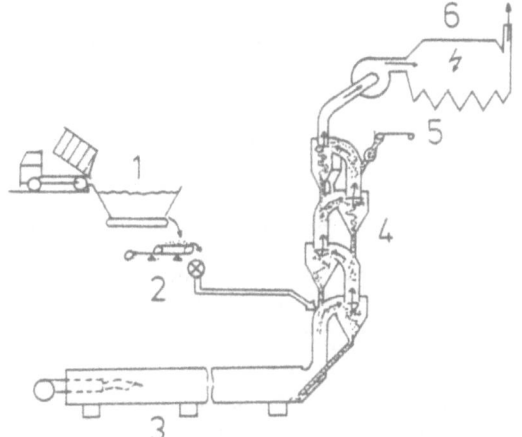

1. Reception of dried sludge
2. Metered feeding
3. Rotary kiln 1500ºC
4. Heat recovery
5. Raw cement meal feed
6. Electrostatic precipitator

Fig. 3. Use and disposal of sludge through cement production.

Keller[11] indicates that the practical limit on using sludge as a fuel supplement in a kiln is 15% of the total heat input. For the case at issue only 3% of the total heat requirements will in fact be met, the cement producer paying for this heat on the basis of equivalent savings in coal. In turn the supplier of dry sludge will be responsible for the installation, maintenance and operation of the sludge feeding system, and will pay both the capital and operating costs of the installation. In addition the sludge suppliers will pay the cement factory a fee for taking (burning) the sludge.

Keller identifies a further potential limit on the use of sewage sludge in this way, this being set by the chlorine contents of the sewage sludge and the raw-meal feed. For rotary kilns with a heat recovery system the total chlorine input, he states, should not in practice exceed 0.02% with respect to final kiln product.

No adverse effects of sludge on cement clinker quality were anticipated. However, no tests had been performed to confirm this view.

6. ELECTRICITY GENERATION

The Hyperion Energy Recovery System (HERS) in Los Angeles was designed to upgrade sludge handling facilities and ultimately provide an energy efficient alternative to the ocean discharge of sludge[12]. The basic process consists of dewatering digested sludge with solid bowl low speed centrifuges, drying the cake by the Carver Greenfield Process ie multi-stage evaporation, combustion of the dried material in fluid bed incinerators and landfilling of the incinerator ash.

Electrical power is generated from two sources, digester gas and heat recovery from the incinerators. Digester gas is compressed and scrubbed for hydrogen sulphide removal and burned in gas turbines to generate electricity. Incinerator off-gases are passed through waste heat boilers which generate high pressure steam which in turn drive steam turbines for power generation. Both turbine systems generate low pressure steam which is used to heat the digesters and to drive the Carver Greenfield Process.

The projected total electrical energy production from the plant is 25 megawatts. Approximately 15 megawatts will be utilised by the liquid and solids processing systems at the plant, leaving 10 megawatts available for sale to a local utility.

The HERS system is capital intensive and complex but it has been designed as a complete integrated system in order to maximise the utilisation of all available energy streams. The success of this method of sludge treatment and cost recovery ie the sale of electrical power, will depend on a number of factors. The first is plant size. Some of the components of the HERS system such as the fluid bed incinerator and Carver Greenfield Process are very costly and can only be justified when the economy of scale at larger plants is achieved. In the case of Los Angeles the design capacity is 366 tonnes of raw sludge per day. Another consideration in any thermal process is the efficiency of the sludge dryer. Although capital intensive, the Carver Greenfield Process is a very efficient method of drying sludge. More conventional dryers such as rotary drum dryers and paddle dryers require 3 to 4 MJ/kg of water removed but the multiple evaporator effects of the Carver Greenfield Process reduce the energy requirement to less than 1 MJ/kg. Probably the most important consideration in a project of this nature is whether there is a market for the electrical power and at what price. In order to stimulate energy recovery, federal law in the United States guarantees that excess energy will be purchased by the local utility at the premium rate which they charge their customers. This ensures that a market for excess electrical power is always available. This situation is not true in many other countries. In many places the local power authority is under no legal obligation to buy power from individual generators. In this case a long term contract negotiated with the local utility would be a prerequisite for considering this method of sludge disposal and energy recovery.

7. FUEL PRODUCTION THROUGH LOW TEMPERATURE CONVERSION

Enormous efforts have been made world-wide in recent years to research and develop alternatives to the use of fossil fuels, and it is important that the use of sludge as a fuel source is seen within this context. Thermochemical conversion of biomass to fuel products is just one route which has been examined, and there are three principle process classes - pyrolysis, gasification and liquefaction. These processes have been reviewed elsewhere[13]. It was concluded that for sewage sludge treatment and disposal low temperature conversion would have the most immediate potential.

The basic concept of low temperature conversion of sewage sludge to produce fuel products has been known for many years[14]. Recently, German researchers have made significant advances in understanding the mechanisms by which sludge is converted to oil[15]. They heated dried sludges to 300-350°C in an oxygen free environment for about 30 minutes. The researchers postulated that catalysed vapour phase reactions converted the organics to straight chain hydrocarbons, much like those present in crude oil. Analysis of the product confirmed that aliphatic hydrocarbons were produced, in contrast to all other processes which produce aromatic and cyclic compounds, whether utilising sludge, cellulose or refuse as the substrate. The German researchers have demonstrated oil yields ranging from 18-27% and char yields from 50-60%. The oil had a heating value of approximately 39 MJ/kg and the char about 15 MJ/kg.

A research program conducted by Environment Canada's Wastewater Technology Centre has generated comparable results[16]. A continuous reactor, see Fig. 4, was designed and operated, in part, to determine the effect of operating variables on product quality and yield. Tests, at a feed rate of 750 g/h, with a number of mixed sludges (primary + waste activated) resulted in yields of 22 to 25% for oil, 50 to 60% for char, 10 to 12% for non-condensible gas (NCG) and 5 to 12% for reaction water. Typical calorific values for the oil, char and NCG were 36.7, 6.2 and 5.8 MJ/kg, respectively. Temperature affected both the yield of individual products and the split between products. As the temperature increased, the yield of oil increased until the optimum temperature had been achieved. At higher temperatures, the oil yield tended to decrease as conditions favoured the formation of increasing quantities of NCG. The calorific value of the NCG increased in direct proportion to the temperature. Oil viscosity and char yield tended to decrease as the temperature increased. The elemental characteristics (C,H,N,O,S) of the oil were not significantly affected by processing conditions. In general, the process proved to be quite stable, in that, while swings in operating parameters such as temperature were reflected in relative yields, the process did not fail nor did the quality of the oil change significantly.

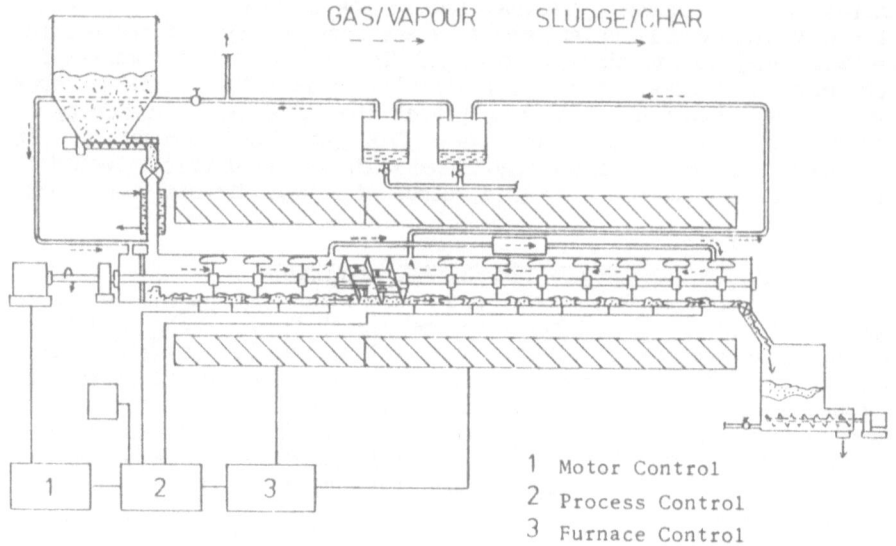

GAS/VAPOUR   SLUDGE/CHAR

1 Motor Control
2 Process Control
3 Furnace Control

Fig. 4. Continuous laboratory equipment for low temperature
         conversion of sludge.

A significant feature of this conversion option is that the major
energy product - oil - is produced in a single step, and that storage and
transportation of the product should be economic. This is in contrast to
many other methods of exploiting the energy content of sludge in which
the recovered energy - methane, steam, hot flue-gas, electricity - is
mainly used in-house or in the immediate neighbourhood of the plant.
     Currently, the least valuable use for the oil would seem to be as a
substitute for No.6 fuel oil. However, the high nitrogen content of the
oil, and the consequent potential for NO_x emissions, is cause for some
concern in this area.   And it remains to be shown that the oil as
produced by this process is non-corrosive and has stable storage
characteristics. Preliminary tests indicate that the oil can likely be
upgraded to an industrial transportation fuel. These are issues critical
to the success of any marketing operation, a fact recognised by the EEC
Commission in its formulation of the 1985-1988 EEC Non-Nuclear Energy R &
D Programme, which calls in part for research proposals to process
pyrolytic oils into suitable fuels.
     Any proposed process has to be examined in relation to the whole
system of which it will form one component, and this systems approach has
been applied by WRc in its preliminary assessment of low temperature
conversion as an alternative sludge disposal method. Since the oil is
formed from the organic and volatile component of a sludge the obvious
choice of feedstock in order to maximise oil production will be raw
sludge. However, many works currently employ anaerobic digestion and it
is wholly feasible that this process would be retained. One potential
advantage of retaining anaerobic digestion in such a disposal route would
be the achievement of complete thermal self-sufficiency, using digester

gas alone, under optimum conditions of solid/liquid separation, dryer operation and heat recovery, see Fig. 5.

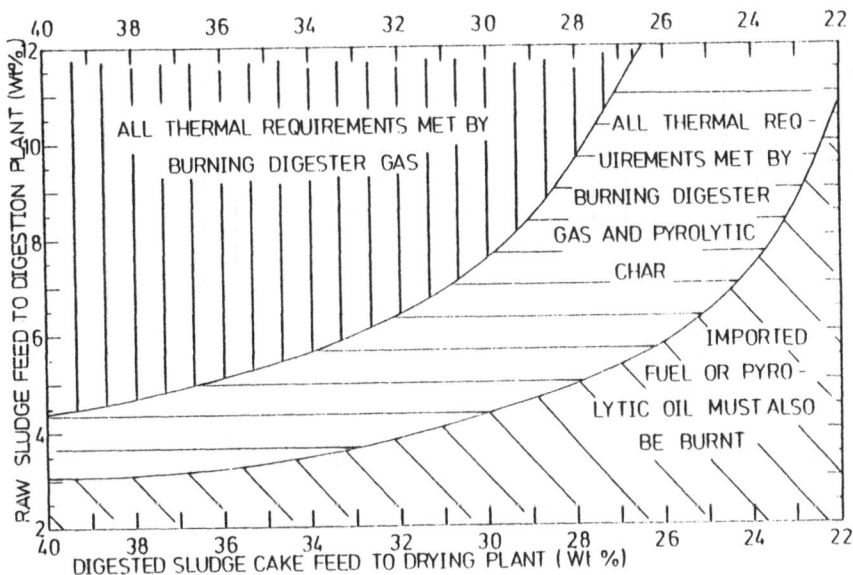

Fig. 5. Domains for thermal self-sufficiency of a route involving low temperature conversion of anaerobically digested sludge (combustion gases used directly for flash drying, waste heat from dryer recovered to provide digester heat, oil marketed externally).

The economics of converting both raw and anaerobically digested sludge have been estimated therefore, and compared with those of the equivalent incineration routes. These comparisons have been made for a raw sludge production of 30 000 tonne DS/year, and have assumed an economic value for the oil equivalent to that of crude petroleum oil (20.2.85). The results of this appraisal are shown in Fig. 6 and indicate that a thermal conversion route could be up to 40% more economic than an equivalent incineration route, regardless of whether raw or anaerobically digested sludge was processed.

This alternative use for sewage sludge clearly has an exciting potential, and Environment Canada's long term plans call for the demonstration of both conversion technology and fuel use at a 25-tonne/d facility. The current schedule projects the demonstration phase of the programme for late 1988 and 1989.

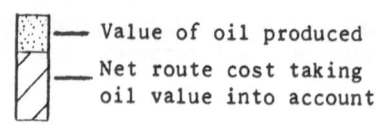

Value of oil produced

Net route cost taking oil value into account

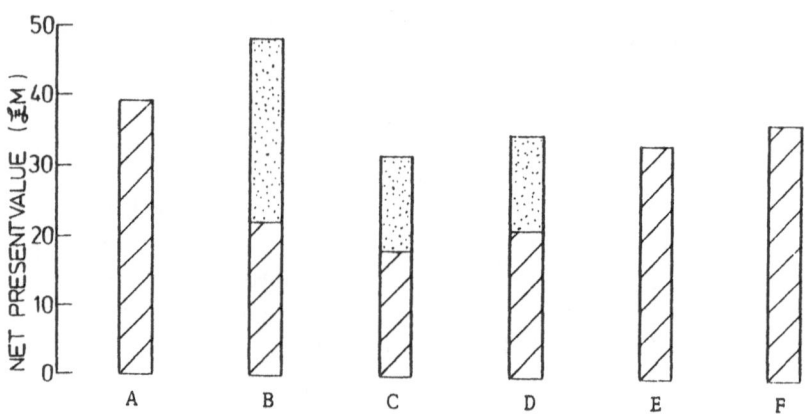

A   Incineration of raw sludge
B   Low temperature conversion of raw sludge
C/D Low temperature conversion of anaerobically digested
     sludge with/without existing digesters
E/F Incineration of anaerobically digested sludge with/
     without existing digesters.

Fig. 6.   Economic comparison of incineration and low temperature
          conversion routes for sludge disposal.

8.   SUMMARY AND CONCLUSIONS

Whilst there is a lot of interest in the potential for using sludge it has to be recognised that most of the proposed uses are not proven. Much research, development and evaluation work needs to be done before this situation can change. If this work is carried out and is successful in proving a number of the more unconventional uses for sludge, it is likely that the influence of local factors will bear heavily in deciding which use is most beneficial and whether it can be justified on economic and financial grounds.

A set of criteria is outlined in Table 3, and it is urged that sludge disposal authorities employ such or similar criteria to assess proposed uses of sludge. It ought to be possible to assess alternatives using these, with ever increasing confidence, from conception to implementation.

REFERENCES

(1)   MONTGOMERY, H.A.C. (1984) Minor uses of sewage sludge. A review prepared for the Water Research Centre. ER 747-M.
(2)   HURLEY, B.J.E. and PETERS, D.C. Potential uses of sewage sludges. Presented at Water Industry '81 Conference, Brighton, England, 19 June 1981.

Table 3. Some criteria for the assessment of proposed uses of sludge

Economic and financial

- Appraisal relative to alternative treatment and disposal routes

- Efficiency of resource utilisation

- Market size and distribution

- Pricing policy necessary to ensure financial viability

- Sensitivity to fluctuating product market value

- Dependence of economic and financial viability on external factors

- Competition from alternative wastes

- Comparison with alternative means of using sludge that would tap the same resource

Operational

- Does the proposed use constitute a disposal outlet for sludge?

- Size of disposal outlet

- Environmental credentials

- Organisation of any processing required

- Product marketability

- Organisation of the marketing operation/s

- Stability of demand for product

- Need for product storage and product storability

- Guarantees on the continued availability of the use and disposal outlet

- Public acceptance

(3) MUMMA, R.O. et al. National survey of elements and other constituents in municipal sewage sludges. Arch. Environ. Contam. Toxicol. (1984), 13, 75-83.

(4) PYNE, R.L. "Preliminary Evaluation of Precious Metals Extraction from Incinerated Sewage Sludge". Report to Environment Canada, 1984.

(5) KATO, H. and TAKESUE, M. (1984). Manufacture of artificial fine lightweight aggregate from sewage sludge by multi-stage stream kiln. Presented at International Recycling Congress, Berlin, 30 October - 1 November 1984.

(6) ARAKAWA, Y., IMOTO, Y. and MORI, T. (1984) Cyclone furnace melting process for sewage sludge. Presented at International Recycling Congress, Berlin, 30 October - 1 November 1984.

(7) Personal communication from Building Research Establishment, England (1985).

(8) OWADA, K. Compacted product from sludges. Jap. Pat. 76123774 (1976). (Chem. Abstr. 1977, 86, 59696.)

(9) ALLEMAN, J.E. and BERMAN, N.A. Constructive sludge management: biobrick. J. Env. Eng. (1984), 110(2), 301-311.

(10) BONOMO, L., BOZZINI, G. and GIUGLIANO, M. (1982). Recycling of industrial sludges in the production of bricks. Presented at International Recycling Congress, Berlin, 1982.

(11) KELLER, U. (1984). A concept for sewage sludge recycling by way of drying-fertilizer-fuel. Presented at International Recycling Congress, Berlin, 30 October - 1 November 1984.

(12) HAUG, R.T. and SIZEMORE, H.M. "Energy Recovery and Optimisation: The Hyperion Energy Recovery System", presented at the International Conference on Thermal Conversion of Municipal Sludge, Hartford, Conn., 1983.

(13) FROST, R.C. and KEARSLEY, C.L. (1984). Oil from sewage sludge - where do we go from here? Water Research Centre, England, ER 304-S.

(14) SHIBATA, S. "Procede de Fabrication d'une Huille Combustible a Partir de Boue Digeree". French Patent 838 063 (1939).

(15) BAYER, E. and KUTUBBUDIN, M. "Low Temperature Conversion of Sludge and Waste to Oil". Proceedings of the International Recycling Congress, Berlin, West Germany (1982).

(16) CAMPBELL, H.W. and BRIDLE, T.R. "Sludge Management by Thermal Conversion to Fuels," presented at an EEC Workshop "New Developments in Processing of Sludges and Slurries" Utrecht, Netherlands, 1985.

DISCUSSION

P. BALMER

Would it be better to use raw or anaerobically digested sludge to produce fuels by the low temperature conversion technique?

R.C. FROST

It is not possible to fully answer this question at the present time. There are conflicting factors. On the one hand, conversion of raw sludge will result in a greater final yield of oil and, hence income, than if digested sludge is converted. However, the conversion of a substantial proportion of raw organic and volatile matter to digester gas will cause the dewatering and drying loads to be less if digested, as opposed to raw sludge, is thermally converted to oil. This may result in a less capital intensive route, and model calculating indicate that the route involving digestion will have a much superior overall energy balance. There is also the prospect that the quality of the oil produced from digested sludge may be superior to that produced from raw sludge and, hence, may be more marketable.

Research is currently being undertaken to resolve these questions.

L. SPINOSA

What are the prospects for the incorporation of sludge into building materials?

R.C. FROST

This would seem potentially to be a good, environmentally acceptable method of disposal of sludge. The high temperatures of cement and brick manufacture will destroy pathogenic organisms and many complex chemical pollutants, whilst the heavy metals will be incorporated into the cementatious or brick matrix. It remains to be seen though whether the cement and brick industries are willing to accept sewage sludge. Studies need to be made to find out how much sludge could nationally be disposed of by this route and at what costs.

SESSION 2 : CHARACTERIZATION OF SLUDGE

Sampling techniques for sludge, soil and plants

Abundance and analysis of PCBs in sewage sludges

Chemical methods for the biological characterization of metal in sludge and soil

Characterization of the physical nature of sewage sludge with particular regard to its suitability as landfill

Microbiological specifications of disinfected sewage sludge

Inactivation of parasitic ova during disinfection and stabilisation of sludge

Epidemiological studies related to the use of sewage sludge in agriculture

Transport of viruses from sludge application sites

Health risks of microbes and chemicals in sewage sludge applied to land - recommendations to the world health organisation

SAMPLING TECHNIQUES FOR SLUDGE, SOIL AND PLANTS

A. GOMEZ*, R. LESCHBER** and F. COLIN***

*   I.N.R.A., Station d'Agronomie, Centre de Recherches de Bordeaux (France)
** Institut für Wasser-, Boden-, und Lufthygiene des Bundesgesundheitsamtes, Berlin (FRG)
*** Institut de Recherches Hydrologiques (I.R.H.), Nancy (France)

Summary

The necessity of analyzing inorganic and organic substances in environmental samples in the micro- and ultramicro-analytical ranges has led to higher requirements for accurate sampling. Although in the field of sludge, soil and plant sampling and investigation, a number of recommendations exists, an overview of the general aspects of sampling, sample pretreatment and preservation is pointed out as a basis for representative results in the subsequent investigation and analysis of sludge, soil and plant material.

GENERAL REMARKS

Modern environmental investigation methods and analytical procedures usually cover the microanalytical range and require highly sensitive analytical apparatus and experienced laboratory staff. In the field of sludge investigation, the fulfilment of these requirements did not seem to be necessary for a long time, due to the relatively high concentrations of pollutants, e.g. heavy metals in sludge as compared with other environmental matrices. Due to the improvement of the knowledge on the negative effects of very low concentrations of some of these pollutants in biological processes, there was an increasing need for analyzing substances in the micro- and ultramicroanalytical ranges in sludges as well. In the field of heavy metal analysis, this was mainly due to the determination of mercury and cadmium especially when transfer reactions in the system sludge-soil-plant had to be studied. During the last few years, in addition, there was an increasing interest in the determination of organic substances which may act as pollutants in the above-mentioned transfer system. In the analytical field, all these requirements were met satisfactorily by the improvement of existing analytical systems e.g. flameless atomic absorption for cadmium determination, cold vapour technique for the determination of mercury etc., or by developing new analytical methods e.g. GC-MS-coupling for the determination of unknown organic compounds. These improvements led to new demands in the field of sampling and sample pretreatment for sludges as well as for soils and plants.

SAMPLING OF SLUDGES

Although reference was made to basic requirements of sampling, in former manuals and handbooks of waste water treatment and analysis, detailed procedures were missing for quite a long time. This changed during the last decade when some countries laid down regulations on the agricultural use of sewage sludge including guide or limit values for heavy metals, thus requiring analytical procedures which were given as part of the statutory text or within separate instructions. Now, there are a number of reports and standards which can be used for proper sampling (1 - 6). In these methodologies or procedures, particular emphasis is placed on how to obtain representative samples. Detailed information on the collection of representative samples from tanks, pipes, channels, heaps and stockpiles is given by the British Standing Committee of Analysts (1) and it is pointed out that all procedures should be carried out by competent trained persons. The same requirements are part of the German Standard existing at present as draft standard (2). Moreover, special attention is paid to the cooperation between the staff doing the sampling and the laboratory staff responsible for the performance of analytical procedures. Single or grab samples are considered to be usually inappropriate due to frequent changes in the composition, water content, etc... of sludges. Therefore, most of the references cited contain instructions on how to accurately collect composite samples.

Sampling of sludges benefits also of the accumulated experience in the field of liquid- or muddy- industrial wastes (8, 9). Different cases have to be considered whether the sludge is liquid or solid. Nevertheless, it is always better to carry out the sampling from a material flux during its flow or handling that from a deposit or a storage capacity (9). In the case of liquid sludges in flow, whether the totality of the flux may be sampled during a part of the time, by cutting the liquid vein, or a part of the flux during the totality of the time. Various methods are described, especially to obtain pumping and transverse homogeneity of the liquid (9). Sampling may be executed directly on rotary drums such as digesters.

Homogenization efficiency of the anaerobic digesters has been measured for various configurations and agitation systems (10). Due to accessibility problems (8), special systems are necessary for sampling in lagoons. The available techniques for solid sludge deposits of relatively low thickness are similar to that conceived for soils (4).

Although rotary drums exist upstream, it is reported that great variations in sludge composition within an operating day occur at plant hauling tank truck batches to disposal sites and at sludge dewatering facilities. From this point of view, a composite sample should be composed of at least three grab samples taken at different times (3).

Automatic samplers have been conceived in order to obtain composite samples of liquid sludges (7). Some of them are marketed.

To perform subsequent analytical determinations, it is necessary to homogenize this sample before dividing it into a number of subsamples of equal composition. This may lead to certain errors. To overcome some of these problems, an American Manuel resulting from the cooperation of a number of Agricultural Experiment Stations in the United States recommends the preparation of dry residue to determine as many parameters as possible from homogenized dry matter (5). For the same reason, the German Regulations on sewage sludge demand the determination of heavy metals from dried, homogenized sludge samples which have to be digested with aqua regia. Interlaboratory comparisons performed within the work of Concerted Action COST 68 have shown that the use of dried sludge samples for analysis leads to reasonable accuracy and reproducibility. Nearly in all cases, however, the sampling of aqueous or wet sludge samples (suspensions, pastes, humid solids) is necessary. This step is of greatest importance for all follow-up procedures. Some aspects have already been mentioned above, and some others have still to be discussed, although not in detail because comprehensive experience is described in the cited references.

The choice of apparatus and material may be very important : there are scoops, hollow bodies to be closed by valves, grips and prickers to take samples and bottels or vessels with wide openings to transport and store sludge samples. They are all made of glass, polyolefine, metal or combinations of these materials to make sure that the samples will neither be contaminated during sampling nor altered during storage, which would interfere with following examination. If changes due to diffusion of substances have to be expected during storage of the sample, only glass bottles or metal containers have to be used. If sludge samples have to be frozen, only plastic bottles and, if possible, metal containers may be used.

For the performance of sampling, special attention has to be paid to the choice of the sampling site in order to obtain a representative sample, to the clear ·description of the site and to an unmistakable marking of the sample. The frequency of sampling depends on the requirements of examination and on the importance of possible changes of the sludge properties. Time of sampling and duration may be of importance in processes with a discontinuous production of sludge. The quantity of the sample usually depends on the examination programme and its extent. Possible inhomogeneities of the material require that an adequate number of single samples of the same size is taken at different sites or at different times and that they are pooled to form a representative composite sample.

Many processes described for the sludge sampling, especially concerning the number and the volume of elementary samples constituting a composite one, are based on empirical considerations, for example, taking composite samples from sludge stockpiles or from fields, a number of 25 cores should be taken to form a composite sample (4).

Recently, important progresses have been carried out (3,9), particularly :
- use of variographic analyse result series, obtained either with the time (chronological series) or with the space (depots), allows to determine the degree of successive sample interaction and to fix the minumum elementary sampling frequency.
- use of multifactorial statistical analytical methods allows to carry out variographic studies on a very restricted number of sludge characteristics which describe as well as possible their variability. Thus, it is possible to determine the deposit structure for a relatively low cost. The disposal structure knowledge allows to adjust sampling operations, not only to know the sludge characteristic average but also the statistical property distribution and extreme characteristics.

To carry out the sampling, manual sampling is the most frequent way because experience has shown that sampling devices used in automatic sampling (e.g. pumps, valves) may easily be blocked by sludge components. If the composition allows mechanical or automatic sampling, it has to be taken into consideration that especially the physical properties of sludge may be changed by mechanical conveying installations and that residual quantities of sludge which remain in the system will mix with the next sample. Small sample quantities may have influence on the results.

As sludge may change rapidly due to biological reactions, these samples have to be examined as soon as possible. Physical variables which can be determined easily e.g. temperature, pH value and conductivity would be determined directly at the sampling site. During further examinations in the laboratory, special importance would be attached to a proper and thorough mixing of the sludge before filling it into the sampling bottles for different investigation purposes.

During the period between sampling and analysis, sludge samples must be kept under conditions corresponding to the original ones. If the sludge is kept cool and in the dark, changes caused by biological and chemical processes can be minimized. If sludge samples have to be preserved for a certain period only, cooling to about 4° C would be applied. Preservation by addition of chemicals cannot be recommended.

For the evaluation of results by the investigating laboratory, a sampling report which has to prepared by the person who takes the samples will be of great advantage. This report would include data on the reasons for sampling, sample type and site, date, time, person, sampling device, local observations and observations relating to the sample, pretreatment, possible preservation and storage conditions of the sample until delivery to the investigating laboratory.

SAMPLING OF SOILS

Soil sampling is commonly used in agriculture. The aim of this practice is to establish the major parameter variations responsible of the fertility in order to bring a correctible fertilization for optimizing the planned cultures (11, 12). Following several steps are used for soil sampling :
- taking off elementary samples in the field;
- pretreatment of samples : drying, grinding, sieving, for example in view of reduction and possible preparation of a composite sample;
- laboratory treatment : possible grinding, reduction of the sample to an aliquot for a representative analysis of the whole.

It is generally considered that the 2 last steps are well under control and the main source of errors is in the field sampling.

The increasing production of sewage sludges and their unquestionable agronomic value have led to consider their use in agriculture. Nevertheless, the often high contents in various mineral elements (such as zinc, cadmium, mercury) or organic compounds (such as PCB), regard to the normal soil contents, put a bridle on their agricultural use.

Various regulations have been established in several countries including France (17) to limit the concentrations of elements capable to be phytotoxic or to contaminate the trophic chain. This makes necessary to strike a mineral – and possible organic – balance of the soil which is intended to receive the sludge. Thus, the sampling methodology and analysis must allow the determination of the total element amounts present in the soil and not, in contrast with agronomic practice, the variations in the fertility parameters.

The extrapolation of results, obtained on a few soil samples to the plot from which they were taken off, implies certainly enormous errors. The only way to establich an exact balance of a plot consists in a sampling which assumes completely its diversity and the element gradients intended to be analyzed.

The plot diversity is understood by the pedologic study of its constitutive soils. Inside each pedologic unit, the composition variation is related to the possible gradient of the elements under consideration. Therefore, the sampling planning will be applied to the sub-plots formed by the different pedologic units. As before, the sampling planning will include the following steps : sample taking off, pretreatment, then laboratory treatment before analysis. The 2 last steps being regarded as well known and low tainted with error, it seems to be important to consider more particularly the sample taking off procedure.

From a general point of view, the sampling procedure is executed using an empirical method and is controlled afterwards by statistical studies. It has appeared to us more interesting to take up the following approach : we establish a sampling planning based on the statistical study of a close-mesh sampling of an homogeneous and low metal polluted soil plot (13). The plot* studied in this paper is located on a "Touya" soil

---

* The French Ministry of Environment had paid for this part of the work.

cultivated in non-irrigated maize monoculture since its breaking up in 1984. The soil, a typically brown acid isohumic one, is developped on recent Quaternary (Würm). It is characterized by a relatively high content in organic matter (5 to 6 %), a texture with mainly fine limon (15 to 20 % of clay) and an initially low pH (4,5 to 5) associated with a notable charge in exchangeable aluminium (about 300 ppm). Regular limings have led to a notable increase in pH values of the soils in this agricultural area and, by this way, the disappearance of mobile aluminium.

The maize monoculture receives yearly an average of 180 kg/ha nitrogen (principally urea), 150 kg/ha phosphoric acid (alternance of superphosphate and scoria) and 150 kg/ha potassium chloride. A liming of about 1500 kg of CaO/ha/2 years is applied (dolomy form) in order to increase the initial level of magnesium.

The plot is a square of 100 m x 100 m where 100 samples have been taken off, using a network of 10,40 x 10,40 m chosen to avoid the superposition of sampling points with maize rows. Some heterogeneities may exist among the seed-rows, due to the intense colonisation of the line by the roots-coma or to a localisation always possible of a starter manure or also of treatment produces. It is clear that this possible heterogeneity is destroyed by next tillage.

Each sample is obtained with one sole drill-blow of 20 cm depth. The sample is immediately put into a tight-closed plastic bag and brought to the laboratory.

The scheme of sample treatment is reported on the document n° 1 added in annexe. The weighing of the whole sample, fresh or aid-dried, is carried out to exhibit a possible sampling heterogeneity due to the operator and, on the other hand, to show in the plot the main differences of texture, organic matter content or existence of wettings which may influence the sample moisture. The following parameters were measured on the set of 100 samples : total carbon (Anne method), granulometry (5 fractions), heavy metal concentrations (Cd, Co, Cu, Ni, Pb, Zn). The metals have been analyzed after mineralization with aqua regia, cadmium in graphite oven and the others by atomic flame absorption.

Beyond these determinations carried out on the whole of 100 samples, composite samples have also been prepared. THe procedure used for that is indicated in the added document (annexe n° 2). For each elementary sample silted at 2 mm, 10 g of soil were sampled and then mixed to constitute a sole sample, from which 3 sub-samples were taken off using the "corning and quatering" technique. This operation has been triplicated. Thus, 9 sub-samples were prepared and 3 separated analysis carried out on them.

The analytical results obtained on the five elementary samples are reported in the document annexe n°3. The variation in weight due to the sampling is relatively slight and approximatively equal to fresh - or dry - weight. This confirms a relative textural homogeneity (visible in the field) and the absence of wettings. The variability in carbon and clay is indeed extremely low (variation coefficient of 5 to 6 %) but, on the other hand, large differences are observed for metals.

To determine the minimum sample number necessary for plot characterization, two statistical methods have been used for interpretation of these results : a classical method using relative precision curves and a geostatistical method working by "krigging". Taking account of the spatial independance exhibited by the geostatistical method, the results of both methods lead to a sample number around 20, following an uniform repartition for a relative accuracy of 10 %. A minimum of 25 samples has finally been chosen, uniformly distributed, so that the accuracy will be below 10 %.

The examination of analytical results obtained on composite samples and reported on table-annexe n°4 indicates that, except for cobalt, the results obtained on a composite sample coming from mixing of 100 elementary samples are very close to the average value of the 100 samples analyzed one after the other. It is thus possible to conclude without excessive error that :
- analysis of 25 samples taken off uniformly from an homogeneous plot must lead to an estimation of the content in different elements, with a precision lower than 10 %; a correct characterization of this plot type is consequently possible;
- analysis of a composite sample of 25 elementary samples with several repetitions is able to give a correct estimation of the plot (with an accuracy of about 10 %).

Concerning the precautions required for sampling, pretreatment and laboratory treatment, one is referred to the study presented at the Brighton Symposium (14) by one of us.

SAMPLING OF PLANTS

Plant sampling benefits of the long agronomic experience. Nevertheless, it must be noticed that, in the normal practice, the aim is the deviation appreciation in relation with an optimal-like situation or the research of one - or several - cause (s) responsible of observed symptoms.

In the case of transfer risks of undesirable elements such as heavy metals, the sample will have to allow the total amount appreciation immobilized by plants and, also, the repartition of these elements in the different plant organs. This will allow the evaluation of trophic chain contamination risks (comestible parts) but also the transport and contamination risks (part exported and then restituted to the soil after composting or transformation into manure).

Consequently, it will be necessary to know exactly the concentration in elements of each plant part for establishing a balance and also the respective weight of each organ under analysis when the whole is too big to be analyzed, or to know the average element percentage when the analysis of the whole plant is possible.

An additive notion will have to be taken under consideration : the plant age or stage conditions the element repartition and concentration in it. For comparative studies (for example, different varieties of the same plant), it will be necessary to locate with a maximum of accuracy the step to which the sampling and then the analysis are referred.

From these general considerations, a sampling planning can be deduced (16) :
a/ definition of the plant number to be sampled - This depends, at the same time, on the size of the plant and of the plot where the sampling is executed. If the plant is small, for example Ray-grass, several individuals have to be taken off in a given point to constitute the sampling unit. In the opposite case (maize in the ear stage for example), one sole plant may constitute the sampling unit. The number of sampling units is dependent of the observed plot heterogeneity. If one is referred to the above paragraph concerning soil sampling, for a same pedologic unit, 20 to 25 sampling units must be taken off to integrate the possible gradients relative to the elements in the soil. Each sampling unit may be either submitted to organ separation (if the different elementary repartitions are wished) or stored in its whole before going through the pretreatment process.
b/ sampling - During the sampling harvest, all precautions would be taken to avoid pollutions such as loam projections by the sampling instrument. If possible, it is preferable to operate always in the same conditions : for

example, same hour of sampling, same climatic conditions. Indeed these factors may influence the elementary composition of the plant and, more certainly, its moisture. Samples will be then put in tight-closed containers (for example, plastic bags) and hold in a cool place to avoid any variation prejudiciable to the next operations. Too important sampling campaigns will be avoided, because they increase too much the delay between the pretreatment and the treatment assuming the sample conservation.

c/ <u>treatment of the sample</u> - Each sampling unit would be submitted to certain operations necessary to preserve its representativity and conservation with the time, until the constitution of an average sample or its analysis. The following operations are distinguished :

. **cleaning** - The sampled plant may have been polluted by loam or compounds such as pesticides or spraying of nutritive solutions. This operation involves the following steps : wiping dry if the sample is very dirty, washing by brief rubbing of the surface with cotton wool or woven impregnated with water, eventually added or liquid detergent or diluted acid solution; then rinse 2 or 3 times by rapid passage of the sample into a water bath, the last of them being constituted by demineralized water, then wringing of the sample.

. **conservation of the fresh sample** - The analyses are relative to fresh material and if it is impossible to do it immediately after cleaning. The samples will be stored in plastic bags thightly closed in a refrigerator. In case of storage in a freezer, particular precautions must be taken to recover the whole sample, because it will be constituted of 3 phases : vegetal fragments, exsudate ice crystals, water of defreezing.

. **dessication** - This operation must be carried out on the whole sample as quickly as possible to stop the biological processes which may alter the sample. Different possibilities are available :

- dessication by heating in a drying oven, made more efficient by aircirculation; for the choice of the working temperature it will be taken into account of possible losses of some elements by volatilisation (case of mercury and selenium). Generally the temperature will never exceed 80° C
- dessication by infra-red heating
- dessication by freeze-drying.

. **grinding** - The aliquot for the analysis being generally smaller than the sample, it is necessary to grind this one. The material used must not bring any pollution (case of stainless steel for the analysis of some metals) and will avoid the contamination of one sample by the precedent one. This operation will always be followed by a rehomogeneization of the obtained powder. One must note that a direct relationship is found between the fineness of the ginding and the weight of the aliquot assuming its representativity.

d/ <u>Constitution of an average sample</u> - For practical and economical reasons, it is not always desirable to analyze each sampling unit; in this case an average sample representative of the whole of the sampling units would participate in an equivalent way of the others to the constitution of the final sample. One of the methods leading to this is : after grinding and rehomogeneization of each sampling unit, a same weight of vegetal material is taken off, to constitute the average sample after rehomogeneization of the aliquots. It is possible to carry out this operation, either on the whole plant or on dry material. To avoid errors due to dessication of the sample, weighing must be done immediately after grinding, when fresh material is used. The material used does not bring pollution or elementary retention (for example, it will be better to avoid the use of a mixer with stainless-steel pieces for the analysis of some metals).

e/ When a long storage precedes the analysis, a rehomogeneization of the vegetal powder would be executed before any analysis procedure. Indeed, a sedimentation of the vegetal particles, due to gravitational force and differences in the grain density in the powder may occur during the storage phase.

GENERAL CONCLUSION

Sampling is a decisive step in the study of pollutants transfers in the system sludge/soil/plant. Many precautions are necessary and the present study cannot be considered as exhaustive. Nevertheless, it has an indicative value, which may be used as a basis in most of the research cases. Only a personnal experience allows to improve the steps reported here because earch research is a specific case.

Résumé

La nécessité d'analyser des échantillons pour le contrôle de l'environnement, dans le domaine minéral ou organique et ceci à des niveaux micro ou ultra-micro-analytiques, conduit à des exigences accrues dans les procédures d'échantillonnage. En ce qui concerne les boues de station d'épuration, les sols et les plantes, il existe un certain nombre de recommandations pour l'échantillonnage et la recherche. Les principaux aspects relatifs à l'échantillonage sont abordés ici. Au long de ces pages, l'importance d'un échantillonage correct, du pré-traitement de l'échantillon et de sa conservation est mise en évidence pour l'obtention de résultats représentatifs dans le domaine de la recherche et de l'analyse des boues, des sols et des plantes.

REFERENCES

(1)   Department of the Environment - The sampling and initial preparation of sewage and waterworks sludges, soils, sediments and plant materials prior to analysis 1977. Report of the Standing Committee of Analysts, 14 pp. Her Majesty's Stationary Office, London, 1978.
(2)   Deutsches Institut für Normung (DIN) - Sampling of sludges; Groups Sludges and Sediments = DIN 38 414 Part 1, Standard draft, unauthorized English translation, Berlin, 1985.
(3)   MONTEITH H.D. and STEPHENSON J.P. - Development of an efficient sampling strategy to characterize digested sludges. Environment Canada Res. Report n° 71, 124 pp. Research Programme for treatment of municipal pollution under provisions of The Canada-Ontario agreement on Great Lakes water quality, Project n° 74-3-16, Ontario Ministry of the Environment, Toronto, 1978.
(4)   COKER E.G. - The design and establishment of field experiments on the agricultural use of sewage sludge. Water Research Centre Techn. Report TR 123, 25 pp. WRC Medmenham, Marlow, Bucks, Oct. 1979.
(5)   Kansas State University, Agricultural Experiment Station (Dir. F.W. SMITH) - Sampling and analysis of soils, plants, waste waters and sludge. Suggested standardization and methodology. Res. Publ. 170, 20 pp., Oct. 1976 preparaed by a Subcommittee of NC - 118 (Chairman R. ELLIS, Jr., Kansas State Univ.).
(6)   AFNOR - Boues des ouvrages de traitement des eaux usées urbaines. Boues liquides. Echantillonnage en vue de l'estimation de la teneur moyenne d'un lot. Pr. U 44-108.

(7) BOLOGNA A. - Automatic waste sludge sampler. Deeds and Data, WPCF, Arpil 1974, pp. D4-D5

(8) PETERSEN T.A. - Link actuated sampler for sampling lagoon sludges. Industrial Wastes, Sept.-Oct., 1983, pp. 16-17.

(9) Bureau de Recherches Géologiques et Minières - Echantillonnage des déchets industriels : approche théorique, recommandations et exemples pratiques. Rapport BRGM n°84 RDM 023 MIN, Orléans, mars 1984.

(10) BAUMANN P.G., HUIBREGTSE G.L. - Evaluation and comparison of digester gas mixing systems. Journal WPCF, Vol. 54 n°8, 1194-1206.

(11) COCHRAN W.G. - Sampling techniques. John Wiley and Sons Inc., New-York, London, Sydney, 1966.

(12) Soil testing and Plant analysis. Part I : Soil testing. Soil Sci. Soc. of Amer., Inc. Publisher, Madison, Wisconsin (U.S.A.), 1967.

(13) BRUN T. - Application de la géostatistique à l'étude des métaux lourds. Document du Service Statistique de l'A.C.T.A Contrat Ministère de l'Environnement : Etude des techniques d'échantillonnage des Sols.

(14) GOMEZ A. - Utilisation des boues de station d'épuration en agriculture : problèmes posés par l'échantillonnage pour la détermination des éléments-traces. 3e Symposium international : Processing and use of sewage sludge, Brighton, 1983, p. 132-139. D. Reidel Publ. Co., Dordrecht-Boston - Lancaster, 1984.

(15) Soil testing and Plant analysis - Part II : Plant analysis. Soil Sci. Soc. of Amer., Inc., Publisher, Madison, Wisconsin (U.S.A.), 1967.

(16) MARTIN-PREVEL P., GAGNARD J., GAUTIER P. - L' analyse végétale dans le contrôle de l'alimentation des plantes tempérées et tropicales. Technique et Documentation, Lavoisier, Paris 1984.

(17) AFNOR - Boues des ouvrages de traitement des eaux usées urbaines. Dénominations et spécification. NF U 44-041.

Annexe 1 –

Sample processing

Fresh weight   ⟶   Air drying   ⟶   Dry weight
(homogeneity of samples)   (Homogeneity of the moisture)

Aliquote sampling   ⟵   Packing   ⟵   Grinding (2 mm)

Grinding (300 μ)   ⟶   Mineralization   ⟶   Solution

C (and S) analysis   N analysis   Analysis P_2O_5 colorimetry
                                                     metals

Granulometry

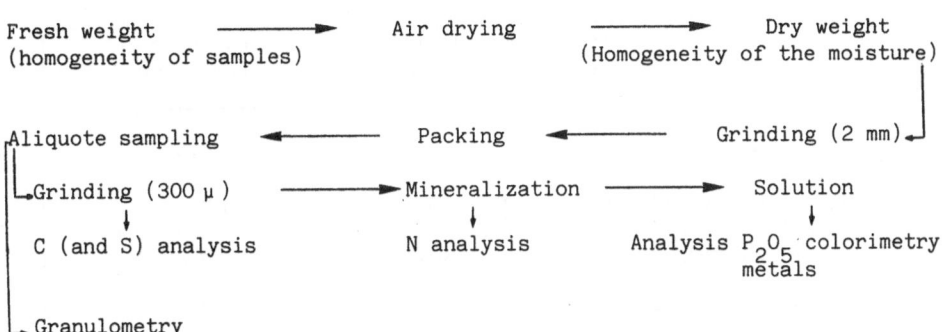

Annexe 2 –

Composite sample

100 samples

10 g/sample

Mixture

Average sample   ⟶   M   A   M   B   M   C

Analysis

x 3
(M_1 M_2 M_3)

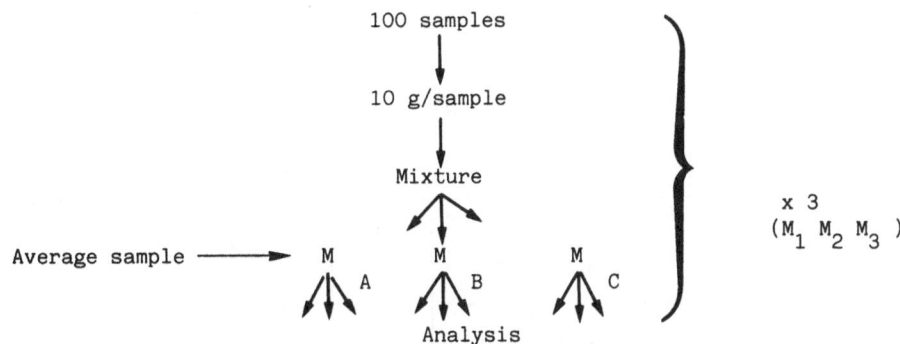

Annexe 3 -

Results of separated analysis on 100 samples

		Mean Average	Standard deviation	Coefficient of variation
C	p. 1 000	27,3	1,73	6,3
Clay	p. 1 000	183	9,4	5,1
Co	mg/g	1,22	0,25	20,2
Cu	"	3,41	1,00	29,4
Ni	"	7,69	0,89	11,5
Pb	"	11,6	1,24	10,7
Zn	"	15,7	2,12	13,5
Fresh weight	g	1021	120	11,7
Dry weight	g	797	94	11,8

Annexe 4 -

Composite sample results (ppm)

	Mean (mg/g)			Standard deviation			Coef. Variat. %		
	M_1	M_2	M_3	M_1	M_2	M_3	M_1	M_2	M_3
Co	2,16	2,07	2,29	0,147	0,314	0,005	6,8	15,1	0,2
Ni	7,59	7,22	7,55	0,287	0,009	0,467	3,8	0,1	6,2
Cu	3,30	3,55	3,44	0,310	0,318	0,165	9,4	9,0	4,8
Pb	11,1	10,8	10,7	0,320	0,556	0,419	2,8	5,2	3,9
Zn	15,0	16,4	17,0	0,440	0,558	0,250	2,9	3,4	1,5

DISCUSSION

The main points to emerge from the discussion were that:
1. It was important to ensure that sampling schemes were practical and feasible and not just based on statistical theory.
2. The details of a sampling programme inevitably depended on a first step of setting out the aims of what you want to achieve.
3. Analysis for some organic contaminants could be very expensive so it was only worth doing on a representative sample even if this in turn was expensive to obtain.
4. Liquid sludge was easier to sample properly than solid sludge.
5. For soil sampling, depth of sample could be important on sludge-treated fields and also the pH value and nitrogen content of the soil should be considered as well as the metal content when assessing the implications of results.
(Contributors to discussion: <u>Hall, Levi-Minzi, Fleming</u>).

ABUNDANCE AND ANALYSIS OF PCBs IN SEWAGE SLUDGES

J. TARRADELLAS*, H. MUNTAU**, H. BECK***

*   : Institut du Génie de l'Environnement
      Ecole Polytechnique Fédérale de Lausanne
      CH-1015 LAUSANNE

** : Commission of the European Communities
      Joint Research Centre Ispra
      I-21020 ISPRA (Varese)

***: Bundesgesundheitsamt
      D-1000 BERLIN 33

Summary

PCBs could be an hindrance for the agricultural use of
municipal sewage sludges. In some cases PCBs from sewage
sludges can be at the origin of the extreme biomagnifica-
tion of these pollutants in agroecosystems. A systematic
control of PCBs content of sewage sludges is needed. The
Program COST 68 ter of the European Communities had promoved an intercomparison project to collect the experiences of European laboratories involved in PCBs-determina-
tion in sewage sludges and to identify the major error
sources of this analysis. The samples submited to analysis
were a set of pure PCBs congeners, a mixture of comercial
PCBs, cleaned and rough extracts from sludges and a dried
sewage sludge powder. The results obtained show that improvements are necessary, mainly in the strictness of
the handling of the samples and the identification of
the most acceptable extraction method, to reach an acceptable coherence between the results of different laboratories.

I. ABUNDANCE OF PCBs IN SEWAGE SLUDGES

The term "PCBs" represents a well-known family of organic micro-pollutants which consists of theorically, 209 different congeners, only 102
congeners of which are found in comercial mixtures. The presence of
these products in municipal sewage sludge had been largely studied in
the last years (ref. 1 to 9). The concentrations of PCBs found in western European sewage sludges varies mainly between 1 to 10 ppm (dry
materiel).
Municipal sewage sludge contain a lot of organic micro-pollutants from
which only PCBs had been well studied (ref. 1, 2, 11, 12, 13). The
main families of organic-micro-pollutants found in sewage sludges are :

. Halogenated aromatics :   . polychlorinated biphenyls PCBs
                                 . polychlorinated terphenyls PCTs
                                 . polychlorinated naphtalenes PCNs
                                 . polychloro-benzenes

. Aromatic amines and nitroso-amines

. Halogenated aromatics containing oxygen :
                                 . phenols
                                 . chlorophenols
                                 . polychloro-diphenylethers
                                 . polychloro-dibenzofuranes
                                 . polychloro-dibenzo-p-dioxins

. Polyaromatic and heteroaromatic hydrocarbons

. Halogenated aliphatics

. Aliphatic and aromatic hydrocarbons

. Phtalate esters

. Pesticides :   . Lindane
                                 . Dieldrin
                                 . DDE
                                 . Organo-phosphorus

In a recent research realized on the sewage sludges of six swiss
towns (ref. 10) the concentrations of PCBs were compared with the
concentrations in PAH (polyaromatic hydrocarbons) phtalates, 4-nonyl-
phenol, cadmium and zinc. The results of this research are presented
on the FIGURE 1

II. EFFECTS OF PCBs FROM SLUDGES ON SOILS, PLANTS AND AGRICULTURAL ECOSYSTEMS

Organic micro-pollutants in sewage sludges, cannot be, generally, so
easily translocated in the plants as heavy metals. That could be one of
the reasons for the existence of only few studies about environmental
problems induced by organic micro-pollutants in soils (ref. 31) and in
sludges. Again, PCBs had been the better studied amongst these products,
partly because they seem to be the less degradable. Some authors (ref.

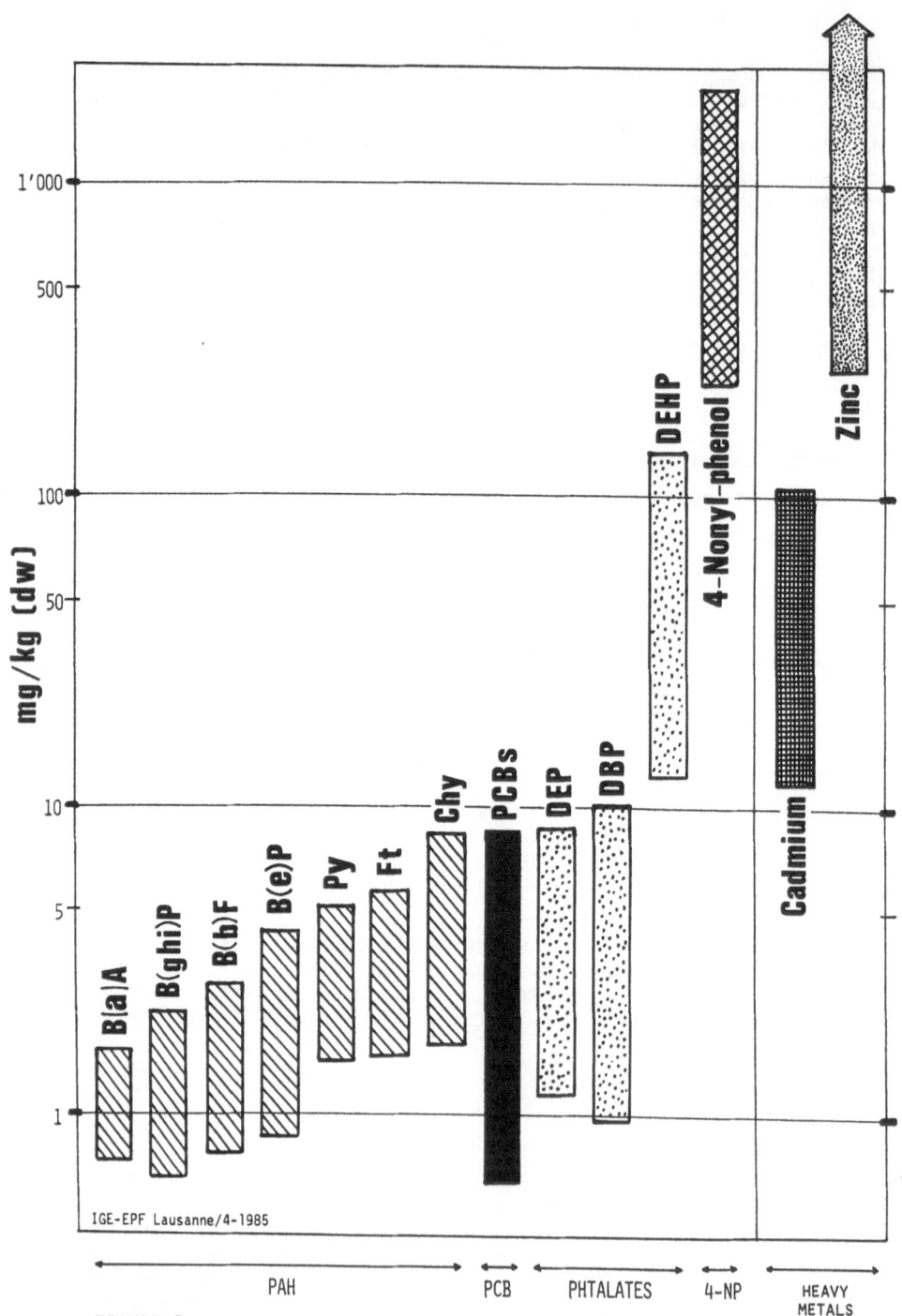

mg/kg [dw]

1'000

500

100

50

10

5

1

B(a)A

B(ghi)P

B(h)F

B(e)P

Py

Ft

Chy

PCBs

DEP

DBP

DEHP

4-Nonyl-phenol

Cadmium

Zinc

IGE-EPF Lausanne/4-1985

PAH   PCB   PHTALATES   4-NP   HEAVY
                                       METALS

FIGURE 1:
MICRO-POLLUTANTS CONCENTRATION IN SEWAGE SLUDGES FROM SWITZERLAND (1983 - 1984)

14, 15, 16) have studied the degradation of PCBs in soils. Results depend on the rate of chlorination of PCBs and the biological activity of the soil. The half-life varies between 2 monthes for 2, 4-dichloribiphenyl to 15 years and up for PCBs with more than 3 chlorine atoms.

FIGURE 2 shows the range of concentrations of PCBs found in natural Swiss soils and soils treated by municipal sewage sludges and composts (ref. 10). The mean magnification factor of PCBs concentration between natural soils and sludges-treated soils is about 5 and between natural soils and compost-treated soils about 20.

FIGURE 2 shows also the range of concentrations of PCBs found in some species of Swiss agro-ecosystems. The role of PCBs brought by sewage sludges and composts to the bio-accumulation of these products in the fauna is clearly underligned by the significative increment of PCBs concentration in the earthworms living in soils treated by sewage sludges and composts. Earthworms are a basic food for a lot of animal species of agro-ecosystems and their PCBs contamination by sludges and composts is probably one of the main reasons for the extremely high concentrations of PCBs found in the eggs of birds of prey.

III. ANALYSIS OF PCBs IN SEWAGE SLUDGES

1) Présentation of the COST-Intercomparison

The results presented above show that PCBs are one of the major organic micro-pollutants to be controlled in sewage sludges. But systematic control of PCBs content of sewage sludges, soils and agricultural products is hampered considerably by the lack of simple, straightforward, rapid and accurate analytical methodologies.

Consequently the Working Party 2 of the COST 68 ter programm had promoved since 1982 an intercomparison project with the view to :

- Collect the experiences of European laboratories involved in PCB-determination in sewage sludge, and establish a realistic picture of the current state-of-the-art.
- Identify the major error sources associated with PCB-determination in sewage sludge.

The first exercise aims at a correct picturetaking of the situation with regard to PCB-determination in sewage sludge and its structure was adequately designed.

This paper presents the results of this exercise.

The following materials was sent out for analysis :

1. Hexane-solution of the PCB-components IUPAC no :
    28 : 2,4,4'-trichlorobiphenyl
    52 : 2,5,2',5'-tetrachlorobiphenyl

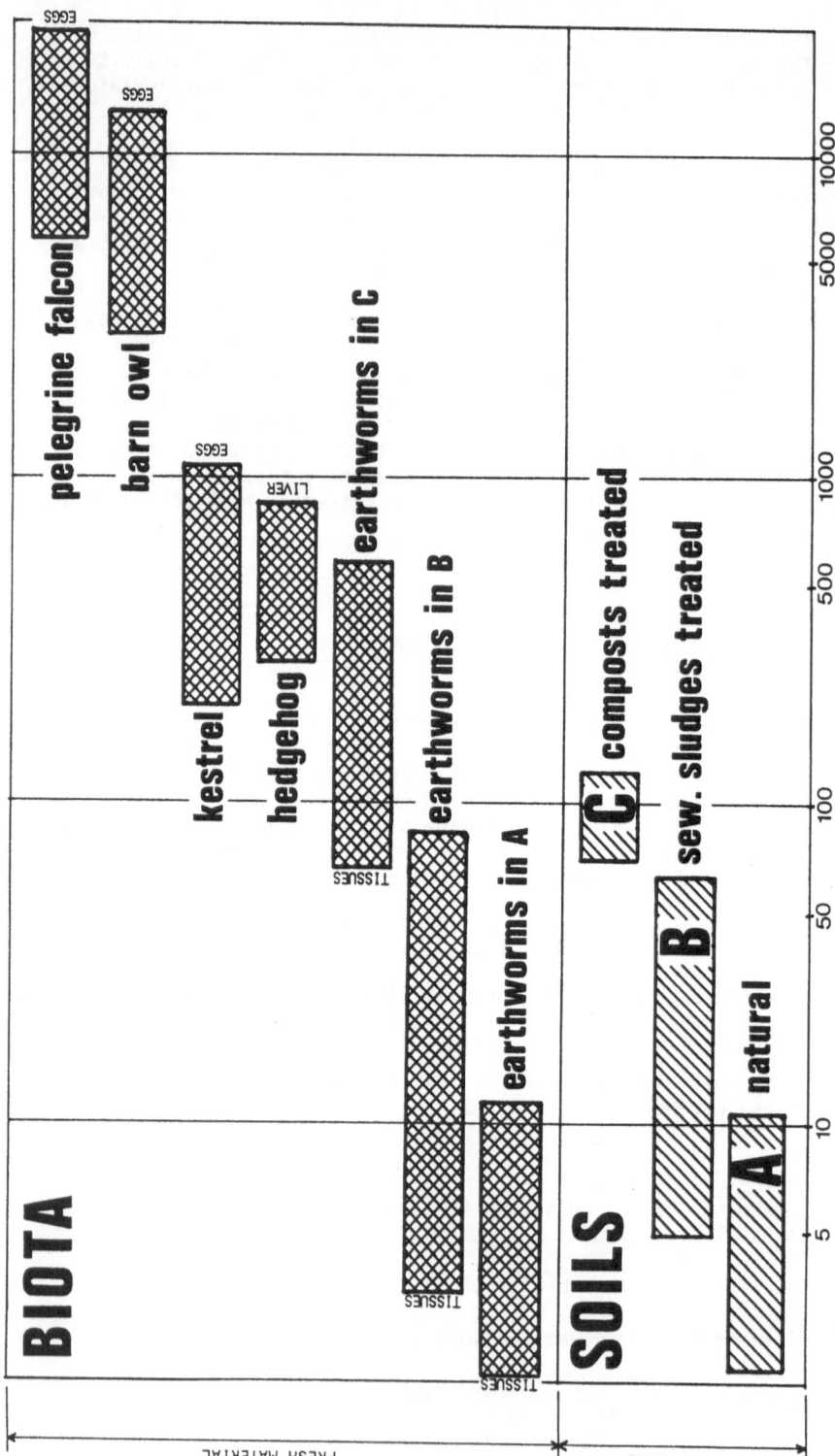

FIGURE 2 : MAGNIFICATION OF PCBs CONCENTRATION IN SWISS AGRO-ECOSYSTEMS

PCBs concentration [µg/kg]

```
 77 :  3,4,3',4'-tetrachlorobiphenyl
101 :  2,4,5,2',5'-pentachlorobiphenyl
138 :  2,3,4,2',4',5'-hexachlorobiphenyl
153 :  2,4,5,2',4',5'-hexachlorobiphenyl
180 :  2,3,4,5,2',4',5'-heptachlorobiphenyl
```

The code of this material was M1.

2. Hexane-solution of technical mixtures Chlophen A30, A40, A60.
   Code M2.

3. Two raw-extracts made from Ispra standard sewage sludge SL3 according to two different extraction techniques. Codes M3A and M3B.
   M3A was extracted by acid-digestion way of the dried sludge during
   24 h in a solution of 2 volumes of 70 % perchloric acid in 3 volumes of glacial acetic acid (ref. 17).
   M3B was extracted by direct stirring of the dried sludge in the
   n-hexane.

4. Purified (cleaned) extract made from Ispra Standard sewage sludge
   SL3. Code M5.
   M5 was obtained cleaning M3A solution by sulfuric acid (ref. 18)
   and Florisil column chromatography (ref. 19) and precipitation of
   sulfur by metallic mercury (ref. 20).

5. Ispra standard sewage sludge SL 3. Code M4.

For the material M1 it was asked to express the concentration of each
the seven congeners (in μg/ml) in the solution. For the other materials the results had to be expessed according to the sheet of results presented on the FIGURE 3 and it was asked to add a description
of analytical methods.

The exercise was developed stepwise :

- First step :   Analysis of M1 and M2. After this step each labora-
                  tory have received the true value corresponding to
                  these materials. These results could be utilized to
                  quantify later on M3, M4 and M5.

- Second step :   Analysis of the extracted sludges materials M3 and
                  M5.

- Final step :   Analysis of the dried sludge material M4.

At the beginning of the exercise, 71 laboratories had indicated an
interest in participating. At least, 39 laboratories have actually
send results. The origins of these laboratories are :

24 from Federal Republic of Germany
 4 from France
 3 from Nederland
 3 from Switzerland
 2 from Finland
 1 from Belgium
 1 from Ireland
 1 from Sweden

SHEET OF RESULTS

SAMPLE M

1) . Concentration of isomer 28[1] in M   :   .... µg/ml

   . Concentration of isomer 52[1] in M   :   .... µg/ml

   . Concentration of isomer 77[1] in M   :   .... µg/ml

   . Concentration of isomer 101[1] in M   :   .... µg/ml

   . Concentration of isomer 138[1] in M   :   .... µg/ml

   . Concentration of isomer 153[1] in M   :   .... µg/ml

   . Concentration of isomer 180[1] in M   :   .... µg/ml

2) Concentration of the isomers 28, 52, 77, 101, 138, 153 and 180[1] together in M   :

                                      .... µg/ml

3) . Concentration of clophen A30[2] in M   :   .... µg/ml

   . Concentration of clophen A50[2] in M   :   .... µg/ml

   . Concentration of clophen A60[2] in M   :   .... µg/ml

                            TOTAL   .... µg/ml

4) Concentration of the sample M in total PCBs[3]:   .... µg/ml

---

[1] According to Ballschmitter numbering.

[2] Or equivalent trade mark (Aroclor, Phenochlor, Kanechlor...).

[3] Please describe how do you proceed to calculate this concentration in total PCBs :
....
....
....

Amongst these 39 laboratories :
- 36 sent results for M1
- 37 sent results for M2
- 21 sent results for M3A and M3B
- 18 sent results for M4
- 20 sent results for M5

About chromatographic columns :
- 37 laboratories have utilized capillary columns
- 2 laboratories have utilized packed columns.

About detection method :
- 38 laboratories have utilized ECD detector
- 1 laboratory has utilized MS detection.

2) Results

TABLE 1 presents the "true" and mean value for all the results. The dispersion of the results is presented by means of figures showing some selected results classified in function of the percent of deviation with regard to the "true" and mean value.

This percent of deviation was calculated utilizing the following formula :

$$\text{Percent deviation} = (\frac{R}{\overline{R}} - 1) \times 100$$

$\overline{R}$: mean or true value for the material
R : value found by the laboratory.

On each figure the true (for M1 and M2) or the mean (M3, M4, M5) value is indicated.

2.1) Results for the directly injectable solutions : M1, M2 ans M5

2.1.1) Results on M1

The FIGURE 4 shows the results obtained for total concentration of the seven congeners.

The FIGURE 5 shows the results obtained for the congener 28.

2.1.2) Results on M2

The FIGURE 6 shows the results obtained in terms of total PCBs in the material M2.

2.1.3) Results on M5

The FIGURE 7 shows the results obtained for the concentration of the congener 28 in the material M5.

TABLE 1			
SAMPLE	CONGENER	TARGET VALUE (μg/ml)	MEAN VALUE
M1	28 52 77 101 138 153 180	2,5 0,9 0,75 0,5 0,45 0,35 0,3	
M2	Clophen A30 Clophen A50 Clophen A60 28 52 101 138 153 180	5,0 2,5 5,0	 0,82 μg/ml 0,35 " 0,35 " 0,62 " 0,54 " 0,42 "
M3 A	Total PCBs 28 52 101 138 153 180		1,41 μg/ml 0,062 " 0,034 " 0,031 " 0,052 " 0,044 " 0,026
M3 B	Total PCBs 28 52 101 138 152 180		2,88 μg/ml 0,14 " 0,07 " 0,065 " 0,14 " 0,13 " 0,073 "
M4	Total PCBs 28 52 101 138 153 180		4,9 μg/g 0,26 " 0,13 " 0,15 " 0,23 " 0,21 " 0,14 "
M5	Total PCBs 28 52 101 138 153 180		1,15 μg/ml 0,055 " 0,028 " 0,034 " 0,055 " 0,048 " 0,028 "

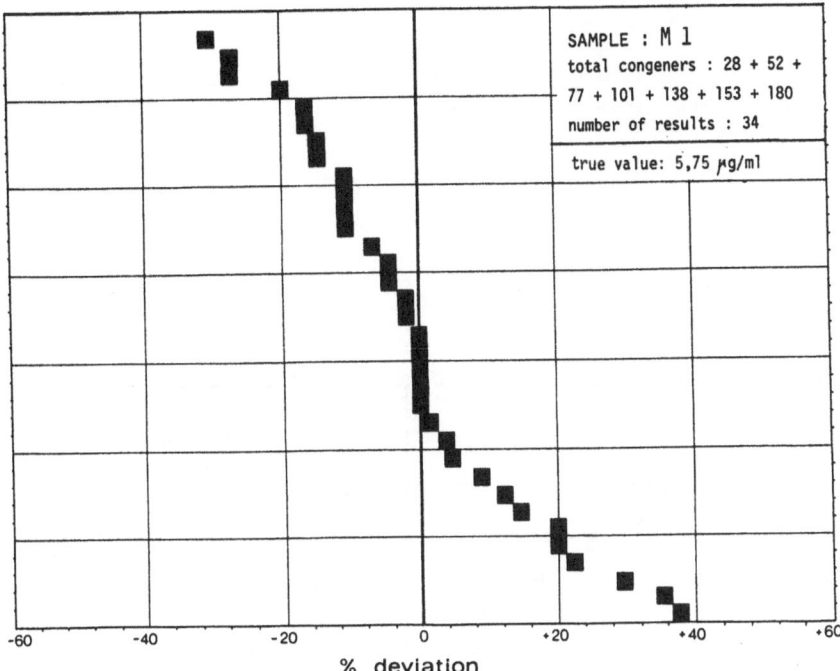

SAMPLE : M 1
total congeners : 28 + 52 +
77 + 101 + 138 + 153 + 180
number of results : 34

true value: 5,75 µg/ml

% deviation

FIGURE 4 : % deviation of the results obtained for the concentration of total of congeners 28 +
52 + 77 + 101 + 138 + 153 + 180 in the sample M 1.

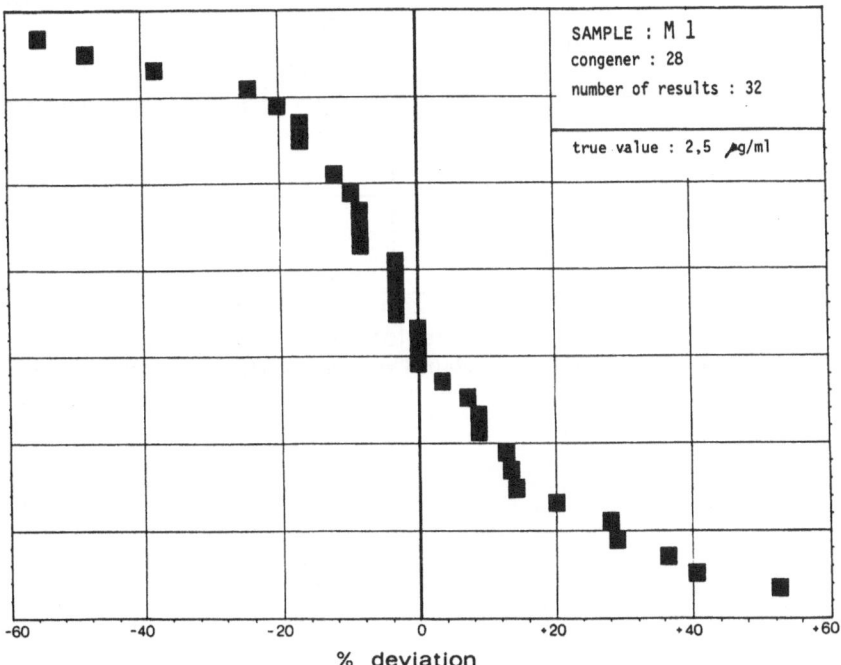

SAMPLE : M 1
congener : 28
number of results : 32

true value : 2,5 µg/ml

% deviation

Figure 5 : % deviation of the results obtained for the concentration of the congener 28 in
the sample M 1.

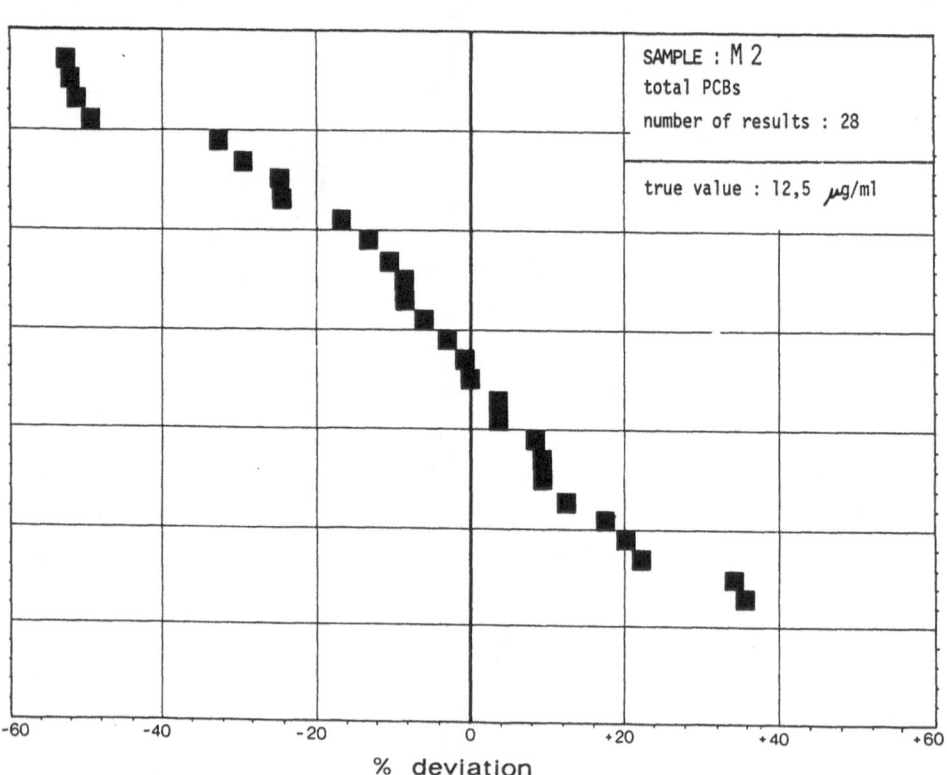

SAMPLE : M 2
total PCBs
number of results : 28

true value : 12,5 μg/ml

% deviation

FIGURE 6 : % deviation of the results obtained for the concentration of total PCBs in the sample M 2.

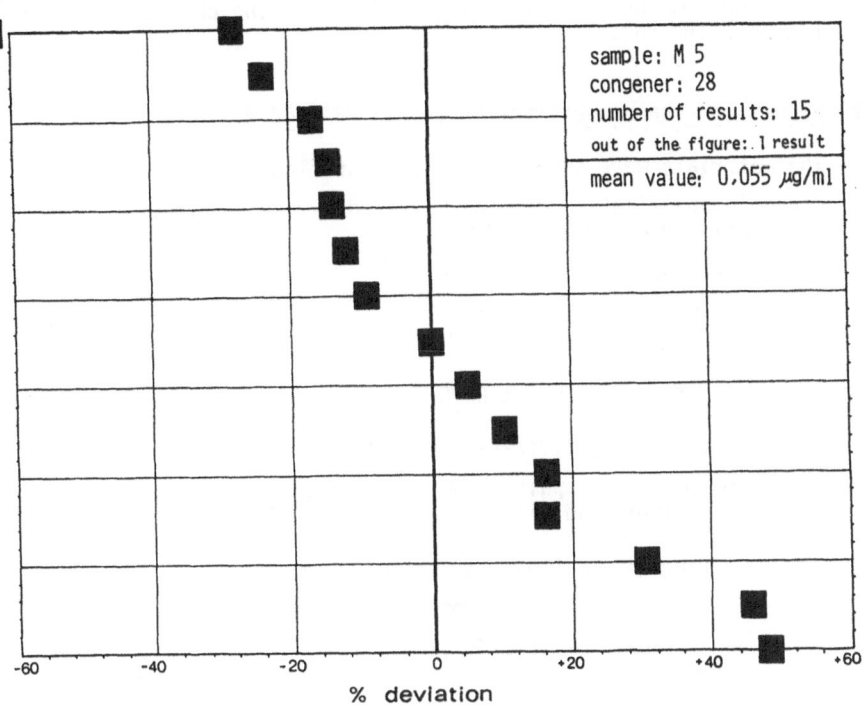

FIGURE 7 : % deviation of the results obtained for the concentration of the congener 28 in the sample M 5.

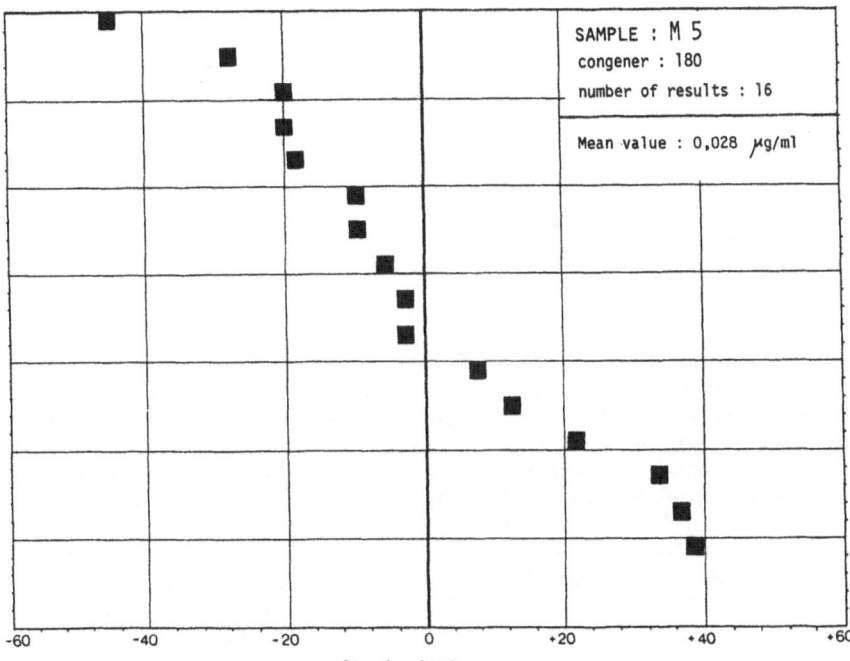

FIGURE 8 : % deviation of the results obtained for the concentration of the congener 180 in the sample M 5.

The FIGURE 8 shows the results obtained for the concentration of congener 180 in the material M5.

2.2) Results on the raw extract materials M3A and M3B

The materials were send to test the purification methods.

The main purification methods utilized by the laboratories having analized M3 are listed below :

- Chemical purification methods :

    -- H_2SO_4 concentrated

    -- H_2SO_4, 7 % SO_3

    -- ethanolic KOH at 60°C

- Chromatographic purification method :

    -- chromatographic columns packed with : Florisil, Alumina and Silicagel.

- Additional purifications for sulfur removal :

    -- precipitation with metallic copper or mercury

    -- chemical reaction with Na_2SO_3.

The results obtained for M3A material about the congeners 28 and 180 are presented on the FIGURES 9 and 10. There was no differences in the quality of the results obtained between M3A and M3B.

2.3) Extraction of the dried sludge, M4

The main target of this exercise was to test the extraction methods. These extraction methods can be classified in two groups:

2.3.1) Direct solvent extraction methods

2.3.1.1) Batch procedures

B-1 : Acetone-petroleum ether extraction (ref. 21)

The reference sample M4 was shaken for one minute. Ten grams of powder were weighted in a centrifuge tube with screw cap, next 20 ml of acetone were added and the tube was shaken for 20 minutes by means of a mechanical shaker.   Next the sample was centrifugated for 10 minutes with a relative centrifugal force of 915 N kg^{-1}. The acetone phase was decanted into a separation funnel and the whole extraction procedure was repeated with another 20 ml of acetone. After combining the acetone phases 5 ml of a saturated solution of sodium sulfite were added to remove sulfur and the solution was shaken for 1 minute. Next 350 ml of

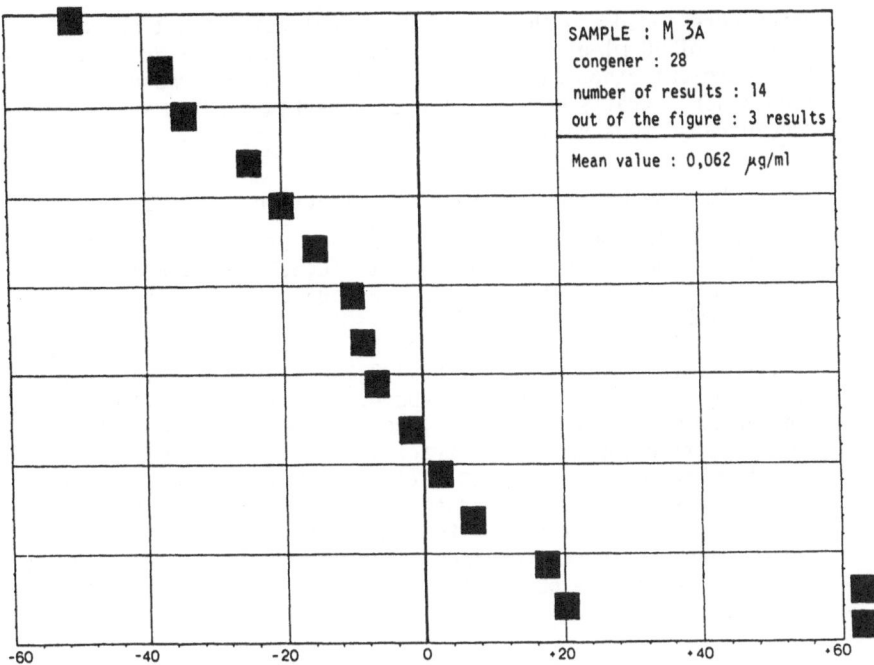

FIGURE 9 : % deviation of the results obtained for the concentration of the congener 28 in the sample M 3a.

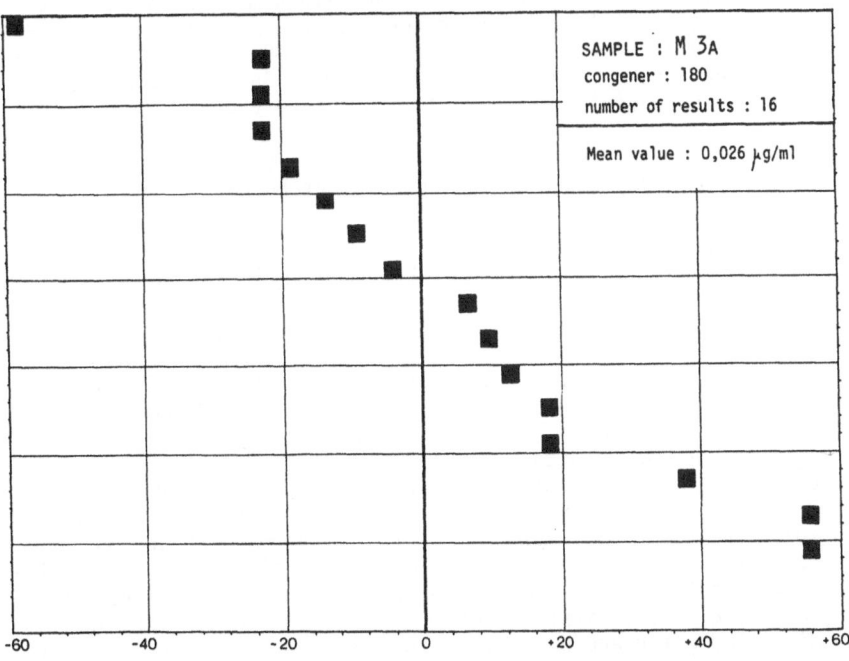

FIGURE 10 : % deviation of the results obtained for the concentration of the congener 180 in the sample M 3a.

water were added and the solution was extracted two times with portions of 50 ml of petroleum ether by shaking for 20 minutes. The petroleum ether phases were decanted and the combined petroleum ether extracts were dried over anhydrous sodium sulfate and concentrated to a volume of 20 ml.

B-2 : Acetone-hexane-diethylether extraction (ref. 22)

10 ml distilled water was added to 2.00 g of the dried sludge in a test tube, mixed well, and centrifuged. The supernatant was removed and saved. 20 ml acetone was added to the wet sludge and the tube rotated overnight.
After centrifugation, the acetone extract was transferred to a separatory funnel containing 20 ml 0,9 % sodium chloride solution plus the supernatant from above. The sludge was re-extracted for 30 min. with 15 ml 2 : 1 hexane : acetone, and after centrifugation the solvent extract was added to the separatory funnel, which was then shaken well. After phase separation, the organic phase was transferred to a pear-shaped flask. The aqueous phase was re-extracted with 10 ml 9:1 hexane:diethyl ether, and the organic phases combined and concentrated to 4 ml under a nitrogen stream in a 60°C water bath.

B-3 : Acetonitril-water and hexane extraction. AOAC general method for chlorinated pesticides (ref. 23, 24).

25 g of Ispra Powder are extracted twice with acetonitrile-water mixture (75:25) in a centrifuge tube with a mixer. The extracts were transferred to 1 l. separatory funnel. Saturated NaCl solution (10 ml), water (600 ml) and hexane (100 ml) were added. The funnel was shaken vigorously for 1 minute. An aliquot (75 ml) of hexane was concentrated to 2 ml with rotary vacuum.

B-4 : Pre-formaldehyde solution and acetone-hexane extraction (ref. 25).

10 g of the sample M4 and 12 ml of a 35 % formaldehyde solution in water were put in an 1000 ml Erlenmeyer flask. This mixture stood without stirring over night. Then 200 ml of an 2:1 hexane acetone solution was added. After shaking for 12 hours the insoluble particels were separated by centrifugation (5 minutes at 2500 rpm) and the liquid part was transferred in a 1000 ml separatory funnel. The insoluble residue in the centrifuge tube was stirred with 50 ml of hexane acetone 2:1, subsequently this mixture was centrifuged, too and the clear solution was also decanted in the separatory funnel. To the united extractes 150 ml distilled water and 50 ml of a saturated sodium chloride aqueous solution were added. After shaking for 1 minute the aqueous phase was let off in a second separatory funnel with 50 ml hexane. This procedure was repeated once more and both hexane extracts were collected in the first separatory funnel and washed with 150 ml distilled water and 50 ml of a aqueous saturated sodium chloride solution. After drying the hexane extract for an 1/2 hour over sodium sulphate and filtration three drops of dodecane were added and the hexane was completly evaporated. The residue was

dissolved in 10 ml of a 1:1 cyclohexane ethylacetate mixture.

B-5 : Direct hexane extraction (ref. 26, 27).

The dried sludge (0,2 g), distilled water (10 ml), and n-hexane (20 ml) were homogenised in a centrifuge tube for 5 minutes with a Silverson mixer. When necessary, the mixture was centrifuged at 3'000 rpm for 3 minutes. The sample was re-extracted twice with n-hexane. The combined n-hexane extracts were concentrated to approximately one ml for direct application to the clean-up column.

2.3.1.2) Soxhlet extraction

Seven laboratories had extracted the dried sludge by the Soxhlet method. The solvents utilized were :

S-1 : 25 % ml acetone, 75 % vol. n-hexane.

S-2 : hexane, acetone, diethyl ether, petroleum ether (2,5 : 7,5 : 1 : 1).

S-3 : pure n-hexane.

The time of reflux varies between 6 to 24 hours.

Below there is the description of one of this extraction method (ref. 28).

Put the empty extraction thimble into the soxhlet, connect reflux condenser and round flask and extract by a mixture of 12,5 ml acetone and 37,5 ml n-hexane until gas chromatogram by ECD detection does not show any PCB peaks or peaks of other disturbing impurities.

This procedure may take one day or even longer. In this case the extraction mixture has to be renewed once or several times until the blank test is satisfactory.

Put an exactly weighed amount of the dried sewage sludge (about 5 g) into the clean extraction thimble and extract in Soxhlet by a mixture of 12,5 ml acetone and 37,5 ml n-hexane for eight hours. The extract solution will become yellow to brown. Concentrate the extract solution by an evaporator until nearly dryness.

Add 20 ml of n-heptane and evaporate until nearly dryness.

Solve the residue in n-heptane and transfer to a 10 ml volumetric flask and fill us by n-heptane the calibration mark.

2.3.1.3) Steam distillation (St).

One laboratory have indicated, without details, to have extract the material M4 by a combined steam distillation-extraction method utilizing pentane as solvent.

2.3.2) Pre-digestion solvent extraction

2.3.1.1) Saponification (ref. 29) (Sp)

Macerate 20,0 g sewage sludge in a macerator (Waring Blendor) with 15 ml 0,2 M NH_4Cl-solution for 5-10 s. Add 100 ml hexane-acetone (1:1 v/v). Macerate 5 min. Transfer the macerate into a 250 ml centrifuge beaker. Centrifugate at 2000 g for about 3 min. Transfer the above solvent over glass wool into a 500 ml separating funnel and transfer the sediment with 50 ml hexane-acetone (1:1 v/v) into the macerator. Add 10 ml 0,2 M NH_4CL to the sediment and macerate 5 min. Transfer the above solvent over glass wool in the separating funnel after centrifugation.
Transfer the hexane extract into a 100 ml flask, add 20 ml of alcoholic potassium hydroxide and saponify on a waterbath for at least 45 min. at 70°C (measured inside the flask). Use a condensor on the flask. Add a few drops of water and mix. If the soap solution turns turbid proceed with saponification. Cool the solution and pour it into a separating funnel (250 ml). Add 30 ml of pentane, 20 ml of water and shake for 30 s. Transfer the pentane layer into another separating funnel. Extract the water layer 3 times with 15 ml of pentane and transfer the pentane layers to the second separating funnel. Wash the combined pentane solutions several times with 15 ml of water until neutral. The pentane solution should now be clear. Pass the pentane solution through a chromatographic column containing anhydrous sodium sulphate. Rinse the column with pentane and concentrate carefully on a rotavapor to 2 ml at 30°C under slight vacuum.

Note - Especially chlorobiphenyls with lower chlorine contents are volatile and may evaporate.

2.3.1.2) Acid digestion (ref. 17) (A)

The dried powder is vigorously shaken during one minute. Approximately 10 g of powder are exactly weighed and then digested by 80 ml of acid solution (i.e : 2 vol.s 70 % perchloric acid in 3 vols glacial acetic acid), for 24 hours over a steam bath. After cooling and doubling the volume by distilled and hexane washed water, the sample is transferred to a separatory funnel and extracted with 3 vols of 20 ml each of n-hexane. The 60 ml of total hexane extract are concentrated to approximately 2 ml.

2.4) Results obtained on the material M4

FIGURES 11 and 12 show the results obtained for the concentrations of the congeners 28 and 180 by the laboratories having performed the total extraction, purification and quantitation of the dried sludge material M4.

On these figures the extraction methods are indicated in front of the points representing the results.

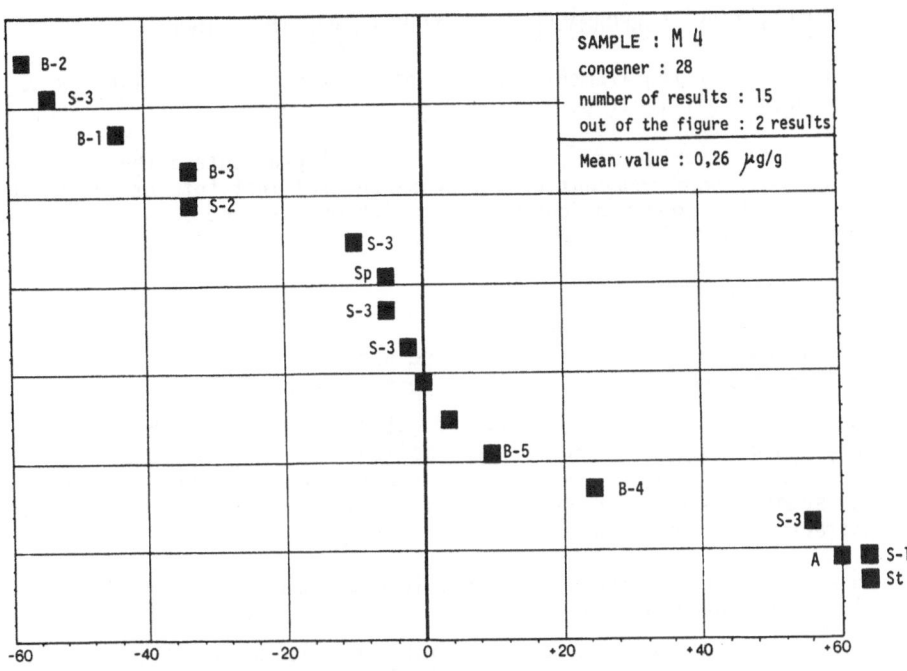

FIGURE 11 : % deviation of the results obtained for the concentration of the congener 28 in the sample M 4.

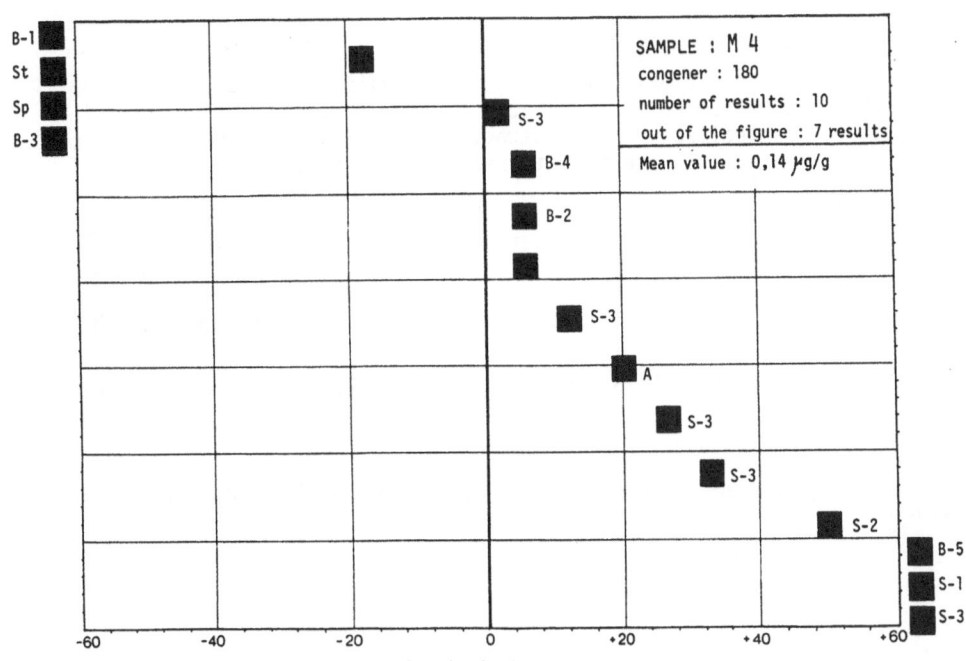

FIGURE 12 : % deviation of the results obtained for the concentration of the congener 180 in the sample M 4

IV. QUALITATIVE PERFORMANCES OF FOUR EXTRACTION METHODS

One of the major requirements in view of a successful extraction is to avoid further difficult purification steps.

Utilizing an other dried sludge from the EEC Ispra Center, the *Institut du Génie de l'Environnement* of the Swiss Federal Institute of Technology of Lausanne had compared four methods which do not need special stirring material or special glass equipment (ref. 30).

These methods were :

- Acid digestion (A), described above.
- Saponification method (Sp), described above.
- A direct hexane-extraction without special mixer :

  The dried sludge (2.0 g), distilled water (50 ml), and n-hexane (80 ml) were homogenised in a 250 ml glass-balloon with a magnetic stirrer for 15 hours. The mixture was centrifugated at 3'000 rpm for 10 minutes. The sample was re-extracted twice with n-hexane. The combined n-hexane extracts were concentrated to approximately one ml.

- An alkaline extraction method abble to extract both PCBs and PAH from sludges (adapted from ref. 36) :

  5 to 10 g of dried sludge are reflux boiled with methanolic-KOH 2 N, during two hours. After cooling, the mixture is transferred to a separatory funnel and extracted 3 times with 50 ml of hexane. The combined extracts are concentrated to approximately one ml.

The other conditions were :

- Clean-Up : The extracts from the four previous methods are purified by the same method on Florisil column chromatography. An additional clean-up was realized on the extracts from Acid digestion and direct hexane extraction to precipitate sulfur by metallic mercury.

- Chromatographic conditions : A Perkin-Elmer Sigma 3 B chromatograph, with an electron-capture ^{63}Ni detector was used with N_2 as carrier gas. Injection in split mode. A fused silica column (30 m., i.d.: 0,25 mm) coated with SE 52 was utilized. Temperature conditions : injector temp.: 275°C., detector temp.: 350°C, oven temp.: 170°C up to 265°C, ramp rate : 2°C/min.

  A Perkin-Elmer Sigma 10 B data station was used as recorder and integrator.

Two extractions were performed which results are presented on the TABLE below :

	PCBs concentration (mg/kg)		
	Extraction 1	Extraction 2	M̄
Acid digestion	3,19	3,17	3,18
Saponification	2,84	2,29	2,56
Direct hexane extract.	4,23	4,48	4,35
Alkaline extraction	4,04	3,94	3,99

The chromatograms obtained after a single Florisil clean-up are presented by FIGURE 13.

The chromatograms obtained show that acid digestion and direct hexane extraction give extracts with a lot of sulfur and consequently these methods need a further purification step by metallic mercury (ref. 20), additional step which is not necessary with the saponification and alkaline methods.

V. DISCUSSION

1) Comments about the quantitation

Most of the researches dedicated to PCBs contamination and most of the regulations about this pollutants give results on total PCBs content. However during the ten last years developments on capillary GLC and consequent higher resolution, had allowed to determine and quantify individual congeners (ref. 32, 33, 34).

PCB residues in the environment and in sewage sludges result from different mixtures of various technical products which, however, may have undergone changes in their distribution pattern due to degradation or metabolization of individual congeners. Depending upon the substrate, distribution patterns of the individual PCB components in the residues may vary. For example, PCB residues in fish are of a different composition than those in milk fat.

Owing to their complex nature of PCBs involving numerous individual congeners, gas chromatographic determination of the PCB content is quite problematical. So far, this has been done mostly by using packed columns with a low separation efficiency. The peaks commonly measured are broad and may consist of numerous individual PCB congeners or even other interfering substances. Since in addition the response of the detector signal (peak area) is different for the individual compounds, identical peak areas may well be consistent with the presence of different amounts of individual congeners. It is not possible to assign the peaks to defined compounds in a qualitative mode and confirm this. For quantitative evaluation, the areas of a number of peaks of a sample are compared with peaks occuring with identical retention time from a commercial product (e.g. Clophen A 60). From this a "PCB content (calculated as Clophen A 60)" is inferred although the non-identity of the PCB distribution pattern of the peak with that of the sample is known.

For some time already, a number of laboratories have used gas chromatography with capillary columns of high separation efficiency for determination so that now most individual PCB components result in separate peaks. However, quantitative determination of all individual congeners to determine the "true" total content of PCBs involves considerable expenditure (calibration by means of approximately 100 single components) which cannot be justified for routine analysis. As an interim solution a few randomly selected peaks from a capillary gas chromatogram were referred to commercial technical products and from these an apparent total PCB content was extrapolated. In such extrapolations, it is entirely possible that the true PCB content is significantly higher than the calculated one; i.e. in fish which contain besides the components of (highly chlorinated) Clophen A 60 also low-chlorinated (i.e. from Clophen A 30) ones. On the other hand, the true content in mother's milk is only ca. 50 % of the extrapolated PCB content because numerous individual congeners of the commercial technical products are no longer contained in mother's milk.

The Federal Health office of the Federal Republic of Germany had developped a concept of PCB residue analysis according to which extrapolation to a fictitious PCB content is replaced by the actually measured contents of selected individual PCB congeners. For this purpose, the following 6 individual congeners have been selected (the figures given in brackets refer to the IUPAC numbering (in accordance with ref. 32).

1) 2, 4, 4'-trichlorobiphenyl	(28)
2) 2,2',5,5'-tetrachlorobiphenyl	(52)
3) 2,2',4,5,5'-pentachlorobiphenyl	(101)
4) 2, 2',3,4,4',5'-hexachlorobiphenyl	(138)
5) 2,2',4,4',5,5'-hexachlorobiphenyl	(153)
6) 2,2',3,4,4',5,5'-heptachlorobiphenyl	(180)

These defined individual congeners are to serve as indicators of the entire PCB distribution pattern in the samples. They represent limiting factors for both the technical PCB products, which are the cause of the world-wide contamination ot fhe environment and for the varying PCB patterns in environmental samples due to metabolization and/or accumulation within the food chain and therefore they are suitable for the establishment of maximum residue limits (ref. 35).

The selected individual PCB components are commercially available both in pure form and as solutions.

For the previously presented reasons, these six particular congeners were those selected for the quantitation in the COST exercise.

In fact some other congeners are of importance in environmental and sewage sludge samples. The TABLE 2 presents the abundance of 14 individual congeners in a mixture of Aroclors, waste-waters and sludges

T A B L E   2

PCB IUPAC no	% content in a mixture of AROCLOR 1242 + 1254 + 1260 (1:1:1)	FRIBOURG (CH) waste-waters (residential district)	MORGES (CH)		BIENNE (CH)		M 4*
			waste-waters	sew.sludges	waste-waters	sew.sludges	
18	1,3	2,9	1,1	1,0	0,8	0,6	5,0
28 + 31	3,5	9,5	4,3	3,8	3,0	2,5	2,6
52	2,0	2,7	2,1	3,4	1,9	1,9	
60	1,8	3,6	2,6	2,7	1,5	2,0	
95	2,3	3,1	2,0	2,5	1,3	2,1	3,1
101	2,6	3,0	2,7	2,9	2,1	2,9	
110	6,4	6,9	11,9	7,6	8,1	8,1	
118	5,0	5,7	10,0	7,7	7,2	7,6	4,6
138	5,4	4,8	6,1	7,5	7,0	8,1	
149	4,0	3,2	4,0	3,4	4,8	3,8	4,3
153	4,8	4,2	6,3	6,2	6,7	6,8	
170	2,1	1,0	1,9	1,4	2,1	2,0	2,8
180	4,5	2,1	3,4	3,8	4,4	4,5	

% content of total PCBs extracted from :

MAIN PCB CONGENERS IN THE WASTE WATERS AND SEWAGE SLUDGES FROM 3 TOWNS OF SWITZERLAND AND IN DRIED SEWAGE SLUDGE M4

*According mean values, see TABLE 1.

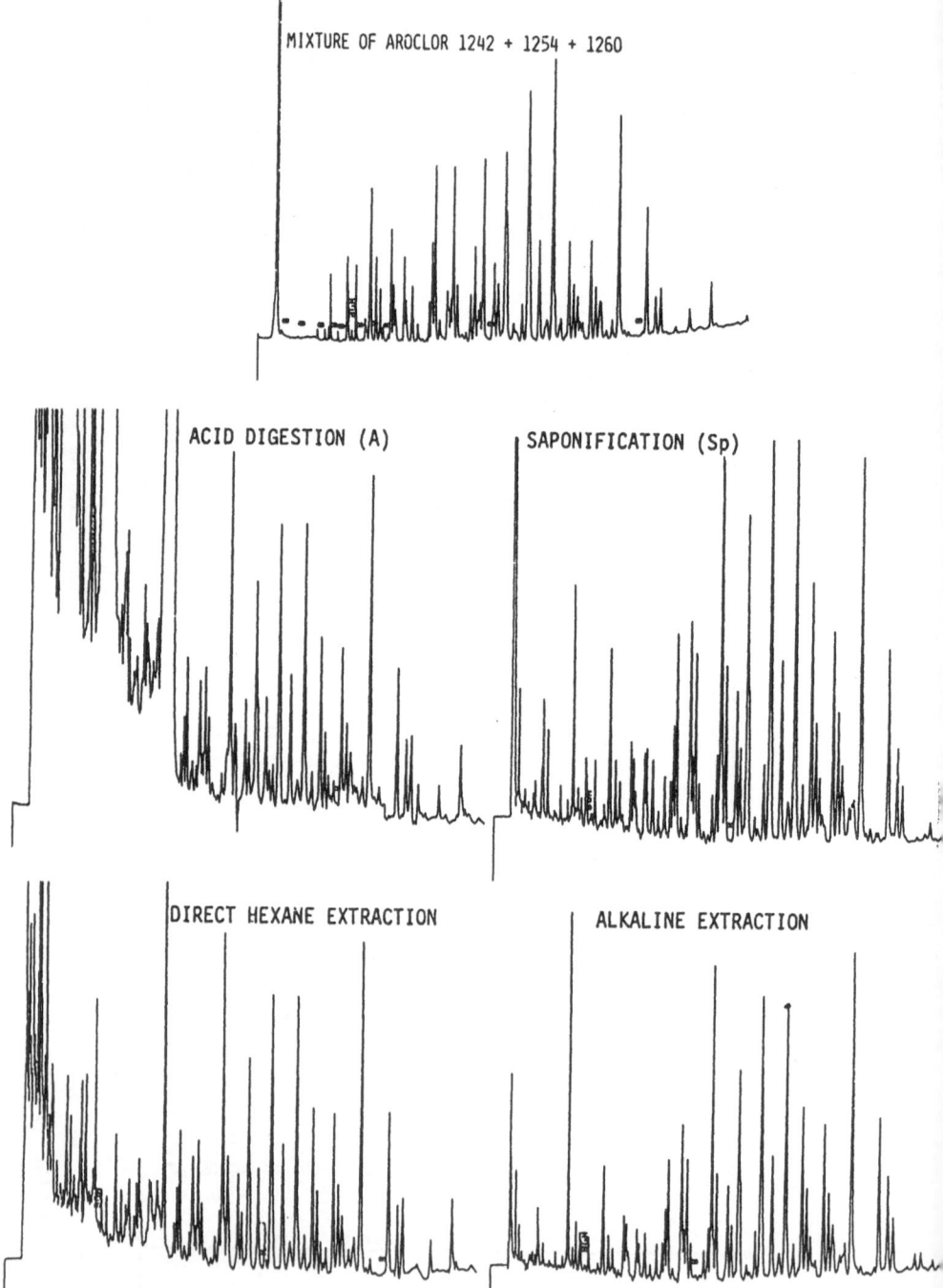

MIXTURE OF AROCLOR 1242 + 1254 + 1260

ACID DIGESTION (A)

SAPONIFICATION (Sp)

DIRECT HEXANE EXTRACTION

ALKALINE EXTRACTION

FIGURE 13: Chromatograms of PCBs extracted*from a EEC-Ispra sewage sludge by the four methods tested, and chromatogram of an Aroclor mixture (*: after a single Florisil purification step)

from three Swiss-towns (ref. 10) and in the dried sludge utilized for the COST exercise. According the results presented in this Table, congeners 60, 95, 110, 118 and 149 should be eventually considered in addition of the six congeners utilized in the COST exercise and selected by the Federal Health Office of the FRG.

2) Quality of the results obtained in the COST exercise

In spite of insufficient number of laboratories having send results the main conclusion of the exercise is that the dispersion of the results obtained increases with the quantity of preparation step of the samples.

For the sample M1 for which it was needed to quantify pure congeners :

. 75 % of the results are in the range of +/- 20 % of deviation for the total concentration of the 7 congeners.

. 68 % of the results are in the range of +/- 20 % of deviation for the congener 28, which quantitation have gave the worst results.

For the sample M2 for which it was needed to quantify a mixture of commercial PCBs :

. 55 % of the results are in the range of +/- 20 % of deviation for the total PCBs quantitation.

For the sample M5 which was a purified extract from a sewage sludge :

. 62 % of the results are in the +/- 20 % range of deviation for the congener 28.

. 62 % for the congener 101.

. 56 % for the congener 180.

For the sample M3A which have needed a purification step realized by each laboratory :

. 53 % of the results are in the +/- 20 % range of deviation for the congener 28.

. 53 % for the congener 101.

. 56 % for the congener 180.

For the sample M4 which have needed a total process of extraction, purification and quantitation from a dried sludge :

. 41 % of the results are in the +/- 20 % range of deviation for the congener 28.

. 41 % for the congener 101.

. 35 % for the congener 180

3) Errors on extraction and purification steps

It appears that extraction and purification are the critical steps to increase the dispersion of the results. These two operations are in fact closely linked.

- If the extraction step is poor the extract is clear and easy to purify but the recovery of PCBs will be low.

- If the extraction step is efficient the rate of recovery of PCBs will be good but the extract will need a difficult purification step in which one can loose the benefits of the extraction.

The results obtained in the COST exercise and the research presented in the chapter IV suggest that :

- Steam distillation and saponification (ref. 29) methods are quite poorly extractive, mainly for highly chlorinated congeners. However, saponification gives very clear extracts.

- Acid digestion (ref. 17) is an extremely strong extraction method (particularly concerning sulfur content of the samples) which needs special attention to purification steps.

- Batch-solvent extractions (ref. 21 to 27), Soxhlet extraction (ref. 28) and alkaline extraction (ref. 36) give both good recovery and quite clear extracts.

Less differences appear between purification methods. Column Chromatography (with Florisil, alumina or silica-gel) seems to be well adapted to sewage sludge extracts. The experience of the operator in the use of these chromatographic materials is the critical point for this operation. Desulfuration can be operated by metallic mercury (ref. 20)

4) Instrumental analysis and quantitation

The COST exercise has indicated some other sources of errors in the analyses of PCBs in sewage sludges :

4.1) Solvents

It must be absolutely ensured that the standard substances and the samples are dissolved in the same solvent because otherwise changes in vapourization profiles may result which in turn could lead to changes in retention times, peak areas and peak heights. For example, using toluene instead of iso-octane as a solvent has resulted in an increase of peak height by 35 % (ref. 35).

4.2) Calibration graphs

Calibration graphs are to be established to ensure that operation takes place in the (approximately) linear range of the ECD. These calibration graphs have to be established utilizing different rates of dilution of the standard solution.

4.3) Internal standard

In many laboratories, the use of internal standards (e.g. isodrin; 1, 2, 3, 4-tetrachloronaphthalene; hexabromobenzene) to reduce the injection error has proved to be useful.

4.4) Computerized interpretation of the chromatographic peaks

Quantitative evaluation is conducted by determination of the peak area or heights for the individual PCB components in the sample extract and comparison with the areas and heights, respectively for a standard solution of known concentration.

In most of the cases this operation is performed by a computerized equipment. The recognition of the peaks is made utilizing relative or absolute retention time. The calculation of peak areas and heights is performed by the computer according some criteria. Amongst these criteria :

- the test area of the peak before confirmation (area sensitivity),
- the slope and the rate of change in the slope of the signal level used by the system to determine wether it is on baseline or not (base sensitivity),
- the criterion used by the system to consider or not a small peak on the trailing side of a big peak (skim sensitivity),
- the baseline treatment (valley to valley, horizontal, etc).

Due to the presence of impurities and the difference of balance between congeners the computerized system can integrate differently the same peak between the standard solution and the sample-extract solution. FIGURE 14 shows an example of this type of errors. The figure shows an enlargement of the area taken as reference by a computerized integrator for the peak corresponding to the congener 180 :

A : in the chromatogram of a mixture of AROCLORS 1242 + 1254 + 1260.

B : in the chromatogram of the extract from an environmental sample (brown trout).

In this particular case, the error induced by the computerized area calculation of this peak is about 16 % in proportion of the comparison of the heights.

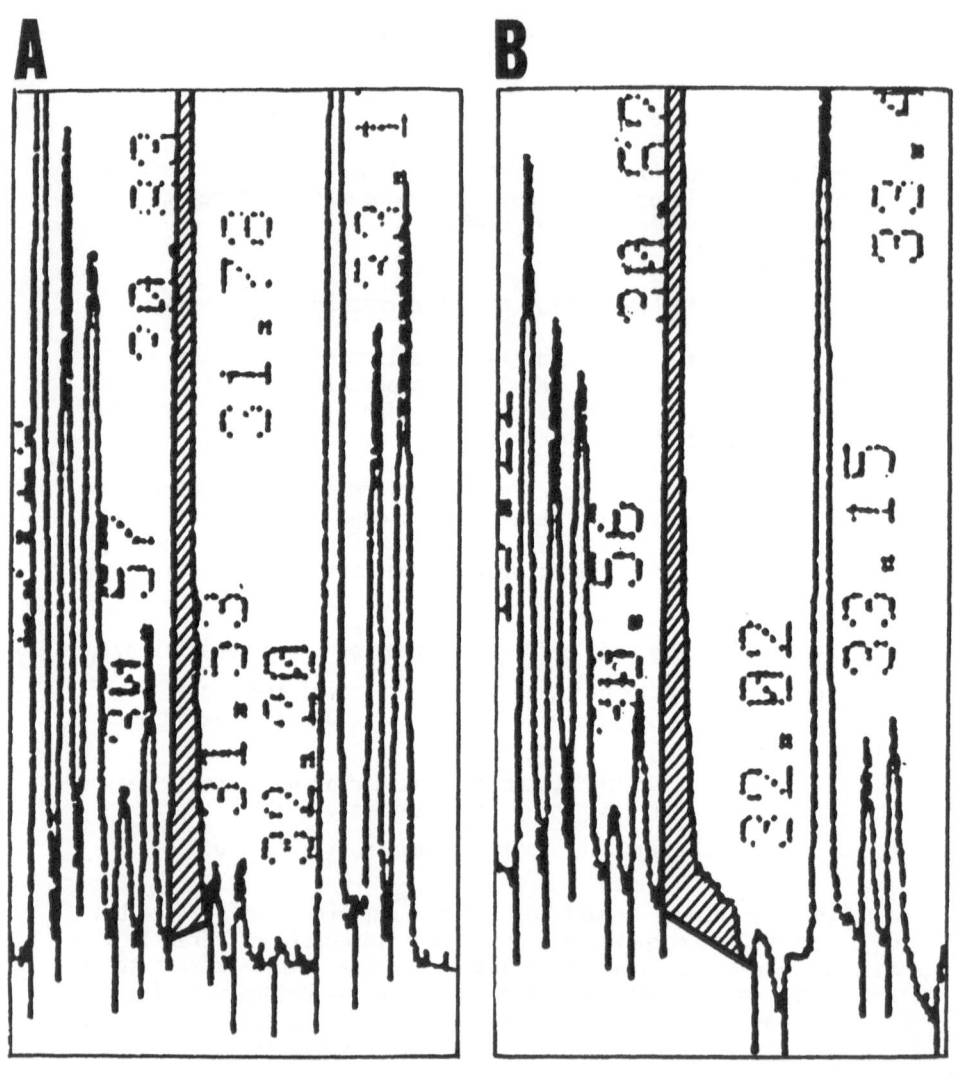

FIGURE 14 : Diferences between the areas taken as reference for the computerized
integration of the same peak (congener 180):

A: in a mixture of comercial Aroclors 1242 + 1254 + 1260

B: in the extract of an environmental sample (brown trout)

VI. CONCLUSION

The intercomparison exercise of the WP-2 COST 681 group has shown that the analyses of PCBs in sewage sludges have to be improved to reach an acceptable coherence between the results of different laboratories.

The two main aspects of improvement are :

- the strictness of the handling (calibration graphs, same solvent for samples and standards, internal standard, quantitation operating on the heights of the peaks)
- identification of the most suited extraction method to be choose amoungst : batch-solvent extraction, Soxhlet extraction or alkaline extraction.

Any future intercomparison program should forms on the improvement of the methodological aspects.

But PCBs are not the only organic-micropollutants in the sludges. To increase the acceptability of the sewage sludges for agricultural use, but also in view of their incineration, it is absolutely necessary to increase our knowledge about the presence and the analysis of all the other main group of organic micro-pollutants which are present in this sludges. For this purpose, extraction procedures allowing to study several types of organic micro-pollutants should be promoved.

ACKNOWLEDGEMENTS

The authors wish to thank Prof. Dr. Ing. Reimar LESCHBER, chairman of the WP-2 COST 681 and M. Pierre L'HERMITE from the Environment Research Programm of the CEC for their support during the study presented in this paper. They are also highly grateful to all the laboratories having participated in the intercomparison presented in this paper.

REFERENCES

1) J.N. Lester
   *Occurence, behaviour and fate of organic micropollutants during waste
   water and sludge treatment process.*
   Paper presented at the EEC concerted action on the characterisation,
   treatment and use of sewage sludge. (Working party 5, may 25-26, 1982
   Stevenage, United Kingdom).

2) T.R. Bridle and M.D. Webber
   A Canadian perspective on toxic organics in sewage sludge.
   Ibid.

3) A.E. Mc Intyre, J.N. Lester, R. Perry
   *The influence of chemical conditioning and dewatering on the
   distribution of PCBs & organochl. insect. in sewage sludges.*
   Environ. Pollut., B. 2, pp. 309-320, (1981).

4) A.E. Mc Intyre, R. Perry, J.N. Lester
   *The behaviour of PCBs and organochlorine insecticides in primary
   mechanical wastewater treatment.*
   Environ. Pollut., B. 2, pp. 223-233, (1981).

5) G. Burgermeister, P. Ammann, J. Tarradellas.
   PCB dans les boues de quelques stations d'épuration de Suisse,
   Characterization, Treatment and Use of Sewage Sludge. D. Reidel,
   publ. pp 264-269 (1981).

6) L.F. de Alencastro, J. Tarradellas
   *Etude de la concentration en PCB des eaux usées dans les stations
   d'épuration.*
   Gaz-Eaux-Eaux Usées, 63, 3, pp. 113-122 (1983).

7) W. Fieggen
   Pesticides and PCBs in sewage sludges
   Paper presented at the EEC concerted action on the characterization,
   treatment and use of sewage sludge. (Groningen. The Netherlands,
   1980).

8) E.E. Shannon, F.J. Ludwig, I. Valdmanis
   Polychlorinated biphenyls in municipal wastewaters
   Environment Canada, research report, 49 (1976).

9) A.E. Mc Intyre and J.N. Lester
   *PCB and organochlorine insecticide concentrations in forty sewage
   sludges in England*
   Environm. Pollut., B, 3, pp. 225-229 (1982).

10) J. Tarradellas, L.F. de Alencastro
The role of analytical technics in the study of PCBs in wastewaters.
15th Annual Symposium on Analytical Chemistry of Pollutants,
Jekyll-Island (USA), 20-22 may 1985.

11) H. Harms, D.R. Sauerbeck
*Toxic organic compounds in town waste materials : their origin con-
centrations and turnover in waste composts, soil and plants,*
Environmental effects of organic and inorganic contaminants in
sewage sludge. D. Reidel, publ. pp 38-51.

12) M. Rocher, A. Copin
*Identification of some organic micro-pollutants in urban sewage
sludges.*
Ibid. pp. 52-58

13) T.R. Bridle, M.D. Webber.
A Canadian perspective on toxic organics in sewage sludge.
Ibid. pp. 27-37

14) I.J. Tinsley
Chemical concepts in Pollutant Behaviour
Wiley Interscience publ., p. 153 (1979).

15) P. Moza, J. Scheunert, W. Klein, F. Korte
*Studies with 2,4',5-trichlorobiphenyl - 14 C and 2,2',4,4',6-pen-
tachlorobiphenyl-14 C in carrots, sugar, beets and soil.*
J. Agric. Food Chem. 27, 1120-1124 (1979).

16) G.F. Fries
*Potential polychlorinated biphenyl residues in animal products from
application of contaminated sewage sludge to land.*
J. Environm. Qual. 11(1), 14-20 (1982.

17) R.L. Stanley, H.T. Lefavoure
*Rapid digestion and clean-up of animal tissues for pesticide residue
analysis.*
J. of Ass. of Official Agricultural Che. , 48, 666-667 (1965).

18) P.G. Murphy
*Sulfuric acid for the clean-up of animal tissues of acid stable
chlorinated hydro-carbons residues.*
J. of Ass. of Off. Anal. Chemists, 55, 1360 (1972).

19) P.A. Mills, B.A. Bong, L.R. Camps, J.A. Burke.
*Elution solvent system for Florisil column clean-up in organochlorine
pesticides residue analysis.*
J. of Ass. of Off. Anal. Chemists, 55, 39 (1972).

20) C.F. Rodriguez, W.A. Mc Mahon, R.E. Thomas
Method development for determination of polychlorinated hydrocarbons in municipal sludge.
Technical Report Data EPA 600/2-80-029 (1980).

21) R.C.C. Wegman, A.W.M. Hofstee
Water Res. 16, 1265 (1982).

22) S. Jensen, J. Mowrer
Dept. of Chemical Environmental Analysis. Stokholms Universitet-Sweden. (Personal Communication to WP 2-COST 681).

23) General method for chlorinated and phosphatated pesticides.
Official methods of analysis of the AOAC, 11th ed., 475, (1970).

24) C. Corvi.
Laboratoire Cantonal de Chimie, Case Postale 109, 1211 Genève (Switzerland) (Personal Communication to WP 2-COST 681).

25) A. Trenkle
Staatliche Landwirtschaftliche Untersuchungs- und Forschungsanstalt Augustenberg. Nesslerstr. 23, D-7500 Karslsuhe 41 (BRD).
(Personal Communication to WP 2-COST 681).

26) A.E. Mc Intyre, J.N. Lester and R. Perry.
Analysis of Organic Substances of Concern in Sewage Sludge.
Final report to the Department of the Environment (Great Britain) for Contracts DGR/480/66 and DGR/480/240. November 1979.

27) J.F. Eades and M. O. Neill.
An Foras Taluntais. Oak Park Research Centre. Carlow (Ireland)
(Personal Communication to WP 2-COST 681).

28) K. Wrabetz,
Bayer AG, Zentrale Analytik, 5090 Leverkusen (FRG)
(Personal Communication to WP 2-COST 681).

29) L.G.M. Th. Tuinstra, A.H. Roos
A Universal method for the Analysis of PCB Congeners : Sewage Sludge.
State Institute for Quality Control of Agricultural Products.
P.O. Box 230. 6700 AE Wageningen (The Netherlands).
(Personal Communication to WP 2-COST 681).

30) L.F. de Alencastro, J. Tarradellas
PCBs in dried sewage sludge : comparison between four extraction methods.
Institut du Génie de l'Environnement, EPFL-Ecublens, 1015 Lausanne (Switzerland).
(Personal Communication to WP 2-COST 681).

31) H. Lorenz, G. Neumeier
Polychlorierte Biphenyle (PCB)
MMV Medizin Verlag München, BGA Schriften 4/83, p. 81-91 (1983).

32) K. Ballschmitter, M. Zell
Analysis of Polychlorinated Biphenyls (PCB) by Glass Capillary Gas-Chromatography.
Fresenius Z. Anal. Chem. 302, 20-31 (1980).

33) E. Schulte, R. Malish
Berechnung der wahren PCB-Gehalte in Umweltproben.
Fresenius Z. Anal. Chem. 314, 545-551 (1983).

34) J.C. Duinker, M.T.J. Hillebrand
Characterization of PCB components in Clophen formulations by Capillary GC-MS and GC-ECD Techniques.
Environm. Sci. Technol., 17, 449-456 (1983).

35) H. Beck, W. Mathor
Analytical Procedure for the Determination of Selected PCB Components in Food.
Bundesgesundheitsblatt, 28, no 1, p.1-12 (1985).

36) G. Grimmer, H. Bohnke, H. Borwitzky.
Gas-Chromatographische profil Analyse der PAH in Klarschlammproben.
Fresenius Z. Anal. Chem. 289, pp.91-95 (1978).

DISCUSSION

It was pointed out that the use and disposal of PCBs in industry was now strictly controlled. Also sludge was spread on only a small percentage of agricultural land. Doubt was expressed about the significance of sludge as a source of PCBs in agroecosystems. Aerial deposition was likely to be more important.
(Contributors to discussion : Davis, Tjell)

CHEMICAL METHODS FOR THE BIOLOGICAL CHARACTERIZATION OF METAL IN SLUDGE AND SOIL

H. HÄNI and S. GUPTA

Swiss Federal Research Station for Agricultural Chemistry
and Hygiene of Environment
CH - 3097 Liebefeld-Berne

Summary

0.1 M NaNO$_3$ is used to assess bioavailability of metals in different
polluted soils. Promising results are obtained for cadmium, zinc and
copper whereas more research is still needed for nickel, to find out why
the plant response is not so well correlated to soluble soil concentrations as in the case of the other three metals. The German colleagues
prefer CaCl$_2$ because this solvent extracts 2 to 4 times more of the metals
than NaNO$_3$.

It is further shown that the NaNO$_3$ extraction method is also suited to
characterize critical metal levels in soils which were "naturally"
contaminated by different pollution sources. The consequences of the
findings for the soil protection against pollution in the Swiss legislation are emphasized.

1. INTRODUCTION

In the Münster Seminar (3) the following was concluded:

- Bioavailable concentrations of metals in soils are required to assess the
  significance of contamination in terms of effects on crops and the food
  chain.
- Neutral salt extractants provide the most promising chemical method to
  reflect the bioavailability of metals from very different soils.
- Progressive acidification and sequential extraction techniques offer as
  research methods a way of characterizing forms of metal in soil at least
  on an empirical basis, and of evaluating the long-term bioavailability of
  metals.
- The soil solution is the interface between the plant root and the soil
  and therefore the concentration of metal in soil solution is likely to
  be closely related to bioavailabitity.

It is the purpose of this paper to substantiate the results of neutral salt
extractants and to give indications that the method, in addition to cadmium
and zinc, can possibly be extended to copper and nickel.

2. NORMAL UNCONTAMINATED SOILS

On analysing normal unpolluted soils the answer one would like to get of
these analyses is to know if the status of nutrients is sufficient for a
proper plant growth.

In the case of trace elements, Sillanpää, e.g. estimated critical deficiency levels for copper and zinc in soil using acid ammonium acetate-EDTA (pH 4.65) as extracting agent (5):

- copper < 0.8 - 1.0 ppm
- zinc   < 1.0 - 1.5 ppm (+ pH correction)

At the moment it remains doubtful if much milder extractants (such as neutral salt solutions) which are proposed for polluted soils can also be used for the so-called normal soils. The very low metal concentrations found in these solvents may cause serious analytical difficulties. However, at the last consultation of the FAO network on trace elements held in Murat-le-Quaire (France, 17-20 September, 1985) it was decided to examine in a final attempt the suitability of various weaker soil extractants to characterize the plant availability of trace elements in normal soils. It has to be repeated that the use of mild extractants was proposed in the light of soil pollution, believing the status of trace element deficiency could furthermore be described with sufficient reliability by the amount dissolved in stronger extractants.

3. CONTAMINATED SOILS

3. 1 Cadmium and zinc

In the Münster Seminar (3) it was shown that neutral salt solutions like 0.1M $NaNO_3$ and 0.05 or 0.1M $CaCl_2$ extracted the amount of cadmium and zinc from very different soils which was reasonably well correlated with the plant content. Sauerbeck (4) prefers the extraction with $CaCl_2$ because the amounts brought into solution by this solvent are 2 to 4 times higher for cadmium, zinc and nickel (comparative measurements for copper are not yet available) than the amounts extracted by $NaNO_3$. Thus, detectable amounts of cadmium can be dissolved from neutral soils of low contamination degrees which is not possible with $NaNO_3$. However, beside the purely analytical aspects the chemical reactions of the anion should also be considered: In contrast to nitrate, chloride has marked complexing properties. Further experiments have to show how these complexing properties influence the soil plant relationship for other metals than cadmium and zinc (e.g. copper and nickel).

The results of a pot experiment with soils which were exposed in the field to different sources of pollution are presented in tables 1 and 2 as well as in figure 1. The plants were analysed for cadmium, zinc, nickel and lead. Because the contents of zinc and cadmium were the only ones which were situated well above the natural background levels in some cases, these two elements have to be regarded as the metals which, considering the investigated pollution sources, can cause serious problems in soils.

It is clearly confirmed by these results that the total content of heavy metals in soil is by no means a good approach to assess bioavailability. This is especially true for soils 4 and 7 in figure 1: Due to unfavorable properties of soil 4 already 51.9 ppm of total zinc give rise to 1.91 ppm of soluble zinc in soil and correspondingly high values in plants, whereas the properties of soil 7 prevent a considerable increase of the soluble fraction even at a total content of 1212 ppm.

Table 1   Influence of zinc in soils - enriched from different sources - on the zinc content and growth of plants (for soil properties s. figure 1)

Nr	Contamination sources	SOILS (1) Total content of Zn (ppm)	Zn conc. in 0.1M NaNO3 (ppm)	RYE GRASS mg Zn/kg dry matter	(cut 1-3) Dry matter yield (g/pot)	SOIL (2) Zn conc. in 0.1M NaNO3 (ppm)	RADISH LEAVES mg Zn/kg dry matter	Dry matter yield (g/pot)	RADIS TUBERS mg Zn/kg dry matter	Dry matter yield (g/pot)
1	Zinc coating plant A	858	0.02	284.8	70.99	0.02	360.4	14.93	114.7	27.97
2	Zinc coating plant B	217	6.17	359.6	68.77	8.59	1633.4	10.77	521.2	11.67
3	Motorway C	84.6	0.01	62.7	81.17	0.1	51.8	14.93	33.7	29.88
4	Motorway D	51.9	1.91	104.6	65.63	1.91	685.5	16.40	181.0	28.85
5	Refuse incineration plant E	118	0.07	90.0	80.97	0.1	70.4	20.60	37.5	35.38
6	Refuse incineration plant F	74	0.01	83.0	67.37	0.37	154.3	19.17	70.1	28.09
7	Compost in vine-growing G	1212	0.15	186.6	80.77	0.22	258.6	17.97	89.7	33.29
8	Sewage sludge H	77.2	0.01	53.4	92.67	0.1	43.7	20.13	30.5	28.12
9	Sewage sludge I	612	0.02	146.9	75.43	0.12	250.6	17.17	69.8	34.69
10	Sewage sludge K	121	0.06	91.9	62.53	0.12	94.5	17.87	67.6	30.60

(1) After the growth of rye grass

(2) After the growth of radish

In soil 2, where the highest concentration of zinc in 0.1 M NaNO$_3$ is found, toxic effects on plants can't be excluded (s. the low dry matter yield of radish tubers in table 1).

However, it has to be mentioned that contrary to soils contaminated by salt solutions some unexpected results were obtained with the "naturally" polluted soils. The zinc contents of plants grown in soils 1, 7 and 9 are too high compared to the zinc concentration in 0.1 M NaNO$_3$. One phenomenon which should further be studied is the kinetics of the dissolution of metals in different polluting compounds. It has to be considered that it might be difficult to understand in all cases the dissolution reaction of metals in the polluting compounds by a single extraction procedure of short duration (e.g. 2 hours) so that the plant reaction can be predicted properly.

According to table 2 the concentrations of cadmium in 0.1 M NaNO$_3$ lie below the critical value of 0.06 ppm in all soils. This is in agreement with the cadmium contents of less than 1 ppm in rye grass and radish tubers. For radish leaves, however, soluble soil concentrations below the value of 0.06 ppm can already lead to contents which are above the zootoxic threshold of 1 ppm.

Table 2 Influence of cadmium in soils - enriched from different sources - on the cadmium content of plants (for soil properties s. figure 1)

SOILS 1 - 10		PLANTS (mg Cd/kg dry matter)		
Total content of Cadmium (ppm)	cadmium conc. in 0.1 M NaNO$_3$ (ppm)	Rye grass (cut 1-3)	Rad.tubers	Rad.leaves
o.14 - 6.1 (crit.value:1)	< 0.06 (crit. value 0.06)	0.2-0.84	0.2-0.88	0.2-3.11

3. 2 Copper and nickel

Figures 2 and 3 show the results of a pot experiment with 6 different soils to which zinc and copper have been added separately in form of their nitrates. Concerning the assessment of bioavailability, the comparison of the two figures is encouraging because 0.1 M NaNO$_3$ seems to be an almost as good extractant for copper as for zinc (1). Much lower correlations were found for NH$_4$OAc and DTPA (s.figures 4 and 5). The same results were obtained for cadmium and zinc (2).

Figure 6 is a presentation of copper contents in vineyard soils. Although total contents often exceed the critical value considerably, the soluble soil concentrations remain within the tolerated range. This harmonizes with the analysis of vine leaves because the measured contents of 11.3 - 17.9 ppm are still below the phytotoxic threshold of 20 ppm.

First results with nickel are promising. However, further research is needed to find out why the plant response is not so well correlated to soluble soil concentration as in the case of cadmium, zinc and copper.

4. EFFECT OF METALS ON MICROORGANISMS

Similar to metal contents in plants and to plant growth soil microorganisms and their activities are related to the metal concentration in the soil solution. The figures in table 3 illustrate this statement.

Table 3   Effect of soluble cadmium concentration on % carbon mineralization

	SOILS	
	Sandy loam pH 5.7	Sandy loam pH 7.0
% carbon mineralization	59	101
Total Cd added (mg/kg soil)	16	16
Soluble Cd (mg/kg soil)	0.61	0.05
CEC (meq./kg soil)	80	124

The investigation of the effect of metals on soil microorganisms is essential for the judgment of proposed guiding values in soils. For copper, e.g. it was observed that the soluble soil concentration (0.1 M $NaNO_3$) which caused 20 % reduction in carbon mineralization was 7 times lower than the concentration which induced phytotoxic contents of 20 ppm in plants (1).

5. CONCLUSIONS

- Due to the importance of bioavailable concentrations of metals in soils in assessing the significance of contamination, soluble metal concentrations in 0.1 M $NaNO_3$ were introduced as guiding values beside those of total contents in the Swiss legislation.
- For safety reasons the guiding values are half the critical values.
- In judging the guiding values the reaction of different plants and of soil microorganisms have to be considered.
- Switzerland started the experiments with $NaNO_3$ and has therefore the greatestexperience with this extractant. CaCl2 is the preferred method in Germany. For the future work it should be tried to decide on a standard method in Europe.

REFERENCES

1. Gupta, S.K., Häni, H. and Rudaz, A. (1985). Relationship between metal ion concentration and biological effects. Proc. of Int. Conf. on Heavy Metals in the Environment. Athens, Sept. 1985. Publ. CEP Consultants Ltd. (in press)

2. Häni, H. and Gupta, S.K. (1985). Reasons to use neutral salt solutions to assess the metal impact on plants and soils. Proc. of a Seminar organized by the Commission of the European Communities, held in Münster, 11-13 April 1984, 42-47.

3. Leschber, R., Davis, R.D. and l'Hermite, P. (1985). Chemical methods for assessing bio-available metals in sludges and soils. Elsevier Applied Science Publishers.

4. Sauerbeck, D.R., and Styperek, P. (1985). Evaluation of chemical methods for assessing the Cd and Zn availability from different soils and sources. Proc. of a Seminar organized by the Commission of the European Comunities, held in Münster, 11-13 April 1984, 49-66.

5. Sillanpää, M. (1982). Micronutrients and the nutrient status of soils: a global study. FAO Soils Bulletin 48.

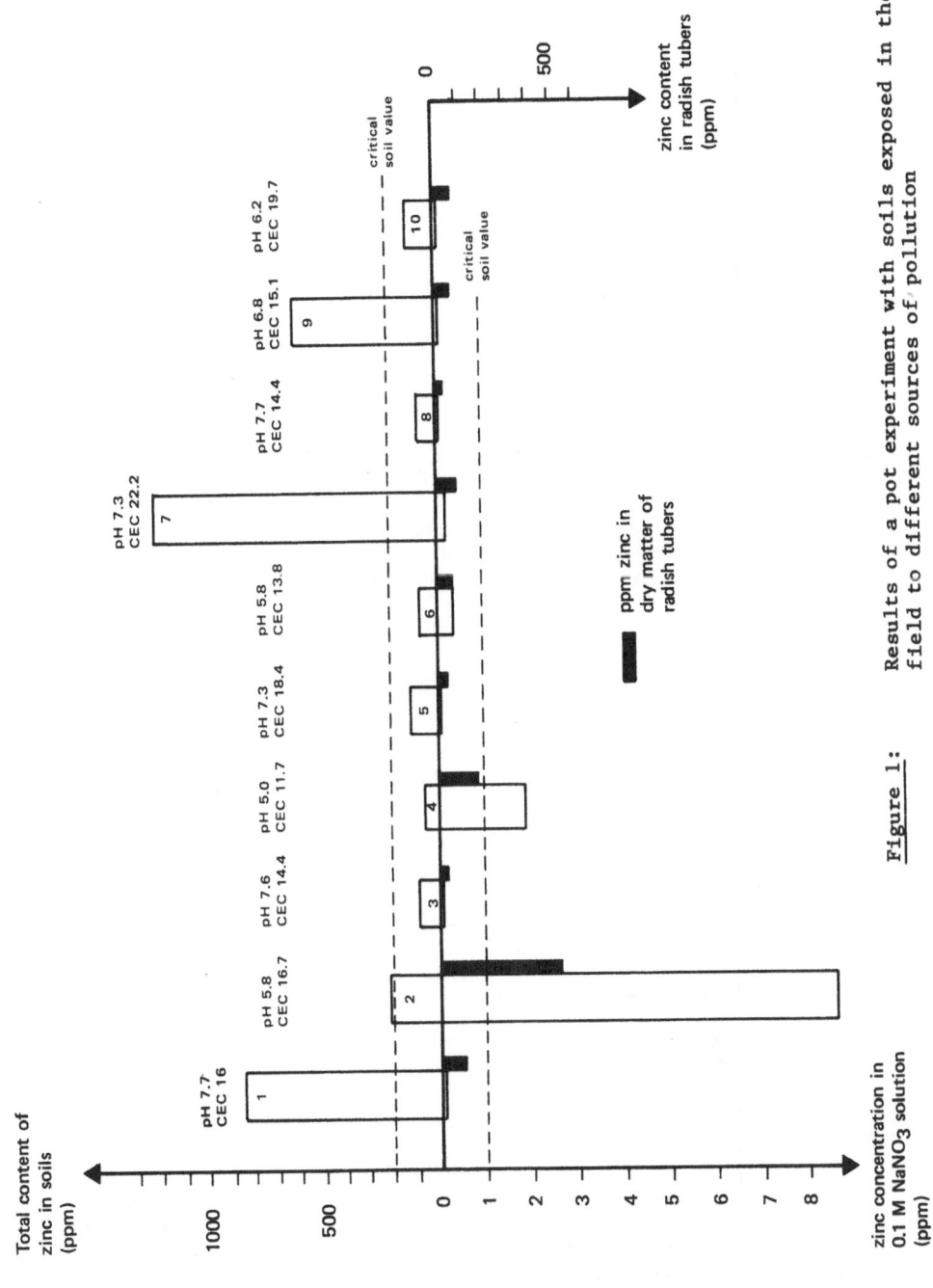

Figure 1: Results of a pot experiment with soils exposed in the field to different sources of pollution

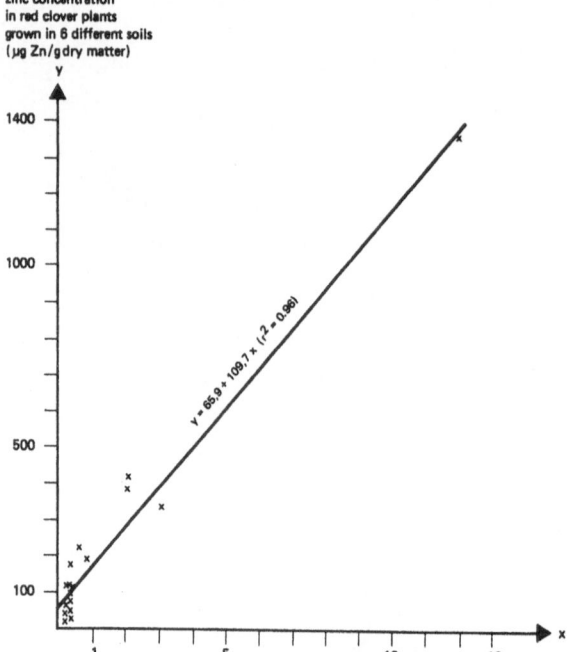

Figure 2:   Results of a pot experiment with six different soils to
              which Zinc has been added in the form of its nitrate

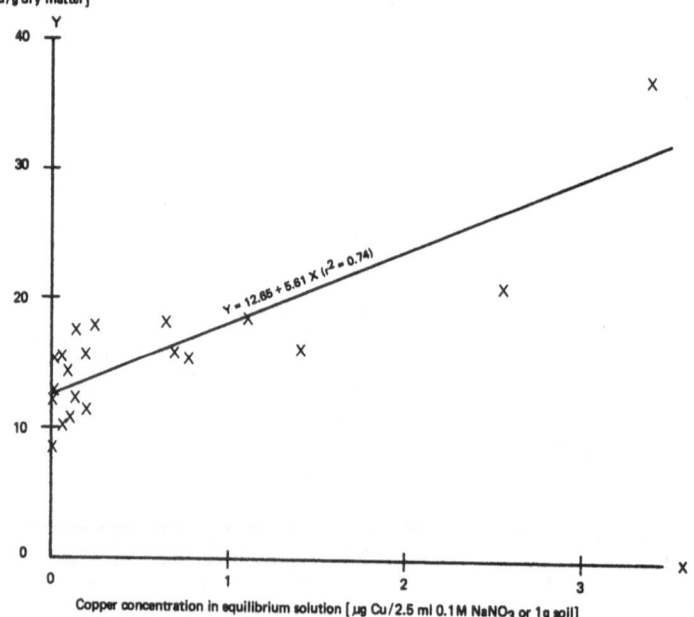

Figure 3:   Results of a pot experiment with six different soils to
              which Copper has been added in the form of its nitrate at
              a concentration of 0 to 3 µg Cu/1g soil

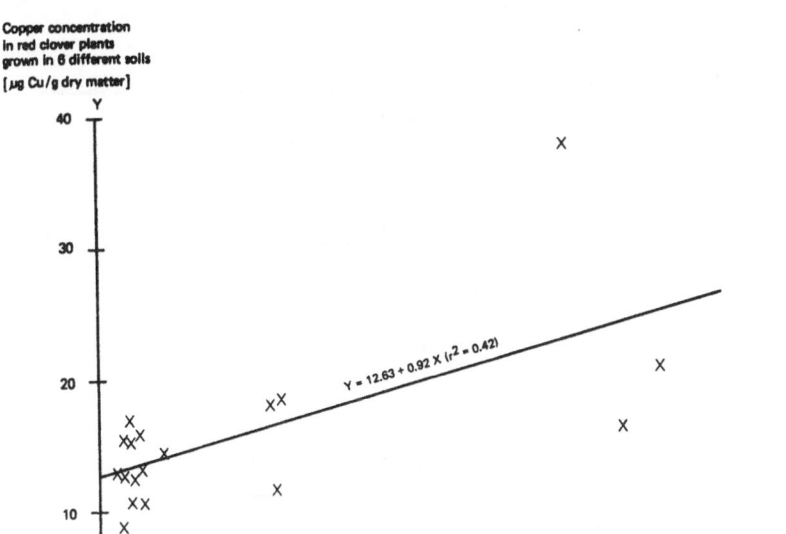

Figure 4: Results of a pot experiment with six different soils to which Copper has been added in the form of its nitrates at a concentration of 0 to 15 µg Cu/1 g soil

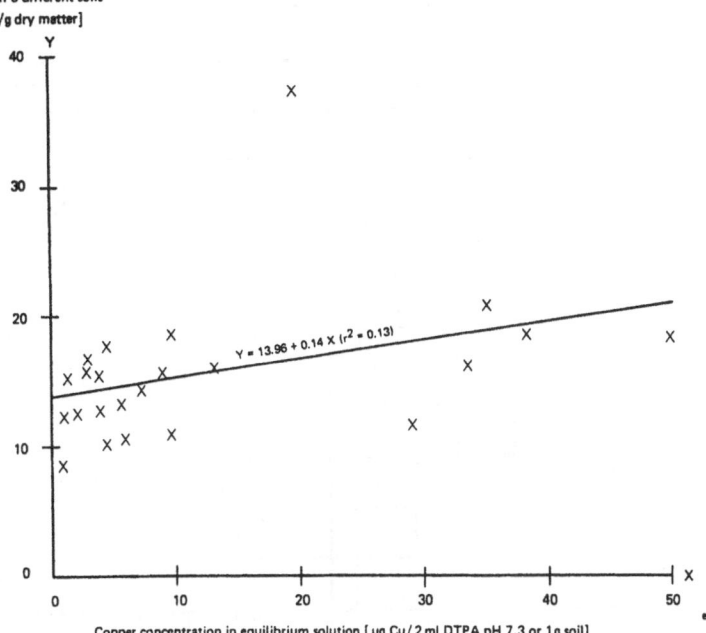

Figure 5: Results of a pot experiment with six different soils to which Copper has been added in the form of its nitrates at a concentration of 0 to 50 µg Cu/1 g soil

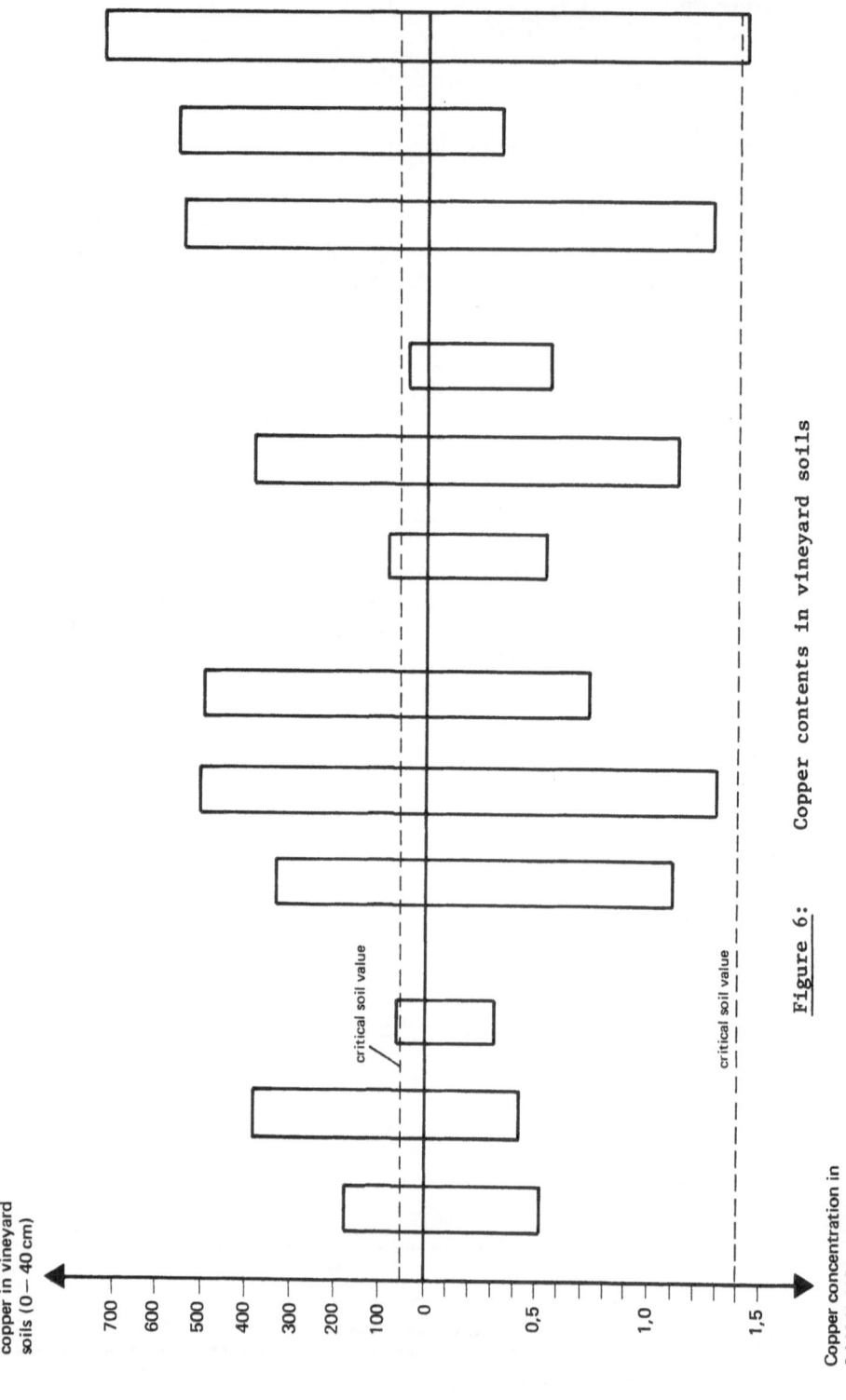

Total content of
copper in vineyard
soils (0 – 40 cm)

700
600
500
400
300
200
100
0
0,5
1,0
1,5

critical soil value

critical soil value

Copper concentration in
0.1 M NaNO₃ solution
(ppm)

Figure 6:   Copper contents in vineyard soils

- 166 -

DISCUSSION

The discussion focussed on the relative merits of calcium chloride compared with sodium nitrate. The important principle was that neutral salts were the best extractants for assessing bioavailable concentrations of metals in soils. The authors preferred sodium nitrate because they were experienced with its use and had an archive of reference results going back five years. Calcium chloride removed more cadmium from soil than did sodium nitrate and therefore analysis of the sample was likely to be easier especially for soils near to background concentrations. On the other hand calcium chloride extracts could pose interference problems and the significance of the formation of chlorocomplexes with cadmium was unclear. (Contributors to discussion : Leschber, Sauerbeck, Matthews, Levi-Minzi)

CHARACTERIZATION OF THE PHYSICAL NATURE OF SEWAGE SLUDGE

WITH PARTICULAR REGARD TO ITS SUITABILITY AS LANDFILL

U. LOLL
Abwasser-Abfall-Aquatechnik
Heidelberger Landstraße 52
D-6100 Darmstadt, FR of Germany

Summary

The most significant property of those sewage sludges which are planned to be deposited is the physical nature. Nevertheless the criteria of dry solid matter after natural and mechanical dewatering of sewage sludge has been mostly used until now. This parameter describes the behaviour of the material placed in the depositing bulk only insufficient. Therefore it is necessary to define parameters for the bearing capacity, the density and the shearing strength that are able to describe the soil mechanical behaviour of sewage sludges which are planned to be deposited. The paper presents new appropriate measuring methods, measuring devices and criteria to determine the depositing capability of dewatered sewage sludges.

1. INTRODUCTION

The physical properties of sewage sludge are especially important for the design of sludge pumps, thickeners and dewatering equipment and for the optimization of conditioning, dewatering, heat treatment and controlled tipping operations (landfill).

For the physical parameters listed in Table 1, there are standard methods which have become widespread internationally. Full details of the tests and the corresponding national and international standard specifications are given in the final report of the COST 68 TER Project (2) or in that of the COST 68 Project (1), with reference to further relevant literature. In addition, Colin (4) has described the most important features involved in characterizing the physical state of sludges.

At present there is no standardized method of determining "suitability as landfill", which is extremely relevant to sludge disposal practice. As land available for tipping becomes ever scarcer, especially in the heavily populated areas of the industrial countries, and the dumping in them of dewatered sludge, either alone or mixed with household or industrial waste, has to be carried out under technically controlled conditions, there is a pressing need for suitable parameters giving information on the bearing capacity, solidity and a shear strength of dewatered sludges. Possible test methods and equipment have thus been tested in Germany in a cooperative project (supported by the Federal Ministry of Research and Technology - BMFT) over the last 3 years, and initial proposals have been made for the establishment of threshold values.

The aim is to develop a standard, internationally-applicable test to determine the suitability of dewatered sludge as landfill on the basis of the results reached together with the other partners in the Euro-COST Programmes.

Table 1: Internationally standardized methods to describe the physical
status of sewage sludge

```
┌────────────────────────────────────────────────────────────────────┐
│  I.)    GRAVIMETRIC                                                  │
│                                                                      │
│         1.)  Total solids (per cent dry matter)                     │
│         2.)  Volatile solids (loss on ignition)                     │
│         3.)  Dissolved solids                                       │
│         4.)  Suspended solids                                       │
│                                                                      │
│  II.)   CONDITIONING AND DEWATERING                                 │
│                                                                      │
│         5.)  Conditionability (coagulant demand)                    │
│         6.)  Effect of shear (floc strength)                        │
│         7.)  Specific resistance to filtration                      │
│         8.)  Capillary suction time                                 │
│         9.)  Compressibility coefficient                            │
│                                                                      │
│  III.)  OTHER PARAMETERS                                            │
│                                                                      │
│         10.) Particle size distribution                             │
└────────────────────────────────────────────────────────────────────┘
```

2. PARAMETERS FOR DESCRIBING THE PHYSICAL STATE OF THE SLUDGE WITH
REFERENCE TO ITS PROCESSING AND DISPOSAL

Colin (in references 3 and 4) defined the categories for characterizing the physical state of sludge as follows:

(a) Liquid state   Sludge which flows under the effect of gravity. This
                    is the best state for handling.

(b) Plastic state   The state in which the material may undergo permanent
                    deformation if the mechanical stress applied to it
                    exceeds a certain threshold. Sludges in this state can
                    be pumped if sufficient pressure is applied.

(c) Solid state with shrinkage This is a solid state in which the volume
                    decreases as drying progresses, i.e. it relates to a
                    medium still saturated with water.

(d) Solid state without shrinkage Below the moisture content level cha-
                    racterizing the limit of the preceding state, the
                    volume of the sludge no longer decreases as drying
                    progresses, i.e. the medium is no longer saturated
                    with water.

In properly run tips, sewage sludge, which is mostly mechanically dewatered, is usually placed in the "plastic" or "solid with shrinkage" state. Sludge in the "solid state without shrinkage" is also sometimes placed in landfill.

In any event, the "solid state without shrinkage" is also fully suitable as landfill, as the material can have a load placed upon it without damage, both during casting and when subsequently placed under load, and it does not have limited solidity or low shear strength.

The threshold of suitability for use as landfill probably lies in the upper range of the plastic state (where a large force must be applied to deform it) or in the solid state with shrinkage.

There is no direct connection between the above definitions and suitability as landfill, which means that efforts at measuring the former are hardly applicable to the problem to be addressed here.

The penetrometer presented by Colin, which is used for describing the plastic state, is most suitable for consideration.

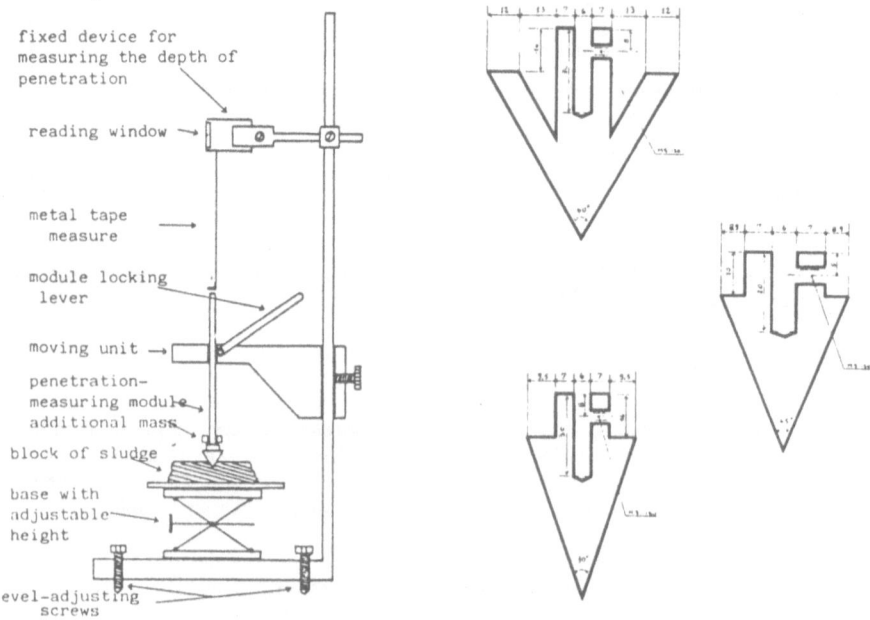

Fig. 1: Diagram of the penetrometer with various diveces with different angles of conicity (after Colin)

Since the tests developed by Colin are only related to sludge transport, storage handling and production techniques, the results obtained cannot be applied to the setting of targets for values appropriate to landfill.

In particular, the way in which the samples are prepared and mounted in the test apparatus does not enable the boundary conditions in the landfill to be simulated.

3. RESEARCH OBJECTIVE

The aim of the studies, which have been planned to cover a fairly long period, is to define the suitability of dewatered sludges for use as landfill on the basis of figures drawn from soil mechanics, and to develop easily manageable measuring techniques for use in sludge dewatering and controlled tipping operations.

In particular, the minimum strength of sludges to be tipped was to be determined, with regard to the way in which they are placed and the way in which they are expected to behave over a long period in the landfill site. In the first cooperative experiments, the strengths of several hundred sewage sludges of various origins, which had been subject to varying stabilization, conditioning, dewatering and post-treatment processes, were studied using the test equipment and procedures described below. After choosing a suitable test procedure, the aim was to promote the international harmonization of determination methods with the recognized standard test on the basis of the results obtained.

The possible influence of sludge treatment, conditioning and, where used, subsequent treatment on changes in its consistency were to be studied to enable sludge treatment sequences to be optimized.

4. TEST APPARATUS AND PROCEDURES

To perform the investigations, the following laboratory apparatus, which is familiar in soil mechanics, being based on the associated test procedures, were used:
- Laboratory vane apparatus to measure the vane shear strength (Fig. 3)
- Falling-cone apparatus for measuring cone shear strength (Fig. 4)
- Plunger apparatus to measure bearing capacity (Fig. 5).

The sample or specimen container shown in Fig. 2 was used for all tests so as to produce constant test conditions.

The sample container has a capacity of 1.5 litres, a height of approx. 150 mm and a diameter of 113 mm.

The dewatered sludge is placed in the sample container in 4 layers, each being compacted with a hand rammer with a surface area of 10 cm^2 until no further compaction is possible and the surface is reasonably even. To provide a defined test surface, the compacted specimen is cut through with a knife under the removable headpiece and the upper part of the specimen discarded.

The vane apparatus for measuring vane shear strength, which is shown in Fig. 3, has an electric motor and can also be operated by hand.

The vane is turned either manually with a handle or using the motor. The vane is connected through a torsion spring to a display which shows the angle of deflection on a 360° scale by means of a trailing pointer.

The torsional moment can be calculated from the angle of deflection shown and the spring constant.

The torsional moment increases as the vane turns and the pointer remains stationary at the reading of maximum moment, when the specimen fails. The experiment is repeated three times for each measurement in order to exclude any accidental readings. The value of the vane shear strength is calculated from the torsional moment at failure of the specimen, the geometrical data on the vane and tables specific to the apparatus.

The falling cone shear strength is defined as the resistance of a soil to penetration by a geometrically defined cone. This 'cone resistance' is determined from the weight and vertex angle of the cone and the depth to which it penetrates. With the apparatus shown in Fig. 4, the cone is so placed at the beginning of the experiment that its point just touches the surface of the specimen. It is held in this position magnetically. After release the cone falls and penetrates the material to be tested. The depth of penetration of the cone is measured after 5 minutes. Five measurements are carried out in each experiment at different points on the same specimen. The falling-cone shear strength can be determined from the depth of penetration of the cone using correlation tables relating the penetration depth of the various cones and the shear strength determined with the vane apparatus.

Bearing capacity is determined using a plunger apparatus with a bearing surface area of 10 cm^2, which is balanced with a counterweight and placed vertically on the surface of the sludge to be investigated. According to the consistency of the sludge, it is loaded to between 0.1 and 10 kg. After 30 seconds under load, the depth of penetration of the plunger is measured. The depth of penetration is plotted against the load in KN/m^2, giving load curves illustrating the behaviour of the material under load. From these curves, the load under which the plunger penetrates to a depth of 4 mm is defined as the bearing capacity of the material under investigation. For each test, the specimen is placed under load three or more times with varying loads to determined this point. The depths of penetration allowed lie between 2 and 6 mm.

5. EVALUATION OF TEST APPARATUS AND METHODS

Although all 3 apparatuses gave the expected scatter of results, partly caused by the nature of the material sampled, they yielded a relatively high degree of uniformity in the sludge strength data obtained, after a staff familiarization period. Interpretation of the results showed that, even with differing types of sludge to be tested, all three test methods performed similarly and indicated comparable sludge strengths. Correlations were obtained for each combination of the three apparatuses described. The correlation coefficients obtained between these, $r \cong 0.9$, showed good comparability of results.

Converting the minimum strengths laid down yielded the following error estimation for the three laboratory test apparatuses used:
- vane shear strength $\quad\quad\quad C_{FL} = 10.0$ KN/m^2 $+$ 2.0 KN/m^2
- falling-cone shear strength $C_{FK} = 12.3$ KN/m^2 $\mp$ 1.3 KN/m^2
- bearing capacity $\quad\quad\quad\quad C_{TF} = 23.5$ KN/m^2 $\mp$ 2.3 KN/m^2

The measuring errors which were determined were partly caused by the test technique, but also often by the lack of homogeneity of the specimens. The primary causes of inaccuracies were uneven compaction of the specimen and fibres and other solid particles.

As a rule, such errors give the impression of a strength higher than it really is. However, these can usually be quickly eliminated by taking multiple readings.

SAMPLE CONTAINER

with 1500 cm³ Volume

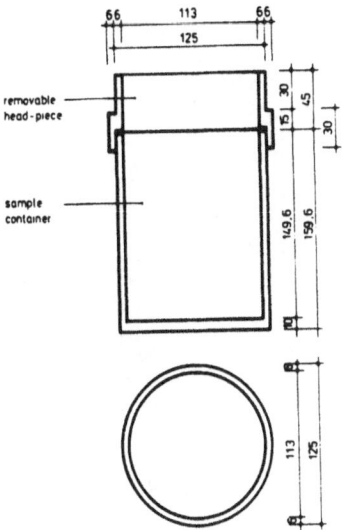

Fig. 2: Sample container

LABORATORY VANE APPARATUS

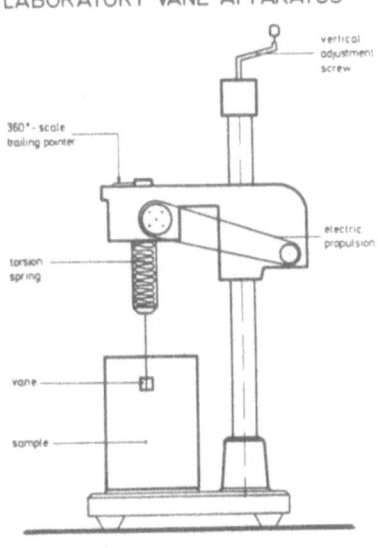

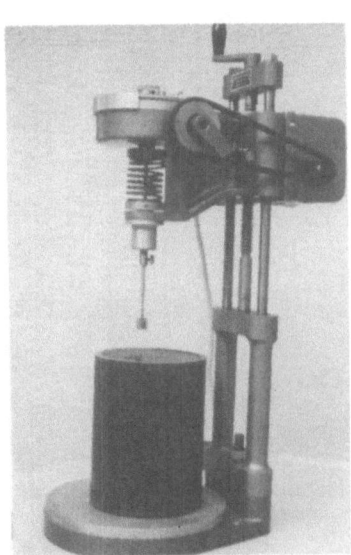

Fig. 3: Laboratory Vane Apparatus (vane shearing strength)

FALL-CONE APPARATUS

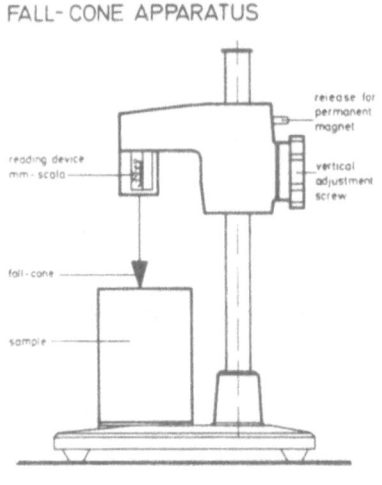

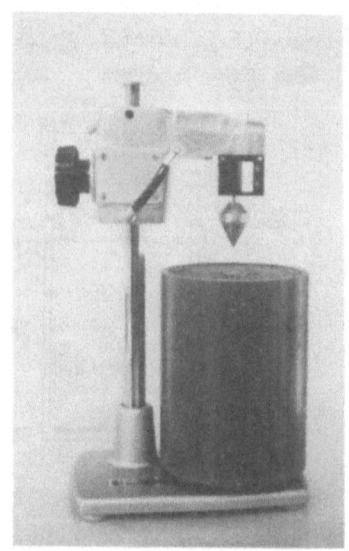

Fig. 4: Fall-Cone Apparatus (fall-cone shearing strength)

LOAD PUNCHEON APPARATUS

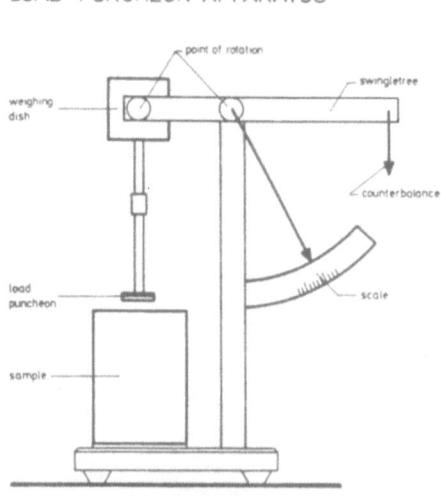

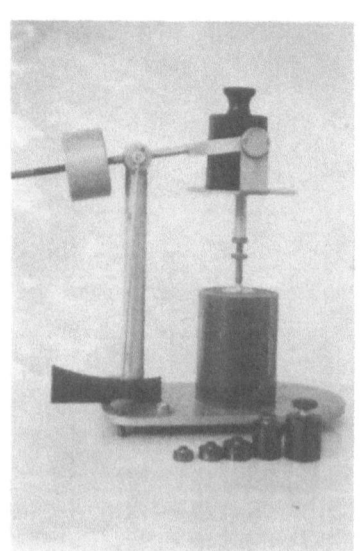

Fig. 5: Load Puncheon Apparatus (bearing capacity)

Drying of the surface of the sample had a differing effect on the different apparatuses. While the vane apparatus was virtually unaffected, the plunger was somewhat affected and the falling cone considerably.

Without going into the reasons in detail, these being destined for inclusion in an extensive research report, the results up to now indicate that the vane apparatus is the most accurate of the three test apparatuses presented.

6.   RESULTS

From a large number of test results, we now proceed to a more detailed description of the correlation between vane shear strength, as determined by the apparatus assessed as best, and the percentage of dry solid matter in the dewatered sludge. In Fig. 6, pairs of results for vane shear strength and solid matter content in the sludge being tested are correlated for a number of differing sludges.

If the possible minimum values for sewage sludge to be suitable as landfill are taken to be 35% dry matter or a vane shear strength of 10 KN/m^2 (see Fig. 6), it can be seen that there is no linear relationship between the two values. This means, in the face of current practice which holds that sewage sludges with more than 35% d.m. are suitable as landfill, that, in the light of their soil mechanics properties, this is not always true. As may be seen in Figures 6 and 7, many sludges did not attain the minimum strength of 10 KN/m^2, which has been determined as the minimum working value, although their d.m. content was over 35%, and that some of the sludges studied had already reached the criterion fixed for minimum strength and thus for suitability as landfill at a dry matter content below 35%. Sludges for which both parameters lie below the minimum value must generally be regarded as not suitable as landfill. Sludges with both properties exceeding the minimum values may be regarded as unrestrictedly suitable as landfill. In general it can be stated that the criterion of dry solid matter (or of water content) in dewatered sludge is not a satisfactory measure of its suitability as landfill. The tests described above represent a significantly better method for determining this parameter, as it relates to real tipping conditions.

The introduction of these new criteria would provide a guarantee for the tip operator that the sludges delivered could be placed without difficulty and without having deleterious effects on the stability of the tip. For operators of sludge dewatering plants, it could mean that they would need to use new sludge stabilization, conditioning, dewatering and subsequent handling techniques in order to comply with the new criteria. In this connection it should be emphasized that a higher final solid matter content in dewatered sludge does not automatically mean that it has a higher strength.

7.   CHANGES IN THE SUITABILITY OF MECHANICALLY DEWATERED SEWAGE SLUDGE

AS LANDFILL ON ADDITION OF QUICKLIME

There are various ways of rendering more suitable as landfill sludges which do not comply with the minimum values for the criteria given. One example is to add quicklime after mechanical dewatering. Fig. 7 shows changes in the dry matter content and vane shear strength for a digested sludge dewatered on a filter belt press. At that stage, without the addition of lime, the sludge had a dry matter content of 20.8% and a vane shear strength of 2.5 KN/m^2. Both final dry matter

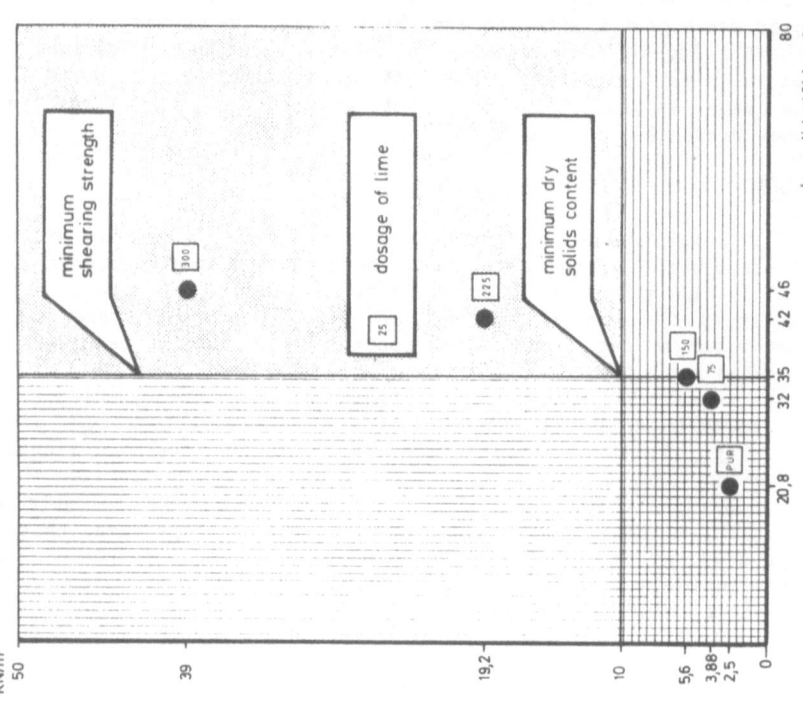

Fig. 7: Increase of dry matter content and shearing strength after different dosages of lime

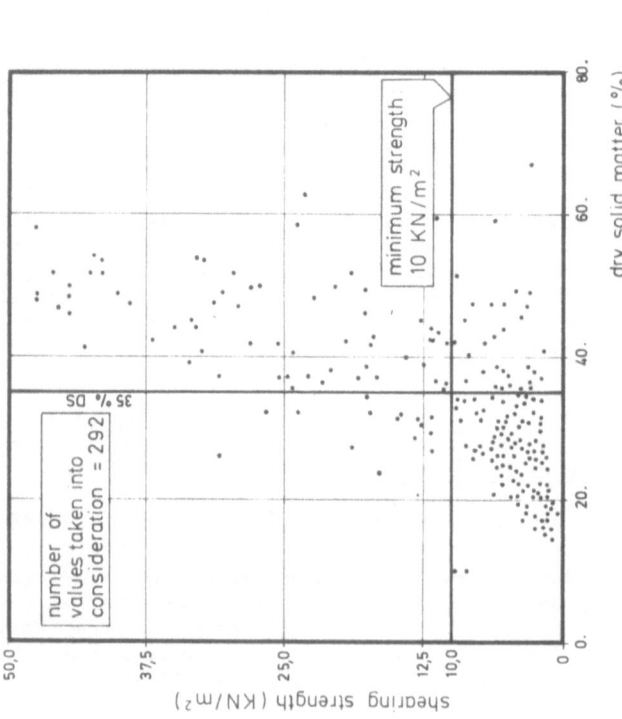

Fig. 6: Correlation between shearing strength and dry solid matter

content and vane shear strength increased with increasing lime dosage. It emerged that whereas after the addition of approximately 75 kg of quicklime per m^3 of dewatered sludge, a final dry matter content of 35% was reached, the vane shear strength still did not exceed the minimum value.

After the addition of approx. 225 kg of lime per m^3, both parameters reached the range in which the sludge's suitability for dumping is undisputed. When 300 kg of lime per m^3 were added in the tests, this was clearly an overdose.

8.   PRELIMINARY RESULTS AND CONCLUSIONS

These studies, on which only a brief report has been possible here, have between them determined that the criterion of final dry matter content is not a satisfactory description of the suitability of dewatered sludge for use as landfill. The test apparatuses presented are in principle suitable for measuring soil mechanics data which describes the behaviour of dewatered sludge when used as landfill. Current and future studies are intended to refine the test methods and in particular to adjust the initial postulated working values of minimum strengths to the real necessities of landfill operations. Attempts are to be made at further development of the proposed test methods in cooperation with more international institutions, with the aim of developing a generally recognized standard of measurement. The work described here was funded by the Ministry of Research and Technology of the Federal Republic of Germany.

REFERENCES

(1)   Commission of the European Communities: COST-ACTION 68, Treatment of sludge, Final report (1975).
(2)   Commission of the European Communities: Concerted action treatment and use of sewage sludge/COST 68 TER, Final report (1984).
(3)   Commission of the European Communities: Methods of characterization of sewage sludge. Proceedings of a workshop in Dublin (1983), Reidel Publishing Company, Dordrecht-Boston-Lancaster.
(4)   COLIN, P and G. AMBROSIO: Etude de la caractérisation de l'état physique des boues. Institut de Recherches Hydrologiques, Nancy (1978).
(5)   MÖLLER, U., KASSNER, W., KOHLHOFF, D., LOLL, U., OTTE-WITTE, R.: Neudefinition der Deponierfähigkeit von Klärschlämmen. Korrespondenz Abwasser, H. 11, S. 928-933 (1984).
(6)   OTTE-WITTE, R., MÖLLER, U., KASSNER, W., LOLL, U., GAY, Ch.G.: New definition of sewage sludge depositing capabilities. Recycling International (1984), Berlin, page 404/409.

MICROBIOLOGICAL SPECIFICATIONS OF DISINFECTED SEWAGE SLUDGE*

D. STRAUCH** and M. DE BERTOLDI***

Institute of Animal Medicine and Animal Hygiene,
University of Hohenheim, Stuttgart, Fed. Rep. of Germany**

Institute of Agricultural Microbiology, University of Pisa;
Study Center for Soil Microbiology, C.N.R., Pisa, Italy***

Summary

Some selected literature dealing with indicator organisms for control of disinfection measures for sewage sludge is discussed. The
question whether bacteria or bacteriophages can be used as indicators for pathogenic viruses is still undecided and must be further
investigated. Also still uncertain is the suitability of certain
groups of bacteria as reliable indicators for other pathogenic bacteria. Enterobacteriaceae and fecal streptococci seem to be suitable
to a certain extent. Further comparative investigations are still
necessary.
Since no standardized microbiological methods in this field are
available internationally a pragmatic proposal of the requirements
for an efficient disinfection of sewage sludge with various methods
is presented for discussion. This seems necessary because in several
countries the authorities are in need of such definitions to prepare legal regulations for public health reasons.

1. INTRODUCTION

The task of Working Party 3 has been to exchange scientific information on all aspects of pathogenic microorganisms in sewage sludge, as
they affect human and animal health when sludge is applied to land. Attention has been focused on the definition of an adequate time interval between sludge application and allowing cattle to graze, sludge disinfection, indicator organisms and epidemiological studies. The working party
has discussed these subjects in its meetings on the basis of reviews by
some of the members and has produced summary reports representing the
general opinion of the members which does not exclude that in certain
points different opinions still do exist.

As far as indicator organisms are concerned the following conclusions have been drawn for the Final Report (Part II) of COST 68ter Treatment and Use of Sewage Sludge (1).

Indicator bacteria have only a limited use in the field of sewage
sludge hygiene. In many cases it is advisable to analyse for or seed with
pathogens directly. Two possible uses of indicator organisms have been
identified: monitoring disinfection processes and checking for regrowth
after disinfection. Fecal streptococci appear to be most useful for critical monitoring of disinfection processes. E. coli or other related counts
(coliforms, enterobacteriaceae) have a general low resistance and would
only detect gross under-processing. Bacteriophages (especially male-spe-

---

*This paper reflects the personal opinion of the authors and not
 necessarily that of Working Party 3

cific) and mycobacteria are potentially useful, but there is too little
data to further define their use. To check for regrowth in disinfected
sludge, enterobacteriaceae and most probably also total coliforms can be
used.

In the following paper we do not intend to repeat all the discussions on indicator organisms with pro and contra. We rather would like to
discuss the problem on the basis of some recent developments in that
field and finally present a pragmatic proposal for the hygienic evaluation
of some disinfection methods as in some countries it may be needed for political decisions.

2. BACTERIA AND BACTERIOPHAGES AS INDICATORS FOR VIRUSES

Scarpino has recently discussed indicator systems for monitoring the
virological quality of potable water, wastewater, solid waste, shellfish,
fish and crops (2). For drinking water he directed concern at the bacterial
indicator system that is routinely used to approve its safety. Based on
reported viral isolations, the coliform standard used may itself be in-
herently defective or the bacterial sample quantity and timing may be the
difficulty. In addition, consideration must be given to the wide range of
susceptibility to chlorine disinfection displayed by different viruses
and the more recent disclosure that there may be disinfection rate differences between stock virus strains commonly used in laboratory disinfection studies and the apparently more resistant indigenous viruses. Studies
have shown how low levels of viruses may pass through the water treatment
procedure and be later isolated from drinking water samples while the
coliform indicators apparently do not survive the rigors of water treatment.

It has also been established that fecal coliforms, fecal streptococci
and total coliforms are not always adequate indicators of viruses in disinfected primary sewage effluents. Viruses were found to survive in all
the effluents even though the fecal coliform counts were within the limits
permitted in USA for recreational waters and for approved shell-fish-gro-
wing waters. It was emphasized that even if the total coliform counts had
been used instead of the fecal coliforms, the bacterial counts would still
have been within the acceptable limits of safety - with pathogenic viruses
still present.

Concerning bacteriophages as an indicator of water contamination the
author expresses guarded optimism that under certain circumstances bacteriophage indicators can be used to detect viral presence and inactivation
but not in all situations.

Studies with solid wastes suggest that Salmonella are rather stable
microorganisms which could be a better indicator for the potential presence
of viruses in sewage than currently used fecal streptococci or fecal coliforms. Studies of viruses and bacteria under identical conditions show
that viruses survive longer than bacteria under most conditions. This of
course would cast doubt on using bacteria as indicators of virus presence.
Scarpino draws the conclusion in his study that in the solid waste environment it would appear that the only reliable indicator of virus presence is finding the virus itself.

For the indication of viruses on or in crops he thinks that although
fecal coliforms have been used to evaluate public health risk in regard
to crops, the indicators for virus presence seem justifiable at present.
Also in this case there does not appear to be a substitute for direct evaluation of pathogenic viral presence.

Scarpino has drawn these conclusions: "Certainly we should continue to strive to develop a surrogate system or model that would enable us to replace the direct determination of pathogenic viruses. However, this has not always been possible, but the vision does not need to fade; the same and other candidates need to be re-examined or discovered. Perhaps we should accept the fact that indicators need not to be 100 % successful in every case - and that direct pathogenic virus determination is the best way to safeguard public health. The indicators could provide us with assistance in reducing public risk, but should never replace direct observation. Again, perhaps the best indicator for pathogenic virus presence is the virus itself." (2)

3. INDICATOR ORGANISMS FOR PATHOGENS OTHER THAN VIRUSES

3.1 The US-Environmental Protection Agency has initiated a literature review on density levels of indicator and pathogenic organisms in municipal wastewater sludge (3). This report is based on the assumption that "sludges originating from municipal wastewater treatment plants harbor a multitude of microorganisms, many of which present a potential health hazard. Risk of public exposure to these organisms is possible when sludges are applied to land as a means of disposal. In recognition of this problem, and as required by Section 405 of the Clean Water Act of 1977 8PL 95-217), criteria for the control of infectious disease in the land application of sewage sludge and septic tank pumpings were issued by the U.S. Environmental Protection Agency (EPA) in 40 CFR Part 257 (Federal Register Vol. 44, No. 179, September 13, 1979).

The "Part 257 criteria" specify what minimum treatment of municipal wastewater treatment plant sludges is required prior to land application of the residue. Acceptable treatment methods, termed "Processes to Significantly Reduce Pathogens," are as follows:
- Aerobic digestion - Agitation of sludge in aerobic conditions at residence times ranging from 60 days at 15°C to 40 days at 20°C, with a volatile solids reduction of at least 38 %.
- Air drying - Draining and/or drying of liquid sludge on underdrained sand beds, or on paved or unpaved basins in which the sludge is at a depth of 9 inches (22.9 cm). A minimum of three months is needed, two months of which temperature average on a daily basis is above 0°C.
- Anaerobic digestion - Maintenance of sludge in the absence of air at residence times ranging from 60 days at 20°C to 15 days at 35°C to 55°C, with a volatile solids reduction of at least 38 %.
- Composting - Using the withinvessel, static aerated pile or windrow composting methods, the sludge is maintained at minimum operating conditions of 40°C for five days. For four hours during this period, the temperature exceeds 55°C.
- Lime stabilization - Application of lime to sludge in quantities sufficient to produce a pH of 12 after two hours of contact.
- Techniques demonstrated to be the equivalent of the above on the basis of pathogen removals and volatile solids reduction.

An additional category of treatment processes, termed "Processes to Further Reduce Pathogens", was designated in Appendix II of 40 CFR Part 257 as required if (1) affected land is to be used within 18 months of sludge application for the cultivation of food crops and (2) the edible portion of the crop is likely to be exposed to the sludge. These additional processes are:

- High temperature composting
- Heat drying
- Heat treatment
- Thermophilic aerobic digestion
- Irradiation."

    For the review the following organisms, categorized into four groupings, were emphasized:
- Indicators - Total coliform, fecal coliform, and fecal streptococcus bacteria; Clostridium perfringens (welchii); bacteriophage
- Pathogenic bacteria - Salmonellae, Shigellae, Pseudomonas sp., Mycobacterium spp., Candida albicans, Aspergillus fumigatus
- Enteric viruses - Enterovirus and its subgroups (polioviruses, echoviruses and coxsackieviruses), reovirus and adenovirus
- Parasites - Entamoeba histolytica, Ascaris lumbricoides, Taenia spp., Schistosoma spp., and others.

    In addition to reporting density levels in raw sludge and septage, and the effectiveness of conventional sludge treatment processes in reducing density levels, this review also identified design and operating variables that affect process efficiency, compared results of laboratory pilot-scale studies to those of full-scale plants, and contrasted survival of indicator organisms to that of pathogens. Methods used by each researcher to enumerate organisms were also described, and brief summaries were provided of related citations that were encountered but were not actually used in this report.

Density levels in raw sludge

    Levels of bacteria, viruses and parasites in raw sludge are presented in Table 1. Note that the densities of pathogenic organisms are several logs less than indicator organisms. Also, there is a noticeable lack of information on the densities of select pathogenic organisms in raw sludges and septages (i.e., lack of parasite organisms data in septages).

    The following sludge treatment methods were evaluated in their effect to reduce the numbers of indicator organisms and of pathogens: anaerobic digestion, mesophilic aerobic digestion, "mesophilic" composting (45°-65°C or higher), lime stabilization, drying beds, sludge storage /lagooning, sludge conditioning/ mechanical dewatering.

    The review results in the following conclusions and recommendations. "Because a large body of literature containing comparable data is not available, it is recommended that additional research be conducted on the effectiveness of sludge treatment processes in reducing density levels of organisms. It is recommended, further that researchers document carefully all pertinent aspects of their experimental design.

    The following conclusions appear to be valid based on the literature reviewed.

    Anaerobic digestion and lime stabilization consistently produce reductions of about 1 to 2 logs in densities of indicator and pathogenic bacteria and, in the case of anaerobic digestion, in densities of viruses as well. At a minimum, effectiveness depends on the processes being carried out under the conditions specified in 40 CFR Part 257. Neither sludge stabilization process appears to be particularly effective for inactivating parasite organisms.

    Conditions of mesophilic composting may inactivate common indicator and pathogenic bacteria and viruses, provided that specified temperatures are attained uniformly throughout the compost mass for over the specified time period. The pathogenic fungus Aspergillus fumigatus thrives under conditions of mesophilic composting, however, and parasite ova appear to survive this process.

Density reductions of bacteria by aerobic digestion are variable and of relatively small magnitude. However, there is a lack of data on the performance of this process and also of air drying in reducing densities of microorganisms.

Sludge lagoons can achieve 1-log reductions in densities of bacteria and viable parasite ova, but, depending on conditions, storage of one month to more than three years may be required.

Mechanical dewatering of sludge, with or without the use of chemical conditioners, has little reliable effect on densities of pathogens.

Few of the laboratory-scale studies reviewed could be related to results obtained at full-scale treatment plants. Operating parameters used in laboratory experiments differed radically from those at full-scale plants. For this reason, comparing density levels was seldom possible. In addition, laboratory studies often used seeded bacteria, viruses, or parasites and it is doubtful whether their behavior mimics that of naturally occurring organisms.

No single indicator organism (either bacteria or bacteriophage) was found to maintain a density level of a constant relative value to that of pathogenic organisms. The data available made it impossible to determine whether this inconsistency is due to the inability of current techniques to enumerate pathogenic bacteria and enteroviruses accurately, or to the fact that densities actually vary.

Of the traditional indicators, fecal streptococci appear to be the most conservative indicator of both the density levels of pathogenic bacteria and enterovirus in raw sludge and of their inactivation during sludge treatments. Additional research is required to identify other indicator systems, both bacterial and viral, whose numbers better reflect both density and reduction of density levels of pathogenic organisms.

A wide variety of methods were used to enumerate all of the organisms considered in this review. Although standard methods are available for quantifying the coliform and streptococcus bacteria and for Salmonella sp., there are no standard techniques for other pathogens, enteroviruses, or parasites. It is recommended that this area be addressed so that comparable data can be produced in future studies."

3.2 In extensive microbiological studies a variety of composting systems were investigated by de Bertoldi et al. (4, 5) and the isolated numbers of Salmonella spp., fecal coliforms and fecal streptococci compared (Table 2 and 3).

Based on their results in Table 2 and 3 as well as on results published elsewhere and on unpublished data de Bertoldi et al. (5) did evaluate the composting methods tested. Their conclusions are:

"Several factors affect pathogen survival during treatment. In composting, heat is one of the primary factors contributing to pathogen inactivation. Most pathogenic microbes are destroyed by heating for several hours to temperatures above 50°C. Microbial competition is another important factor in controlling pathogen diffusion during composting. A variety of saprophytic microorganisms participates in the composting process; these microorganisms might be considered the indigenous or natural microflora of the compost system. Municipal sludge contains a second microbial population, the pathogens, which represent a numerically insignificant fraction of the total microbial population. Hence competition comes into play when the community is heterogeneous and the population density is high relative to the supply of any limiting feature of the en-

vironment. The indigenous saprophytic population has a highly distinct
competitive advantage over the other population; composting material is
not the natural environment for pathogenic microorganisms, therefore in
this ecosystem competition will tend to result in the elimination of the
less fit rival.

The choice of the most suitable composting process depends on differing local situations and conditions; nevertheless only processes which
guarantee good quality and sanitized end-products should be chosen.

All the experimental data reported in this paper indicate that static
composting systems provide a better control over pathogens than does turning.

In the closed systems, horizontal reactors seem to guarantee better
control over pathogens than vertical continuous reactors. This may be
mainly due to the fact that masses over three meters high are difficult to
control during the composting process. In particular oxygenation of the
mass is not homogeneous, resulting in anaerobic zones with lower temperatures.

Since the term "composting" encompasses numerous processes which may
create physiologically disparate physico-chemical environments for pathogens in the sludge being composted, it can not be tacitly assumed that
composting renders municipal sludge pathogen-free. However, well-conducted
processes with prolonged periods of high temperatures do seem to provide
a sanitized product.

Finally, there is a need for a better definition of the term "sanitized compost". At present, there is no legal specification of what should
be a disinfected sludge.

For this reason, and because it is essential for compost to have specifications for pathogens, too, it is of primary importance to lay down not
only which microorganisms are to be monitored, but also their maximum permitted level in a sanitized product."

It is therefore suggested to use as indicators enterobacteriaceae
with a limit of 5×10^2/g and fecal streptococci with a limit of 5×10^3. These
figures derive from numerous experiments which were carried out by de
Bertoldi and co-workers in Pisa on many different composting processes.
When these indicators are lower in number than the limits given above,
Salmonella spp. is always absent.

For parasites the eggs of Ascaris suum are usually considered as the
representative indicator or test organism. Their thermosensitivity is in
the range between 50° - 55°C so that they are covered by tests performed
with microorganisms like fecal streptococci or f2-phages with a higher
thermostability.

The figures for fecal streptococci mentioned above present a contradiction to a statement of Havelaar (6) who proposed for the hygienic evaluation of the aerobic-thermophilic stabilization (ATS) of liquid sludge
the fecal streptococci as indicators with an upper limit of 10^1/g. Further
investigations under practical con- ditions are necessary to disentagle
this difference in opinion.

There is a further contradiction in the opinions as far as the hygienic evaluation of ATS is concerned. The evaluation of ATS is based on the
assumption that ATS is performed in only one reactor (6). Recent investi-
gastions in practice have shown that ATS in one reactor may result in a
hygienically safe product when all necessary parameters for pathogen kill
are favorable, which never can be guaranteed under the hard conditions of
sewage works practice (7). Therefore a prerequisite for the production of
a hygienically reliable product is the two or three stage operation of such
ATS plants combined with a continuous registration of the temperatures in
the reactors with liquid sludge (8, 9, 10). Under these conditions the

proposed continuous monitoring of each batch of the end-product of ATS plants based on indicator organisms like fecal streptococci (6) is superfluous. Moreover a batch-wise control of the end-product of a single-stage ATS plant is practically not possible because none of the ATS plants to date is operated as batch system for economic reasons; all of them are continuous or semi-continuous systems.

3.3 In a very thorough study in three German plants where municipal refuse is composted in different systems:
- Prat-system: refuse + refluxcompost + water + ammonium compounds
- DANO-system: refuse + sludge
- Hazemag-system (aerated static pile): refuse + sludge,
the authors inoculated the raw material directly with liquid cultures of
  Staphylococcus aureus   ATCC 6538
  Klebsiella pneumoniae   ATCC 4532
  Streptococcus faecalis   ATCC 6057
  Candida albicans   ATCC 10231
These bacterial strains are the same which are used in the Federal Republic of Germany as test strains for testing the antibacterial effect of chemical disinfectants (11).

The results of this study are very important for the evaluation of composts from municipal refuse. In none of the composts investigated the outer compost layers were considered to be hygienically safe. Even if the material of these layers was put into the inner core of the windrows by mechanical means the authors did not always succeed in destroying all of their test microbes. But also in these experiments Streptococcus faecalis proved to be a reliable indicator organism.

Based on their results the authors doubt that the results of studies in other composting systems which are published in the literature and indicate that these other systems are producing composts free of pathogens, are appropriate. They ask the question whether it is really necessary to make such rigid demands on the hygienic quality of refuse composts and tried to answer it by other investigations.

They made microbiological analyses of composts of the three plants tested which were sold to the public and compared the results to analyses of organic fertilizers, garten molds, peat which were bought in shops and soils of private gardens, sand from children play grounds and top-soil of forests. Three typical results are shown in Figures 1-3. From the results of this study the authors draw the conclusion that the refuse composts usually had higher densities of germs than the fertilizers or soils. This seems to be very natural since the composts are produced by microbiological conversion processes and are used from bacteria and fungi as substrates for nutrition. They consider all 18 of the detected bacterial species as pathogenic or facultative pathogenic. From these they found in compost and in the other materials investigated: Citrobacter, Enterobacter, E. coli, Klebsiella, Pseudomonas, Aspergillus, Salmonella. Only from compost were isolated: Achromobacter, Proteus, Serratia, Enterococcus, Streptomyces, Mucor, Rhodotorula, Penicillium, Geotrichum candidum.

Since these microbes may become pathogenic only under certain conditions they do not raise the hygienic risk of the compost so that no aggravating differences between the composts on the one hand and the sum of the other substrates analyzed could be seen. The question of the real risk of the utilization of refuse or refuse/sludge composts in agriculture, landscape gardening or private gardening can only be answered by a thorough analysis of risk which could not be done within this study. As long as the utilization of the substrates like organic fertilizers, garden molds, peat, soils etc. is not regularized the authors do not have objections against the utilization of refuse composts for the purposes for which they are destined (11).

3.4 Another approach to the problem of indicator organisms and monitoring pathogen destruction was made by a research group in the USA (12, 13). Although several studies showed the potential for destruction of pathogens by composting, the authors thought a method for determining the capability of a particular system was needed.

"A straightforward method of determining the success of a process in destroying pathogens would be, of course, monitoring for the pathogens themselves or for an indicator organism. The utilization of pathogens is not practical because, for most, the procedures are too difficult for routine application. The use of a specific pathogen is ruled out not only because of the difficulty, but to be effective one would have to know the concentration of the pathogen at the initiation of composting. A specific pathogen could be low enough in numbers or sensitive enough to destruction by heat so that its absence would not guarantee that all others had been destroyed.

The use of the indicator organisms (fecal coliforms) were ruled out because although these organisms may be destroyed during the composting process, they are capable of repopulating the compost if they are reinoculated and reinoculation is impossible to prevent in most composting processes as they are presently conducted.

To overcome these problems it was decided to use a temperature-by-time monitoring system with the relatively heat-resistant bacterial virus f2 as a standard. A search of the literature showed that the resistance of f2 to heat was high enough so that if it was destroyed all enteric pathogens would be also. The D values of a number of enteric pathogens as compared with f2 are shown in Table 4. A D value is the amount of time required to cause a ten-fold (one log) reduction in population numbers. The temperature must be specified. A plot showing the number of logs of f2 killed for different temperatures with time is shown in Figure 4. As a reasonable goal for a composting operation it was decided to use 15 logs of f2 inactivation. At 55°C this could be achieved in 2.5 days. In the canning industry, the time at a particular temperature to inactivate 11 logs of bacillus spores is considered adequate, but even with the control realizable in the canning process for uniform heating the theoretical spore reduction often is not obtained. Therefore, in a much less controllable composting system, the use of a 15 log criterium did not seem to be an unreasonable value to insure safety.

To determine the applicability of the approach to the Beltsville procedure and to determine the most suitable monitoring sites, 15 composting piles were monitored taking random probings associated with three pile regions: the center, the lateral portions, and the lower portions at the pile end (pile toes). Consistently the latter area was found to be at the lowest temperatures. Nevertheless it was found that one could expect to achieve at least 55°C in this portion of the pile and therefore in all portions of the pile for 10.6 days of the 21 day composting period with a confidence level of 95 percent. Therefore it appeared reasonable to monitor the pile in this zone.

Studies have shown that salmonellae are capable to some degree of repopulating sludge and compost. In sterilized compost growth can be extensive, but in sludge and compost that has not been sterilized, the population increase over the inoculation levels is usually only a log or so with a reduction to undetectable levels in a few days. The experience at Beltsville has been that during periods when windrow composting of raw sludge was interrupted by heavy rains and cold, one was able to find salmonellae in stored compost and on the site pad. But monitoring monthly for one and one-half years while composting by the aerated-pile method, one was unable to detect salmonellae on the pad or in stored compost. As a result of this experience it is recommended that in addition to temperature monitoring the finished compost be monitored monthly at the site for salmonellae by an agency independent of the organization operating the site and doing the temperature monitoring.

As of March 18, 1982, the Dickerson sewage sludge composting site in Montgomery Country, Maryland had been in operation for 15 months and had been using the temperature-by-time criteria and salmonella monitoring since starting. Once the six monitoring points have maintained a minimum temperature of 55°C for at least three consecutive days then only biweekly temperature monitoring is required. In addition to temperature monitoring, a monthly sampling of the 21 day old aerated piles, screened compost storage piles, and unscreened compost storage piles was conducted. Samples were taken at a point one-foot below the pile surface and analyzed for salmonellae. As yet no samples positive for salmonellae have been found."

3.5 In own experiments according to the Swiss regulations for the determination of enterobacteriaceae (EB) after disinfection of sludge we found that in quite a considerable number of cases EB could be cultivated on selective solid medium and identified even when the Mossel-enrichment broth was negative for EB.

We also could confirm the statement of Scarpino (2) that there may be differences in disinfection rate when using laboratory stock virus strains versus indigenous virus. We found certain differencies in the temperature sensitivity between laboratory strains of Salmonella spp. when seeded into the sludge and indigenous Salmonella spp. strains isolated from the same sludge. The same was also true for strains of Streptococcus faecium (ATCC 6057) and strains indigenous of the same sludge.

In other own experiments it could be proved that indigenous fecal streptococci (FS) have a considerably lower pH-sensitivity than Salmonella spp. thus indicating that it may become necessary to choose different indicators for the hygienic control or evaluation of different sludge treatment systems.

3.6 In summarizing the literature quoted and other literature as well as own experiences it must be concluded that the problem of monitoring methods to render sewage sludge free of pathogens by looking for indicator organisms is not solved for a long time yet. Several approaches were described but all of them still have to prove their efficacy under practical conditions and for more than only one system of sludge treatment. The concerns of Scarpino (2) and others about the use of bacteria or bacteriophages as indicators for pathogenic viruses have to be considered as well as the conclusions and recommendations of Pederson (3) and the demands of de Bertoldi et al. (5) to better define the term "sanitized compost" and to lay down not only which microorganisms are to be monitored, but also their maximum permitted level in a sanitized product. For this evaluation also the results of Schrammeck and Sauerwald (11), which are of practical importance, have to be drawn into consideration.

In following the conclusions of Working Party 3 that it is in many cases advisable to analyse sludge for pathogens directly and use them to seed sludge to test the efficacy of treatment systems the described observations have to be considered that under certain circumstances differences in sensitivity of indigenous pathogens and the seeded laboratory strains may exist which can distort the results of such studies. This question has especially to be investigated in connection with the recommendation of f2-phages as indicator organisms (12, 13).

Much more practical work is to be done in this respect before final decisions can be made and standardized methods are internationally available. To achieve this easier the recommendation of Pederson (3) should be better observed in the future that researchers document carefully all pertinent aspects of their experimental design.

Since the authorities in several countries are preparing regulations for the safe utilization of sewage sludge in agriculture, gardening and possibly in forestry the politicians need definitions of parameters for hygienically safe sludges which are relating to practice, which can be controlled by the authorities and which can easily be changed when new scientific results indicate that there is a need for revision. Therefore in the following chapter we are presenting for discussion a proposal of the requirements for efficient disinfection of sewage sludge with various methods.

4. PROPOSAL FOR REQUIREMENTS OF DISINFECTING SLUDGE

I. At the time of delivery of hygienically safe sludge by the sewage treatment plant to the customer the sludge should meet the following requirements:
- the sludge should not contain more than 5×10^2 Enterobacteriaceae and 5×10^3 Fecal Streptococci per gram or milliliter,
- the sludge should not contain salmonellas (pre-enrichment) in 100 g,
- the sludge should not contain infective parasite ova.

II. For efficient disinfection of sludge the technologies used should meet the following requirements:

1. Pasteurization

Pasteurization of sewage sludge is only appropriate if raw sludge is treated prior to digestion (pre-pasteurization). Pasteurizing of digested sludge (post-pasteurization) should be abundoned because it almost inevitably results in a recontamination with regrowth of enterobacteriaceae, salmonellas and other bacteria.

Pasteurization should generally be done for 30 minutes at at least 70°C which includes a safety margin. Higher temperatures can also be applied with appropriate exposure times. Pasteurization between 60°-70°C for at least 30 min is only acceptable in combination with subsequent stabilization processes like e.g. anaerobic digestion.

2. Composting

a) Windrow (Stack)-Composting

To ensure a hygienically safe product the composting process must be operated for at least three to four weeks. During this time the temperature should exceed 65°C for at least one week.

It is most important that all of the compost mass in the windrow must reach this temperature for the recommended period of time. These requirements can be met by operating a program of regular turning of the piles or by covering them with a layer of composted material. Thus a more uniform temperature profile in the composting material can be achieved. To further improve the hygienic quality of the compost the composting process should be followed by curing the material in stacks for another three or more weeks.

b) Bioreactor-Composting

In bioreactors more uniform temperature profiles in the composting material can be attained and the temperature can be influenced by the aeration rate and the technique of feeding. For a hygienically safe product the temperature of the compost in the reactor should reach 65° - 70°C and the material should not pass this temperature range faster than in 48 hours. The flow time of the material through the reactor(s) should be in the range of two weeks or above. A subsequent curing time of at least three weeks will help to improve the technical and hygienic quality of the compost.

In bioreactors the temperature must closely be monitored and in case of disturbaces immediate counteractions by the operator are necessary. That material which has not reached the required temperataure either must be used as reflux material for another passage through the reactor or must be stored in a windrow for at least three weeks with at least one turning.

3. Aerobic-thermophilic stabilization

Aerobic-thermophilic treatment of liquid raw sludge as the sole stabilization process, combined with the aim to disinfect the sludge, should, as a rule, always be operated at least as a two-stage system with two reactors. The mean hydraulic detention time in the system should at least be five days and the minimum detention time in each of the reactors one day.

Disinfection under these technical conditions can be achieved if in the course of the whole process a minimum temperature of 55°C for at least 48 hours with a pH about 8 or 60°C for at least 24 hours with a pH distinctly above 8 can be ensured.

With single stage aerobic-thermophilic stabilization an inactivation of pathogens in sewage sludge can be achieved under certain conditions and observance of certain parameters. But it is nearly impossible in the daily practical operation to constantly maintain these necessary parameters and control them. Moreover hydraulic short-circuits resulting in still infectious effluent cannot be excluded.

Even in the two-stage aerobic-thermophilic stabilization process one has to reckon with a breakthrough of pathogens in reactor I for similar reasons as in the single-stage process which can be counterbalanced by the thermical and biochemical activities in reactor II. Therefore aerobic-thermophilic stabilization of raw sludge with the purpose of eliminating pathogens basically should only be operated as a two or more stage system.

4. Aerobic-thermophilic treatment with subsequent anaerobic digestion

When a single-stage aerobic-thermophilic treatment is used as a kind of "pre-pasteurization" for the following anaerobic mesophilic digestion process, the requirements are less stringent than those for aerobic-thermophilic stabilization because it results in a more effective and more rapid decomposition of the sludge and of the pathogens in the subsequent mesophilic anaerobic stabilization process.

For these reasons the single-stage aerobic-thermophilic treatment should be operated in the temperature range of about 55°C for 24 hours and during this time a temperatuare of 55° - 60°C should be maintained for at least one hour in order to eliminate viable parasite ova.

5. Lime treatment

a) Lime as CaO

Unslaked lime is used to solidify dewatered sludge. This results in an exothermic chemical reaction by which temperatures between 55°C and 70°C should be reached. In the inner parts of the stacks these temperatures remain for about 24 hours, thus resembling a prolonged pasteurization process. The initial pH of the sludge-lime mixture should be at 12.5 or above. For the hygienic success of the whole process the intensity and speed of mixing lime with sludge, the method of conditioning and degree of dewatering the sludge as well as the CaO concentration and quality are critical factors which have carefully to be observed.

Disinfection in this temperature and pH range is achieved within a few hours. For optimizing the process it is recommended to use an insulated and closed vessel for storage of the lime-sludge mixture.

b) Lime as Ca(OH)$_2$

Slaked lime is frequently added to liquid sludge as a conditioning step before dewatering. This will result in an increase of pH, depending on the lime dose and the sludge characteristics.

Vegetative bacterial cells and the hygienically relevant viruses are destroyed by liming of the sludge with an initial pH of 12.5 or better above. The limed sludge should be kept at the treatment plant for at least 3 days before further utilization. It does not matter whether the lime is used for dewatering sludge or just for mixing with liquid sludge as a disinfecting measure. Thorough mixing of lime and sludge is most important.

As far as parasite ova like those of <u>Ascaris</u> and <u>Taenia</u> are concerned it is possible to reduce the epidemiological risk considerably when after liming a pH of 12.5 or above was reached and a minimum storage time of two months is observed.

Limed raw sludge can also be used for anaerobic digestion without technical disturbances in the digester, even if the initial pH is above 12.5. In this case the method can be compared in its hygienic effect with "pre-pasteurization".

A dosis of 0.2 kg lime/kg DM is provisionally recommended.

6. Thermal Drying

Thermal drying generally is performed at temperataures slightly over 100°C for a long enough time to destroy vegetative bcterial cells, viruses and parasite ova. Certain bacterial spores, which are not relevant for human or veterinary epidemiology, may survive this process in reduced numbers.

Generally thermically dried sludges can be considered to be hygienically safe.

7. Irradiation

The recommendations issued by Working Party 3 of the CEC-Concerted Action "Treatment and use of sewage sludge" - COST 68 bis for the required irradiation doses for sludge disinfection are:

     500 Krad for irradiation of liquid sludge
     1.000 Krad for dewatered or dried sludge.

III. Newly developed technologies for the disinfection of sludge should be tested before they are introduced into practice and, according to the results, be approved by national or local authorities. For these tests the system should be controlled for its efficacy by using known and representative pathogens like salmonellas, parvovirus and Ascaris eggs. If this is

not feasible for certain reasons also fecal streptocci can be introduced
as bacterial test organisms. For testing the disinfecting efficacy on viruses instead of parvovirus also the f2-phage should be used to add more
practical experience to the results described by Burge et al. (12, 13).

Disinfection processes should result in a complete inactivation of
salmonellas and infective ascaris eggs, a reduction of 4 logs of parvovirus or a reduction of 4 logs of fecal streptococci (that not more than
5×10^3 g/ml can be isolated) and 15 logs of f2-phages. All the results obtained in following these recommendations of part III should be subject
to international discussions and be tessera for a final international
standardization of monitoring methods.

5. LITERATURE

(1) Concerted action treatment and use of sewage sludge. COST 68 ter; Final
report of the Community - Cost Concertation Committee; II. Scientific
Report; CEC SL/94/83 - XII/ENV/44/83, Brussels 1983
(2) SCARPINO, P.V. (1982). Selection of practical indicator systems for
monitoring the virological quality of potable water, wastewater, solid
waste, shellfish, fish and crops. 11th Conference of the International
Association on Water Pollution Research, Post-Conference Seminar 2:
Water Virology, Pretoria/SA, 5./6. April 1982.
(3) PEDERSON, D.C. (1981). Density levels of pathogenic organisms in municipal wastewater sludge - a literature review. Natl. Techn. Inform.
Service, 5285 Port Royal Rd., Springfield, VA 221 61, USA. Order No.
PB 82-102 286.
(4) DE BERTOLDI, M., COPPOLA, S. and SPINOSA, L. (1983). Health implications in sewage sludge composting. In: Bruce, A.M., Havelaar, A.H.,
L'Hermite, P. (Eds.), Disinfection of sewage sludge-technical, economic
and microbiological aspects, 165-178. D. Reidel Publ. Co., Dordrecht/
Holland.
(5) DE BERTOLDI, M., FRASSINETTI, S., BIANCHIN, L. and PERA, A. (1985).
Sludge hygienization with different compost systems. Joint Seminar CEC-
German Veterinary Medical Society: Inactivation of microorganisms in
sewage sludge by stabilization processes, Hohenheim, 8-10 Oct. 1984.
Strauch,D. and L'Hermite, P. (Eds.). In print. Elsevier Appl. Science
Publ., Barking/Essex, England.
(6) HAVELAAR, A.H. (1985). Mikrobiologische Spezifikationen für die Desin-
fektion von Klärschlamm (Microbiological specifications for the disinfection of sewage sludge). Paper to be presented at 119. Seminar of
FGU Berlin 'Umwelthygiene in der Abfallwirtschaft - Stand und Beurtei-
lung (Environmental hygiene in waste management - state and assessment).
Berlin 29-30 April 1985, (canceled).
(7) HAMMEL, H.-E. (1983). Hygienische Untersuchungen über die Wirkung von
Verfahren zur Kompostierung von entwässertem Klärschlamm und zur aerob-
thermophilen Stabilisierung von Flüssigschlamm (Hygienic investigations
on the efficacy of methods for composting dewatered sludge and for
aerobic-thermophilic stabilization of liquid sewage sludge). Veterinary
Medical Thesis, Univ. Giessen/Germany.
(8) STRAUCH, D., HAMMEL, H.E. and PHILIPP, W. (1985). Investigations on
the hygienic effect of single stage and two-stage aerobic-thermophilic
stabilization of liquid raw sludge. Loc. cit. No. 5.

(9) RÜPRICH, W. and STRAUCH, D. (1984). Technologische und hygienische
     Aspekte der aerob-thermophilen Schlammstabilisierung - System Fuchs
     (Technological and hygienic aspects of aerobic-thermophilic sludge
     stabilization - system Fuchs). Korrespondenz Abwasser, 31(11) 946-952.

(10) LANGELAND, G. and PAULSRUD, B. (1985). Aerobic thermophilic stabili-
     zation. Loc. cit. No. 5.

(11) SCHRAMMECK, E. and SAUERWALD, M. (1984). Gutachten über die hygieni-
     sche Beschaffenheit von Müll- bzw. Müll-Klärschlammkomposten aus den
     Werken in Ennepetal, Duisburg und Lemgo (Expertise on the hygienic
     quality of compost from municipal refuse or refuse/sludge produced
     in the plants of Ennepetal, Duisburg and Lemgo/Germany). Hygiene-In-
     stitut des Ruhrgebietes, Rotthauser Straße 19, D-4650 Gelsenkirchen.

(12) BURGE, W.D., COLACICCO, D. and CRAMER, W.N. (1981). Criteria for
     achieving pathogen destruction during composting. J.Water Poll. Contr.
     Feder. 53(12) 1683-1690.

(13) BURGE, W.D. (1983). Monitoring pathogen destruction. BioCycle March/
     April, 48-50.

Table 1. Density levels of organisms in raw sludge and septage
(Average geometric mean of organisms per gram dry weight) (3)

Organism	Primary	Secondary	Mixed	Septage
Total coliform bacteria	1.2×10^8	7.1×10^8	1.1×10^9	1.4×10^8
Fecal coliform bacteria	2.0×10^7	8.3×10^6	1.9×10^5	1.2×10^6
Fecal streptococci	8.9×10^5	1.7×10^6	3.7×10^6	6.6×10^5
Bacteriophage	1.3×10^5	NR	NR	NR
Salmonella sp.	4.1×10^2	8.8×10^2	2.9×10^2	5.1×10^{-1}
Shigella sp.	NR	NR	ND	NR
Pseudomonas aeruginosa	2.8×10^3	1.1×10^4	3.3×10^3	2.6×10^1
Parasite ova/cysts (total)	2.1×10^2	NR	$<5.0 \times 10^1$	NR
Ascaris sp.	7.2×10^2	1.4×10^3	2.9×10^2	NR
Trichuris trichura	1.0×10^1	$<1.0 \times 10^1$	0	NR
Trichuris vulpis	1.1×10^3	$<1.0 \times 10^1$	1.4×10^2	NR
Toxocara sp.	2.4×10^3	2.8×10^2	1.3×10^3	NR
Hymenolepis diminuta	$6. \times 10^0$	2.0×10^1	0	NR
Enteric viruses*	3.9×10^2	3.1×10^3	$3.6 \times 10^2{**}$	NR

NR = No data available
ND = None detected
* = Plaque forming units per gram dry weight (PFU/gdw)
** = $TCID_{50}$ = 50 percent tissue culture infectious dose

Table 2. Effect of sludge composting on the recovery of salmonellas, fecal coliforms and fecal streptococci (5)

MICROORGANISMS	SLUDGE + SUW		SLUDGE + RICE HULL		
	sludge	compost	sludge	compost	
Salmonella sp	2.4×10^1	absent	2.4×10^1	absent	A
Fecal coliforms	4.8×10^5	5.6×10^4	5×10^6	3×10^5	
Fecal streptococci	2.4×10^5	2.0×10^5	1.8×10^5	3×10^6	

	SLUDGE + SUW		SLUDGE-WOOD CHIPS		SLUDGE-INERTS		
	sludge	compost	sludge	compost	sludge	compost	
Salmonella sp	2.4×10^1	absent	1.0×10^5	absent	5.0×10^4	absent	B
F. coliforms	4.8×10^5	4.1×10^2	2.0×10^7	5.0×10^1	7.0×10^7	1.0×10^1	
F. streptococci	2.4×10^5	7.0×10^2	7.0×10^6	3.5×10^1	7.0×10^5	1.0×10^1	

	SLUDGE + SUW		SLUDGE + CORK		SLUDGE + STRAW		
	sludge	compost	sludge	compost	sludge	compost	
Salmonella sp	2.4×10^1	absent	3.1×10^1	absent	1.2×10^2	absent	C
F. coliforms	4.8×10^5	1.2×10^3	9.3×10^5	5.9×10^2	8.5×10^6	7.8×10^3	
F. streptococci	2.4×10^5	8.2×10^3	6.0×10^6	8.1×10^3	4.8×10^6	3.0×10^3	

A = Composting in turned piles (30 d)
B = Composting in static pile aerated by vacuum induced pressure (suction)
C = Composting in static pile aerated by forced pressure ventilation (blowing)
SUW = Solid urban waste

Table 3. Effect of sludge composting in different types of bioreactors on the recovery of salmonellas, fecal coliforms and fecal streptococci (5)

MICROORGANISMS	STARTING MATERIAL	AFTER REACTOR PASSAGE	
Salmonella sp.	absent	absent	
F. coliforms	1.4×10^7	2.5×10^6	A
F. streptococci	3.4×10^5	9.0×10^6	
Salmonella sp	absent	absent	
F. coliforms	2.2×10^8	7.8×10^3	B
F. streptococci	8.5×10^6	1.9×10^9	
Salmonella sp	absent	absent	
F. coliforms	3.1×10^8	2.5×10^5	C
F. streptococci	4.7×10^7	7.6×10^6	
Salmonella sp	1.3×10^2	absent	
F. coliforms	5.7×10^7	2.5×10^2	D
F. streptococci	8.0×10^6	1.7×10^2	
Salmonella sp	8.1×10^2	absent	
F. coliforms	7.9×10^8	9.2×10^1	E
F. streptococci	4.5×10^6	1.9×10^2	

A = Vertical reactor (Weiss). Mixture sludge + organic fraction of SUW (14 d)
B = Vertical reactor (Weiss). Mixture sludge + organic fraction of SUW (21 d)
C = Vertical reactor (BAV). Mixture sludge + organic fraction of SUW (15 d)
D = Horizontal reactor (Yokohama/Japan). Mixture sludge + composted sludge (10)
E = Horizontal reactor (Paygro/USA). Mixture sludge + sawdust + composted sludge (14 d)

Table 4. Time required for a 10-fold reduction of enteric pathogen by heat as compared with that for f2 bacteriophage (13)

Organism	D Value (Minutes)	
	55°C	60°C
Adenovirus, 12 NIAID	11	0.17
Poliovirus, type 1	32	19
Ascaris ova	-	1.3
Histolytica cysts	44	25
Salmonella*	80	7.5
Bacteriophage, f2	267	47

* Serotype senftenberg 775 W

Figure 1.
Qualitative microbiological analysis of refuse compost samples

Probe	Wassergehalt %	pH-Wert (in Eluat 1+10)	Total germ count Koloniezahl g⁻¹ (Gelatine)	Total germ count Koloniezahl g⁻¹ (Agar)	"Coliforme Keime" g⁻¹	"E.coli" g⁻¹	Achromobacter	Citrobacter	Enterobacter	E.coli	Klebsiella	Proteus mirabilis	Proteus vulgaris	Serratia	Salmonellen	Enterokokken	Gasbrand-Gruppe	Pseudomonaden	Streptomyces	Mucor	Rhodotorula	Aspergillus	Penicillium	Geotrichum candidum	kein Wachstum
Edelkompost I Werk Duisburg	25	8,4	40.000.000	70.000.000	23.000	32.000																			
Edelkompost II Werk Duisburg	30	7,9	6.000.000	80.000.000	300.000	260.000																			
Edelkompost III Werk Duisburg	29	8,2	5.000.000	25.000.000	25.000	11.000																			
Edelkompost IV Werk Duisburg	35	7,9	30.000.000	35.000.000	10.000	10.000																			
Lagerkompost I Werk Duisburg	40	8,6	7.000.000	800.000.000	13.000	16.000																			
Lagerkompost II Werk Duisburg	38	7,9	1.000.000	90.000.000	12.000	9.000																			
Lagerkompost III Werk Duisburg	29	7,7	22.000.000	45.000.000	300.000	50.000																			
Lagerkompost IV Werk Duisburg	25	7,8	2.400.000	6.000.000	7.000	4.000																			
Lagerkompost V Werk Duisburg	31	8,0	2.400.000	1.900.000	20.000	1.000																			
Sackware I Werk Duisburg	29	7,5	10.000.000	300.000.000	500	300																			
Sackware II Werk Duisburg	20	8,1	690.000	55.000.000	70	120																			

Probe	Wassergehalt %	pH-Wert (in Eluat 1+10)	Total germ count Koloniezahl g^{-1} (Gelatine)	Total germ count Koloniezahl g^{-1} (Agar)	"Coliforme Keime" g^{-1}	"E.coli" g^{-1}
Rinderdüng, getrocknet I Hersteller A	52	8,2	4.000.000	160.000.000	0	0
Rinderdüng, getrocknet II Hersteller A	49	8,5	2.000.000	1.000.000	0	0
Guano I Hersteller B	12	8,8	140	0	0	0
Guano II Hersteller B	12	8,9	10	0	0	0
organ. Blumendünger I Hersteller C	7	8,3	550	7.000	10	0
organ. Blumendünger II Hersteller C	10	7,9	1.000	2.000	150	0
organ. Gemüsedünger I Hersteller D	14	8,5	6.700	15.000	20	0
organ. Gemüsedünger II Hersteller D	n.b.	8,5	27.000	30.000	10	10
Hornmehl und -späne I Hersteller D	7	6,2	88.000	40.000	0	0
Hornmehl und -späne II Hersteller D	8	6,7	14.000	400	0	0
Bintkompostiert Hersteller D	n.b.	8,0	10	61	10	0

Qualitative columns (organisms): Achromobacter, Citrobacter, Enterobacter, E.coli, Klebsiella, Proteus mirabilis, Proteus vulgaris, Serratia, Salmonellen, Enterokokken, Gasbrand-Gruppe, Pseudomonaden, Streptomyces, Mucor, Rhodotrouls, Aspergillus, Penicillium, Geotrichum candidum, kein Wachstum

Figure 3. Qualitative microbiological analysis of soils, sand and garden compost

Probe	Wassergehalt %	pH-Wert (im Eluat 1+10)	Total germ count Koloniezahl (Gelatine) g⁻¹	Total germ count Koloniezahl (Agar) g⁻¹	"Coliforme Keime" g⁻¹	"E.coli" g⁻¹
Gartenerde Wattenscheid ohne Hilfsstoffe	15	7,9	250.000	9.000	12	10
Gartenerde Marl ohne Hilfsstoffe	10	8,1	100.000	10.000	0	0
Gartenerde Recklinghausen ohne Hilfsstoffe	20	8,0	400.000	15.000	950	0
Gartenerde Duisburg ohne Hilfsstoffe	12	8,4	570.000	230.000	0	0
Gartenerde Duisburg nach Klärschlamm- und Mistdüngung	10	8,4	160.000	950.000	300	200
Gartenerde Duisburg nach Klärschlamm- und Kompostgabe	15	7,9	160.000	126.000	7	0
Gartenkompost Wattenscheid einjährig	20	7,4	2.100.000	1.800.000	950	11
Gartenkompost Wattenscheid zweijährig	15	7,9	200.000	300.000	99	10
Gartenkompost Wattenscheid dreijährig	21	7,9	100.000	79.000	96	0
Gartenkompost Marl einjährig	14	7,8	900.000	950.000	30	0
Gartenkompost Marl zweijährig	17	8,0	250.000	1.200.000	0	0

Organism columns (presence indicated by filled squares): Achromobacter, Citrobacter, Enterobacter, E.coli, Klebsiella, Proteus mirabilis, Proteus vulgaris, Serratia, Salmonellen, Enterokokken, Coeobrand-Gruppe, Pseudomonaden, Streptomyces, Mucor, Rhodotroula, Aspergillus, Penicillium, Geotrichum candidum, kein Wachstum

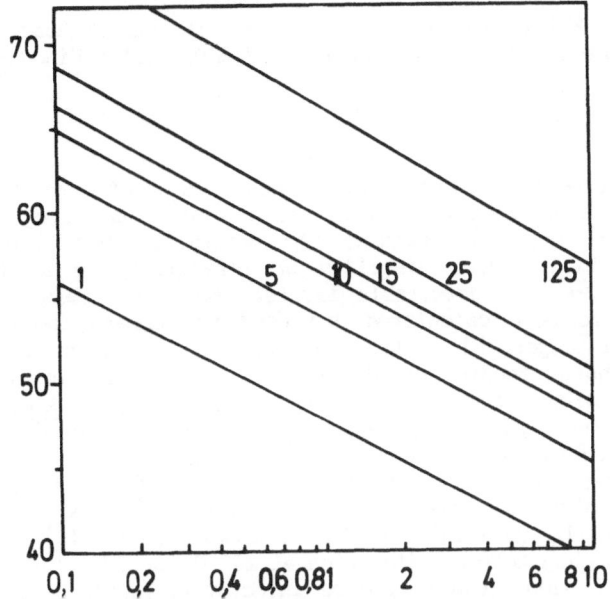

Figure 4. Curves showing the time - by - temperature regimes necessary for the inactivation of a desired number of logs of f2 bacteriophage(12)

INACTIVATION OF PARASITIC OVA DURING DISINFECTION AND STABILISATION OF SLUDGE

E B Pike and E G Carrington
Water Research Centre, PO Box 16, Henley Road,
Medmenham, Marlow, Bucks, SL7 2HD, United Kingdom.

Summary

Quantitative information upon the destruction of eggs of Ascaris spp. and Taenia saginata by thermal disinfection and during stabilisation of sludge has been examined to determine those processes which significantly reduce viability or infectivity by at least 90% or reduce them to undetectable levels. Where possible specific decay rates have been calculated and used to compare the efficacies of processes operated either batchwise or with continuous or discontinuous feeding of sludge. At temperatures above 50 $^{\circ}$C, eggs of T. saginata are more resistant than those of Ascaris spp. and complete destruction requires treatment in excess of 60 $^{\circ}$C for 3h, as in pasteurisation, treatment with quicklime or in bioreactor composting. Significant reductions in the infectivity of T. saginata ova for calves are given by anaerobic mesophilic and accelerated cold digestion, by lagooning for 28d, treatment at 52-60 $^{\circ}$C for 3h and by thermophilic oxidative digestion at 47 $^{\circ}$C for 6d. Lime treatment at pH12 will significantly reduce infectivity of T. saginata but will require at least 7d to reduce Ascaris viability significantly.

1. PARASITES IN SLUDGE AND THEIR CONTROL

The eggs of enteric helminth parasites and protozoan cysts are passed by active cases and carriers in the community and so are found in sewage. Because of their small dimensions and similar density to that of water they are only partially removed in sedimentation processes, except in sewage treatments which are largely quiescent, such as oxidation ponds. Nevertheless, they appear in greatly concentrated numbers in primary sludges compared with sewage and treated effluent. Hence, sludge will appear on a volume to volume basis to present the greatest hazard to the health of man and animals in its disposal, although the real risk will be determined by the actual numbers ingested and by such factors as the immunity of the recipient host animal and the infectivity of the parasite.

A World Health Organisation (WHO) Working Group considered that the risks to human health from agricultural use of sludge was greater for Taenia saginata and salmonellae than for the other pathogens identified. (1).

Man is the definitive host for the adult beef tapeworm, T. saginata and for the large roundworm Ascaris lumbricoides. Sewage sludge and other faecal contamination are the means by which infection is transmitted, in the former case to cattle and the latter to man. Roundworm infection by Ascaris suum is very common in intensively reared pigs but is not usually transmissible to man and infection in pigs is normally by direct contamination on the farm. With A. suum, Echinococcus species, Toxocara species and the sporozoan parasite, Toxoplasma gondii, domestic, farm and other animals are the normal or definitive hosts, although man can acquire infection. The role of sludge is therefore unimportant since the input of these parasites to the sewerage system must normally be small. The Working Group noted that the parasites for which sludge was a significant vector were T. saginata, Ascaris species, Toxocara species and sarcosporidia (1).

More recently, a COST 681 Workshop on Epidemiological Studies of Risks associated with Agricultural use of Sewage Sludge concluded that sludge spread on land could act as a vector of T. saginata and Ascaris species and that its role in the spread of Giardia, Sarcocystis, and Toxoplasma was less clear and required further study (2).

It is obvious that disinfection of sludge by methods sufficiently intense to destroy the infectivity of the distributive stages of these parasites will completely remove any hazard from agricultural use of sludge. The practical problem lies in defining a degree of treatment that will guarantee disinfection, with an adequate margin of safety, and which can be carried out reliably at an acceptable cost. Another more widely acceptable, but pragmatic, approach is to identify the ability of commonly used processes for stabilising sludge to destroy the infectivity of parasites and then to place suitable restrictions on the use to which the land receiving sludge may be put, in order to allow the infectivity to decay to an extent posing negligible risk. Such an approach is implicit in many national guidelines or laws. Since incidences of infections in the community, agricultural practices, climate and other local factors will influence the risk, it is to be expected that such recommended practices and restrictions will vary from region to region. The features of these two approaches to disease control - disinfection and stabilisation with controls on disposal - have been discused (3).

2. DETERMINING THE EFFECTS OF TREATMENTS ON INFECTIVITY OF PARASITES

The distributive stages of intestinal parasites found in sludge and sewage are either eggs in the case of the helminths considered or cysts for the protozoans. These are resting stages with much greater powers of resistance to disinfection, drying and other treatments than the adult worms or trophozoites which inhabit the body.

Ideally, the effect of disinfection and other treatments upon infectivity should only be obtained by observing the ability of the treated resting stage to cause infection in test host animals. This is usually impossible, particularly if the definitive or only host is man or very expensive if a large animal, such as cattle. Most investigators have resorted to using systems modelling infectivity. These have included microscopic examination to determine cellular integrity or disorganisation, or attempts to initiate outgrowth and hatching of infective stages from the resting stage by incubation in acid conditions resembling those in the stomach and small intestine - as with the bile/trypsin hatching test with T. saginata ova (4). The eggs of Ascaris spp. are fertilised in utero and undergo cell division subsequently after release from the female worm. After about three weeks at normal ambient temperatures, development will have proceeded as far as the infective, L2 larval stage, in which the embryonic motile worm can be seen inside the shell. Incubation of treated eggs in acidified water for three weeks is often used as the basis of assaying viability of Ascaris eggs (4).

Great care must be taken in interpreting the results of experiments to determine the 'viability' or infectivity of parasitic eggs and much of the published work is suspect. Thus comparative work has shown that the bile/trypsin hatching and activation test for viability of T. saginata ova greatly over-estimates their ability to infect calves (5,6) and is not even correlated with it (6). The results of animal infectivity tests are influenced by the innate or acquired immunity of the test animals, although the latter can be obviated by segregation of the animals from all likely sources of infection prior to test and by using animals of the same breed and age. There is a considerable degree of variation in the number of cysts produced in seemingly identical animals in a batch fed with identical numbers of T. saginata eggs (6,7) and the response is not proportional to the numbers of Taenia spp. fed to host animals (8).

The reason for these anomalies is probably because infection is not a single or even simple process. Apart from considerations of the immunity of the host animal and its state of health, the infection process begins when eggs are ingested. Enzymic activities in the gut bring about dissolution of the protective shell of the eggs, followed by attachment of the embryo to the gut wall and its migration through the animal body. If large numbers of eggs are swallowed it is likely that an immune response will result so reducing the final degree of infection.

These are analytical problems, similar in essence to those encountered when attempting to estimate numbers of bacteria by viable counting methods. Other difficulties arise when laboratory-scale models are used to determine the decay of pathogens in treatment processes and when 'surrogate' marker organisms are used instead of pathogens to indicate the efficacy of treatment. These problems will be described in later sections of this paper.

Another indirect way of determining the efficacy of treatment is by intensive surveillance of disease and by studies of case histories of outbreaks. In the general conclusions and recommendations of the EC COST Workshop on epidemiology it was noted that ..."improved surveillance and reporting of infections and diseases in human and animal populations exposed to sludge are necessary to identify risk associated with presently accepted sludge disposal strategies" (2).

3. PRINCIPLES OF DISINFECTION

The efficacy of a disinfection process depends ultimately upon two factors, the time and the intensity of exposure. It is completely unknown for all organisms exposed to disinfection to die simultaneously, hence there is no such concept in reality as 'survival time'. The term is often used, wrongly, to indicate the period of exposure, after which the analytical methods failed to detect any survivors.

To determine and to predict the degree of disinfection needs a death rate to be applied to a model of disinfection so that the effect of period of exposure can be described. 'Survival times' cannot be used to calculate death rates unless the initial population density and the analytical limit of detection are known.

The simplest disinfection model, that of exponential decay, assumes that all the pathogens are equally susceptible to disinfection and that they are equally and randomly exposed to the agency, so that a constant fraction x_t/x_o of the original population, x_o, is killed in any defined interval of time $(t - t_o)$

$$x_t/x_o = e^{-k(t-t_o)} \qquad\qquad 1$$

where k is the exponential death rate constant. Deviations from this law will occur when organisms are clumped, are of differing resistances, or when infectivity or viability are comprised of a number of vital properties each differing in sensitivity.

Another important feature, often ignored when comparing mechanical processes is the configuration of the reaction vessel or plant. There are two ideal types of reactor, the batch and the continuous stirred tank reactor (CSTR).

In the batch reactor, the material to be treated is retained for a finite process time, before being emptied. It should also be stirred so that all elements of the material are equally exposed to treatment. The effect of disinfection is directly described by Equation 1, if the special conditions apply. If the treatment is carried out in continuously-fed narrow channels or pipes, or in a large number of stirred tanks in series the treatment process is termed 'plug-flow'. If back-mixing is unable to occur, so that each element of the material receives identical exposure, then Equation 1 equally holds, if the term $(t-t_o)$ is replaced by the retention period.

In the CSTR, material is fed continuously and overflows after treatment at the same rate, expressed as the 'dilution rate', D, expressed as volume changes per unit time. The mean retention period is obviously $1/D$. The CSTR is assumed to be completely mixed so that all elements have an equal chance of being removed in any interval of time. There is thus a statistical distribution of residence times, such that the fraction of elements with a residence time equal to or greater that t is expressed as e^{-Dt}. It can be shown that after one volume change has occurred in a CSTR (the mean retention period), 63.2% of the elements originally present will have overflowed and therefore have received a less complete treatment than they would if exposed in a batch reactor for a similar process time. It follows that the CSTR provides a much less effective degree of disinfection than the batch or plug-flow reactor and, in general, CSTR's are completely unsuitable for processes where disinfection is the primary aim.

The disinfection performance of a CSTR can be described, if it is assumed to be in steady operation so that a number balance can be applied to the pathogens entering, leaving and dying.

$$\text{Rate of output} = \text{Rate of input} - \text{Rate of death}$$

$$Dx = Dx_o - kx \qquad\qquad 2$$

where x_o and x are the populations entering and leaving the reactor respectively. Since the mechanisms and the kinetics of death are unchanged by the type of reactor under ideal conditions, the death rate constant, k, has the same meaning in Equations 1 and 2. It is therefore permissible to determine k experimentally in one type of reactor and apply in to predict the efficiency of another.

In practice, continuous reactors, such as anaerobic digesters are fed discontinuously, have unmixed or dead spaces, perhaps filled with grit, which increase their effective dilution rate and may be poorly mixed, so that Equation 2 does not hold completely.

If the reactor is fed batch-wise so that a certain proportion is replaced at periodic intervals, then its efficiency in disinfection will be intermediate between that of the batch reactor and the CSTR. The more frequently it is fed, relative to the mean retention period, then the more closely its efficiency will approach that of the CSTR. If the fraction of the reactor volume replaced at each feeding is R and an equivalent volume is removed before feeding, then the fraction x/x_o of the original population, x_o, surviving after treatment is given by

$$x/x_o = R/(e^{kR/D} -1+R)$$

In practice, stabilisation processes, such as anaerobic mesophilic digesters, are fed batch-wise either at daily intervals, as at smaller works, or semi-continuously, as at larger works. Thus, their disinfecting effect is best described by Equation 3 or Equation 2 respectively. It will be obvious that the disinfecting effect will be reduced with batch-feeding, if fresh sludge is fed before an equivalent amount is withdrawn. Sludge pasteurisation plants always treat separate batches of sludge or are continuous with plug-flow characteristics ensuring that each element of sludge receives a defined degree of treatment, eg 30 minutes at 70 ^{o}C. In this case, Equation 1 will apply. Composting in lower processes, or windrows, lime treatment, stockpiling of sludge and cold anaerobic digestion processes will also be described by Equation 1. Where sludge is lagooned, the retention characteristics are usually poorly defined unless the lagoon is in the form of tanks in series or a long channel preventing mixing and short-circuiting.

The effects of various sludge treatment processes upon the eggs of Ascaris and T. saginata will now be considered, with the aim of predicting how efficient full-scale processes should be at destroying them. Not all published work can be used predictively, either because unsuitable, inaccurate methods have been used to measure decay or because the data is not quantitative.

There remains the problem of interpretation of risk, because risk cannot be directly related to numbers of parasites surviving. The United States Environmental Protection Agency has categorised sludges and the way in which disposal to land is regulated, according to the degree of pathogen reduction expected during treatment. Thus, a 'process to significantly reduce pathogens' must achieve 90% reduction in pathogens (type unspecified) or 99% reduction in faecal coliform bacteria and the operating conditions necessary for such processes to achieve this are specified. Greater degrees of disinfection are achieved by other defined processes 'to further reduce pathogens' which presumably are able to reduce pathogens to undetectable levels (9). These definitions are convenient, but arbitrary and will be used in this paper for interpreting published data.

4. ASCARIS

The eggs of the pork roundworm (Ascaris suum) and of the human roundworm (Ascaris lumbricoides) cannot be distinguished microscopically. Both are usually, but not always, specific for their respective host. However both species can be present in sludge, particularly if there is drainage from pig-rearing units, markets and abattoirs. The fertilised eggs, released by the female worm and passed with faeces, undergo cellular development to the L2 larval stage in the external environment and can only re-infect the host upon ingestion once this stage is reached. The eggs remain infective for many months in soil and water and are very resistant to chemical disinfection.

Table 1 shows that only processes in which the eggs are subjected to heat of 50 ^{o}C or greater can reduce viablility to below detectable limits. The temperature is critical. Thus, exposure of eggs to 45–55 ^{o}C in the Bauart-Weiss bioreactor for 72h failed to destroy viability completely (10). Pasteurisation of eggs in sludge at 45 ^{o}C for 3h had no significant effect on viability, whereas it was destroyed after 15 min at 55 ^{o}C (11). Indeed if the data of Carrington (11) are examined further by a plot of the natural logarithm of the specific death rate (ln k) upon reciprocal absolute temperature (Arrhenius plot) it can be deduced that the energy of activation for the disinfection reaction (the slope of the curve) progressively decreases with increasing temperature. A similar phenomenon was found for loss of viability in vitro of T. saginata eggs (4). While the theoretical interpretation is speculative (eg that death is caused by starvation at lower temperatures and chemical denaturation of cell components at higher temperatures), the practical significance is clear – that temperature control must be precise over the whole reactor volume and must allow an adequate safety margin.

The processes shown in Table II are those which significantly reduce (ie by at least 90%) viability. Lime treatment gave a progressive loss in viability, significant at pH 12.5 and 11.6, but not at 7.7 and 10.6 in the experiments of Schuh et al. (17) and in the experiments of Polprasert and Valencia (18) at pH 9, 10, 11 and 12 – in which eggs of A. lumbricoides in infected faeces were used. Only at 12.5 was the rate of disinfection, corresponding to 90% loss in viability in 7.4d, high enough to make lime stabilisation an effective method for disinfecting sludge containing Ascaris eggs.

All the processes in Tables I and II are batch or plug-flow processes with the exception of anaerobic digestion at 49 ^{o}C, which is not a practical proposition, since gas production and control is poor.

5. TAENIA SAGINATA

Although the in vitro test for viability of Taenia eggs is relatively simple and rapid compared with tests for infectivity towards calves, which involve slaughter and dissection of carcases, 2–3 months after infection, viability bears no simple relationship to infectivity and always over-estimates infectivity (5,6). Apart from cost, infectivity experiments present several difficulties. In countries where tapeworm infection is rare, supply of eggs for experiments is difficult. Only those proglottides voided by cases or a few proglottides at the distal end of the worm contain infective ova and, since eggs are continually expelled from 'ripe' proglottides, few contain a full complement of eggs (50–80 thousand).

During storage at 4 ^{o}C, infectivity is progressively lost, so that fresh eggs must be used and that where the sludge treatments involve exposure of the eggs for several weeks, as with anaerobic digestion, control animals must be infected at the start and conclusion of each experiment in order to separate the effect of the experimental treatment from that of decay in storage. Thus during experiments conducted by Severn-Trent Water Authority (STWA) and the Institute for Research on Animal Diseases (IRAD) (see Reference and Acknowledgements), storage of eggs for 25, 28 and 50 days respectively caused a decline in numbers of cysts produced from a given number of eggs to 70, 76 and 0.96% of those produced in initial control infections. This represents a specific decay rate of 0.18d^{-1} (or 90% loss in infectivity in 13 days).

The relationship between numbers of eggs fed to susceptible animals and the count of cysts produced is not linear and even with the same doses to animals of identical breed and ages, is extremely variable

Table I. Processes giving complete destruction of the viability of
Ascaris eggs*

Process and conditions	Authors (and reference)
Aerobic thermophilic stabilisation, Fuchs system; 49.5 °C, 57h, pH 7.5	Strauch (12)
Aerobic, thermophilic stabilisation, Thieme system; >53 °C, 3-4d.	Strauch, Hammel and Philipp (13)
Bioreactor composting: Kneer system; 50% sludge, 25% sawdust, 25% recycled compost, 75 °C max., 24h; Bauart-Weiss system; sludge 20% sawdust 30%, recycled compost, 50%, 67 °C, 24h.	Strauch, Berg and Fleischle (10)
Pasteurisation: batch system; 60 °C and 65 °C, 30 min; 70 °C, 5 min.	Strauch and Berg (14)
Pasteurisation: batch, 55 °C, 15 min	Carrington (11)
Quicklime treatment (calcium oxide): to produce 62 °C in first 2h, pH 13.4.	Strauch and Berg (15)

* Eggs of Ascaris suum added to sludge (11) or in sealed ampoules or
   'germ carriers' into the treatment process. Viability estimated by
   ability to embryonate on incubation and/or ability to produce L3
   larvae in livers of mice.

Table II Processes significantly reducing the viability of Ascaris
eggs.

Process and conditions	Survivor ratio*	Specific death rate,* k (d^{-1})	Authors (and reference)
Anaerobic digestion; 4-1 digester fed daily, 49 °C:			
Retention 10d	<0.0043	>1.7	Carrington
16d	0.0043	1.3	and Harman (16)
20d	0.0043	1.1	
Pasteurisation: batches, 53 °C, 1h	0.011	4.5	Carrington (11)
Lime treatment; batches, 29 °C pH 12.5, 7.4d	0.10	0.25	Schuh, Philipp and Strauch (17)

* Survivor ratio is fraction of treated eggs recovered from sludge able
   to embryonate compared with untreated eggs; k calculated from Equation
   3 (batch-fed) or Equation 1

(6,8). Analysis of data relating yield of cysts (y) to the number of eggs given (x), STWA/IRAD experiments (6,7), yielded the relationship

$$y = 0.3886x^{0.7075}$$

This shows that production of a single cyst would require ingestion of 6 eggs and that 10^5 eggs would yield 1300 cysts on average. It therefore follows that generalised infections with cysticerci must have resulted from ingestion of one or more proglottides. Similarly, although a threshold for infection exists it is low, compared, for example, with the thousands of salmonellae required to set up clinical disease.

Few infectivity experiments have been carried out quantitatively. Silverman and Guiver's (20) experiment showed that eggs in sludge withdrawn after batch digestion at 35 °C for 5, 10, 15 and 20 days, pooled and fed to a single calf were not infective. Shafai showed that aliquot samples of 20 000 eggs treated in water (not sludge) for 15 min at 60 °C, 70 °C for 5 min and 80 °C for 2.5 min respectively were rendered non-infective, compared with controls(5).

Table II summarises the results of infectivity experiments carried out by STWA and IRAD. The experimental conditions have been described (6,7). Results for those processes which did not significantly reduce infectivity - anaerobic mesophilic digestion for 24h at 35 °C (survivor ratio, 1.0) and aerobic stabilisation at 7 °C in an oxidation ditch 'package' plant (0.16) are omitted. The eggs in these experiments were exposed to the treatments for specified times, so that if first-order kinetics were obeyed, disinfection would be described by Equation 1. Some of the processes are carried out full-scale with continuous or periodic feeding. As explained in section 3, this considerably reduces the efficiency of disinfection, as shown by the calculation of survivor ratio assuming the reactor to be a CSTR. Many processes are fed only daily and if the treated liquor is withdrawn before feeding, the efficiency is slightly improved, but only when the retention period is short and the death rate high. The difference between the calculated survivor ratios is appreciable only for thermophilic oxidation, eg 0.11 for the CSTR mode and 0.059 for daily feeding, with 6 days retention in each case.

The difference between treatment at 60 °C in water, in which infectivity was destroyed in 15 min (5) and at 60 °C in sludge in which 1.1% of infectivity remained (albeit for only 4 cysts per animal for a dose of 4850 eggs) emphasises that experiments should be carried out in sludge to be representative.

The data of Table III on analysis show, as noted in Section 4 for Ascaris eggs, that the specific death rate is not linearly related to temperature. The Arrhenius plot of ln specific death rate on reciprocal absolute temperature shows that the energy of activation, associated with disinfection, decreases as temperature rises. This confirms the need for careful control of temperature during pasteurisation.

6. CONCLUDING DISCUSSION

Comparison of Table I-III shows that the death rates of Ascaris eggs are much lower than those of T. saginata at temperatures below about 50 °C, whereas the latter are more resistant above 50 °C. Thus it must be assumed that for a thermal process to kill both organisms, the aim must be to kill T. saginata. This would appear from the evidence given to require conditions in excess of 60 °C for 3 hours, although the

Table III. Processes significantly reducing the infectivity of
T. saginata ova for calves*.

Process and conditions	Survivor ratio observed	Specific death rate, k (d^{-1})	Survivor ratio for CSTR
Accelerated cold digestion; 3 °C, 50d.	0.076	0.052	0.27
Anaerobic mesophilic digestion, 35 °C:			
15d retention	<0.00059	>0.50	<0.12
30d retention	0.0053	0.17	0.16
10d + secondary digestion 15d	0.0011	-	-
Lagoon storage; 7 °C, 28d	0.018	0.14	
Lime stabilisation, pH12 initally			
24h	0.041	-	-
48h	0.050	-	-
Pasteurisation, 3h:			
52 °C + digestion 35 °C, 1d	0.041	-	-
55 °C	0.014	34.0	
60 °C	0.011	36.0	
Thermophilic oxidative digestion; 49 °C, 6d	0.00036	1.3	0.11

* From results of STWA/IRAD experiments, some published (6,7). Ova
suspended in microporous carriers in reactors. Survivor ratio
observed calculated as ratio of number of cysts produced from treated
eggs to that initially produced from same number of untreated eggs.
Specific death rates calculated from Equation 1. Where the full-scale
process is continuously-fed (CSTR), survivor ratios have been
re-calculated from Equation 2.

experiments of Table III did not determine conditions needed to reduce
infectivity below detectable limits. Table III shows that many processes
commonly used to stabilise sludge will also bring about significant
reductions in infectivity of T. saginata eggs for calves. Particularly
interesting, because of their suitability for use at small sewage works
are lagoon storage, accelerated cold digestion and lime treatment at
pH12. It will be noted that lime treatment has less effect upon Ascaris
eggs and it may be inferred that the initial pH value should not be less
than 12 and that if Ascaris infection is common amongst the human
population, sludge will need storing for longer that 24h, perhaps 7-10d,
to achieve significant reduction. The efficacy of mesophilic anaerobic
digestion in reducing infectivity of T. saginata eggs is confirmed by
Table III.

Lastly, the results presented confirm the views of Working Party 3
expressed at the 3rd International Symposium on Processing and Use of
Sludge on the use of indicators of efficacy of treatment (21). It was

recommended that in the initial development of new processes, tests of efficiency should be carried out with pathogens rather than with indicators. Thus the new processes of accelerated cold digestion and of thermophilic oxidative digestion have been validated in Table III with T. saginata, rather than with Ascaris ova as a more easily obtainable and easier to use indicator. In this report of Working Party 3 an attempt was made to express disinfecting ability in kinetic terms to enable comparisons of resistance to be made. This approach has been continued in this paper to enable comparisons to be made between small-scale batch processes and large continuous and semi continuous processes.

ACKNOWLEGEMENTS

The data of Carrington (4,11) and that of Severn-Trent Water Authority and the Institute for Research on Animal Diseases (6,7) was obtained partly under funding from the UK Department of the Environment (Contracts Nos. PECD 7/7/062, PECD 7/7/084 and PD 77/126-50/84) and the Water Research Centre, Environment Directorate and is published with their permission.

REFERENCES

(1)   The Risk to Health of Microbes in Sewage Sludge Applied to Land. (1981) Report on a WHO Working Group, Stevenage, 6-9 January. EURO Reports and Studies No. 54. World Health Organisation, Copenhagen, 1981, 27 pp.

(2)   HAVELAAR, A.H. and BLOCK, J.C. (1985). Epidemiological studies related to the use of sewage sludge in agriculture. In Proceedings of Fourth International Symposium, Processing and Use of Organic Sludge and Liquid Agricultural Wastes, Rome, 8-10 October, 1985. Commission of the European Communities, in press.

(3)   PIKE, E.B. and DAVIS, R.D. (1984). Stabilisation and disinfection - their relevance to agricultural utilisation of sludge. In Sewage Sludge Stabilisation and Disinfection (ed A.M. Bruce) Ellis Horwood, Chichester, 61-84.

(4)   PIKE, E.B., MORRIS, D.L. and CARRINGTON, E.G. (1983). The inactivation of ova of the parasites Taenia saginata and Ascaris suum during heated anaerobic digestion. Wat. Pollut. Control. 82, 501-509.

(5)   SHAFAI, H. Studies on Some Aspects of Bovine Cysticercus bovis. Ph.D. Thesis, University of Dublin, 1975. Cited by HANNAN, J. In Characterisation, Treatment and Use of Sludge, 2nd Europ. Symp., Vienna, 21-23 October 1980. D. Reidel Publishing Co., Dordrecht, 1981, 330-349.

(6) HUGHES, D.L., MORRIS, D.L., NORRINGTON, I.J. AND WAITE, W.M. (1984). The effects of pasteurisation and stabilisation of sludge on Taenia saginata eggs. In Inactivation of Micro-organisms in Sewage Sludge by Stabilisation Processes ed D. Strauch, A.H. Havellar and P. l'Hermite. Elsevier Applied Science Publishers, London, pp 123-134.

(7) MORRIS, D.L., HUGHES, D.L., HEWITT, R.J. and NORRINGTON, I.J. (1982). The effects of sludge stabilisation and treatment processes on the viability and infectivity of beef tapeworm eggs. Water pollut. Control (in press).

(8) GEMMELL, M.A. (1977). Experimental expidemiology of hydatosis and cysticercosis. Adv. Parasitol. 15, 311-369.

(9) UNITED STATES ENVIRONMENTAL PROTECTION AGENCY (1984). Environmental Regulations and Technology. Use and Disposal of Municipal Wastewater Sludge. EPA 625/10-84-003. US EPA Intra-Agency Sludge Task Force, Washington DC, pp. 21-22.

(10) STRAUCH, D., BERG, T. and FLEISHCLE, W. (1980). Mikrobiologische Untersuchungen zur Hygienisierung von Klärschlamm. 4. Untersuchungen an Bioreaktoren. GWF-Wasser/Abwasser, 121, 331-334.

(11) CARRINGTON, E.G. (1984). Pasteurisation: effects on Ascaris eggs. In Inactivation of Micro-organisms in Sewage Sludge by Stabilisation Processes ed D. Strauch, A.H. Havellar and P. l'Hermite. Elsevier Applied Science Publishers, London, pp 121-125.

(12) STRAUCH, D. (1980) Mikrobiologishe Untersuchungen zur Hygienisierung von Klärschlamm. 7. Weitere Untersuchungen an einem Verfahren zur aerob-thermophilien Schlammstabilisierung. GWF-Wasser/ Abwasser, 121, 552-555.

(13) STRAUCH, D., HAMMEL, H.E. and PHILIPP, W. (1985). Investigations on the hygienic effect of single stage and two-stage aerobic-thermophilic stabilisation of liquid raw sludge. In Inactivation of Micro-organisms in Sewage Sludge by Stabilisation Processes, ed D. Strauch, A.H. Havelaar and P. l'Hermite. Elsevier Applied Science Publishers, London, pp. 48-63.

(14) STRAUCH, D. and BERG, T. (1980) Mikrobiologische Untersuchungen zur Hygienisierung von Klärschlamm. 2. Versuche an Pasteurisierungsanlagen. GWF-Wasser/Abwasser, 121, 184-187.

(15) STRAUCH, D. and BERG, T. (1980). Mikrobiologische Untersuchungen zur Hygienisierung von Klärschlamm. 6. Untersuchungen an einem Verfahren zur Schlammverfestigung mit Branntkalk. GWF-Wasser/ Abwasser, 121, 493-495.

(16) CARRINGTON, E.G. and HARMAN, S.A. (1984). The effects of anaerobic digestion, temperature and retention period on the survival of Salmonella and Ascaris ova. In Sewage Sludge Stabilisation and Disinfection, ed A.M Bruce. Ellis Horwood, Chichester, pp 369-380.

(17) SCHUH, R., PHILIPP, W. and STRAUCH, D.L. (1985). Influence of sewage sludge with and without lime treatment on the development of <u>Ascaris suum</u> eggs. In Inactivation of Micro-organisms in Sewage Sludge by Stabilisation Processes, ed. D. Strauch, A.H. Havelaar and P. l'Hermite. Elsevier Applied Science Publishers, London, pp 100-120.

(18) POLPRASERT, C. and VALENCIA, L.G. (1981). The inactivation of faecal coliforms and <u>Ascaris</u> ova in faeces by lime. <u>Wat. Res</u>, <u>15</u>, 31-36.

(19) JEPSEN, A. and ROTH, H. (1952). Epizootology of <u>Cysticercus bovis</u> - resistance of the eggs of <u>Taenia saginata</u>. Report of the XIth International Veterinary Congress, London, 8-13 August 1949, Vol II. His Majesty's Stationery Office, London, pp 43-50.

(20) SILVERMAN, P.H. and GUIVER, K. (1960). Survival of eggs of <u>Taenia saginata</u> (the human beef tapeworm) after mesophilic, anaerobic digestion. J. Proc. inst. Sew. Purif. <u>3</u>, 345-347.

(21) PIKE, E.B. (1984). Indicators of pollution and efficacy of treatment: significance and methodology. In Processing and Use of Sewage Sludge, ed. P. l'Hermite and H. Ott. D. Reidel Publishing Co., Dordrecht, pp. 213-219.

EPIDEMIOLOGICAL STUDIES RELATED TO THE USE OF SEWAGE SLUDGE IN AGRICULTURE

A.H. HAVELAAR[1] and J.C. BLOCK[2]

[1] National Institute of Public Health and Enviromental Hygiene,
P.O. Box 1, 3720 BA Bilthoven, the Netherlands
[2] Centre des Sciences de l'Environnement, Université de Metz,
1 Rue des Récollets, 57000 Metz, France

Summary

The present knowledge and future research needs for epidemiological
studies related to the use of sewage sludge in agriculture were
discussed in a Workshop, organized by COST 681 Working Party 3, in
Metz (F) from May 21–23, 1985. The discussions considered four
separate themes: bacteria, parasites, viruses and occupational
hazards.   One conclusion, common to all themes, was that there exists
a need for better surveillance of sludge exposed (human or animal)
populations and adequate reporting of surveillance data. These data
would give a continuous flow of information, and could indicate the
need for more detailed epidemiological studies in particular cases or
areas.
Such studies were at present considered necessary only with regard to
parasites, because it was concluded that the guidelines proposed in
the draft EEC directive were not always sufficient to prevent
parasitic infections.
Sufficient knowledge was present with regard to risk assessment and
prevention of bacterial infections.   Little was known about viral
infections or occupational hazards related to land use of sludge, but
epidemiological studies were not recommended because of the complexity
of the design and the small chance to obtain conclusive results.
It was again stressed that adequate prevention of sludge transmitted
infections requires measures that depend on local situations with
regard to such factors as prevalent infections, type of animal
husbandry, availability of land, climate, etc. Guidelines, in
particular those from supranational bodies such as EEC, should take
this into account and local authorities should, where necessary,
require further steps to be taken.

1. INTRODUCTION

Working Party 3 of EEC COST Concerted Action Project 681 concerns
itself exclusively with pathogenic microorganisms in sewage sludge. It has
been well established that a great variety of pathogenic microorganisms
(bacteria, parasites, viruses, fungi, etc.) may be present in sludge, the
actual numbers depending on a great many factors such as country, prevalent
infections among the human population, season, industrial wastes (e.g.
slaughterhouses), etc. Therefore, the use of sewage sludge poses a health
hazard to man and animals. In this context, a hazard [1] is defined as the
potential for adverse health effects.   Whether there indeed is a
significant health risk (i.e. a distinct possibility for infections to
occur) depends on factors such as the nature of the pathogen, its actual

numbers, the sludge disposal practice, the use made of the land and other geographical, climatological and demographic factors. It is the actual risk that defines whether particular sludge disposal practices are acceptable from a hygienic point of view and guidelines should preferably be based on risk assessment.

Risk assessment of sludge use in agriculture has since long been a matter of much debate and as yet no consensus has been reached on desirable preventive hygienic measures. To further discuss and evaluate this problem, COST 681 Working Party 3 organized a Workshop (21-23 May, 1985, Metz, France) under the title: Epidemiological studies of risks associated with agricultural use of sewage sludge: knowledge and needs. This Workshop brought together a multidisciplinary group of 39 scientists from 12 countries with a wide range of expertise in problems on sludge hygiene.

The objectives of the Workshop were to consider carefully available epidemiological studies and case reports, particularly from a methodological point of view and to consider the risks associated with sewage sludge in the overall epidemiology of certain infectious diseases. Furthermore, recommendations were to be made as to which epidemiological studies should be carried out in future.

The presently available knowledge was presented in review papers and subsequently discussed in panel discussions, which set out to answer questions on future research needs. The panel discussions were organized in parallel on four themes: bacteria, parasites, viruses and occupational hazards. The conclusions from the panel discussions will be briefly presented below. Full details of the Workshop can be found in the proceedings, that shall be published early 1986.

2. BACTERIA

Traditionally, concern about pathogenic bacteria in sludge has been almost exclusively directed towards the genus Salmonella and this emphasis was again confirmed in the Workshop. It was concluded that other bacteria were of little concern as far as transmission by sludge was concerned, but that there was ample documentation in the case of salmonellosis. In particular, the following outbreaks were considered.
- Scotland: 23 incidents involving sewage effluents (10 incidents), septic tank effluent (8 incidents), sewage sludge (3 incidents), abattoir effluent (2 incidents), [2].
- Scotland: detailed account of one of the above outbreaks involving 27 cattle and 98 human cases consuming raw milk. Sludge containing slurry from a chicken factory was spread on grassland and cattle allowed to graze almost immediately [3]. Rare S.typhimurium RDNC isolated from cattle, pasture, sludge and a poultry factory.
- U.K.: leakage of septic tank waste from household with S.paratyphi B carrier on to field. Five cattle infected on farm. Contamination of water supply and failure of chlorination; 90 cases in village supplied from borehole in field [4].
- U.K.: 30/90 cattle infected with S.aberdeen. Raw sewage from cracked sewage pipe overflowing on the pasture [5].
- Switzerland: nine cattle infected with rare S.tokoin fed with grass cut from pasture treated 4 weeks earlier with sludge; serotype isolated from pasture 7 weeks after sludging. Cattle did not graze this pasture, so cross-contamination was not involved [6]. The dried sludge remaining on the pasture yielded S.tokoin. Other feedstuffs were negative for salmonellas.
- Switzerland: five other outbreaks of salmonellosis involving use of sludge on grassland.

It was noted however that these outbreaks occurred before the publication of national guidelines and that they might not have occurred if rules such as those proposed in the Draft EEC Directive on the use of sewage sludge in agriculture had been obeyed. Preventive measures in the Directive include the requirement for stabilization, which brings about a significant, albeit variable reduction of viable salmonellae, and a no-grazing interval between sludge spreading and the admission of cattle of at least 6 weeks. Under most circumstances, these measures may be considered to be effective but, depending on local conditions they might be supplemented with national or local guidelines which may be more restrictive. Such guidelines might include the need for disinfection to eliminate positively any hazard.

The Workshop considered the available information on salmonellosis and its control with regard to sludge adequate, and therefore no experimental epidemiological studies were considered necessary. It was emphasized however that there was a high priority need for recording and publishing details of outbreaks, which could eventually lead to amendment of national or international guidelines.

3. PARASITES

Contrary to the situation with bacterial infections, the spectrum of parasites to be considered in relation to sludge use in agriculture is very broad and fundamental information is largely absent. Parasites of concern include the nematodes of the Ascaris spp., the trematode Taenia saginata, the coccidia Sarcocystis and Cryptosporidium and possibly the protozoans, Entamoeba, Giardia and Toxoplasma. It was agreed that the only absolutely safe practice was disinfection of the sludge before application to land by processes such as pasteurization or composting, provided proper control on the efficiency of the process was exercised. On the other hand, improperly handled sludge has been documented as a vector of such parasitic infections as ascariasis and taeniasis in the past and might still be so in the future. The proposed EEC Directive was not considered to be safe from the parasitological point of view. In particular, concern was expressed on the exclusion of treatment plants of a small processing capacity, on the effectiveness of soil injection in eliminating the risk of transmission of parasitic infections and on the highly variable effect of different stabilization processes on viability and infectivity of parasitic stages. In order to gain more information and possibly recommend further safety measures, the following was recommended.
- Development of guidelines for parasitological investigation of sewage sludges throughout Europe with particular reference to the types, concentrations and origin (human or animal) of the various parasites.
- Epidemiological studies on the role of sludge in the epidemiology of parasitic diseases.
- Surveillance of any parasitic diseases eventually associated with the use of sludge under the proposed Directive.

4. VIRUSES

Though much is known about the occurrence of viruses in sewage or sludge, many questions remain unanswered. One major obstacle is the inadequacy of present methods to accurately measure and identify viruses in environmental samples, in particular the lack of detection methods for epidemiologically relevant viruses like the hepatitis-A virus or gastro-enteritis viruses (rota, calici, astro, parvo and the group of Norwalk agents). Presently detectable viruses (predominantly in the group of enteroviruses) are of little epidemiological concern in the field of sewage and sludge hygiene and should be regarded as indicator viruses for human faecal pollution. It is questionable however in how far the resistance of these indicator viruses is relevant for other viruses, and consequently what value should be attributed to their inactivation during sludge processing and application.

While the role of sewage as a vector of viral infections by various routes (shellfish, drinking water, recreational water) has been amply documented, little information is available on sludge. Only two case-histories were identified, one on milk-borne hepatitis-A in Czechoslouwakia originating from pollution of process water with cesspool contents [7] and one on hepatitis-A among sludge spreaders [8].

It was considered practically impossible however, to design and imply meaningful epidemiological studies with regard to viruses in sludge. Firstly, there is no specific association between infection and disease (low percentage of clinically manifest cases, wide range of symptoms for one virus and same symptoms for different viruses). Secondly, such studies might be immensely costly and even then lead to inconclusive results.

It was emphasized that an effective system of surveillance and reporting should be developed and preferably be organized on an international basis. For the time being, preventive measures should be based on studies regarding the nature and survival of culturable viruses in sludge, in particular the efficiency of various sludge treatment processes.

A particular point of concern for the virologists in the Workshop was the increasing use of commercially available (sludge derived) compost for private gardens, without any regulation with regard to its hygienical quality.

5. OCCUPATIONAL HAZARDS

The panel on occupational hazards considered the possible adverse health effects of microbiological origin for people who by their work come into contact with sludge. It was realized however, that in some situations (e.g. at a sewage treatment plant) it was difficult to discriminate the effects of sludge and wastewater. Health risks identified and documented in the literature were infections with hepatitis-A virus [8] among sludge spreaders, and Aspergillosis infections among compost workers. In the latter case, the infections are more likely to be connected with the composting process per se than with sludge. In a study in Ohio (USA) no difference was detected in the health status of farming families using sludge on land or controls. This implies that a health risk was either absent or too small to be detected by the study as designed.

In view of the complex design of meaningful studies in this field, these were not considered of high priority. Alternatively, an effective system of surveillance of the health status of the exposed population was recommended, together with the implication of a code of good hygienic practice. Surveillance data might indicate the need for further, more detailed epidemiological studies.

6. GENERAL CONCLUSIONS

The participants in the Workshop have not been able to give quantitative estimates of actual risks related to sludge use in agriculture. Available epidemiological data were either on incidents where relevant guidelines were not yet effective or not followed, or - in the case of prospective studies - often inconclusive. Nevertheless, remarkable consensus was obtained on the definition of obviously hazardous and safe procedures, and it should be possible to base risk management of sludge disposal techniques on the opinions as expressed in the panel discussions.

Improved surveillance and reporting of infections and disease in human and animal populations exposed to sludge are necessary to identify risks associated with presently accepted sludge disposal strategies. Criteria should be developed for designing and implementing appropriate surveillance systems. Surveillance may then indicate specific situations for epidemiological studies.

Presently, the Workshop has identified a need for expertly designed epidemiological studies only in the field of parasitic infections. For other pathogens (e.g. salmonellae) it was felt that adequate knowledge existed or, if this was not so, the designing, conducting and interpreting of prospective studies might not be possible and/or the cost might be prohibitive.

Reduction of pathogens in sewage sludge before disposal may decrease risks because it reduces exposure. However, there are many other transmission routes of pathogens and the disease incidence associated with sludge disposal is not known. Therefore, it cannot be predicted whether or not specific treatment and disposal guidelines will result in a decrease of disease incidence at a national level.

The primary aim of sewage sludge disposal guidelines is to provide a product which is safe for use in agriculture. The present EEC directive is not considered adequate to prevent spread of infectious microorganisms under all circumstances. Therefore, in certain regions additional precautions or treatment may be required and further studies under local circumstances are needed to define these needs. Such studies would concern fate and transport of pathogens in the environment including groundwater and health effects on exposed populations (human or animal).

ACKNOWLEDGEMENTS

Although the opinions expressed in the paper are those of the authors, the presentation has been based largely on the reports of the panel discussions prepared by E.B.Pike (United Kingdom), J.Hannan (Ireland), E.Lund (Denmark) and J.C.Block (France).

REFERENCES

[1] The risk to health of microbes in sewage sludge applied to land. Report on a Working Group. World Health Organization, Regional Office for Europe, Copenhagen, 1981, p. 8.

[2] REILLY, W.J. et al. (1981). Human and animal salmonellosis in Scotland associated with environmental contamination, 1973–1979. Vet. Rec., Vol. 108, 553–555.

[3] BURNETT, R.C.S. and MAC LEOD, A.F. (1980). An outbreak of salmonellosis in West Lothian. Communicable Disease Scotland, Vol. 14, 6–8.

[4] GEORGE, J.T.A. et al. (1972). Paratyphoid in man and cattle. Br. Med. J., Vol. 3, 208–211.

[5] BICKNELL, S.R. (1972). Salmonella aberdeen infection associated with human sewage. J. Hyg. Camb., Vol. 70, 121–126.

[6] HESS, E. and BREER, C. In: Proceedings of symposium Radiation for a Clean Enviroment, Munich, 1973. International Atomic Energy Agency, Geneva, pp. 203–208.

[7] RASKA, K. et al. (1966). A milk-borne infectious hepatitis epidemic. J. Hyg. Epidemiol. Microbiol. Immunol., Vol. 10, 413,427.

[8] TIMOTHY, E.M. and MEPHAM, P. (1984). Outbreak of infectious hepatitis among sewage sludge spreaders. Comm. Dis. Rep., 20th January, 3.

TRANSPORT OF VIRUSES FROM SLUDGE APPLICATION SITES

P.H. JØRGENSEN and E. LUND
Department of Veterinary Virology and Immunology
Royal Veterinary and Agricultural University of Copenhagen
13 Bulowsvej, 1870 Frederiksberg C, Denmark

Summary

The information on virus in sludge is reviewed. Sludge, even digested sludge, can contain probably up to 10^4 - 10^5 infectious units of enteric viruses. The viruses that may be present are especially the enteroviruses (polio-, coxsackie-, echo- and infectious hepatitis virus), adenoviruses, and presumably rotaviruses. The enteric viruses are relatively resistant towards inactivation in the environment and may persist for several months in sludge. They do not withstand severe drying. Usually the virus is adsorbed by the upper soil layers but under extreme conditions they can pass through to the groundwater, e.g. when heavy rains cause water to pass through sand at a high rate. Virus adsorption is a reversible process influenced by pH, ionic strength, temperature, soil characteristics and type and amount of virus. Adsorption has been investigated in laboratory models and confirmed in the few natural systems that have been investigated. It is concluded that more investigation on the transport of virus through soil is needed, because of the potential public health problems involved in groundwater contamination.

1. INTRODUCTION

With the increasing land disposal of sewage sludge it has become important to evaluate the fate of indigenious human and animal viruses in the sludge. Most types of urban wastewater and wastewater sludge may be contaminated with a wide range of mammalian enteric viruses. Little is known about the content of viruses from domestic animals in sewage and sewage sludge due to the lack of investigations concerning these types of viruses. Such viruses can however be present in sewage effluent from abattoirs etc. as had been shown earlier (29). Animal and human viruses are in all likelyhood similar with respect to inactivation rate and persistance in the environment, when corresponding virus types are compared.

2. HUMAN VIRUSES INDIGENIOUS IN SEWAGE SLUDGE

The occurence of human enteric viruses in water and sewage sludges have been reported in numerous publications e.g. (4) (8) and (45). The most commonly studied group has been the picornaviridae, which includes polioviruses, coxsackieviruses A and B, echoviruses, enteroviruses 68-71 and as a newly accepted member enterovirus 72 (the etiologic agent of hepatitis A). Also adenoviruses and reoviruses have been isolated from domestic sewage. Hepatitis A virus and Norwalk type agents have been responsible for water-borne and especially sea-food borne epidemics traced to sewage contamination. Hepatitis A virus can only be grown in the laboratory with great difficulties, and the Norwalk type agents have not yet been cultivated. Reoviruses have also been isolated from sewage, but their role in human disease is not understood. An important group of reoviruses are the rotaviruses which appear worldwide to be the major cause of childhood diarrhea. Human rotaviruses are difficult to cultivate in the laboratory in connection with water and sludge examinations; in fact indigenous rotavirus has not yet been demonstrated.

Some enteric viruses are excreted in concentrations as high as 10 log units per gram feces and more than 5 log units of infectious virus per litre have been detected in raw sewage (31). Mechanical treatment by sedimentation combined with biological treatment reduces the concentration of enteroviruses in wastewater by 1-2 log units (1) (25). Viruses removed from the water will concentrate especially in the primary sludges. This transfer takes place because the virus particles adsorb to particular matter in the wastewater. The concentration of enteric viruses in raw sludges may vary between 3 and 6 log units per 1 under European and North American conditions (1). This is considerably more than in the raw sewage (26).

The digestion of the sludges is carried out mainly to produce odourless and stable sludges and to reduce the bulk of the material. As the digesters usually are run continuously the detention time is an average measure and part of the recently added escapes when the next cycle of fill and draw is carried out. In extreme cases a significant fraction of the new sludge may be withdrawn after only a few hours treatment, even though the average treatment time is measured in weeks. Under such conditions virus reduction in the digester will be much less than in a batch digester with the same (average) detention time. Mesophilic digestion of sludges before disposal reduces the number of infectious virus particles with approximately 1-2 log unit under normal conditions (1) (28). In addition to the influence of the detention time, it is known that high values of pH, temperature, drying and certain chemical and biological factors may greatly increase the inactivation rate of viruses (28).

The amount of virus present in the sewage is highly variable and depends on factors such as the prevalence of infection in the community and climatic factors. In temperate climates peak levels of enteroviruses occur in the late summer and early fall, although they may be isolated all year round with sufficiently careful methods. In warmer climates the seasonal variation in the virus concentrations are less pronounced.

It seems reasonable to estimate the concentration of indigenious human enteric viruses in mesophilic digested sewage sludge to be up to 4-5 log units per 1 before land application.

3. TRANSPORT AND TRANSMISSION OF VIRUSES FROM SLUDGES

Viruses indigenous in sludge may after land disposal be either inactivated on the application site or infectious virus may be transported with horizontal or vertical waterflows. Also transport by aerosols may occur depending on the application method. Viruses are also inactivated during transport through the soil. Even passive adsorption and desorption may provide time enough to significantly reduce the amount of infectious virus. The rate of this inactivation will depend on specific conditions and the mode of transport and can be very slow.

As viruses are inactivated in the environment no matter what route of transport they may follow, the initial virus concentration, the inactivation rate, the transport distance and time and the infective dose are important parameters, when viral hazards associated with land disposal of sewage sludge are evaluated.

4. AEROSOLS

Viruses may be transported from land application sites of sludges by means of aerosols. Transport via this route is restricted to the very act of application. If the sludge is sprayed onto the soil surface, aerosols are created. The content of dry matter of the sludge may reduce the possible virus transport by aerosols. Viruses are likely to be associated with the solids, and aerosol droplets containing sludge particles will not move for long distances because of their size. As land disposal sites normally are situated in remote areas, this transport route mainly seems to concern the people actually doing the job.

Several investigations, reviewed in US EPA (41) have dealt with a possible association between exposure to wastewater aerosols and illness of different population groups, e.g. sewer workers, inhabitants of wastewater treatment plant surroundings etc., but they have all failed to find causal relations of statistic significance.

A retrospective epidemiological study was carried out on the association between enteric disease incidence and wastewater utilization in 79 kibbutzim in Israël having a population of 32.672 (39). However, no evidence of excess risk associated with effluent irrigation was found, contrary to what was indicated in a preliminary work (15).

The literature concerning aerosol transport of viruses from land application of sludge is very sparse. Nevertheless the problem should not be ignored. The negative results of the investigations may partly be due to the lack of sensitivity of the epidemiologic methods.

5. INACTIVATION OF VIRUSES IN THE ENVIRONMENT

The inactivation rate of indigenous viruses in sewage sludge disposed on land is of importance for both horizontal and vertical transport of infectious viruses. Various factors influence the inactivation rates. High temperature, drying of the sludge and ultraviolet inactivation from sunlight will all increase the inactivation rate. The inactivation rate is so strongly dependent upon temperature that for many purposes only the exposure to the higher end of the temperature range contributes significantly to the loss of infectious virus.

Damgaard-Larsen et al (3) found that coxsackievirus mixed in surface applicated sewage sludge on the top of lysimeters was reduced from 9 log units per 1 of sludge to undetectable levels during a 23 week period under Danish winter conditions (temp. range : just below zero to 10°C. Tierny et al. (40) found that the concentration of poliovirus 1 dropped from 3 log units per gram of soil to undetectable levels within a period of 96 days during the winter (temp. range : − 14°C/+27°C) and 11 days in summer peirods (temp. range : +15°C/+34°C). The virus was added to the soil with inoculated wastewater and sludge. Nielsen and Lydholm (34) isolated indigenous coxsackievirus B5 from anaerobic mesophilic digested sludge 4 months after field application during the spring/early summer period in Denmark. Bitton et al. (2) investigated inactivation of poliovirus 1 inoculated in sewage sludge. The sludge was applied to the top of soil columns, which were placed under out-door conditions. The virus inactivation was monitored during a warm, dry period (soil temperatures in the range of +18°C/+27°C) and in a warm, humid period (soil temperature in the range of +24°C/+29°C). The virus was inactivated from 7 log units of PFU per gram dry matter to undetectable level during a period up to 21 days in the dry period under the former conditions and during a period of more than 35 days in the humid period. The results thus confirm, that the relative humidity of the atmosphere influences the inactivation rate. Jørgensen and Lund (13) isolated indigenous viruses (poliovirus type 1 and

2, coxsackievirus B1 and B3, adenovirus type 2 and 5) up to 21 weeks after disposal on land from anaerobic mesophilic digested sewage sludge. The sludge was applied on a sandy soil in a forest plantation in Denmark during the month of September. The temperature varied in the range 0°C/13°C during the investigation period. The concentration of virus was estimated to vary between 1.7 and 4 log units per litre. These results are in accordance with the results of Damgaard-Larsen (3) and Nielsen and Lydholm (34).

6. HORIZONTAL TRANSPORT

Horizontal transport of viruses may lead to contamination of surface waters. Viruses present in such waters may reach susceptible hosts where surface water is employed for drinking water production. Also transmission via recreational water, consumption of shellfish, harvested from contaminated water or consumption of crops, grown in surface water fertilized fields are associated with potential viral hazards. Many surface waters are contaminated with human enteric viruses (1) (31) (45). The most prevailing contamination route is via treated or untreated sewage discharge directly. The literature concerning horizontal transport of virus from land application of sludge is very sparse. The best studied case seems to be the report from Czekoslovakia 1966 (4, 36) where an outbreak of infectious hepatitis was traced to illegal use of cesspool sewage on farm land in the early spring when the ground was frozen. The results was gross fecal pollution of the stream supplying potable water, but the actual outbreak became milk-borne.

7. VERTICAL TRANSPORT

Virus concentration of ground waters has become a matter of serious concern especially in connection with land treatment of wastewater. Several reliable reports of vertical virus transport have been published (16). The public health hazard is obvious, as ground water is widely used for production of drinking water. The treatment of ground water only rarely aims at removing viruses from the water and infectious virus particles present in the groundwater supply may therefore pass to drinking water.

Occurence of human enteric viruses in natural groundwater has been reported by several authors. Wellings et al, Schaub and Sorber (37), Koerner and Haws (18) isolated virus from groundwater below irrigation or slow rate infiltration wastewater land treatment, and Vaughn et al. (42) reported occurrence of viruses in groundwater below high rate wastewater land infiltration. In all cases the soils were sandy and highly permeable. One report (44) describes detection of up to 78 PFU of enteroviruses in 190 l samples collected beneath a wastewater irrigation site. The soil was pure sand and heavy rainfalls preceeded the sampling.

Damgaard-Larsen et al. (3) could not detect coxsackievirus B3 in leachates collected from lysimeters. On the top of 4 lysimeters with different soil types, sewage sludge seeded with coxsackieviruses B3 to a concentration of 9 log units TCID5o per 1 was displaced. The percolates were collected 125 cm below the soil surfaces during a 5 months Danish winter period. Tritium marking of the sludge indicates that the water went through after 2 months.

Farrah et al. (7) examined groundwater samples from areas with landfill disposal of sewage sludge. Although viruses could be isolated from sludge samples none of the ground water samples contained detectable amounts of virus. Jørgensen and Lund (13) detected coxsackievirus B3 and poliovirus 2 in one out of 16 groundwater samples taken from sandy soil plantation area with soil surface application of anaerobic mesophilic digested sewage sludge (25 tons sludge dry matter per hectare). Some of the negative samples were collected from areas without sludge disposal. The volumes of the samples varied between 5 and 100 1, the volume of the positive sample was 30 litres. The soil of the area is a highly permeable diluvial sand, and the distance between the soil surface an the groundwater level varies between 1.5 and 10 m. The positive sample was collected 11 weeks after the disposal of sludge, and the period preceeding the collection was characterized by increased precipitation compared to the normally registered average of the region.

8. EXPERIMENTS EMPLOYING SOIL COLUMNS

The number of investigations of vertical virus transport from surface deposited sludge are relatively few. The obvious reason for this is the great practical and economical difficulties encountered in studies of groundwater. As a consequence a number of reports have appeared where experiments under artificial conditions employing soil columns in the laboratory have been carried out. The problems envolved in these studies are essentially connected with their acceptability as good models for the real world situation. Such factors as proper dimensions and packing of the soil in the columns to avoid seeping at the sides and canalization must be controlled. Other factors like flow rate, water quality and composition are more easily managed.

Many investigators have performed soil column experiments of different design concerning virus transport in soils. Many factors influence the possible transport of virus in the columns, e.g. pH and ionic strength of the water, whether the flow is saturated or unsaturated, the available adsorption surface of the soil, competing organic materials in the water, the chemical composition of the soil matrix, and the water flow rate and the virus strain. Adsorption of viruses to the surface of soil or sludge particles is a reversible process which does not influence the infectivity of the virus (27), but in general the virus stays with the top soil.

Lance et al. (19) and Duboise et al. (6) found in soil column experiments, that the adsorption of poliovirus 1 was positively correlated with the ionic strength of the percolating water. In column experiments with highly permeable sandy soil Landry et al. (23) found that different strains of enteroviruses apparently adsorbed in different ways. The highest adsorption was seen in soil previously percolated with wastewater. Wang et al. (43) could not confirm significant strain differences, but reported inverse proportionality between retention of virus and flow rate of the percolating water. Funderberg et al. (10) noticed proportionality between adsorption of poliovirus 1 and reovirus 3 and three soil characteristics namely the CEC (cation exchange capacity), the content of organic carbon and of clay. Lance et al. (20) did not register significant differences between adsorption of echovirus 29, echovirus 1 and poliovirus 1 below 40 cm depth in soil columns. Schaub et al. (38) detected indigenous enteroviruses from 52 % of 29 percolate samples collected 170 cm below the surface of soil lysimeters percolated with primary and secondary treated wastewater.

In Berlin Filip et al. (9) examined soil and ground water samples from fields irrigated with wastewater for year long periods. From (8 %) of 87 soil samples poliovirus and coxsackievirus were isolated. Only one of the positive samples was taken below 60 cm's depth. Virus was not recovered from ground water. Bitton et al. (2) detected no virus in percolates collected from the previously mentioned unsaturated soil columns. Lance and Gerba (21) found, that the mobility of poliovirus 1 in soil columns was fastest with deionized water, intermediary with wastewater and slowest with tap water. Presence of organic matter in the water reduced the adsorption. The virus adsorption was also registered to increase with increasing concentration and increasing valence of cations. Dizer et al. (5) found maximum adsorption of 4 different enteroviruses in sandy soil columns when suspended in ground water or tertiary treated wastewater. Surprisingly adsorption to soil materials suspended in distilled water exceeded adsorption when secondary wastewater was used. These authors found that hydrophobic interaction between virus particles and soil matrix enhanced retention of virus. Jørgensen (14) carried out experiments with 2 soil types in 100 cm unsaturated columns. On the top of each column sewage sludge seeded with adenovirus 1 and coxsackievirus B3 was placed. The columns were kept in the dark and percolated with artificial rain water for a 6 months period in accordance with the natural precipitation of Denmark. No virus could be detected in any of the collected percolates though the applied sludge was seeded to concentrations of approximately 8 log units TCID50 per 1 of each virus type. Furthermore no virus could be detected in soil samples taken from below 2.5. cm in the columns at the end of the period though both viruses were detectable above this depth in the columns. Wang et al. (43) showed in soil column experiments that inverse proportionality existed between virus adsorption and the flow rate of the percolating water. Several investigations have indicated that the mobility of viruses is greater under saturated compared to unsaturated conditions. This has been shown in experiments (2) (19) (22) and rendered probable under natural conditions (9). Flow rates of water in unsaturated soil are very much slower than in saturated soil.

9. STUDIES ON VIRUS ADSORPTION

The nature of the adsorption mechanisms between virus particles and soil material has been studied in several batch experiments. These experiments provide information about adsorption mechanisms under saturated conditions similar to deep soil horizons. It remains difficult to evaluate how far such laboratory experiments are relevant for natural conditions.

Moore et al. (32) (33) and Lipson and Stotzky (24) found in different bath experiments that approximately 90-99 % of human enteric viruses (reovirus and poliovirus) adsorbed to a wide range of minerals and soils. Lipson and Stotzky (24) found that adsorption of reovirus increased with increasing surface and with increasing CEC of clay minerals. This finding could not be confirmed by Moore et al. (32) (33), but the methods employed were not the same in the different experiments. Moore et al. (33) and Lipson and Stotzky (24) also found that the virus adsorption increased with the concentration and the valence of cations. This was confirmed by Lance and Gerba (22) in soil column experiments.

Presence of organic matter in the soil reduced the adsorption of virus in batch experiments (11) (32) (33). This could not be confirmed by Funderberg et al. (10) and Landry et al. (23), who showed that high content of organic matter in soil columns enhanced virus adsorption. In other soil column experiments high concentration of dissolved or suspended organic matter in the percolating water reduced the adsorption of viruses (22). These different results may reflect differences in the methods employed, or differences in the organic matter. It could probably be concluded that when organic matter is dissolved or suspended, it may compete with virus in the adsorption process. Immobilized organic matter, integrated in the soil matrix, apparently tends to enhance adsorption of virus and adsorbed virus is more resistant to thermal inactivation than free virus. If the proportions between suspended and adsorbed virus can be measured, the retention factor, which is the proportion between the water and virus transport velocity, theoretically can be calculated (30) (35). Calculated retention factors values of 500 are reported for poliovirus in soils with high concentration of clay minerals and water with high cation concentration. This factor is so high that for practical purposes it means that virus does not move, before it is inactivated.

10. STABILITY OF VIRUS UNDER GROUNDWATER CONDITIONS

Keswick et al. (17) examined the inactivation of coxsackievirus, poliovirus and simian rotavirus SA 11 in McFeters perfusion chambers. The temperature of the water was 13-15°C, and the pH was 7.8. They found that poliovirus and coxsackievirus were inactivated with 1 log unit per 5 days, while the same inactivation of rotavirus SA 11 occured within 2.7 days. Jørgensen and Lund (13) examined the stability of coxsackievirus, adenovirus and echovirus under simulated groundwater conditions in the laboratory. We found, that these viruses were inactivated with 1 log unit per 15-25 weeks at +6°C. All three viruses adsorbed readily to soil particles in the experiment, and the adsorption generally tended to protect virus against inactivation.

11. CONCLUSIONS

Viruses from man and animals may be present in sludge and in the environment after sludge disposal. They are also present in treated sewage. Their survival in the environment depends on their type and the amount of virus and on local conditions, essentially the climatic situation. Some viruses like rotaviruses and the virus of infectious hepatitis are important food- and waterborne viruses, but are much more difficult to demonstrate than enteric viruses like poliocoxsackie-and echoviruses. The enteric viruses are quite resistant in the environment but are inactivated by severe drying. When digested sludges are disposed on land virus may be demonstrated in top soil samples several months after the disposal. Infectious viruses may be transported through the soil to groundwater but this happens only under extreme conditions with heavy precipitation passing through sand or when canalization through the soil occurs. In general virus is retained in the uppermost soil layers but the adsorption is a reversible process influenced by ionic strength, pH, flow rate of water, nature of the soil and even the type of virus. This has been confirmed in model experiments employing various types of soil columns in the laboratory under varied conditions. If groundwater becomes contaminated with human enteric viruses serious public health problems may result, because the virus persist for very long periods under groundwater conditions and conventional water treatment may not be sufficient for their removal.

REFERENCES
(1) Berg, G. (1984). Monitoring for Virus Disseminated in Water. Monographs in Virology, 15, 17-29.

(2) Bitton, G., Pancorbo, O.C. and Farrah, S.R. (1984). Virus Transport and Survival After Land Application of Sewage Sludge. Applied and Environmental Microbiology, 47, 905-909.

(3) Damgaard-Larsen, S., Jensen, K.O., Lund, E. and Nissen, B. (1977). Survival and movement of enterovirus in connection with land disposal of sludges. Water Research, 11, 503-508.

(4) Dean, R.B. and Lund, E. (1981). Water Reuse. Problems and Solutions. Academic Press 1981.

(5) Dizer, H., Nasser, A. and Lopez, J.M. (1984). Penetration of Different Human Pathogenic Viruses into Sand Columns Percolated with Distilled Water, Groundwater or Wastewater. Applied and Environmental Microbiology, 47, 409-415.

(6) Duboise, S.M., Moore, B.E. and Sagik, B.P. (1976). Poliovirus Survival and Movement in a Sandy Forest Soil. Applied and Environmental Microbiology, 31, 536-543.

(7) Farrah, S.R., Bitton, G., Hoffmann, E.M., Lanni, O., Pancorbo, O.C., Lutrick, M.C. and Bertrand, J.E. (1981). Survival of Enteroviruses and Coliform Bacteria in a Sludge Lagoon. Applied and Environmental Microbiology, 41, 459-465.

(8) Feachem, R., Garelick, H. and Slade, J. (1981). Enteroviruses in the Environment. Tropical Diseases Bulletin, 78, 3, 185-230.

(9) Filip, Z., Seidel, K. and Dizer, H. (1983). Distribution of Enteric Viruses and Microorganisms in Long-Term, Sewage Treated Soil. Water Science and Technology, 15, 129-135.

(10) Funderberg, S.W., Moore, B.E., Sagik, B.P. and Sorber, C.A. (1981). Viral Transport Through Soil Colums under Conditions of Saturated Flow. Water Research, 15, 703-711.

(11) Gerba, C.P., Goyal, S.M., Cech, I.C. and Bogdan, G.F. (1981). Quantitative Assesment of the Adsorptive Behavior of Viruses in Soils. Environmental Science and Technology, 15, 940-944.

(12) Gerba, C.P. Keswick, B.H., Dupont, H.L. and Fields, H.A. (1982). Isolation of Rotavirus and Hepatitis A Virus from Drinking Water. Paper presented at the International Symposium on Enteric Viruses in Water Herzlia, Israel, Dec. 1982. In Melnick, J.L. (ed) (1984) : Enteric Viruses in Water. Monographs in Virology, 15, 119-125.

(13) Jørgensen, P.H. and Lund, E. (1984). Detection and Stability of Enteric Viruses in Sludge, Soil and Ground Water. Presented at IAWPRC's 12th Biennial International Conference on Water Pollution Control, 17-21 September 1984, Amsterdam.

(14) Jørgensen, P.H. (1984). Examination of the Penetration of Enteric Viruses in Soils under Simulated Conditions in the Laboratory. Presented as poster N° 16 at IAWPRC's 12th Biennial International Conference on Water Pollution Control, 17-21 September 1984, Amsterdam.

(15) Katzenelson, E., Buium, I. and Shuval, H.I. (1976). Risk of Communicable Disease Infection Associated with Wastewater Irrigation in Agricultural Settlements. Science, 194, 944-946.

(16) Keswick, B.H. and Gerba, C.P. (1980). Viruses in groundwater. Environmental Science and Technology, 14, 1290-1297.

(17) Keswick, B.H., Gerba, C.P., Secor, S.L. and Cech, I. (1982). Survival of Enteric Viruses and Indicator Bacteria in Groundwater. Journal of Environmental Science and Health, A17, 903-912.

(18) Koerner, E.L. and Haws, D.A. (1979). Long-term effects of land application of domestic wastewater : Vineland, New Jersey, Rapid infiltration Site. US EPA-600/2-79-072.

(19) Lance, J.C., Gerba, C.P. and Melnick, J.L. (1976). Virus Movement in Soils Columns Flooded with Secondary Sewage Effluent. Applied and Environmental Microbiology, 32, 520-526.

(20) Lance, J.C., Gerba, C.P. and Wang, D.-S. (1982). Comparative Movement of Different Enteroviruses in Soil Columns. Journal of Environmental Quality, 11, 347-351.

(21) Lance, J.C. and Gerba, C.P. (1984a). Effect of Ionic Composition of Suspending Solution on Virus Adsorption by a Soil column. Applied and Environmental Microbiology, 47, 484-488.

(22) Lance, J.C. and Gerba, C.P. (1984b). Virus Movement in Soil During Saturated and Unsaturated Flow. Applied and Environmental Microbiology, 47, 335-337.

(23) Landry, E.F., Vaughn, J.M., McHarrell, Z.T. and Beckwith, C.A. (1979). Adsorption of Enteroviruses to Soil Cores and Their Subsequent Elution by Artificial Rainwater. Applied and Environmental Microbiology, 38, 680-687.

(24) Lipson, S.M. and Stotzky, G. (1983). Adsorption of Reovirus to Clay Minerals : Effects of Cation-Exchange Capacity, Cation Saturation and Surface Area. Applied and Environmental Microbiology, 46, 673-682.

(25) Lund, E., Hedstrom, C.E. and Jantzen, N. (1969). Occurrence of Enteric Viruses in Wastewater after Activated Sludge Treatment. Journal Water Pollution Control Federation, 41, 2, 169-174.

(26) Lund, E., (1976). Disposal of sludges. In Berg, G. et al. (eds). Viruses in Water. American Public Health Association, 196-205

(27) Lund, E. (1981). Methods for Virus Recovery from Solids. In Goddard, M. and Butler, M. (eds). Viruses and Wastewater Treatment, Pergamon Press, 189-199.

(28) Lydholm, B. and Nielsen, A.L. (1981). The Use of a Soluble Polyelectrolyte for the Isolation of Virus from Sludge. In Goddard, M. and Butler, M. (eds) : Viruses and Wastewater Treatment, 85-89, Pergamon Press, Oxford.

(29) Malherbe, H.H., Strickland-Cholmley, M. and Geyer, S.M. (1967). Viruses in Abattoir Effluents. In Berg, G. (ed). Transmission of Viruses by the Water Route, Interscience Publishers, 347-354.

(30) Matthess, G. and Pekdeger, A. (1981). Concepts of a survival and transport model of pathogenic bacteria and viruses in groundwater. The Science of the Total Environment, 21, 149-159.

(31) Melnick, J.L. (1984). Etiologic Agents and Their Potential for Causing Waterborne Virus Disease. Monographs in Virology, 15, 1-16.

(32) Moore, R.S., Taylor, D.H., Sturman, L.S., Reddy, M.M. and Fuhs, G.W. (1981). Poliovirus Absorption by 34 Minerals and Soils. Applied and Environmental Microbiology, 42, 963-975.

(33) Moore, R.S., Taylor, D.H., Reddy, M.M.M. and Sturman, L.S. (1982). Adsorption of Reovirus by Minerals and Soils. Applied and Environmental Microbiology, 44, 852-859.

(34) Nielsen, A.L. and Lydholm, B. (1980). Methods for the Isolation of Virus from Raw and Digested Wastewater Sludge. Water Research, 14, 175-181.

(35) Pekdeger, A. and Matthess, G. (1983). Factors of Bacteria and Virus Transport in Groundwater. Environmental Geology, 5, 49-52.

(36) Raska, K. et al. (1966). A Milk-Borne Infectious Hepatitis Epidemic. Journal of Hygiene, Epidemiology, Microbiology and Immunology, X, 413-427.

(37) Schaub, S.A. and Sorber, C.A. (1977). Virus and bacteria removal from wastewater by rapid infiltration through soil. Applied and Environmental Microbiology, 33, 609-619.

(38) Schaub, S.A., Bausum, H.T. and Taylor, G.W. (1982). Fate of Virus in Wastewater Applied to Slow-Infiltration Land Treatment Systems. Applied and Environmental Microbiology, 44, 383-394.

(39) Shuval, H.I., Fattal, B. and Wax, Y. (1984). Retrospective Epidemiological Study of Disease Associated with Wastewater Utilization. US EPA-600/S1-84-006.

(40) Tierney, J.T., Sullivan, R. and Larkin, E.P. (1977). Persistence of Poliovirus 1 in Soil and on Vegetables Grown in Soil Previously Flooded with Inoculated Sewage Sludge or Effluent. Applied and Environmental Microbiology, 33, 109-113.

(41) United States Environmental Protection Agency (U.S. EPA) (1980). Wastewater Aerosols and Disease. EPA-600/9-80-028.

(42) Vaughn, J.M., Landry, E.F., Baranosky, L.J., Beckwith, C.A., Pahl, M.C. and Delihas, N.C. (1978). Survey of human virus occurrence in wastewater recharged groundwater of Long Island. Applied and Environmental Microbiology, 36, 47-51.

(43) Wang, D.-S., Gerba, C.P. and Lance, J.E. (1981). Effect of Soil Permeability on Virus Removal Through Soil Columns. Applied and Environmental Microbiology, 42, 83-88.

(44) Wellings, F.M., Lewis, A.L. and Mountain, A.W. (1974). Virus survival following wastewater spray irrigation of sandy soils. In Malina, J.F. and Sagik, B.P. (eds) : Virus Survival in Water and Wastewater Systems. Center for Research in Water Resources, Austin, Texas, 253-260.

(45) World Health Organization (1979). Human viruses in water, wastewater and soil. Report of a WHO Scientific Group. WHO, Geneva.

HEALTH RISKS OF MICROBES AND CHEMICALS IN SEWAGE SLUDGE APPLIED TO LAND - RECOMMENDATIONS TO THE WORLD HEALTH ORGANISATION

Dr.P.J.Matthews, *+   E.Lund,*++ and R.Leschber,***+

* European Water Pollution Control Association (EWPCA) and Anglian Water Huntingdon UK.

** Royal Veterinary and Agricultural University, Copenhagen, Denmark.

*** Institute of Water, Soil and Air Hygiene, Federal Health Office, Berlin 33,FRG.

+ Member of WHO Working Group on Risks to Health of Chemicals in Sewage Sludge Applied to Land;

++ Member of WHO Working Group on Risks to Health of Microbes in Sewage Sludge Applied to Land.

These authors prepared and presented this paper in their capacities as temporary advisers to the WHO Regional Office for Europe, Copenhagen.

SUMMARY

The paper summarises the findings of two WHO Advisory Working Groups. These were convened to make recommendations on the risks of pathogens and chemicals present in the sewage sludges applied to land. Whilst the Groups did not favour universal operational guidance and criteria, they identified the main features of concern and made recommendations on the main points which should be taken into account when deriving guidance on safe practices. The Groups also gave some consideration to further work which is needed to develop criteria which are available now.

1. INTRODUCTION

In 1981 and 1984 the Regional Office for Europe World Health Organisation (WHO) convened two Advisory Working Groups to discuss and report on the risk of sewage sludge applied to land. One dealt with the risks of pathogens, and the other dealt with the risk of chemicals present in sludge. In both instances the Working Groups drew heavily on European and North American experiences. This paper summarises the main elements of the two reports.

Providing adequate sanitary facilities is a major responsibility facing communities throughout the world. Hence the WHO has a legitimate and extensive role in providing support and guidance on sewage treatment as an important contribution to the good health of Europe and the rest of the world.

As communities shift in balance from being predominantly rural to being predominantly urban and as social expectations rise, the principal method of dealing with human wastes shifts from essentially domestic, such as privvies and septic tanks, to municipal in which the water-borne wastes are taken away in sewers and treated. The last two or three decades has seen massive investment in sewage treatment facilities in many countries and hence sludge production has risen rapidly; further increases in many countries are still expected.

In Western Europe, current annual production is about 5.5 million tonnes (expressed as dry solids) and this is expected to rise by about 25-50% (EWPCA 1984). In the USA, the flow receiving at least primary sedimentation treatment will rise from 133.4 x $10^3 m^3$/d in 1982 to 161.5 x 10^3/d in 2000 and the volumes receiving only primary treatment will fall from 11.5 x $10^3 m^3$/d to 0.8 $10^3 m^3$/d (US EPA 1983). In Japan, it was estimated that sludge production for disposal would rise from 0.9M dry tonnes in 1980/81 to 1.56M dry tonnes in 1985 (Kurihara 1981).

The safe and economic disposal of these sludges constitutes a problem which is being given urgent attention by many countries. National policies are being developed. The principal options available are agricultural utilisation, marine dispersal, landfill and incineration. The WHO Working Groups were charged with consideration of agricultural utilisation and similar disposal methods, such as use on vegetable plots and gardens.

In disposing of a sludge, all the available disposal and environmental options should be considered. After taking account of any statutory limitations and necessary measures to protect public health and the environment by using differing forms of environmental impact analysis, each option is appraised economically. The cheapest option should then be selected. This approach is known as the "best practicable environmental option". In doing the economic appraisal, the short and long-term costs and benefits should be evaluated.

At the present time the often-made decision to use a particular sludge as a fertiliser or soil conditioner is usually determined within national policies, on the basis that by reference to criteria for safe disposal it is the most economically feasible method of dispersal. The available low cost options of landfill and marine dispersal may not be available for a particular sludge due to geographical factors or specific policies which preclude their use. Although incineration is expensive and destroys what could be a valuable resource, it may be necessary if local conditions prohibit the application of sludge to land.

Sludge may be treated prior to disposal in order to reduce disposal costs and to render it more suitable for disposal. Hence, the choice of treatment, if at all, depends on acceptability of the product and economics.

2. PRINCIPLES OF SAFE UTILISATION PRACTICES

If sewage sludge is to be disposed satisfactorily by use as a fertiliser or soil conditioned it must be wanted, even needed, by customers, particularly farmers. It is a source of nutrients particularly nitrogen and phosphate, organic matter, lime (especially where this has been used in the treatment of the sludge) and even water. The services provided by the disposer must fit in with good agricultural practice to the extent that there is mutual benefit to disposer and farmer. Sludge only has a modest contribution to make in terms of national fertiliser needs; for instance in the UK with about 80% of the population served by sewage treatment, if all the sludges were to be used for their nitrogen and phosphate content only about 5% of N and 10% of P_2O_5 maximum of national needs could be supplied.

The inorganic fertiliser industry has, therefore, nothing to fear from this practice. However, the contribution on a local basis around sewage treatment works can be substantial.

Farmers and communities will want to be reassured that the use of sludge is safe. Even with good practice, this presents a difficult public relations problem for disposers because necessary hygiene training at an early age causes a psychological aversion to all feacal matter. This aversion is strengthened by the fact that sewage sludge is the "sink" for most of the water-borne waste of the urban community.

Sewage sludge contains a variety of pathogens, including bacteria, viruses, parasites, and fungi, reflecting the presence of these agents in the human and animal population contributing to the sewage. Consequently treatment may be required prior to disposal.

The chemical composition of municipal sludge varies widely, particularly in terms of minor constituents. The strictly domestic sources of metals and organic pollutants include a background from faecal matter, as well as contributions from detergent, cosmetics, insecticides, paints and other materials used in house care. Contributions from corrosion of the buildings and substances applied to gardens, may be included together with aerial deposits from smoke and automobile exhausts if urban runoff is combined with domestic sewage.

Municipal sewage also includes contributions from restaurants, food handlers and other institutions whose sewage closely resembles that from entirely domestic sources. In addition, there are contributions from small business - painters, printers, dry cleaners, jewellers, dentists, garages, auto workshops, electroplaters, school laboratories, etc. - which can greatly increase the concentration of one or another substance, depending on local conditions.

Industrial effluents may make substantial contributions according to the nature of the industrial processes. In industrialised areas, the concentrations of several metals may be elevated by discharges from metal processing industries, such as electroplating, leather processing, battery and paint manufacture, etc. Very high concentrations can arise because of discharge from one or more factories.

Industries may discharge large quantities of specific waste, either by historical permission or because of local inattention to pollution problems. In some cases, although pollutants are known to be discharged, there is reluctance to interfere with companies supplying the source of local payrolls.

It is important, therefore, that industrial effluent control is exercised by the statutory authorities and that manufacturing processes are managed and the industrial effluents are pre-treated to produce acceptable qualities with respect to the sewage effluent and sludge disposed by the receiving sewage treatment works. This is effective in reducing the excessive loads by individual discharges and in reducing the loads discharged by industry, in general, within a catchment area.

The summary of the two Working Groups' views was that good practice for sludge utilisation should have the following features:

(i)   Chemical quality should be controlled in such a way that the receiving soil does not become contaminated so as to cause crops to be hazardous to eat or for reduction of their yields. Clearly irresponsible practice can cause insidious long term problems. Care should be exercised so that pollution of surface and groundwaters does not occur.

(ii)   Microbiological quality should be taken into account so that the spread of human animal and plant diseases is prevented. It is not possible to control the microbiological quality in quite the same way as chemical quality (except by the control of animal derived industrial effluents). Hence the objective of prevention of the spread of disease is achieved by a combination of rigorously controlled utilisation practices and treatment of the sludge on the sewage treatment works as required.

(iii)   Aesthetic considerations should be taken into account and the operation should not cause nuisance or offence.

(iv)   The product and service should be acceptable to farmers.

3. CHEMICAL QUALITY

The Working Group met in Malta in October 1984 (WHO 1985). It considered elemental, particularly metallic, and organic chemical residues in sludge and in soils which have received sludge; it also gave some consideration to the way in which water may become polluted by sludge use and, in this context, nitrate received the most attention.

The Group concluded that for heavy metals the following points were of importance. Sludge from areas that have strict discharge controls has lower heavy metal concentrations than sludge from less controlled areas. Heavy metals in sludge, except cadmium, are not expected to affect human health through accumulation in food and fodder plants. There would appear to be little risk to health of applying sludge containing cadmium to forest land or to land producing fodder and seed crops. Tobacco grown on land receiving sludge causes an additional risk to human health from the inhalation of cadmium in tobacco smoke. The uptake by crops of lead from land treated with sludge does not contribute an appreciable amount of lead to the human food chain.

With respect to organic compounds the Group concluded that information on identified organic pollutants in sludge and their pathways to man is limited. The concentrations reported in the literature appear to be low, and almost always below 10 mg/kg dry solids. The most significant route or organic pollutants from sludge to man is through ingestion of soil and sludge by grazing animals, transfer through food crops is negligible. The contribution to total human intake of identified organic pollutants resulting from sludge application to land is minor and is unlikely to cause adverse health effects.

From the discussions and conclusions the Group were able to make the following recommendations.

General

Both the short- and long-term health risks costs, and benefits should be evaluated when selecting a method of sludge disposal. When applying a sludge to land, management practices should consider concentrations of organic and inorganic chemicals in the sludge, maximum allowable accumulation levels in soils, rates of accumulation, local soil conditions (pH, soil organic matter and cation exchange capacity), climatic conditions, and topography. Sludge should be spread in a manner that meets the needs of the plants and preserves the quality of soil, ground an surface water. Sludge utilisation should be supported by effective industrial effluent control practices and policies. The implications for sewage treatment and sludge utilisation should be evaluated before a new chemical is considered for wide use. Sludge should not be applied to growing food crops, such as vegetables and fruits, or in such a way that would contaminate drinking water sources. Sludge should not be applied to land that is used by the public (parks, playgrounds, sports fields) unless adequate precautions are taken to ensure the protection of human health. The lead content should not be above 1000 mg/kg dry solids to protect children against ingestion of lead in soil. Exposure studies should be conducted on worst-case population groups, such as exposed workers, and on adults and children, using home-grown vegetables in acidic soils that have been treated with sludge. Special emphasis should be given in pathway studies to the long-term bioavailability of the different chemical pollutants (both heavy metals and organic compounds).

Cadmium

Concentrations of cadmium in sludge applied to land should be as low as practicable to decrease the risk of its transfer to man through food. Additions of such sludge should be assessed in the light of other cadmium sources, such as fertilisers and atmospheric deposition. Sludge that is excessively contaminated with cadmium should not be used on light and acidic soils because uptake of cadmium is likely to be highest in vegetables grown on these areas. Sludge application rates that are higher than those based on cadmium limitations may be acceptable on land producing plants not used for food, provided there is no risk of contaminating the groundwater or surface water. Tobacco should not be grown on land receiving sludge because of excessive cadmium accumulation by this crop. Further dietary and faecal cadmium studies should be undertaken to improve knowledge of the long-term individual intake and retention of cadmium. Studies should be undertaken to establish (a) how cadmium concentrations in sludge affect potential cadmium uptake by plants and (b) the persistence of cadmium availability to crops. A group of experts should re-evaluate the "provisional tolerable weekly intake (of cadmium) per individual" set in 1972 by a Joint FAO/WHO Expert Committee on Food Additives.

Organic Compounds

More information should be obtained on the organic pollutant content of sludge, especially where local point sources of industrial pollutants exist. More information should be obtained on the fate and behaviour of persistent organic chemicals of significance to health in sludge and soil. Improved method should be developed for the extraction and preparation for analysis of organic pollutants of health significance in sludges, with a view to undertaking systematic surveys of these compound in sludge to provide basic information.

More studies should be carried out to quantify the relative importance of the pathways of organic pollutants of health significance from sludge-treated soil to man. Broad-spectrum tests related to effects on human health (such as enzyme inhibition and mutagenicity tests) should be developed to overcome the practical problems of analysis posed by the large number of organic pollutants in sludge.

Nitrate

When applying sludge to land, account should be taken of both the total and soluble nitrogen in the soil, because both can contribute to groundwater pollution.

4. **MICROBIAL QUALITY**

The Working Group met in Stevenage, UK in January 1981 (WHO 1981). It discussed the implications of the presence of a wide range of micro-organisms in sewage sludge and concluded that the use of sludge results in the distribution of pathogens in the environment.

However a number of other sources and routes of transmission of pathogens contribute to the total risk to human and animal health. The additional risk of a given sludge disposal practice must therefore be considered against this background. It is however, extremely difficult and expensive to measure the risk by microbiological and epidemiological methods. The public health risk from the disposal of sewage sludge on land may nevertheless be reduced by appropriate treatment and use of the land. It would be unrealistic to expect a significant or early reduction in the incidence of disease if other important sources of pathogens remain unaffected.

Acceptable levels of risk in given communities depend on a number of different factors. These include: the health status of the local population; the nature of the soil; the temperature, humidity, precipitation, and groundwater table; the nature of the agriculture and animal husbandry; and the way in which the sludge can be safely transported and spread on the land. Such acceptable levels cannot be expressed in absolute terms, but the public health risk may be controlled by the use of appropriate guidelines, which would vary from place to place according to local circumstances.

Salmonellae responsible for food poisoning represent one of the risks to human health that may be increased by the spreading on land of sewage sludge containing these organisms. Although there is no epidemiological evidence available for this yet, it has been shown that cattle exposed to such sludge may become carriers of Salmonellae to a greater extent than controls. T. saginata is a specific parasite of man, and where infection occurs the eggs are excreted in human faeces. They may be disseminated by sewage sludge and thus infect cattle, the subsequent development of cysticerci presenting an infection risk for man. Salmonellae and eggs of T. saginata in sewage sludge can be eliminated or reduced by various forms of heat treatment, by ionizing radiation, or by long-term storage. In contrast to the Salmonellae, the eggs of T. saginata are resistant to chemical disinfectants.

The risk to public health from other pathogenic agents appears to be less than those from Salmonellae and Taenia in most areas studied. However, the possible risk from viruses and from parasites such as Sarcocystis has not been adequately evaluated.

Measures should be taken to effect a substantial reduction in the concentration of pathogens in sewage sludge before it is allowed to come into contact with crops such as fresh vegetables and fruit that are brought into the kitchen raw. Sludge containing pathogens that may multiply in or contaminate meat, poultry and dairy products should not be spread on land where food animals are raised unless an adequate interval of time elapses between spreading the sludge on the land and allowing the animals to graze.

The following measures should be taken for the disinfection of sewage sludge so that its use can be relatively unrestricted:

(a)   heat treatment such as 70°C for 30 minutes, thermophilic composting, or heat drying;
(b)   ionizing radiation with at least 5kGy;
(c)   extended batch storage for a time that is inversely related to the temperature;
(d)   treatment with chemical agents that destroy the organisms in the sludge enviroment.

For restricted use, where direct contact with fresh food that may be brought raw into the kitchen is not involved, anaerobic digestion and other stabilisation processes should be utilised if they are carried out in such a way as to minimise recontamination. Uniform regulations should not be imposed over large regions without due regard to local conditions. Carefully planned epidemiological investigations into the relationship between sludge spreading and the public health should be encouraged.

5.   SUMMARY

Land application is a cost effective method of using sludge and results in minimal risk to health when good management is practiced. Good management practice should be adaptable to new developments and should incorporate monitoring of the sludges, sludged soils, effects on crops and effects on crop consumers as and when appropriate. The WHO Working Groups, particularly that for microbial risks, did not favour universal recommendations since depending on local circumstances, different measures are needed in different localities and situations to reduce the problem to an acceptable size. The formulation and application of suitable control measures must be a local decision. Several countries have taken the decision to formulate national statutes or guidelines (WHO 1981, EWPCA 1983, EPA 1984). The WHO Working Groups summarised the main features which control measures should take into account.

In the case of chemicals, which may be present in the sludge, it is important that there should be integration or close liaison between management responsible for sewage sludge disposal and utilisation and that responsible for industrial effluent control. In determining what can be allowed in discharges, amongst a wide range of toxicological data available to the control authorities should be data on the hazards arising from dispersion on agricultural soils.

The WHO has not attempted to produce comprehensive guidance on the principles and main features of all aspects of sludge utilisation. Good practice in sludge utilisation is an important contribution to public health. It assists in the good management of sewage treatment as well as preventing problems in the essential disposal of the sludge.

The EWPCA recommended that it would be appropriate for one consolidated report to be produced which would include not only the precautionary guidance, prepared by WHO, but also advice on agricultural benefits. This will encourage the establishment of safe practices in recycling a valuable and natural resource. However, it may well be that it would be beneficial to extend this report to include conditions outside those in Europe and North America.

6. **REFERENCES**

Each of the reports contains numerous references upon which the conclusions and recommendation are based. These are not reproduced here.

EUROPEAN WATER POLLUTION CONTROL ASSOCIATION (EWPCA) in HOUSE OF LORDS SELECT COMMITTEE ON THE EUROPEAN COMMUNITIES; Report: Sewage Sludge in Agriculture. London HMSO July 1983.

EWPCA EUROWATER - The Water Industries of the EWPCA. Thunderbird Enterprises, Rickmansworth, May, 1984

KURIHURA H. Treatment of sewage sludge. Annual Report of Studies on Sewage Treatment. Kankyo Gijutsu Kenkyu Kai (Japan) 1981.(contact Japan Sewage Works Association)

US, EPA. Environmental Regulations and Technology. Use and Disposal of Municipal Waste Water Sludge. Washington. September 1984.

US ENVIRONMENTAL PROTECTION AGENCY. (US, EPA) The 1982 Needs Survey: Conveyance, Treatment and Control of Municipal Waste-Water Combined Sewer Overflows and Stormwater Runoff. EPA 43019-83-002 Washington DC. June 1983.

WORLD HEALTH ORGANISATION (WHO). The Risk to Health of Microbes in Sewage Sludge Applied to Land. Report of a WHO Working Group, Stevenage, January 1981. EURO Reports and Studies. No.54 WHO Regional Office for Europe. Copenhagen.

WHO. The Risk to Health of Chemicals in Sewage Sludge Applied to Land. Report of a Working Group. R.B.Dean and M.J.Suess Editors. Waste Management and Research 1985, 3,251-278.

8. **ACKNOWLEDGEMENTS**

The views expressed in this paper do not necessarily represent the decisions or the stated policy of the World Health Organisation.

The authors acknowledge gratefully the work done by the Working Groups which they merely summarised.

SESSION 3 : MANAGEMENT CONSIDERATIONS IN SLUDGE AND EFFLUENT DISPOSAL

Recommendations on olfactometric measurements

Manure Nutrient composition : rapid methods of assessment

Efficiency of utilisation of nitrogen in sludges and slurries

Effect of a long term sludge disposal on the soil organic matter characteristics

Contamination problems in relation to land use

Long-term effects of contaminants

Effect on a long term sludge disposal on cadmium and nickel toxicity to a continuous maize crop

Trace metal regulations for sludge utilization in agriculture; a critical review

RECOMMENDATIONS ON OLFACTOMETRIC MEASUREMENTS

M. HANGARTNER
Institut für Hygiene und Arbeitsphysiologie, Zürich

J. HARTUNG
Institut für Tierhygiene, Hannover

J.H. VOORBURG
Government Agricultural Wastewater Service, Arnhem

SUMMARY

A special 'odours' sub-group of COST 681 Working Party 1 has made an inventory of existing guidelines for olfactometric-methods. These were discussed at a recent FAO/EEC Workshop on odours and a group of experts ('Rosewarne Group') agreed a series of recommendations formulating minimum conditions for conducting odour measurements with the aim of producing more consistent results and improved inter-laboratory comparisons. The recommendations cover (1) Sampling, including materials suitable for use, volumes required, transport and storage. (2) Olfactometer design, including minimum flow rates at sniffing ports, choice techniques, calibration and cleaning. (3) Panels, including the required background conditions, panel selection and panel size. (4) Operational procedures including method and numbers of presentations, size of dilution steps, and the minimum interstimulus interval. (5) Presentation of results including the method of calculation of the group threshold value. The recommendations are put forward for comment which could lead to improvements. It is proposed that the work of the special COST 681 sub-group on odours should continue with particular attention being paid to the measurement of odour offensiveness.

1. INTRODUCTION

In the EEC working party on sludge treatment (COST 681) the emission of odours is one of the problems being studied. In order to focus more attention on odour measurement a special subgroup was formed. This subgroup made an inventory of the existing guidelines and prescriptions for olfactometric odour measurements. The results of this inventory are reported in the papers by Hartung (1), Hangartner (2) and Voorburg (3) given at a recent FAO/EEC Workshop on odours.

During the Workshop sessions devoted to odour measurement, the experiences of experts were discussed and evaluated. As a follow-up of the discussion a small group of experts, meeting under the name of the "Rosewarne Group", agreed with these recommendations. The members of the Rosewarne Group are:

A.M. Bruce, Water Research Centre, Stevenage (UK).
M. Hangartner, Institut für Hygiene und Arbeitsphysiologie, Zürich (CH).
J. Hartung, Institut für Tierhygiene, Hannover (FRG).
P. l'Hermite, C.E.C., Brussels.

E.P. Koster, Vakgroep Psychische Funktieleer, Utrecht (NL).
R.L. Moss, Warren Spring Laboratory, Stevenage, (UK).
M. Paduch, Verein Deutscher Ingenieure, Dusseldorf (FRG).
H.M.J. Scheltinga, Staatstoezicht op de Volksgezondheid, Arnhem (NL).
M.F. Thal, Commissariat a l'Energie Atomique, Fontenay aux Roses (F).
V. Thiele, Landesanstalt fur Immissionsschutz, Essen (FRG).
J.H. Voorburg, Government Agricultural Wastewater Service, Arnhem (NL) (Chairman).
H. Wijnen, Ministry of Housing, Physical Planning and Environment, Air Directorate, Leidschendam (NL).

These recommendations are produced to formulate minimum conditions for research workers, organisations and industries, who wish to have consistent results and measurements which can be compared with odour measurements in other laboratories and other countries. Laboratories beginning work on odour measurements are advised to start at least at this level.

Improving an olfactometer into a more consistent and more sensitive one means that the results of earlier measurements lose part of their value.

In all research work, the design of experiments, the sampling technique and the analytical methods used require careful consideration and much commonsense. This is especially true for olfactometric measurements.

Olfactometers are used to assess concentrations of odour emission and immission. An olfactometer is a device which dilutes the odorous air to be tested, with odour-free air. These diluted samples are offered to a panel of people whose ability to detect a smell enables the odour threshold to be determined. Because the nose is the detector, several limitations are imposed by the 'human factor'. In this connection, a warning must be given against the measurement of samples containing harmful substances.

To understand these recommendations, some aspects of the human factor should be mentioned. The sensitivity of a human being is quite variable and influenced by e.g. weather conditions and hormonal changes. Even with an accurate olfactometer, one should realise that measurement of one compound by the same person over some days may give a 3-fold variation. Referring to Koster (4) special attention should be paid to adaptation, habituation and response bias. Adaptation to an odour means loss of sensitivity to the odour. Exposure to clean air for one minute is needed for recovery of sensitivity.

Habituation is a reduced sensitivity caused by loss of interest due to long and monotonous experiments.

Response bias is the capacity to distinguish a stimulus. It is important to understand the difference between a 'yes/no' choice and a forced choice. With a 'yes/no' technique the panelist tests the air from one single sniffing port and has to decide whether the air smells or not. A good panel member is motivated and wants to give correct answers. So he only says: "yes" if he is sure that he can smell something. In the case of a forced choice technique, the panelist is testing the air from two or more tubes. He knows that one of the tubes contains polluted air. At concentrations near the threshold he selects the tube that is the most likely to contain the odour. This results in a threshold at a much higher dilution rate than that determined with a 'yes/no' technique.

Another very important fact is that it is not possible to predict the effect of mixing odours on the concentration in odour units of the mixture; it is never the sum of the individual odour concentrations. The olfactometer should therefore be constructed from materials which do not produce or adsorb odours. Sometimes a part of the odour is not emitted as a gas but adsorbed on dust particles. In measuring odour emissions, the dust particles may block the olfactometer. Olfactometers with high flow rates and large orifices and valves have less problems with dust particles). These recommendations do not, however, provide a solution to the dust problem.

2. DEFINITIONS

Odour concentration expressed in odour units per cubic meter (O.U. m-3) is the number of dilutions to the detection threshold. (The odour concentration of an undiluted sample which is at threshold level is 1 O.U. m-3.)

Individual odour threshold is that concentration which is preceived by the subject in 50% of the cases in which it is presented to him.

Group threshold is the concentration that is perceived by 50% of the panel members.

3. SAMPLING

The materials to be used should be inert so that absorption and desorption are avoided. There should be no reaction with the sample, and the material should not change the odour. Some examples of materials to be used for the probe, the tubes and the valves are:- PTFE (polytetrafluorethylene)
- stainless steel
- glass.

Preflushing is recommended. In the case of long tubes, at least half an hour is required.

Predilution should be provided in order to prevent condensation.

The sample should be representative taking into account whether it is a fluctuating or a constant source. In the case of static sampling, at least two samples should be taken.

Sampling by concentration on an absorbent is not recommended unless further research proves it to be reliable.

In the case of static sampling, the material of the bag should have the same properties as given above and, moreover, should suffer no sample losses by diffusion. Good experiences have so far been made with PFTE, TEDLAR and Polyamid bags.

One should be careful with the re-use of the bags.

The volume of the sample will be determined by the air flow rate to the olfactometer, the number of presentations and the concentration of the sample.

Transport and storage time should be as short as possible with a maximum of 24 hours. The bag should be kept in the dark at a temperature sufficient to prevent condensation.

In the case of new types of odour emission, the reliability of the bag should be tested. The same holds for new bag materials.

If a dust filter is required, it should be upstream of the olfactometer and not upstream of the sampler. Experiments should be carried out to demonstrate that the properties of the odour have not been changed by the filter. Glass fibre filters are in use in Germany (LIS, Essen) and in the Netherlands (MT/TNO, Apeldoorn).

4. OLFACTOMETER

Only dynamic systems should be used. The materials to be used should have the same properties as given under "sampling". The construction should prevent losses of odorants and uncontrolled intake of clean air due to leakage.

The clean air should be odour-free as judged by each member of the panel. We recommend synthetic air or ambient air filtered with charcoal followed by a dust filter, to remove charcoal particles.

The tube or sniffing ports in use have various designs. The main condition is that the panelist should be supplied with the minimum flow required for breathing and that he or she should not inhale air from outside the sniffing port. In present practice, flow rates of between 16 and 170 1. min^{-1} (0.96 - 10 m^3. h^{-1}) are used.

Forced-choice techniques (two or more tubes, one containing the odour and the others only odourless air) and yes/no techniques (one tube) are used. The forced-choice technique is more sensitive. The results obtained should be corrected for guessing by applying the following formula:

$$Pcorr = \frac{Pobs - Pchance}{100 - Pchance} \times 100$$

in which Pcorr   = Pcorrected.
      Pchance   = the percentage correct observations obtained
                 by mere guessing (50% in the case of 2
                 tubes).
      Pobs   = Percentage observed.

Calibration of the flow rate in the olfactometer should be done regularly. The frequency depends on:
- construction of the olfactometer,
- frequency of transport of the olfactometer,
- degree of pollution of the sampled air,
- working situation (dirt, humidity, etc.).
Calibration can be done with a tracer gas, e.g. methane, Once a year, a standard experiment should be carried out with a large panel and H$_2$S as well as n-butanol as odorants.

Cleaning is done with warm air, steam or, if possible, by washing after dismantling. Special attention should be paid to cleaning after working under dirty conditions.

5. PANEL
The following background conditions are important:
- The test area should be odour-free.
- The room should be well ventilated (at least 6 times per hour).
- The tests should be carried out at room temperature and with
  normal humidity (40 to 70%).
There should be comfortable surroundings with no external stimuli such as smoke, noise, perfume, etc. No tests should be carried out within half an hour after a meal.

The test procedure:
There should be an independent panel leader. During the actual assessment there should be no communication among the panel members, and the communication between leader and panel should be very restricted.

Selection of panelists:
Panelists should screened and trained. They should be familiar with the test procedure. This means that at least one series with H$_2$S should be carried out. If possible more extended tests with more components are recommended.

Only individuals between 16 and 50 years of age who have a normal sense of smell and can follow simple instructions should be included in the panel. Persons with an erratic judgement have to be excluded from the panel.

The size of the panel:
The following minimum numbers of panelists are required:
a) 16 panelists to measure a representative threshold,
b) 8 panelists for all practical measurements,
c) 4 panelists in case of comparative judgements.
The panelist should be motivated (interested in the job).
The panelist should not be informed about his performance until after the test.

6. OPERATIONAL PROCEDURE

Fatigue should be prevented. In every hour, the panelist should always have a rest of at least 15 minutes.

The presentation can be at random or in steps of ascending concentrations. Series of descending concentrations should not be used (risk of adaptation).

The size of the dilution steps, i.e. the ratio between two adjacent dilutions, should be between 1.5 and 3.

The maximum exposure time to the stimulus should be 15 seconds.

The interstimulus interval should be sufficient to avoid adaptation of the panelist to the odour and sufficient to allow the olfactometer to provide the new stimulus. This can be attained with a cycle time of 1 minute (cycle time = exposure time plus interval).

The number of presentations (dilution steps) made to panelists in any one series should be at least 5. The range of dilution should be such that responses of less than 16% detection and more than 84% detection are obtained from the panel.

The number of test series should be at least two (one replicate).

7. PRESENTATION OF RESULTS

The group threshold should be calculated as the geometric mean of the observations or by graphical evaluation. Standard mathematical procedures should be followed (5).

Geometric means: the dilution at which the response changes from no perception to perception has to be determined for each test series. After that the geometric mean of the dilution so defined has to be calculated.

Graphical evaluation: the distribution of the frequency of perceptions as a function of dilution has to be determined. The evaluation can be performed using frequencies before or after probit transformation.

8. CONCLUDING REMARKS

The authors are interested in comments on, and experiences with, these recommendations, in order to improve them. Furthermore they propose that the work of the COST 681 sub-group "odour measurement" should be continued with particular attention being paid to the measurement of the offensiveness of odours.

REFERENCES
(1) Hartung, J. (1985). Sampling procedures for olfactometric measurements. Proceedings of the Joint FAO/EEC Workshop on Odours. Silsoe. (In the Press.)

(2) Hangartner, M. (1985). Selection and treatment of panelists for determination of odor thresholds. Proceedings of the Joint FAO/EEC Workshop on Odours. Silsoe. (In the Press.)

(3) Voorburg, J.H. (1985). Standardization of olfactometric measurements. Proceedings of the Joint FAO/EEC Workshop on Odours. Silsoe. (In the Press.)

(4) Koster, E.P. (1985). Limitations imposed on olfactometric measurement by the human factor. Proceedings of the Joint FAO/EEC Workshop on Odours. Silsoe. (In the Press.)

(5) Sachs, L. (1984). Applied statistics. Springer series in statistics, 2nd edition. Springer-Verlag, Berlin, Heidelberg, New York, Tokio.

DISCUSSION

P J Matthews

Did you consider condensation techniques? These have been tried and found successful for sewage odour. Secondly, it is a difficult task to measure odour offensiveness, more complicated than intensity, and is a psychological problem resulting from public health training. Overcoming such problems with the public largely depends on whether they are town or country people.

J H Voorburg

At the Silsoe workshop, Professor Koster from Utrecht played an important role on these psychological issues. The problem in analysing the air to be tested is that from the number of compounds present, the odour quality of the mixture cannot be predicted.

R Levi-Minzi

I have had some experience of testing odours but not as a researcher. The researcher has difficulty himself in understanding limits making this a difficult technique to use objectively. It is better not to know the source of the odour to avoid psychological problems.

J H Voorburg

The measurement of odours is very dependent on the conditions of testing. The panel members need to agree a standard for level and intensity.

MANURE NUTRIENT COMPOSITION : RAPID METHODS OF ASSESSMENT

HUBERT TUNNEY
Agricultural Institute, Johnstown Castle Research Centre
Wexford, Ireland

Summary

This paper deals with a number of rapid methods for estimating the dry matter and fertiliser value of sewage sludge and liquid agricultural wastes. The Slurry Meter is a hydrometer calibrated in percent dry matter and is based on a highly significant straight line relationship between dry matter and specific gravity of these wastes. The fertiliser value can then be estimated from the dry matter. The Nitrogen Meter is based on the release of nitrogen gas after the oxidation of ammonia by calcium hypochlorite. The mixing takes place in an enclosed chamber where the pressure is proportional to ammonia content and is measured by a pressure gauge. The ammonia electrode is extensively used in laboratories for determining ammonia content of waste and waste waters. Water analysis field kits can also be used for estimating the ammonia and phosphate content of manures. The suitability of these rapid tests for estimating the value of wastes under farm conditions is discussed.

1. INTRODUCTION

Results of analyses, of organic sludges and liquid agricultural wastes (slurry), from the literature show that there can be a wide variation in nutrient composition. This variation in composition is a major obstacle to effective integration of these valuable sources of plant nutrients into a farm fertiliser programme.

A simple field test to give a rapid estimate of the dry matter and nutrient content at the time of spreading has obvious advantages. It would enable the farmer or his advisors to estimate the nutrients in the manure and thereby decide the correct rate of application. It would enable an on the spot check of the quantities of N, P and K in a tank or tanker of waste and would put organic manures into a category similar to chemical fertilisers where the composition is shown on the outside of the bag.

A simple reliable rapid field test of composition would permit the correct rates of application and this in turn would help to reduce the risks of soil and water pollution from excessive rates. It would also facilitate the movement of organic manures from sewage treatment plants and intensive livestock farms to surrounding land where the nutrients are required.

The idea of a rapid test for estimating the nutrient content of manures is not new. Over 70 years ago, there were a number of publications on a hydrometer method to estimate the nitrogen content of liquid manures. (Dussere, 1915; Vogel, 1916; Lagers, 1918).

In the 1920's a paper was published showing that sodium hypochlorite could oxidize ammonia in slurry and relese it as nitrogen gas. (Tovborg Jensen, 1929). This is the basis of an instrument called a Nitrogen Meter for estimating available nitrogen in manure.

This paper describes recent work on the hydrometer method or Slurry Meter and the Nitrogen Meter, in addition to other methods which may be used for assesing the nutrient composition of organic manures.

Most of the work to date has been with liquid agricultural manures. However, work is commencing on the suitability of these methods for sludges. Results of an American study on rapid methods for determining fertiliser value of livestock manures has recently been published (Chescheir and Westerman, 1984).

2. SLURRY METER

2.1 Animal Manures.

This method has been developed at Johnstown Castle Research Centre in recent years. Studies on the composition of cattle and pig slurry showed a statistically significant correlation between dry matter and nutrient content (Tunney, 1975). It was evident from these studies that a simple test for dry matter would give a good estimate of the fertiliser value of slurry. The relationship between dry matter and nitrogen content of cattle and pig slurry is shown in figure 1 (Tunney, 1979).

As a result a number of methods for obtaining a rapid field test for dry matter were investigated. These included conductivity, colorimetery and density measurements. The most promising method was based on a highly significant linear relationship between dry matter and specific gravity of slurry (Tunney, 1979). This relationship is reproduced in Figure 2.

Based on this relationship a hydrometer, known as a Slurry Meter and calibrated in per cent dry matter can be used to estimate the dry matter of organic manures. The instructions for the Slurry Meter contains a table showing the relationship between dry matter and the N, P and K content for cattle slurry and pig slurry. The background to this method and practical experience with its use has already been presented at E.E.C. Workshops. (Tunney, 1984; Tunney, 1985).

A copy of instructions for use with the Slurry Meter is shown in Appendix A. The relationship between dry matter and nutrient content will of course depend on the diet of the animal. Ideally the relationship between dry matter and nutrient content should be calibrated for each country with major differences in animal diet.

In addition to Irish work on this method it has also been studied and reported on in Canada (Dube, 1982), America (Chescheir and Westerman, 1984) Denmark (Kjellerup, 1985) and France (Bertrand and Smagghe, 1985) and other studies are currently in progress.

A summary of the equations of the relationships between dry matter and specific gravity adapted from the literature is shown in Table 1. The equations are adapted so that a specific gravity of one corresponds to a dry matter of zero.

The results in table 1 show a good agreement between the dry matter and specific gravity obtained in the different studies.

It is clear that there is a good straight line relationship between dry matter and specific gravity of manures.

Figure 3 shows the relationship between specific gravity and total solids (dry matter) obtained by Chescheir and Westerman (1984) for cattle and pig manure. Dube (1982) in Canada and Chen (1982) in U.S.A. confirmed a similar relationship for pig slurry and cattle slurry respectively.

Table 1. Dry matter versus specific gravity equations adapted from the literature.

Equations	R^2	Reference		Manure[*]
S.G. = 1 + .0039 (% D.M.)	.96	Tunney	1979	P, C
= 1 + .0041	.72	Dube	1982	P
= 1 + .0036		Chen	1982	C
= 1 + .0043	.90	Cheschier	1984	P, C
= 1 + .0036	.92	Tunney	1985	H
= 1 + .0042	.91	Kjellerup	1985	P, C
S.G. = 1 + .004 (% D.M.)	Mean			

[*] P = pig, C = cattle, H = hen manure

Figure 4 shows that there is a similar relationship for manure from laying hens (Tunney, 1985). This relationship is interesting as the poultry manure is solid rather than slurry.

The specific gravity measurements were made by four fold dilutions to obtain a slurry, followed by a hydrometer reading and multiplying the results obtained by 4 to get the final values shown in Figure 4.

Figure 5 shows a diagramatic representation of the slurry meter being used to estimate the dry matter of two slurries. After estimating the dry matter the N, P and K content of the slurry can be read off from a table as shown in Appendix A.

Several hundred Slurry-Meters have been manufactured to date by a hydrometer manufacturing company in London. Feed back from people who have used the Slurry-Meter indicates that they found it easy to use. It is made from glass which is fragile for farm use, however it is resistant to corrosion by slurry. Slurry Meters have also been manufactured in metal, they are not as streamlined as glass and hence accuracy may be reduced with high dry matter or viscous slurries. Plastic Slurry Meters may have advantages, however, a demand for over one thousand would be necessary to justify tooling up costs.

American work (Chescheir and Westerman, 1984) has shown that there is a better correlation between nutrient content and specific gravity than between dry matter and nutrient content. This offers the possibility of improving the accuracy of the Slurry Meter for estimating the nutrient content of manures.

Chescheir and Westerman (1984) calculated the percentage of samples that gave a result with less than $\pm$ 25% error using the direct correlation between specific gravity and nutrient content of slurry, this is summarised in Table 2.

Table 2. Per cent of estimates for nutrients and total solids within ±
25% error (Chescheir and Westerman, 1984).

	Dairy	Swine
Nitrogen	94	93
Phosphorus	88	72
Potassium	76	82
Total Solids	100	86

2.2 Sewage sludge

Sewage sludge was collected from 20 sewage treatment plants in Ireland
in May 1985 and analysed for dry matter and specific gravity. The specific
gravity was determined by weighing 500 mls of the liquid sludge in a 500 ml
volumetric flask. With solid sludges appropriate dilutions were made to
enable accurate measurements. The results of this study are summarised in
Figure 6. This indicates that sewage sludge obeys the same relationship
already established for animal manures.

There was a statistically significant correlation between dry matter
determined by the Slurry Meter and by oven drying. However, estimating the
dry matter of sewage sludge with the Slurry Meter is more difficult because
of the high viscosity, even at low dry matter, of some sludges. In this
study the relationship between dry matter and nutrient content was also
statistically significant.

The sludges tested were from extended aeration treatment plants with
the exception of one from an anaerobic digestion treatment plant. Some of
the sludges had polyelectrolyte added. A number of the high dry matter
sludges had been mechanically dewatered after the addition of
polyelectrolyte.

This work indicated that specific gravity can be used as a rapid
estimate of sludge dry matter. However, the use of the Slurry Meter to
obtain an accurate estimate of sludge dry matter requires further work and
physical or chemical pretreatment of the sludge to reduce viscosity would
be necessary. The presence of a high content of gas bubbles trapped in the
sludge can reduce the accuracy of the specific gravity determination.

3. NITROGEN DETERMINATION

There are three possible rapid methods for estimating the nitrogen
content, or more specifically, the ammonia content of manure. The choices
are (1) Nitrogen Meter, (2) Ammonia Electrode and (3) Water Analysis Field
Kit.

3.1 Nitrogen Meter

Tovborg Jensen (1929) showed that mixing sodium hypochlorite with
manure caused the release of nitrogen gas by oxidation of ammonia. The
pressure of the released gas is proportional to the ammonium content of the
manure.

In 1983 a Sewdish company, Agros, developed the Nitrogen Meter which
is similar in principle to the method of Tovborg Jensen but uses calcium

hypochlorite (Ca (OCl)2) instead of sodium hypochlorite (Na OCl).

The Nitrogen Meter consists of a stainless steel reaction chamber with a lid, incorporating a pressure gauge with three measuring scales. There are three measuring cups, small, medium and large to correspond with the three scales on the pressure gauge. Measurements start with the smaller cup and the final choice of cup depends on the ammonium content of the manure. The slurry is measured from the cup into the chamber and a constant amount of the calcium hypochlorite reagent is added to a tipping tray in the chamber. The air tight lid is sealed on the chamber and the reagent is tipped into the slurry. The reaction takes place and the ammonium content is estimated from the gauge. Care should be taken to ensure that the pressure does not exceed the scale on the gauge.

A comparison of the Nitrogen Meter and standard labortory analyses (Technicon, 1974) for ammonium was made using inorganic salts by Chescheir and Westerman (1984). Their results are shown in Table 3.

Table 3. Comparison of Nitrogen Meter with standard laboratory analyses using inorganic salts (Chescheir and Westerman, 1984).

Inorganic Compound	Standard Solution	Laboratory Analyses	Nitrogen Meter	Difference %
		NH4 - N mg/L		
NH4 Cl	1,000	1,010	1,200	19
NH4 Cl	5,000	5,000	5,400	8
NH4 Cl	10,000	9,930	11,000	11
(NH4)2 SO4	1,000	1,020	1,200	18
(NH4)2 SO4	5,000	5,180	5,150	-0.6
(NH4)2 SO4	10,000	10,020	10,100	0.8

The results in table 3 show good agreement between the two methods. The same authors found satisfactory agreement between ammonium in manure measured by laboratory analyses and the Nitrogen Meter. They concluded that "the Nitrogen Meter is an excellent device for measuring ammonium nitrogen and easily oxidised organic nitrogen in manures, lagoon effluent and sludges".

They noted, however, that the Nitrogen Meter gave much higher values for stacked or scraped poultry manure and scraped pig manure than laboratory analyses and this was attributed to the measurement of urea and perhaps some other nitrogen compounds. The results of one aerobically digested municipal sludge gave a value of 520 mg/L ammonium nitrogen with laboratory analyses and 1,100 with the Nitrogen Meter. (Chescheir and

Westerman, 1984).

Recent work in Denmark (Kjellerup, 1985) on 128 samples of cattle and pig slurry, with a range of 0.5 to 6.0 kg ammonium nitrogen per tonne of manure showed good agreement between ammonium nitrogen determination by distillation and the Nitrogen Meter. The relationship was

NH4-N (laboratory) = 0.5 + 0.84 NH4-N (Nitrogen Meter)

with R^2 = 0.94.

The conclusion was that "under practical conditions the Agros Meter will give a good estimate of the ammonium content".

3.2 Ammonia Electrode

Orion Research Incorporated have developed an ammonia electrode (Orion, 1972). This ammonia selective electrode has been used successfully in many laboratories to measure the ammonia content of organic manures and waste waters.

In one of the earliest studies on cattle and pig slurry (Byrne and Power, 1974) the ammonia electrode was found to give excellent results when compared with conventional analyses for ammonia.

There are several reports of studies with the ammonia electrode for manures and the results indicate that it is an accurate method for ammonia determination (Sievers and Brune, 1978; Hills, 1980; Georgacakis et al. 1982; Le Duy and Samson, 1982; Sievers et al. 1983).

Figure 7 shows the results of a comparison of a colorimetric method with the ammonia electrode for estimating the ammonia content of cattle and pig slurry (Chescheir and Westerman, 1984). The results in Figure 7 show that the ammonia electrode overestimates the ammonia content, and this is in contrast with results found by Sievers et al. (1983). This difference may be attributable to the quantity of sodium hydroxide used. The ammonia selective electrode is used in conjunction with a pH meter and measurement is made after the addition of sodium hydroxide to raise the pH so that ammonium ions in the waste are converted to ammonia. Chescheir and Westerman (1984) stated that the ammonia electrode was very good for determining the ammonia nitrogen in animal slurries; they suggested, however, that a higher amount of NaOH than recommended in the instructions was necessary.

3.3 Water Analysis Field Kits

The ammonia content of pig slurry was estimated by a Nesslers field kit (Steward, 1969) with good accuracy. More recently this method has been tested by Chescheir and Westerman (1984) and they found reasonable agreement with conventional analyses. The results of this work are summarised in Figure 8. These authors found that large dillutions of slurries were required for analyses with field kits. They also used field kits to estimate the ortho-phosphate content of cattle and pig slurry. In summary they stated that the kit was adequate for measuring ammonia in pig slurry but not in catle slurry. In contrast the kit was adequate for phosphorus measurement in cattle slurry but not in pig slurry.

The field kits are based on colorimetric measurements after the addition of chemical reagents to the diluted slurry.

4. DISCUSSION

There is a growing interest in recent years in rapid methods of estimating the fertiliser value of liquid wastes as evidenced by the publications cited.

For a rapid test to be acceptable for farm use it should be simple to use and inexpensive. More elaborate rapid tests, requiring some laboratory facilities e.g. dilution, could be used by advisory services but would not be ideally suitable for general farm use.

The cost of the rapid test equipment from the lowest to the highest is (1) Slurry Meter, (2) Water Analysis Kits, (3) Nitrogen Meter, (4) Ammonia Electrode.

Chescheir and Wasterman (1984) stated that the Slurry Meter and Nitrogen Meter would be suitable for farm use while the Water Analysis Kits and Ammonia Electrode could be used on farms but they would require more skill in operation and in making dilutions.

The Slurry Meter is the simplest to use and the least expensive, costing about the same as the fertiliser value of one tanker load of slurry. It therefore could be recommended as the first choice for a rapid test on the farm. It also has the advantage of being the most accurate rapid test for estimating the dry matter content of slurry.

For more accurate determination of ammonium content of slurry on the farm the Nitrogen Meter would be suitable. The cost is relatively high and it would be best suited for use on larger intensive farms, by farm advisors or co-operatives.

The Ammonia Electrode is already widely used in many laboratories at present, but would not be suitable for farm use except in exceptional circumstances.

The Water Analysis Kits are perhaps of more limited value but could be considered where they are already available, for example in sewage treatment plants.

All the methods would benefit from further study to determine their suitability in practical use and to improve their accuracy.

In conclusion it would seem that any of these rapid tests for slurry would be an improvement on the present situation where application rates are based on average composition.

In the final analyses their major contribution would be as an educational tool so that farmers and their advisors would be convinced of the fertiliser value of these materials and how they can be used to reduce farm costs. In this way it should be possible to reduce excessive application rates and the associated risks of soil and water pollution.

5. REFERENCES

(1) BYRNE, E. and POWER, T. (1974). Determination of ammonia nitrogen in animal slurries by an ammonia electrode. Comm. in Soil Science and Plant Analysis, 5(1), 51-65.
(2) BERTRAND, M. and SMAGGHE, D. (1985). Mise au point methodes rapides d'appreciation de la valeur fertilisentante azotee et potassique de lisiers de porcs. Journees Rech. Porcine en France 17, 327-338.
(3) CHEN, Y.R. (1982). Engineering properties of beef cattle manure. Amer. Soc. Agric. Eng. Paper No. 82-4085, presented at ASAE Summer meeting, Univ. Wisconsin, June 1982. Publ. ASAE, St. Joseph, Michigan.

(4) CHESCHEIR, G.M. and WESTERMAN, P.W. (1984). Rapid methods for determining fertiliser value of livestock manures. Amer. Soc. Agric. Eng. Paper No. 84-4082, presented at ASAE Summer meeting, Univ. Tennessee, Knoxville, June 1984. Publ. ASAE, St. Joseph, MI.

(5) DUBE, A. (1982). Methode rapide de determination de la valeur fertilisante des lisiers de porcs a la ferme. Agriculture 39:1, 32-35.

(6) DUSSERE, C. (1915). Composition des purins provenant d'exploitations agricoles de la Suisse Romande. Ann. Agric. Suisse, 83-88.

(7) GEORGACAKIS, D., SIEVERS, D.M. and IANNOTTI, E.L. (1982). Buffer stability in manure digesters. Agricultural Wastes 4:427-441.

(8) HILLS, D.J. (1980). Methane gas production from dairy manure at high solids concentration. Transactions of the ASAE 23: 122 - 126.

(9) KJELLERUP, V. (1985). Agros Nitrogen Meter for estimation of ammonium nitrogen in slurry and liquid manure. Proc. C.E.C. Meeting on Efficiency of Land Use of Sludge and Manure, Askov (DK), 25-27 June. In Press.

(10) LAGERS, G.H.G. (1918). Het soortelijk gewicht en het stikstofgehalte van gier. Verslagen Landbouwkundige Onderzoekingen 9-18.

(11) LE DUY, A. and SAMSON, R. (1982). Testing of an ammonia ion selective electrode for ammonia nitrogen measurement in methanogenic sludge. Biotechnology Letters 4: 303-306.

(12) ORION RESEARCH, (1979). Instruction manual for ammonia electrode model 95-10. Cambridge, MA.

(13) SIEVERS, D.M. and BRUNE, D.E. (1978). Carbon/nitrogen ratio and anaerobic digestion of swine waste. Transactions of the ASAE, 21:537-541.

(14) SIEVERS, D.M., DOYLE, K. and PORTER, J.H. (1983). Ammonia measurement in high organic wastes. ASAE Paper No. 83-4064. ASAE, St. Joseph, MI.

(15) TECHNICON INDUSTRIAL SYSTEMS, 1974. Industrial methods for NH3 -N and TKN extract analyses. No. 328-74A, Tarrytown, N.Y.

(16) TOVBORG JENSEN, S. (1929). Undersogelser over ammoniakfordampning i forbindelse med kvaelstoftab ved udbringning af naturlig godning. Tidsskrift for Planteavl, 35, 68-71.

(17) TUNNEY, H. and MOLLY, S. (1975). Variations between farms in N, P, K, Mg and dry matter composition of cattle pig and poultry manures. Ir. J. Agric. Res. 14: 71-79.

(18) TUNNEY, H. (1979). Dry matter, specific gravity, and nutrient relationships of cattle and pig slurry. In: Engineering Problems with Effluents from Livestock. Ed. J.C. Hawkins. Proc. C.E.C. Seminar, Cambridge, England. Publ. C.E.C. Luxembourg. 430-447.

(19) TUNNEY, H. (1984). Slurry Meter for estimating dry matter and nutrient content of slurry. Proc. of C.E.C. meeting on Long Term Effects of Sewage Sludge and Farm Slurries Applications, Pisa, 25-27 Sept., In press.

(20) TUNNEY, H. (1985). Slurry Meter - Rapid field test for estimating dry matter and fertiliser value of slurry. Proc. C.E.C. meeting on Efficiency of Land Use of Sludge and Manure, Askov (DK), 25-27 June 1985. In Press.

(21) VOGEL, F. (1916). Versuche mit stallmist und Jauche. Jaherb. DLG. 31, 109-116.

APPENDIX A - SLURRY METER INSTRUCTIONS

1. Collect representative sample of animal manure slurry.

2. Put slurry in a plastic bucket and stir well.

3. Place slurry meter in slurry and read dry matter immediately.

4. Read corresponding N, P and K content from Table 1 below.

5. Cattle slurry higher than 5% dry matter (d.m.) and pig slurry higher than 8% is normally too viscous for accurate measurement. In such cases, mix slurry with an equal volume of water, stir well and take d.m. reading with Slurry Meter. Double the reading obtained to get the correct d.m. of the undiluted slurry. If necessary make further dilutions.

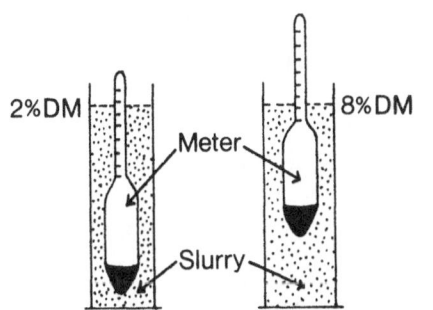

% Dry Matter	Concentration
0 - 5	Low
5 - 10	Medium
10 - 15	High

Note: depth of slurry in bucket should be 25 to 30 cm.

Table 1. Total N, P and K content of cattle and pig slurry in kg/tonne (kg/m³).

Cattle Slurry				Pig Slurry		
N	P	K	% Dry Matter	N	P	K
1.5	0.2	1.6	- 2 -	2.5	0.5	1.2
2.5	0.4	2.2	- 4 -	4.5	1.0	1.7
3.5	0.6	2.7	- 6 -	5.5	1.5	2.1
4.2	0.8	3.1	- 8 -	6.0	2.0	2.3
5.0	1.0	3.5	- 10 -	6.5	2.5	2.5
5.5	1.2	3.7	- 12 -	7.0	3.0	2.6
6.0	1.4	3.9	- 14 -	7.2	3.5	2.7

The higher the dry the higher the fertiliser value. If a slurry sample containing mostly urine is collected the N and K values will be higher and P values lower than shown.

It can be assumed that the N value shown in Table 1 will be half as effective and the P and K will be as effective as their chemical fertiliser equivalents.

Consult your agricultural advisor for information on the correct rates to apply and how you can reduce fertiliser costs by recycling animal manures.

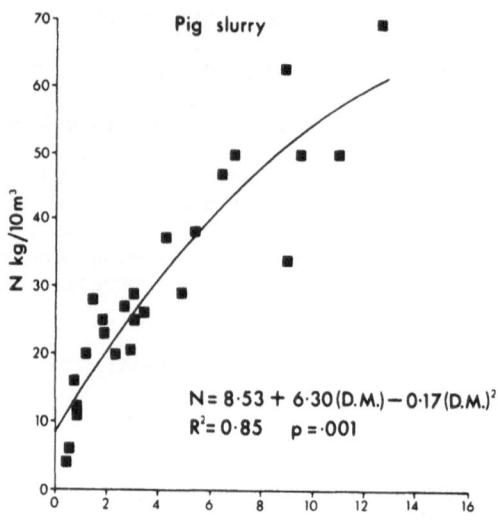

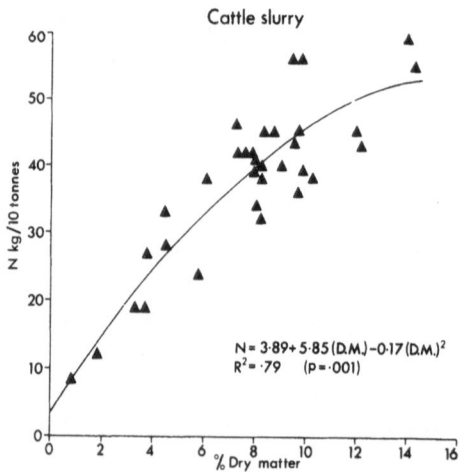

Figure 1. Relationship between dry matter and nitrogen content of cattle and pig slurry (Tunney, 1979).

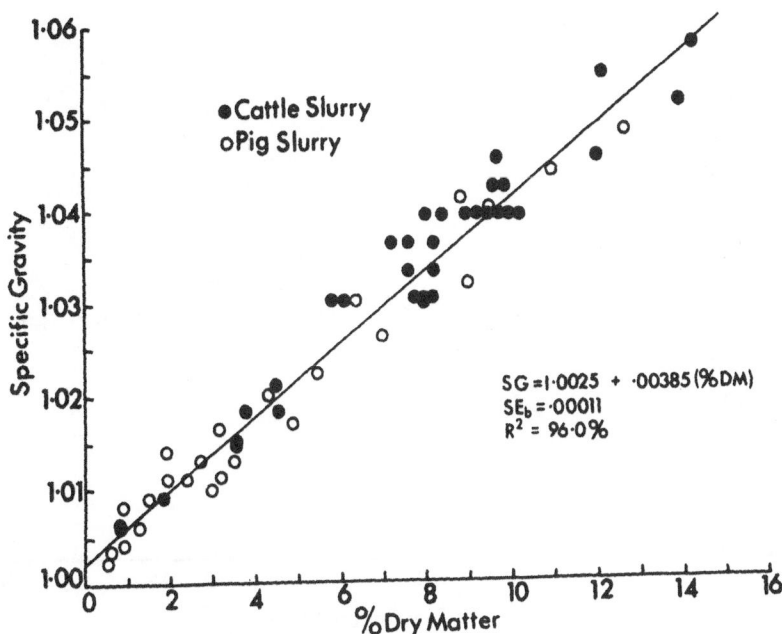

Figure 2. Relationship between dry matter and specific gravity of pig and cattle slurry (Tunney, 1979).

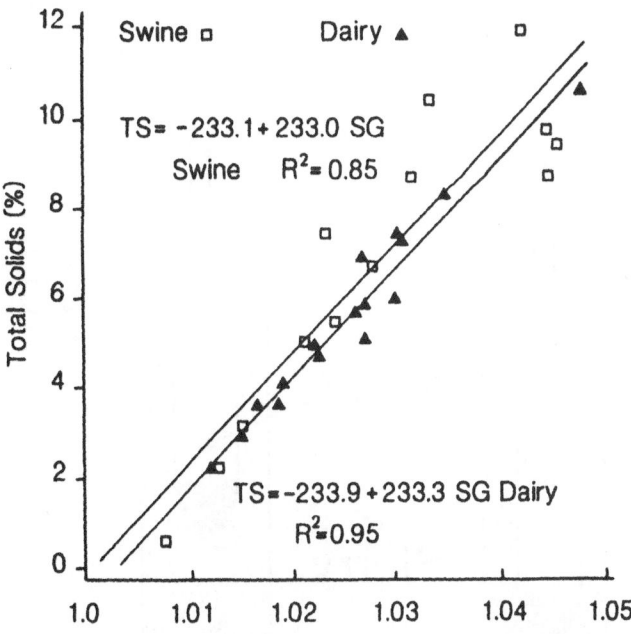

Figure 3. Relationship between total solids (dry matter) and specific gravity from an American study on dairy and swine manure (Chescheir and Westerman 1984).

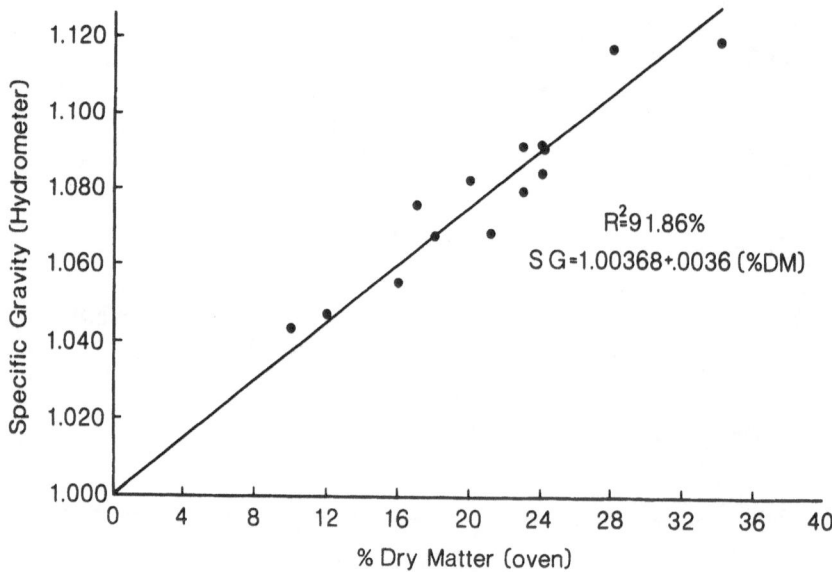

Figure 4. Relationship between dry matter and specific gravity of poultry manure from laying hens (Tunney, 1985).

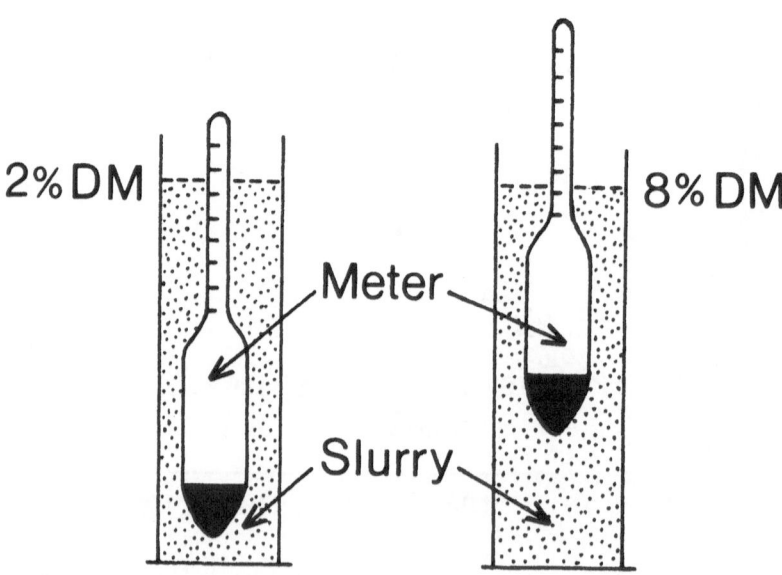

Figure 5. Diagramatic representation of Slurry Meter being used to measure low (left) and medium (right) dry matter slurry.

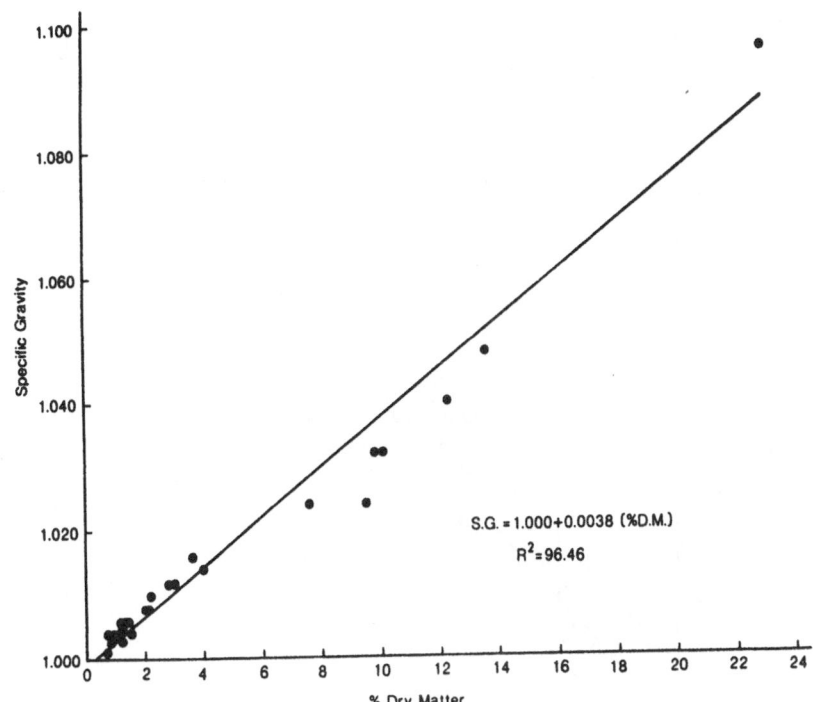

Figure 6. Relationship between dry matter and specific gravity of sewage sludge.

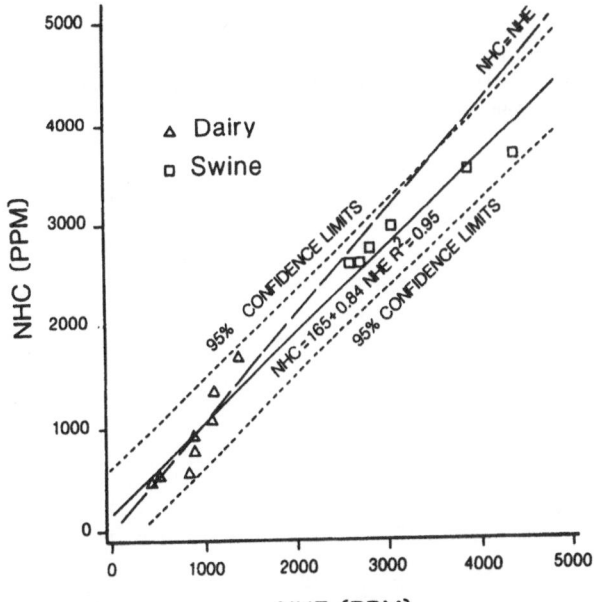

Figure 7. Ammonia nitrogen measured by colorimetric method (NHC) versus ammonia electrode (NHE) in dairy and swine manures (Chescheir and Westerman 1984).

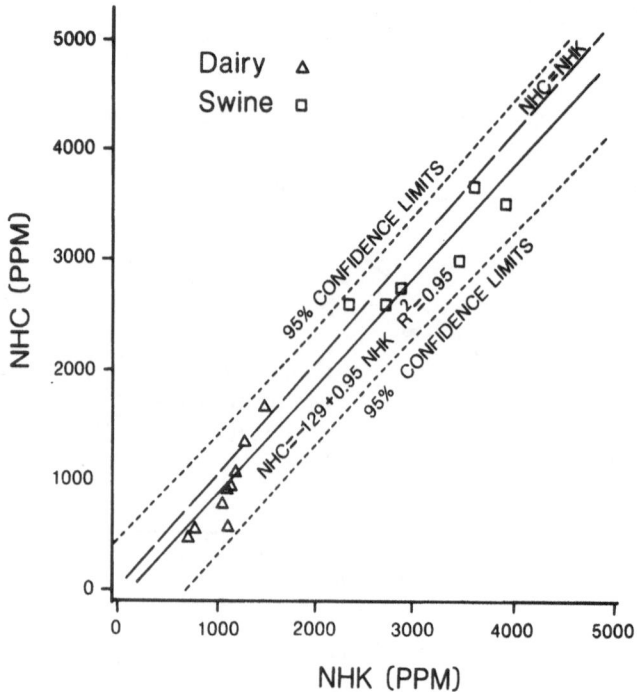

Figure 8. Ammonia nitrogen measured by cholorimetric method (NHC) versus water analysis kit (NHK) in dairy and swine manure (Chescheir and Westerman 1984).

DISCUSSION

D Sauerbeck

This technique is very appealing because it is simple but deviation will be large at high dry matter contents. It could be used for animal slurries but may not be possible for sewage sludge. Do you feel that in the normal range of 4-10% dry solids for animal slurry that the accuracy is good enough?

H Tunney

Advice to farmers is currently on average analyses for slurry which is not very good as there can be a ten-fold variation in actual analysis. We reckon on an accuracy for the meter of $\pm 25\%$ which is a very good improvement on current practice.

P J Matthews

Any method to allow the use of waste to be more efficient is welcomed as it may reduce pollution problems. For sewage sludge, there are good operational reasons for knowing dry solids content to assess the efficiency of sludge transport for instance.

Does the relationship between nutrients and dry solids indicate available or total nutrients? The hydrometer may not be so useful for sewage sludge due to its viscosity - will it meet our operational requirements? There may also be different densities and nutrient contents for different sludges.

H Tunney

The relationship of dry solids to nutrients measures total nutrient content. It can be assumed that total P and K are the same as available but it is more complicated for N. The correlation of nutrients with specific gravity is in fact better than for nutrients and dry solids.

EFFICIENCY OF UTILISATION OF NITROGEN IN SLUDGES AND SLURRIES

J H WILLIAMS and J E HALL
Ministry of Agriculture, Fisheries and Food, Wolverhampton,
West Midlands, UK.
Water Research Centre, Medmenham, Bucks, UK

Summary

The importance of utilising sludges and farm slurries on
agricultural land with maximum benefit to the farmer and at
minimal risk to the environment is now widely recognised. This
paper is concerned with the efficienct use of the nitrogen
contained in these manures concentrating on possible methods of
improving the efficiency of the nitrogen. Rate, timing and
method of application are of prime importance and such manures
need to be applied according to their NH_4-N content and the N
requirements of the crop. On arable land, the efficiency of the
total N in sludges or slurries relative to that of fertiliser -N
appears to be between 50-70% when applied in spring and worked
into the soil. Autumn or early winter applications can reduce
the efficiency to 20-30%. On grassland, the average nitrogen
efficiency is about 50% when surface applied in spring and can be
as low as 30% from autumn application. Efficiency can be
enhanced by soil injection in spring and on grassland, the method
reduces odours, prevents smothering of grass making it more
acceptable and hygienic for grazing stock. Methods of treatment
to improve the efficiency of nitrogen utilisation such as
anaerobic digestion, addition of chemical additives to inhibit
nitrification to reduce nitrate leaching and ammonia losses are
discussed for arable and grassland in relation to greater
flexibility of time of application without detriment to the
environment.

1. INTRODUCTION

The efficient use of sewage sludge and animal slurries in
European countries without detriment to the environment is now
considered to be of high priority. Some 40% of the total sludge
produced in the UK is applied to agricultural land compared to 30% in
the rest of the EEC Community. On average this represents some 20,000
tonnes N which is only 2.5% of the total inorganic fertiliser N
applied in the UK and only 1% in terms of "available" nitrogen.

In contrast nearly 1 million tonnes of total N are contained in
animal wastes produced annually in the UK and, of this, nearly 50%
will require handling and spreading on agricultural land. About half
of this again, ie 250,000 tonnes N will be applied as liquid slurry
and the other half as solid FYM. Until the 1940's farm manures
provided the greater part of the nitrogen used on most farms in the UK
but by 1957, farm manures only supplied about 20% of the nitrogen

applied to crops. By 1982, the Survey of Fertiliser Practice revealed
that only 28% of the area of crops and grass in England and Wales
received organic manures. This apparent decrease reflects a large
increase in the use of inorganic fertiliser and decline of mixed
farming systems. At the same time, livestock numbers increased and,
with intensification, came a change from solid to liquid slurry
systems. In the 1960's and early 70's environmental aspects were less
important, fertilisers were cheap and farmers and advisers were only
concerned with the disposal of slurry. Attitudes have now changed and
the emphasis is on the efficient utilisation of sludges and slurries
and their incorporation into the overall fertiliser policy of the
farm. Losses of nitrogen can occur at any stage from collection to
spreading on land and can continue after application. With the
Control of Pollution Act in the UK there are constraints which require
that farmers do not cause pollution or a nuisance. Safe and efficient
handling and utilisation of all wastes as organic manures is an all
important part of waste management practice and from which a
significant financial gain can be achieved. This paper considers
possible ways and means of improving the fertiliser efficiency and of
obtaining more consistent results from sludges and slurries.

2. NITROGEN CONTENT AND VALUE

2.1 SEWAGE SLUDGES

It is the amount of ammonia present and the degree of stability
of the organic matter that results in differing nitrogen values for
different types of sewage sludge. The proportion of ammonia ranges
from 5% of the total N in unstabilised sludges to more than 70% in
anaerobically digested sludges. Most of the N value of unstabilised
sludges comes from mineralisation of its organic N in the soil and is
thus a slow release source of N but the digestion process, converts
this easily degradable organic N to ammonia thus making liquid
digested sludge a quick acting source of N leaving organic matter that
is relatively resistant to mineralisation.

Dewatered sludges lose their soluble nitrogen and may be regarded
as slow release N sources, their rate of N mineralisation depends on
whether   they were digested or not prior to dewatering.

The nitrogen availabilities and predictive equations for
different sludges have been described by a number of workers and these
have been reviewed (1, 2, 3, 4). Table 1 summarises average nitrogen
contents for different sludge types and gives percent availability and
the average N available in sludge as spread, in the first cropping
year (5, 6). When dewatered sludge is applied or regular applications
of liquid sludge are made, residual effects from the organic N may be
expected in subsequent years; such cumulative and residual effects are
fairly predictable as the value of the organic N from a single
application will decrease by half each year(7).

2.2 FARM SLURRIES

Farm slurries, like sewage sludges are useful sources of nitrogen
if stored, handled and applied correctly. Field experiments to
determine their nitrogen value have been undertaken for some 25 years.

They still continue to provide variable results arising from
variations in the N and dry matter contents and the climatic and soil
conditions under which they were tested. The effectiveness of slurry
N has been shown by some workers to be closely related to the soluble
N content (24 hour extraction in cold 0.1N HCl) and more recently in
Northern Ireland to its NH_4 -N content (8).

Typical content of nitrogen in cattle, pig and poultry slurry are
shown in Table II (9). Content varies with type of feed and age of
stock. Availability of the nitrogen to the plants depends on many
factors. The readily soluble forms and the easily mineralisable
organic fractions will be available in the short term. Of the total N
some 43, 66 and 66% will on average be present as NH_4 -N in cattle,
pig and poultry slurry respectively on the basis of 10% dry matter.
Availability also depends on cropping, method and time of
applications. When spread in autumn or early winter, in particular on
arable land, a significant proportion of the available nitrogen may be
lost through leaching and/or denitrification depending on soil
temperature and moisture conditions.

3. IMPROVING EFFICIENCY OF UTILISATION

The reasons for the variability in the efficiency of the nitrogen
in sludges and slurries are numerous and include:-

i. rate of application and dry matter content;
ii. NH_4-N content and biodegradability of the residual organic
matter;
iii. method and time of application, the cropping situation, soil and
weather conditions.

Dilution of the slurry and frequent small dressings have been
shown to be more effective than a single application on grassland.
Reduction in physical smothering of the sward is obviously a factor to
be considered and any growth response to such treatments is probably
the net effect of the positive and negative benefits. The mechanical
separation of slurry solids might be expected to improve the
efficiency of N in the liquid fraction by increasing the proportion of
NH_4-N and reducing the smothering effect on grass. A number of
farmers have installed aerobic treatment systems in attempts to
overcome odour problems and a few have also built anaerobic digestion
plants. The increased nitrate content of aerobically treated slurry
or the high NH_4-N content of anaerobically digested slurry would be
expected to give improved N utilisation. Farmers have claimed
benefits in terms of consistency of product, slurry handling and
reductions in smothering of grass after such treatment of farm
slurry.

3.1 RATE AND TIME OF APPLICATION

3.1.1 Sewage Sludges

The amount of sludge or slurry applied should be related to the
nutrient requirements of the succeeding crop. Whilst the US EPA
recommended rates of sludge application 1.5 to 2 times the full
nitrogen requirement of the crop (10) there is concern in the UK that

application rates should be limited to 50 m^3/ha at any one time (11, 12). Whilst this incurs higher costs to the sludge disposal authorities it also appears to improve the nitrogen efficiency of sludge applied to grassland as shown by Table V.

It is often necessary for operational reasons to apply sludges and slurries at times of the year when conventional nitrogen fertiliser would not be applied. This relates particularly to applications made over the autumn and winter period when warm, moist soil conditions and high rainfall could lead to large losses of nitrogen by leaching and denitrification.

In grassland, such losses of N tend to be low but are related to rate, timing, rainfall and soil type. Hall and Williams (13), described a series of grassland trials in the UK on sites of differing soil texture with rainfall ranging from 500 to 1000 mm per year which compared sludge applied over the winter period. This is summarised in Table III which gives yields relative to the highest yield (100%) in each trial and indicates that grass yields from the autumn application can vary from only one third to being equal to the spring application.

Winter losses of sludge N from arable soils are not so well researched but are accepted as being much higher particularly those resulting from the period of autumn cultivation and early growth stages of winter sown crops when any nitrate formed is at risk of being leached. Table IV summarises UK cereal trials on a similar basis to Table III and appears to indicate that losses from autumn applications may not be as great from sludge as hitherto thought. The average rainfall for the six month period from October was less than 300 mm so that these results are probably not representative of Europe as a whole but information is lacking on this point.

3.1.2 Farm Slurries

The results from field trials on rate and time of application of cattle slurries on grassland have been very variable ranging from 50-80% efficiency of slurry N applied in late winter or early spring. Early work at the Hannah Research Institute in Scotland showed that applications of dilute liquid manure applied over the growing period commencing in February were between 80 and, 90% as effective as fertiliser N. (The November application was only 35% as effective as the February applied slurry (14, 15, 16).

Conflicting results were reported from Northern Ireland, Scotland and ADAS in England and Wales with efficiencies of cattle slurry N ranging from 16-80%, being highest when applied as a diluted slurry frequently over the grass growing season (17, 18, 19, 20). The experimental results up to the mid-70's were considered by ADAS who suggested an average value of 50% for the availability of cow slurry N following a late winter/early spring application. A further series of field trials were carried out over England and Wales starting in 1977. These are listed together with a summary of the findings when slurry applied to grass in Table VI. A comprehensive review of the UK experiments has recently been published by Smith and Unwin (21).

On arable crops, work in Northern Ireland by Stewart (22) suggested that slurry nitrogen for barley is almost as effective as fertiliser-N if applied in winter or spring. In 1974-76, Pain

investigated the effects of cattle slurry on spring barley (23).
Markedly improved grain yields were obtained from the slurry
treatments compared to fertiliser -N but had little effect on grain
quality in terms of its N content. In 1965-67, experiments on the
fertiliser value of pig slurry for barley were carried out in East
Yorkshire, UK (18). When applied at up to 190 kg/ha total N to the
seedbed, slurry N was 50-70% effective and at higher rates, 25-50%.
Mid-winter applications were 20-35% effective. Recent experiments in
Lincolnshire, UK on spring barley showed even lower efficiencies of
pig slurry N applied in winter - only 10% from December applications
and 20-25% from January applications (24).

The effect of different rates of slurries (pig, cattle and
poultry) on arable crops and grass was studied at Gembloux (25). The
main arable crops studied were winter wheat and winter barley.
Slurries were applied immediately before ploughing in September or
October and compared with increasing quantities of fertiliser N
applied at the recommended time in the spring. Examples of the
results obtained expressed as % efficiency of slurry N are shown in
Table VII. Efficiency of the slurry N applied in the autumn was, on
average, 50% on the deep loamy soil for winter wheat and decreased
with increasing quantity of slurry N applied. On the shallow soil,
efficiencies were much poorer for winter barley, ranging from 6-22%
depending on quantity of slurry applied. It was concluded that the
application of 120 kg/ha slurry N would seem to be optimum for winter
cereals, particularly wheat, which, given the mean composition of
slurry, corresponds to 20-25m^3/ha of pig slurry and 30-35 m^3/ha of
cattle slurry.

Other workers on the continent found a fair measure of agreement
for the efficiency of slurry N for arable crops, applied at different
times between autumn and spring. Their results are summarised in
Table VIII abstracted from the review by van Dijk and Sturm (9). They
are in broad agreement with results of UK field trials and show that
efficiency depends very much on the type of slurry, its composition
particularly its cpntent of readily available nitrogen, soil type and
the method of application.

The data provides ample evidence of reduced slurry N efficiencies
when rates are increased, if applied in late winter/early spring or if
applied later in the season after the first cut of grass. Dilution of
the slurry is obviously another explanation for the variation in N
efficiency as demonstrated by Stewart (26) who showed that several
small dressings on grass were more effective than a single application
and that diluting the cow slurry 1:6 increased N efficiency. Dilution
would reduce the smothering effect on the sward. Loss of N by ammonia
volatilisation during and after applying slurry in the April/May/June
period could account for decreased N efficiency. It has been
suggested by Kolenbrander that losses by this means could be as high
as 30% (27).

3.2 VALUE OF ANAEROBIC DIGESTION

Unstabilised sludges and manures have wide C:N ratios, from 14 to
50, and during anaerobic digestion carbon is metabolised and a
proportion of the nitrogen is mineralised reducing the C:N to about
10. Whilst evidence of nitrogen immobilisation has been observed by
Chaussod (28, 29) in incubation studies with unstabilised sewage

sludges and other organic wastes, this has not been a quantifiable effect under field conditions. Nevertheless, anaerobic digestion does have a significant effect on the nitrogen fertiliser value of sludges. This has been briefly reviewed by Hall (3) and Demuynch (4) who showed that the reported N availabilities for anerobically digested sludge ranged from 20 to 84% with an average value of about 50% whilst the availability of nitrogen in undigested sludge in the year of application was only around 35%.

Figure 1 compares grass yields following the application of stored digested and undigested sludges to a clay loam soil in the UK. Whilst this data is just an example, it does illustrate the point that, at a given application rate, liquid digested sludge not only provides more nitrogen but it is also more available.

In terms of the agronomic benefits, it is very doubtful whether the setting up of anaerobic digestion plants on the farm can be justified purely on economic grounds. Digested slurries will have higher ammonia contents and will be more readily available to the crop but there are very real risks of losses during storage, during and after spreading. However, digestion does improve the quality of the wastes, it resolves the odour problems encountered during spreading, the digested effluents are less sticky and more homogenous which means easier handling and fewer blockages; the smothering effect on grass will be reduced with fewer weed problems leading to improved efficiencies of slurry N in many cases and lastly a reduction in plant pathogens, bacteria and parasitic cysts.

Air temperature is one of the meteorological variables which affects ammonia volatilisation of the most. Losses can be as high as 60-70% of the NH_4-N if applied to almost bare ground followed by dry windy conditions (30, 31). The losses can be avoided by applying digested wastes on crops which fully cover the ground and by incorporation immediately after applications if applied to bare ground.

The nitrogen effect of anaerobically digested and undigested cattle and pig slurry was studied in field experiments with barley and beet at Askov in Denmark (32). The results for spring barley, averaged over 17 trials during 1979-81 and 9 trials during 1982-84, are given in Table IX. Results for 18 trials on fodder beet are shown in Table X. In the cattle slurries used in these trials the NH_4-N was approximately 50% and in pig slurries about 66% of the total N. Based on the amount of total N in the slurry the digested cattle slurries had a slightly better effect on average than raw slurry but not significantly; with pig slurry efficiency was variable but with a tendency to smaller differences between raw and digested than for cattle slurry.

Studies by Suess and Wurzinger in Germany showed slurry from a biogas plant to be more effective than fresh slurry for winter wheat grown on a low yield potential soil (33). The results obtained for winter wheat grown on a fluvio-glacial soil are shown in Table XI. Chemical analysis of the fresh and biogas slurries showed a decrease from 9.7 to 7.3% dry matter and an increase from 3.4 to 3.7 g/litre of NH_4-N after digestion.

In the absence of fertiliser N applied, the yield is higher after the use of biogas slurry even though there is a lower amount of total N applied which suggests a higher availability of the nitrogen in biogas slurry applied to the growing crop. Where 80 kg/ha fertiliser N was applied, differences in grain yield were not statistically significant and no difference between slurry treatments.

3.3 ALTERNATIVE METHODS OF APPLICATION

3.3.1 Soil Injection

Soil injection of liquid manures (particularly sewage sludge) is becoming an increasingly attractive option for both farmers and sludge disposal authorities. Whilst capital and running costs tend to be higher compared to surface spreading, the agronomic benefits (reduced ammonia loss, soil loosening, pasture hygiene) and the environmental benefits (control of odour and surface run-off, compliance with proposed EC Directive (34) are sufficient to outweigh the disadvantages where these benefits are important and can result in an overall reduction in operating costs (ie injection of liquid undigested sludge may be a cheaper option to anaerobic digestion dewatering). These aspects of soil injection have been reviewed by Hall (35, 36).

Soil injection of sludge and slurry improves the efficiency of nitrogen utilisation by preventing ammonia volatilisation. This can have significant effects of crop yields bearing in mind that a high proportion of the total N present in slurry and digested sludge is ammonia and that significant amounts can be lost following surface spreading (30, 37, 38). However, the improved N efficiency through soil injection may be limited to some extent by soil conditions and injector design, particularly on grassland where the quality of finish is important and damage to the sward surface and roots must be minimised.

On arable soils there is generally no such constraint except when injecting into a growing crop. The tines can be positioned closer together than for grassland so that their areas of disturbance interact and produce a greater cultivation effect with improved distribution of slurry under the soil surface. Table XII summarises data from a number of sources where equivalent rates of slurry were surface applied and injected; increases in yields have been expressed relative to fertiliser response. It is clear that in all cases bar one, higher yields resulted from the injected slurry than from the surface applications and this is very largely attributable to the prevention of ammonia loss by volatilisation as indicated by Table XIII.

Larsen (39) has shown with spring barley and fodder beet that close injector tine spacings of 30 cm gave higher yields than spacings of 60 and 75 cm at equivalent application rates. Kemppainen (40) showed, under Finnish conditions, that injection in the autumn before ploughing gave lower barley yields than spring injection but was nonetheless better than an autumn surface application by a factor of two. This work also showed that injecting into an emerging barley crop in early summer can result in serious crop damage. The optimum time for surface application or injection was just before sowing in May or at emergence and injection gave the better yields at this time.

Injection into growing crops is generally only successful in row crops, particularly maize. Beauchamp (41) found that injecting maize when the crop was 15-20 cm high resulted in the same yield as from injection prior to sowing. Larsen (42) compared surface application, placement between rows and injection between rows at different crop heights; the results given in Table XIV indicate that injected slurry produced the highest yields at each growth stage tested with the greatest response at the 15 cm crop height.

On grassland, higher yields following injection have been observed when compared to equivalent surface applications (Table XV). However, the benefits of preventing ammonia volatilisation are reduced to a variable extent by damage to the sward surface and roots. Injection has increased grass yields by up to a factor of three (43) but under difficult soil conditions the yield may be significantly depressed compared to the equivalent surface application (44). It is clear that there is an interaction between soil type and soil moisture, and therefore the prevailing weather conditions at the time of injection.

Injection in the spring in moist, light textured soils generally results in improved yields compared with surface applications (40, 43, 45, 46) but can be improved further by irrigation (30 mm) following injection (40). On heavy textured soils, such irrigation benefitted only the surface application (presumably partly due to preventing NH_3 loss) and generally yields following injection on such soils are lower than for surface applications due to the localised damage around the injection time.

Injection in the autumn appears to allow the sward to recover before growth starts in the spring as yields are higher than from spring injection (44). Generally, less damage is done by injection in the autumn to grassland compared with summer injection as there is a greater likelihood of subsequent rain and surface conditions tend to be more moist. However, in late spring and early summer when the soil is entering a drying phase, with soil moisture deficits over 40 mm (47), increasing amounts of grass die-back may be expected due to soil-root disturbance.

Much of the relative yield reduction caused by injection on grassland is short-lived and often only limited to the first cut of grass following injection. For example in Table XV, the injection of digested sludge into a clay soil in the spring resulted in a very poor initial response. However, the residual effects were such that by the first grass cut the following year, the injected treatment had out yielded the surface application (44). This initial depression in yield was caused by persistent anaerobic conditions in a wet clay soil and it is likely that nitrogen would be lost by denitrification under such circumstances. Ryden found denitrification of about 2 kg N/ha over 8.5 days following injection of cattle slurry into a loam soil in July when denitrification losses from the surface application were only one tenth of this (48). Such losses could perhaps be controlled by nitrification inhibitors, but trials reported by De la Lande Cremer showed only a small benefit from Didin in the first harvest following spring injection and was not worthwhile overall as shown by Table XV (49).

Good tine design is important in minimising sward damage and this has been discussed by Hall (36). Larsen showed on a loam soil that 30 cm spacing (3.5 cm width tine) produced a higher yield than at 60 cm with the same tine or 75 cm with 37.5 cm winged tine. However, on a sandy soil, tine spacing did not affect yields (46) except when 3 passes were made. This is in agreement with tine spacing trials (25 cm winged tine) in the UK (44) on a loamy sand (Figure 2) but on a clay soil, optimum spacing was found to be 65 cm; the lower yield at 50 cm was due to the increased amount of sward damage (ie more tine passes per unit area) although it was found that under wet soil conditions, injection at 50 cm produced the more even growth response.

Overall the data presented in Tables XII and XV show that surface spread slurry is about 25% more effective on grass than on arable land and whilst injection is 1.5 times more effective than surface spreading on arable soils through the prevention of ammonia volatilisation, it is only 95% as effective, on average, as surface spreading on grassland due to sward damage.

Incidental to improving the efficiency of slurry N to grass but important in improving grazing value, is the effect of injection on grass palatability as the surface application of slurry can lead to rejection of herbage by ruminants. Apart from the hygienic aspects of preventing access by cattle to any sludge or slurry-borne pathogens, the intakes of herbage have been shown to be higher on grassland which has been injected compared with surface spreading as indicated by Table XVI and will lead to improved pasture efficiency.

3.3.2 Application to Growing Crops

With the increasing acreages of winter sown cereal crops, sludge disposal authorities in purely arable areas face increasing difficulties i finding suitable land on which to spread sludge over the winter period.

The use of mobile irrigators provides an opportunity to apply sludge effluent or separated slurry liquid to growing crops in the spring period, for example to cereals, potatoes and to maize crops. Diluted liquid manures (2-4% dry matter) will give an improved efficiency of N use. Stage of growth of the crop and dry matter content of the sludge or slurry are important factors to consider if crop damage is to be avoided. Soils are drier and ground conditions generally more favourable for liquid application at this time of the year. Adequate on-farm storage for large volumes of liquid effluents and slurries are essential if such a system is to be successful. More evidence on the agronomic benefits to be achieved is required before a firm recommendation can be made for such a system with the high capital costs that could be involved.

3.4 USE OF NITRIFICATION INHIBITORS

Recently a number of nitrification inhibitors have appeared on the market. The efficacy of some of these have been tested in vegetable growing soils in an attempt to reduce the nitrate content of leafy vegetable crops and economise on the use of nitrogen. There were obviously several factors which affected the efficiency of these inhibitors, that is their stability, persistence under different soil and climatic conditions such as moisture and temperature.

Workers in Holland, Belgium and Denmark have recently investigated some of the inhibitors in an attempt to reduce the rate of nitrification of NH_4-N in slurries when applied to soil thus improving the efficiency of slurry N use and reduce the risk of nitrate leaching and/or denitrification losses. In Belgium for example, Destain and his co-workers recently used dicyandiamid (DCD) as an inhibitor to try and improve the efficiency of N in pig slurry applied to winter wheat grown on loam soils at Gembloux. The DCD was mixed into the pig slurry and applied immediately to the soil in August prior to drilling in the autumn at a rate of 20 kg DCD ha^{-1}. The results obtained were inconclusive but the DCD treatment which provided 135 kg slurry N ha^{-1} gave identical yields to pig slurry applications supplying 172 and 210 kg N ha^{-1} without added DCD. In winter barley there was a small positive response in yield to the inclusion of DCD in the pig slurry at 190 kg N ha^{-1} applied (50).

At the Askov Experimental Station in Denmark three types of inhibitors (Didin, Dwell and N Serve 240E) were tested in field experiments on spring barley and beet (51). The rates of cattle slurry applied were 50 t/ha for barley and 100 t/ha for beet. There were good effects in spring barley in 1983 following the application of all 3 inhibitors in the slurry applied in December 1982 (an additional yield of 1 t/ha of grain compared to non-treated slurry. There was no effect of the inhibitors following slurry application in September or April. In 1984, there was a good effect from the inhibitor applied in September 1983 (+ 1.6 t/ha grain) but not from the December or April treatments. Temperatures were lower in the autumn of 1983 than in 1982 - higher temperatures cause degradation of the inhibitor. By December 1983, soil temperatures were so low (less than 5°C) that, even without inhibitors, there was very little nitrification taking place.

3.5 FIELD METHODS OF ASSESSING NUTRIENT COMPOSITION

Studies on the composition of animal slurries have shown that there is a good correlation between dry matter and the nutrient content. Based on the fact that there was a straight line relationship between dry matter and specific gravity a hydrometer calibrated in percent dry matter was patented by Tunney in Ireland (52). It can be used under field conditions to obtain a rapid estimate of the dry matter content of the slurry and the corresponding total N, P and K content from an accompanying table. It is particularly valuable for on the farm estimation of the slurry dressing required to provide the necessary nutrients.

Complementary to the slurry hydrometer, there has also been produced the Agros Nitrogen Meter in Sweden which permits a rapid and reasonably accurate determination of the ammonium-N content of a liquid slurry. This is a particularly useful parameter on which to base slurry dressings which, if coupled with the hydrometer for dry matter estimation can provide a good estimate of major nutrients present. At Askov Experimental Station, comparisons of measurements by the Agros Nitrogen Meter and the traditional laboratory distillation procedure for ammonia have been made for pig and cattle slurry samples. There was a significant correlation (R^2=0.94) between the two values with comparable absolute NH_4-N contents (53). To overcome variability in composition it is very useful to have a simple method which quickly gives an approximate estimate of nutrient content.

4. AVOIDANCE OF POLLUTION

4.1 ATMOSPHERIC POLLUTION

The application of sludge and slurry to land can give rise to odours (although when digested, these materials are not nearly so offensive); aerosols and ammonia volatilisation may also be of concern. The problem has two stages; firstly the action of spreading, whilst of short duration, can produce aerosols and odours that can drift several kilometers and is related to the design of the spreading equipment (Table XVII). Secondly, there is the longer term problem with manure lying exposed on the ground surface generating odours and losing NH_3 for several days following application, unless immediately cultivated into the soil.

Whilst aerosols are unpleasant, there do not appear to be any significant health risks associated with them (54). However, half of the farm-related odour complaints in the UK are concerned with spreading manure. As previously discussed, NH_3 losses by volatilisation can be high, particularly under drying conditions, and concern has been expressed about possible environmental effects of NH_3 in relation to acid rain (55).

It is clear that spreading equipment that will deliver large droplets close to the ground, such as dribble bars, minimise the problems during spreading but, unless cultivated in, there will still be a persistent odour problem. Potentially, soil injection is the only promising technique which will significantly reduce ammonia loss into the atmosphere (48, 56) and the production of odours (57).

4.2 WATER POLLUTION

Inputs of nitrogen into the system can be large in intensive grass/arable situations. The greater the input, the higher will be the leaching losses particularly where large dressings of slurry or sewage sludge are applied immediately after cereal harvest in the autumn. Following long dry periods, mild and moist weather conditions in the autumn will enhance the degree of nitrification from soil and slurry sources and presents a very serious risk to water quality. Lysimeter studies with sewage sludge and farm slurry have been conducted in Denmark (58). Following an application of 700 kg/ha sludge N (dewatered digested sludge with 27% dry matter) each year for 8 years, on average, 100 kg/ha N was lost through leaching. Similar studies were carried out using cattle slurry for a crop rotation of beet, barley, italian ryegrass and barley. After applying 50 or 100 t/ha of cattle slurry, containing 225 and 450 kg slurry N, to spring barley, losses were high and of the order of 60 and 100 kg/ha N respectively. Leaching was substantially lower after sugar beet and lower still after Italian ryegrass. The difference is due to the longer period of uncropped land from August onwards after spring barley compared to beet or grass. It has been demonstrated by Vetters and Steffens (59) that catch cropping reduces N leaching quite markedly when slurry is applied in the autumn after cereal harvest.

Lysimeter studies in the UK in 1984/85 reported by Unwin (60) demonstrated however, that with all the lysimeters sown to winter barley in October, high leachate nitrogen levels were still being found after applications of 100 and 200 m^3 ha^{-1} of pig slurry.

Following winter barley the losses were 215 and 239 kg/ha N and, after spring barley, 225 and 315 kg/ha N after the low and high rate of slurry applications. The results indicate a considerable pollution risk when slurry is applied at high rates in the cereal stubble situation.

In grass lysimeters, despite a high NH_4-N content, N losses were very low over 3 years even when 1000 kg/ha slurry N were applied during the growing season. A ten year study with grass lysimeters and cattle slurry on sandy loam and calcareous silt loam soils showed that the quantity of nitrate leached from both soils at nil and 125 kg/ha N as ammonium nitrate or slurry has been similar, in the range of 10-14 kg/ha NO_3-N per year. Only 3 leachate samples out of 120 in the last 6 years have exceeded 20 mg/l NO_3-N. Losses from the high slurry treatment (500 kg slurry N/ha) have been less than from 500 kg/ha N as ammonium nitrate which may be due to ammonia losses from slurry through volatilisation. From both these treatments, appreciable amounts of nitrate were leached each year and mean leachate concentrations have been in excess of 13 mg/l for the past 6 years. The increased leaching from the 500 kg/ha N rate is related to seasona weather conditions. In 1981/82, for example, the effect was magnified when fertiliser and slurry were applied during a dry spell in July and August which was followed by the wettest winter in 10 years. Losses reached 240 kg/ha NO_3-N on thee sandy soil with mean leachate concentrations of 46 and 55 mg/l for the fertiliser-N and slurry treatments respectively.

Whilst the risks to ground water quality are probably no more or no less if sludge or slurry is injected provided injection is in the rooting zone and above the depth of any gravel backfill that may be present over the field drains. Injection into dry cracked soils or into drained land with shallow gravel backfill is inadvisable due to the risks of sludge running directly into land drains and watercourses.

Pollution of watercourses through surface run-off can occur following excessive rates of application, heavy rainfall after spreading and on sloping ground. A grass crop or rough cultivation can greatly reduce the risk and if the conditions are right, soil injection can reduce it still further. Winged and inclined injector tines, because they give a better shattering of larger volumes of soil and hence of voids, are much better than simple tines in preventing run-out from injection slots on sloping ground.

5. CONCLUSION

The fertiliser value of organic manures is not being fully exploited at the present time. It would be reasonable to assume that the P and K contained in sludges and slurries can be efficienctly utilised irrespective of time of application. The variability reported in the efficiency of the nitrogen from sludges and slurries can be ascribed to a number of factors which include:-

1. Dry matter and soluble nitrogen content
2. Rate, time and method of application
3. Soil type and depth
4. Cropping
5. Weather conditions over the period between application and sowing of the spring crop or the start of grass growth.

In attempting to maximise the efficiency of the nitrogen in these liquid manures the additional costs incurred, such as better facilities for treatment, collection, storage and spreading can far outweigh the value of the nitrogen. However, many farmers are prepared to accept such costs on the basis that it gives them a more manageable system and one which must be more environmentally acceptable. Having accepted such a system, it becomes more important to use the manures to maximum benefit and, at the same time, reduce the environmental impact to a minimum. To achieve this, soil injection of sludges and slurries offers a very real possibility but will be more costly than surface spreading and may take longer under certain soil and weather conditions especially in grassland. The technique offers a means of reducing odours, ammonia losses and the risk of infection. Results from the use of chemical additives in liquid wastes to slow down nitrification have been variable and whilst show some promise, a firm recommendation for their use cannot be made at the present time.

Whilst anaerobic digestion gives very much improved efficiency of the nitrogen in sewage sludges, in the case of farm slurries it is of doubtful advantage probably because the latter already contains a high content of NH_4-N (about 50% of total N). Digested sludges and farm slurries with such relatively high contents of NH_4-N may suffer losses of ammonia when spread on the surface; such losses can be reduced by:-

1. Late winter and early spring applications when air temperatures are lower.
2. Soil injection.
3. Smaller, split applications which spreads the risk.
4. Applying more dilute slurries to ensure more rapid soil penetration.

The latter would apply particularly on grassland and to growing crops in the early spring. Additives such as aluminium salts or zeolites have shown some promise in reducing ammonia losses by increasing the acidity of the waste.

Rapid methods for assessing the nutrient composition of farm slurries just prior to spreading can provide the farmer with a reasonably good estimate of the dry matter content using a simple hydrometer and from that, its N value. The use of the Agros nitrogen meter can also provide a good estimate of NH_4-N content of cattle and pig slurries which correlates well with the laboratory determinations. Increasing environmental constraints are being placed on farming practices and if farmers are being forced into more costly slurry storage and handling systems it becomes increasingly important to obtain maximum benefit from nitrogen in sludges and slurries in order to be able to reduce fertiliser-N costs.

REFERENCES

1. COKER, E.G. (1981) Sludge utilisation and the availability of nitrogen
    in sewage sludges. Water Research Centre report 94-M.

2. CATROUX, G., CHAUSSOD, R., GUPTA, S., DE HAAN, S., HALL, J., SUESS, A.
    and WILLIAMS, J.H. (1982) Nitrogen and phosphorus value of sewage
    sludges. A state of knowledge and practical recommendations.
    Commission of European Communities.

3. HALL, J.E. (1983) Predicting the nitrogen value of sewage sludges. In
    Processing and use of sewage sludge. Proceedings of the Third
    International symposium, Brighton, UK.

4. DEMUYNCK, M. (1984) Utilisation in agriculture of anaerobically
    digested effluents. Commission of European Communities.

5. WATER RESEARCH CENTRE (1985) The agricultural value of sewage sludge,
    A Farmer's Guide.

6. MINISTRY OF AGRICULTURE, FISHERIES AND FOOD. The use of sewage sludge
    on agricultural land. Booklet 2409. In revision.

7. HALL, J.E. (1984). The cumulative and residual effects of sewage
    sludge nitrogen on crop growth. In Long-term effects of sewage sludge
    and farm slurries applications. Proceedings of CEC COST 681 WP4
    seminar, Pisa, Italy.

8. MCALLISTER, J.S.V. (1981) Responses to slurry N in Northern Ireland.
    In Nitrogen losses and surface run-off from land spread manures. Ed
    J.C. Brogan 389-393. Nijhoff/Junk CEC.

9. VAN DIJK, T.A. and STURM, H. (1983) Fertiliser value of animal
    manures on the continent. Proceedings No 220 of the Fertiliser
    Society (London).

10. US ENVIRONMENTAL PROTECTION AGENCY (1977) Municipal sludge management.
    Environmental Factors Technical Bulletin. UK Federal Register 42,
    211.

11. THE SCOTTISH AGRICULTURAL COLLEGES (1981) Disposal of sewage sludge on
    agricultural land. Publication No 76.

12. MINISTRY OF AGRICULTURE, FISHERIES AND FOOD (1985) Code of Good
    Agricultural Practice. HMSO.

13. HALL, J.E. and WILLIAMS, J.H. (1983). The use of sewage sludge on
    arable and grassland. In Utilisation of sewage sludge on land: rates
    of application and long-term effects of metals. Proceedings of a
    seminar held at Uppsala, Sweden.

14. CASTLE, M.E. and DRYSDALE, A.D. (1962). Liquid manure as a grassland
    fertiliser. 1. The response to liquid manure and to fertiliser. J
    Agric Sci Camb 58. 165-171.

15. DRYSDALE, A.D. (1963). Liquid manure as a grassland fertiliser ii. The response to winter applications. J Agric. Sci Camb 61 353-360.

16. DRYSDALE, A.D. (1965). Liquid manure as a grassland fertiliser iii. The effect of liquid manure on the yield and botanical composition of pasture and its interaction with nitrogen, phosphate and potash fertilisers. J Agric Sci Camb 65 333-340.

17. DAVIES, H.T. (1970). Experiments on the fertilising value of animal surries. Part 1-The use of cow slurry on grassland. Experimental Husbandry, 19 49-60.

18. DAVIES, H.T. (1970). Experiments on the fertilising value of animal slurries. Part II - The use of pig slurry on spring barley. Experimental Husbandry 19 61-64.

19. APPLETON, M and RICHARDSON, S.J. (1976). Cow slurry management with particular reference to Bridgett's EHF. ADAS Quart. Rev 23, 294-305.

20. UNWIN, R.J. (1973-1976). Slurry application to grassland. In Soil Science Reports of Experiments in SW Region of MAFF 1973-1976.

21. SMITH, K.A. and UNWIN, R.J. (1983). Fertiliser value of organic manures in the UK. Proceedings No 221 of Fertiliser Society (London).

22. STEWART, T.A. (1970). Studies on the use of animal slurries to manure barley. 1. Effect of age and dilution of cow and pig slurry when applied at various rate before and after sowing. Rec Agric Res, Ministry of Agriculture, N Ireland 18 125.

23. PAIN, B.F. et al (1978). The effects of slurry and inorganic nitrogen fertiliser on the yield and quality of spring barley. J. Agric Sci. Camb., 90, 283-289.

24. JOHNSON, P.A. and PRINCE, J. (1983). Fertiliser value of pig slurry nitrogen for spring barley. Internal MAFF Record of Investigations (East Midland Region, Shardlow, UK).

25. LECOMTE, R. et al (1979). The influence of agronomic application of slurry on the yield and composition of arable crops and grassland and on changes in soil properties. Proceedings of an EEC seminar on "Effluents from Livestock" held at Bad Zwischenahn, 2-5 Oct, pp 139-183.

26. STEWART, T.A. (1968). The effect of age, dilution and rate of application of cow and pig slurry on grass production. Record of Agric Res 17(1), 68-90 (Ministry of Ag, NI).

27. KOLENBRANDER, G.J. (1981). Limits to the spreading of animal excrement on agricultural land. In "Nitrogen losses and surface run-off from land spread manures". Edited by J C Brogan, Nijhoff/Junk, CEC, The Hague.

28. CHAUSSOD, R., SANCHEZ, C., DUMET, M.C. and CATROUX, G. (1983). Influence d'une digestion anaerobic (methanisation) prealable dur l'evolution dans le sol du carone et de l'azote des dechets organiques. DGRST - Action Concertee 'Valorisation energetique des dechets agricoles' - Aide no. 78-7-2909 INRA, Dijon.

29. CHAUSSOD, R., CATROUX, G. and JUSTE, C. (1985). Effects of anaerobic digestion of organic wastes on carbon and nitrogen mineralisation rates: laboratory and field experiments. Presented to COST 681 WP4 meeting 'Efficient land use of sludge and manure. Askov, Denmark, June 1985. (In press).

30. BEAUCHAMP, E.G. et al (1978). Ammonia volatilisation from sewage sludge applied in the field. J Environ Qual 7, 141-146.

31. COKER, E.G. (1978). The utilisation of liquid digested sludge. Paper 7 in WRC Conference on "Utilisation of Sewage Sludge Land". Water Research Centre, (Stevenage, Herts, UK).

32. LARSEN, K.E. (1985). Fertiliser value of anaerobically treated cattle and pig slurry applied for barley and beet. In Proceedings of EEC Seminar on "Efficiency of land use of sludge and manure" (Askov, DK) (In press).

33. SUESS, A. and WURZINGER, A. (1985). The effect of anaerobic digestion on the nutrient value of farm manure. In Proceedings of EEC Seminar on "Efficiency of land use of sludge and manure" (Askov, DK) (In press).

34. COMMISSION OF THE EUROPEAN COMMUNITIES (1982). Proposal for a Council Directive on the Use of Sewage Sludge in Agriculture.

35. HALL, J.E. and DAVIS, J.M. (1984). Sewage sludge injection into agricultural land. Report of a workshop held at the Institution of Civil Engineers, London. November 1983, Water Research Centre report 714-M.

36. HALL, J.E. (1985). Machinery spreading: soil injection as a barrier to odour dispersion. Presented to a FAO/EEC joint workshop 'Odour prevention and odour control of organic sludges and livestock farming', Silsoe, UK.

37. BEAUCHAMP, E.G., KIDD, G.E. and THURTHELL, G. (1982). Ammonia volatilisation from liquid dairy cattle manure in the field. Can. J. Soil Sci., 62, 11-19.

38. SHERWOOD, M. (1981). Fate of nitrogen applied to grassland in animal wastes. In Proc XIV Int Grassland Congress, Lexington USA. A Smith and HAYS. Ed by Westview Press, USA (1983).

39. LARSEN, K.E. and KELLER, P. (1985). Injection of slurry to barley and fodder beet. Tiddsskr. Planteavl 89, 11-17.

40. KEMPPAINEN, E. (1985). Effect of cattle slurry injection on the quantity and quality of barley and grass yield. Presented to COST 681 WP4 meeting 'Efficient Land Use of Sludge and Manure', Askov, Denmark, June 1985. (In press).

41. BEAUCHAMP, E.G. (1983). Response of corn to nitrogen in preplant and sidedress applications of liquid dairy cattle manure. Can. J. Soil Sci. 63(2), 377-386.

42. LARSEN, K.E. (1985). Injection of cattle slurry to barley, beet, grass and maize. Presented to COST 681 WP4 meeting 'Efficient Land Use of Sludge and Manure, Askov, Denmark, June, 1985.

43. LUTEN, W., GEURINK, J.H. and WOLDRING, J.J. (1983). Yield response and nitrate accumulation of herbage by injection of cattle slurry in grassland. In Efficient Grassland Farming. (Ed. Corrall, A.J.) British Grassland Society 14, 185-191.

44. HALL, J.E. (1985). Soil injection research in the UK. Presented to COST 681 WP4 meeting 'Efficient Land Use of Sludge and Manure', Askov, Denmark, June 1985 (In press).

45. TUNNEY, H. and MOLLOY, S.P. (1985). Comparison of grass production with soil injected and surface spread cattle slurry. Presented to COST 681 WP4 meeting 'Efficient Land Use of Sludge and Manure', Askov, Denmark, June 1985 (In press).

46. LARSEN, K.E. and KELLER, P. (1985). Injection of slurry to grass. Tidsskr. Planteavl. 89, 19-24.

47. GODWIN, R.J. and WARNER, N.L. (1984). Soil injection of sewage sludge. Contract report to Water Research Centre.

48. HALL, J.E. and RYDEN, J.C. (1985). Current UK research into ammonia losses from sludges and slurries. Presented to COST 681 WP4 meeting 'Efficient Land Use of Sludge and Manure', Askov, Denmark, June 1985 (In press).

49. DE LA LANDE CREMER, L.C.N. (1985). Dutch experience with slurry injection. Presented to COST 681 WP4 meeting 'Efficient Land Use of Sludge and Manure', Askov, Denmark, June 1985 (In press).

50. DESTAIN, J.P. (1985). Use of dicyanamid as a nitrification inhibitor to improve slurry nitrogen efficiency. In Proceedings of seminar on 'Efficiency of land use of sludge and manures' held at Askov, Denmark (In press).

51. KJELLERUP, B.Sc (1985). Nitrogen effect of slurry mixed with nitrification inhibitors - field experiments. In Proceedings of seminar on 'Efficiency of land use of sludge and manures held at Askov, Denmark, (In press).

52. TUNNEY, H. (1984). Slurry-Meter for estimating dry matter and nutrient content of slurry. In Proceedings of CEC WP4 Seminar on "Long term effects of sewage sludge and farm slurries applications" (Pisa, Italy) (In press).

53. KJELLERUP, B.Sc Agro V. (1985). Agros nitrogen meter for estimation of ammonium nitrogen in slurry and liquid manure. In Proceedings of CEC WP4 Seminar on "Efficient land use of sludge and manure", (Askov, DK) (In press).

54. JAKUBOWSKI, W. (1985). US EPA-sponsored epidemiological studies of health effects associated with the treatment and disposal of wastewater and sewage sludge. Presented to COST 681 WP2 meeting, Metz, May 1985.

55. VOORBURG, J.H. (1985). Odour research and ammonia volatilisation. Presented to FAO/EEC joint workshop 'Odour Prevention and Odour Control of Organic Sludges and Livestock Farming', Silsoe, UK, April 1985.

56. HOFF, J.D., NELSON, D.W. and SUTTON, A.L. (1981). Ammonia volatilisation from liquid swine manure applied to cropland. J. Env. Qual. 10 (1), 90-95.

57. NOREN, O. (1985). Swedish experiences with soil injection. Presented to a FAO/EEC Joint Workshop entitled "Odour prevention and odour control of organic sludges in livestock farming. Silsoe, UK.

58. DAM KOFOED, A. and SONDERGAARD-KLAUSEN, P. (1985). Leaching of nutrients from sewage sludge and animal manure. In Proceedings of EEC Seminar on "Efficiency of land use of sludge and manure" (Askov, DK) (In press).

59. STEFFENS, G. and VETTER, H. (1983). The leaching of nitrogen after slurry application with and without catchcropping. Landwirtsch. Forsch. 36, Kongre Bband.

60. UNWIN, R.J. (1985). Leaching of nitrate after application of organic manures. Lysimeter studies. In proceedings of EEC Seminar on "Efficiency of land use of sludge and manure" (Askov, DK) (In press).

61. LANDE CREMER, L.C.N. de la (1978). Bewerkung der Wirtschaftsdunger in den Niederlanden. Lecture delivered at the International Colloqium "Dungerberatung" in Gumpenstein, Austria (In press).

62. MINISTRY OF AGRICULTURE, FISHERIES AND FOOD (1982). Profitable Utilisation of Livestock Manures. Booklet 2081(HMSO).

63. VETTER, H. (1983). Sind 3 Dungeinheiten oder 3 Dungergrossviehein - heiten zu wenig? Landiw-Blatt, Weser-Ems 130(30), 15-19.

64. HALL, J.E., CARLTON-SMITH, C.H., DAVIS, R.D. and COKER, E.G. (1983). Field investigations into the manurial value of liquid undigested sewage sludge. Water Research Centre report 652-M.

65. COKER, E.G., HODGSON, D.R. and SMITH, A.T. (1984). The effects of undigested primary sewage sludge on the growth and nitrogen uptake of barley and permanent grass. Water Research Centre report 698-M.

66. HALL, J.E., CARLTON-SMITH, C.H., DAVIS, R.D. and COKER, E.G. (1983). Field investigations into the manurial value of lagoon-matured digested sewage sludge. Water Research Centre report 510-M.

67. EDGAR, K.F., FRAME, J. and HARKESS, R.D. The manurial value of liquid, anaerobically digested sewage sludge on grassland in the west of Scotland. I. Nitrogen value of sludge applied during the winter months. (In press).

68. HALL, J.E. and COKER, E.G. (1981). The effect of time of application of lagoon-matured digested sewage sludge on the yield and composition of winter wheat. Water Research Centre report 146-M.

69. HALL, J.E. (1983). The effect of sewage sludge on the growth and composition of winter wheat. Water Research Centre reports 386-M and 615-M.

70. GRACEY, H.I. (1983). Efficiency of slurry nitrogen as affected by the time and rate of slurry application and rate of inorganic nitrogen. In Efficient Grassland Farming. Proceedings of European Grassland Federation Conference, Reading 1982.

71. PAIN, B.F. and SANDERS, L.T. (1980). Effluents from intensive livestock units – fertiliser equivalent of cattle slurry or grass and forage maize. In "Effluents from Livestock" Ed J K R Gasser, (Applied Science Publications).

72. UNWIN, R.J., PAIN B.F. and WHINHAM, N. The effect of rate and time of application of nitrogen in cow slurry on grass cut for silage. Agricultural Wastes (In press).

73. PAIN, B.F., SMITH, K.A. and DYER, C. Factors affecting the response of cut grass to nitrogen in dairy cow slurry. Agricultural Wastes (In press).

74. KOLENBRANDER, G.J. and LANDE CREMER, L.C.N. de la (1967). Stalmest en gier, waarde en mogeliijkheden. Veenman, Wageningen, pp188.

75. SLUIJSMANS, C.M.J. et al (1978). The spreading of animal excrement on utilised agricultural areas of the Community. I. Scientific basis for the limitation of quantities and criteria for rules thereon. CEC Info on Agriculture No 47, pp154.

76. VETTER, H. and STEFFENS, G. (1979). Gullewirkung mit und ohne erganzende Stickstoff-dungung. Landwirtsch. Forsch. 33(1980), Sonderheft 36, 365-373.

77. NEMMING, O. (1982). Stigende maengder fast Svinegodning og svinegylle til byg. Tidsskr. Planteavl 86, 127-132.

78. SAFLEY, L.M., LESSMAN, G.M., WOLT, J.D. and SMITH, M.C. (1981). Comparison of corn yields between broadcast and injected applications of swine-manure slurry. In Livestock Waste: A renewable resource. Am. Soc. Agric. Engnrs., 178-180.

79. PAIN, B.F. and BROOM, D.M. (1978). The effects of injected and surface-spread slurry on the intake and grazing behaviour of dairy cows. Anim. Prod. 26, 75-83.

TABLE I – AVERAGE NITROGEN CONTENT OF DIFFERENT SEWAGE SLUDGES AND ITS AVAILABILITY
FOR CROP GROWTH.

	Dry Solids (%)	N Content (% ds)	Availability (%)	Total N (kg/m^3)	Available N* (kg/m^3)
Liquid Undigested	5	3.5	35	1.8	0.6
Liquid Digested	4	5 $(3+2)^{**}$	100 NH_4 –N +15 organic N	2.0	1.2
Undigested Cake	25	3	20	7.5	1.5
Digested Cake	25	3	15	7.5	1.1

* Available in first cropping year

** (Ammonia + organic N).

TABLE II – AVERAGE NITROGEN CONTENTS OF CATTLE, PIG AND POULTRY SLURRY – MEAN AND RANGE UNDER
DUTCH CONDITIONS (61).

	Cattle Slurry	Pig Slurry	Poultry Slurry
On dry matter basis (%)	5.0(4.0–8.0)	8.25(5.4–14.4)	7.8(4.24–11.25)
On wet matter basis (kg/tonne or m^3)	5.0(4.0–8.0)	8.25(5.4–14.4)	10.9(5.9–15.75)

For Comparison (kg/m^3)

UK (MAFF) (62)	5.0	6.5	13.8
Germany (63)	4.5	7.0	10.0

TABLE III - RELATIVE YIELDS OF RYEGRASS RESULTING FROM APPLYING SLUDGE AT DIFFERENT TIMES OVER THE AUTUMN-SPRING PERIOD

	Oct	Nov	Dec	Jan	Feb	Mar	Apr	Reference
Undigested								
Sandy loam	-	81	-	-	100	-	-	(64)
Silty loam	-	67	-	-	100	-	-	(64)
Clay loam	-	100	-	-	96	-	-	(64)
Loam	-	90	-	-	100	-	-	(65)
Silty clay loam	85	-	-	-	-	100	-	(45)
Digested								
Sandy loam	-	54	-	-	100	-	-	(66)
Silty loam	-	100	-	-	100	-	-	(66)
Clay loam	-	43	-	-	100	-	-	(66)
Calcareous loam	-	100	-	-	97	-	-	(66)
Clay loam	-	33	-	95	100	-	95	(67)
Sandy clay loam	-	88	-	100	88	-	72	(67)
Silty clay loam	96	-	-	-	-	100	-	(45)

TABLE IV - RELATIVE YIELDS OF CEREALS RESULTING FROM APPLYING SLUDGE AT DIFFERENT TIMES OVER THE AUTUMN-SPRING PERIOD.

	Sept	Oct	Nov	Dec	Jan	Feb	Mar	Apr	References
Undigested									
Spring barley (3 trials)	-	82	-	-	-	-	-	100	(65)
Digested									
Spring barley (11 trials)	-	-	100	-	97	-	85	-	(31)
Winter wheat									
1 trial	86	-	-	93	-	100	-	99	(68)
3 trials	92	-	100	-	-	-	-	-	(69)

TABLE V - MEAN NITROGEN VALUES OF UNDIGESTED SLUDGE APPLIED ANNUALLY TO GRASS IN AUTUMN
AND/OR SPRING, 1978-81 (65).

	Rate of application (m^3/ha)	Fertiliser replacement value (kg N/ha)
Autumn	60*	46
Spring	60	51
Autumn and Spring	30 + 30	60

*$1m^3$ contained 6.4 kg total N.

TABLE VI - SUMMARY OF FIELD TRIALS CONCLUDED SINCE 1977 AND EFFICIENCIES OF SLURRY N ON GRASS.

Year	Country of Origin	Type of Slurry	Rate Applied	N Efficiency	Reference
1983	N Ireland	Cattle	52 kg/ha of slurry N	March - 86% Jan - 80% Nov - 80%	(70)
1980	NIRD in England	Cattle	174-210 kg/ha slurry N in single dressings	25-30% Over whole season	(71)
1979	NIRD and ADAS, England	Cattle	148-370 kg/ha slurry N	Mid-winter - 15% Early spring - 30% After 1st cut - 25%	(72)
1982	NIRD and ADAS, England (27 trials)	Cattle	80 and 160 kg/ha slurry N	Early spring - 38% (low rate) Early spring - 23% (high rate) After 1st cut - 24% (low rate) After 1st cut - 17% (high rate)	(73)

TABLE VII - % EFFICIENCY OF SLURRY N ON ARABLE CROPS (25).

Type of Slurry	Crop	Location	Total N in Slurry	Efficiency
Deep loamy soil				
Pig	W Wheat	Marcq	121	50
Pig	W Wheat	Marcq	247	24
Cattle	W Wheat	Marcq	108	90
Cattle	W Wheat	Marcq	207	48
Poultry	W Wheat	Marcq	142	49
Poultry	W Wheat	Marcq	288	42
Shallow Soil				
Cattle	W Barley	Braibant	91	22
Cattle	W Barley	Braibant	157	13
Cattle	W Barley	Braibant	181	17
Cattle	W Barley	Emptinne	88	11
Cattle	W Barley	Emptinne	152	7
Cattle	W Barley	Emptinne	172	6

TABLE VIII - EFFICIENCY OF SLURRY N FOR ARABLE CROPS (EXPERIMENTS CONDUCTED ON THE CONTINENT) (9)

Year	Country	Type of Slurry	Crop	Efficiency of Slurry N (%)	Reference
1967	Netherlands	Cattle	Arable	Spring - 50 Autumn - 25	Kolenbrander and Dela Lande Cremer (74)
1978	Netherlands	Cattle	Arable	Spring - 50 Autumn - 30	
		Pig		Spring - 55 Autumn - 30	Sluijsmans et al (75)
		Poultry		Spring - 65 Autumn - 30	
1979	Germany	Pig	Arable	Oct/Nov Dec/Jan Feb/Mar Sandy Soil 10-20 30-50 60-70 Silty Clay Soil 20-30 30-50 60-80 Deep Loam Soil 30-50 40-60 60-80	Vetter and Steffens (76)
1982	Denmark	Pig	Barley	Dec/Jan -35	Nemming (77)

- 280 -

TABLE IX - SPRING BARLEY - YIELDS FROM RAW AND DIGESTED SLURRIES (t/ha) (85% DM)-(32)

	1979-81 (Average of 17 trials)	1982-84 (Average of 9 trials)	
Control - Nil N	2.49	2.50	
Fertiliser (CAN)		Yield Increases	
40 kg/ha	1.06	2.20	
80 kg/ha	1.66	2.20	
120 kg/ha	1.68	2.46	
Slurry	Cattle	Cattle	Pig
Raw - 80 kg N/ha	0.92	1.08	1.55
- 160 N/ha	1.25	1.70	2.30
Digested - 80 kg N/ha	0.94	1.29	1.55
- 160 kg N/ha	1.35	1.91	2.26

TABLE X - FODDER BEET - YIELDS FROM RAW AND DIGESTED SLURRIES (CROP UNITS/HA) (1 Crop Unit = 100 kg barley grain) - (32)

	1979-81 (Average of 12 trials)	1982-84 (Average of 6 trials)	
Control - Nil N	74.0	85.0	
Fertiliser (CAN)		Yield Increases	
75 kg N/ha	27.4	28.3	
150 kg N/ha	42.0	31.4	
225 kg N/ha	51.8	32.8	
Slurry	Cattle	Cattle	Pig
Raw - 120 kg N/ha	27.0	24.0	27.1
- 240 kg N/ha	38.1	36.0	40.9
Digested - 120 kg N/ha	30.4	30.3	28.9
- 240 kg N/ha	43.6	35.0	36.1

TABLE XI – EFFECT OF BIO-GAS SLURRY ON THE YIELD OF WINTER WHEAT (t/ha) – (33)

	No Fertiliser N	80 kg/ha Fertiliser N
	(LSD = 0.6 t/ha (P = 0.05)	
No slurry applied	3.90	6.36
30 m^3 fresh slurry (11 March – 156 kg N)	5.30	7.49
30 m^3 bio-gas slurry (11 March – 108 kg N)	6.18	7.08
30 m^3 fresh slurry (19 April – 168 kg N)	4.62	7.12
30 m^3 bio-gas slurry (19 April – 148 kg N)	5.30	7.20

TABLE XII - INCREASE IN YIELDS OF ARABLE CROPS FOLLOWING SURFACE APPLICATION OR INJECTION OF LIQUID MANURE APPLIED AT EQUIVALENT RATES RELATIVE TO N FERTILISER RESPONSE (100)

Crop	Manure	Soil type	Fertiliser	Manure		Tine Spacing (cm)	Rate of application (m³/ha)	Time of application	Reference
				Surface applied	Injected				
Maize	Pig	silt loam	100	80	100	-	-	early spring	(78)
Maize	Cattle	silt loam	100	33	60	75	23-93	spring	(41)
Arable	Slurry	-	100	80	90	-	-	spring	(74)
Barley	Cattle	sand, sandy loam and loam	100	66	109 101 92	30 60 75	20,40	spring	(39)
Fodder	Cattle	sand, sandy loam and loam	100	44	98 76 79	30 60 75	40,80	spring	(40)
Barley	Cattle	sandy clay	100	23 82 61 21	52 91 71 2	50 50 50 50	50 50 50 50	Sept/Oct before ploughing May - before sowing May - sprouting 1 month after sowing	(40)
		Mean	100	54	78				

TABLE XIII - LOSS OF AMMONIA FOLLOWING SURFACE APPLICATION OR INJECTION OF SLURRY (%).

Cumulative loss of NH_3

Slurry	Application rate	Sampling period (h)	Surface applied	Injection	Reference
Pig	90	84	14.0	2.6	(56)
	135	84	12.2	-	
	180	84	11.2	2.6	
Cattle	40	18	10.7	1.0	(48)
		45	14.0	1.2	
		65	14.8	1.2	
		88	15.2	1.2	
		116	15.6	1.2	
		204	16.1	1.2	

TABLE XIV - EFFECT OF DIFFERENT METHODS OF APPLICATION OF CATTLE SLURRY (60 t/ha) TO GROWING MAIZE. INCREASES IN YIELD OF COB AND STALK (tdm/ha). MEAN OF 4 EXPERIMENTS AND 2 SOILS-(42).

Slurry applied at crop height (cm)	1	15	40
Surface application	0.8	1.6	0.6
Placed between rows	1.9	2.1	2.2
Injection	2.4	4.3	3.1

TABLE XVA - INCREASE IN YIELD OF GRASS FROM SURFACE APPLICATION OR INJECTION OF LIQUID SLURRY RELATIVE TO N FERTILISER RESPONSE (100)

Manure	Soil type	Fertiliser	Manure Surface applied	Manure Injected		Tine spacing (cm)	Rate of application (m³/ha)	Time of application	Reference
					Irrigation				
Cattle	sand	100	30	63	0	50	50	June	(40)
			70	84	30 mm	50	50		
	silty clay		52	55	0	50	50		
			60	53	30 mm	50	50		
Cattle	loam	100	45	50		60	25	April	(45)
			63	67		60	37.5		
			52	75		30	50		
			62	70		60	75		
Cattle	sand	100	10	34		50	20	Early spring	(43), (40)
					No. of passes				
Cattle	sand	100	57	74	1	30	80	Spring	(46)
				61	3	30	80		
				76	1	60	80		
				76	3	60	80		
				74	1	75	80		
				54	3	75	80		
Cattle	loam	100	45	58	1	30	80	Spring	(46)
				34	3	30	80		
				49	1	60	80		
				36	3	60	80		
				42	1	75	80		
				15	3	75	80		
					Didin				
Cattle	-	100	42	57	- 1st harvest		20,40,60	Spring	(49)
		85	38	81	+				
		100	28	51	- 5 harvests				
		95	21	49	+				

TABLE XVB - INCREASE IN YIELD OF GRASS FROM SURFACE APPLICATION OR INJECTION OF SEWAGE SLUDGE RELATIVE TO N FERTILISER RESPONSE (100)

Digested loamy sand	100	84	30	1st harvest	65	64,115	March (44)
sludge		103	77	1st year total			
Digested clay	100	89	29	1st harvest	65	136	May (44)
sludge		87	46	1st year total			
		87	107	1st & 2nd year			
Digested silty clay loam	100	135	134		65	140	October (44)
sludge		124	64			140	March
Undigested silty clay	100	73	57		65	140	October (44)
sludge loam		126	16			140	March

TABLE XVI - THE RELATIVE PALATABILITY OF GRASS TREATED WITH CATTLE SLURRY.

Fertiliser	Surface applied	Injected	Ruminant	Reference
100	70	98	Sheep	(40)
100	78	114	Cows	(79)

TABLE XVII - EQUIPMENT DESIGN AND RELATIVE RISKS OF ODOUR AND AEROSOL PRODUCTION

Means of discharge	Droplet size	Trajectory	Distribution device	Odour risk
Vacuum	small	high	gun	high
pump			splash plate	
auger	large	low	dribble bar	
gravity			injector	low

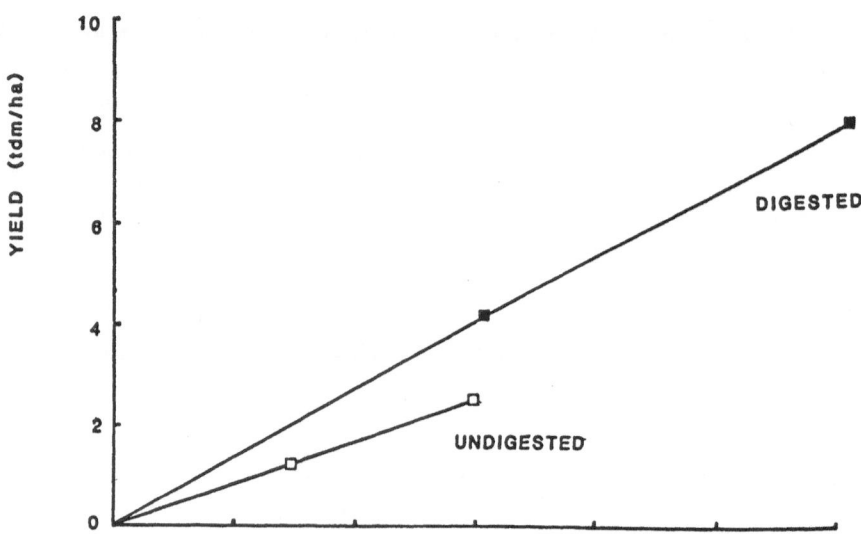

FIGURE 1. Increase in grass yields following the application of liquid digested sludge (■) and liquid undigested sludge (□) at 70 and 140 m³/ha. (Example from reference 63)

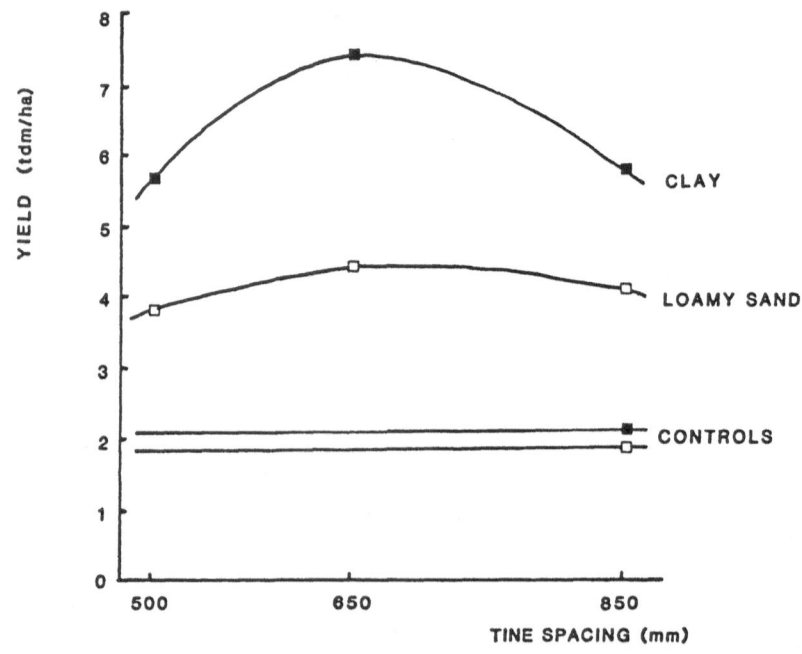

FIGURE 2. The effect of injector tine spacing on mean grass yield. Digested sludge at 140 m³/ha.

DISCUSSION

C Juste
    Amongst the work programme for Working Party 4 it was planned to examine micropollutant effects on nitrogen mineralisation in sewage sludge.

J H Williams
    We have not considered this.

D Sauerbeck
    I was not certain for the basis of calculating efficiencies - total or mineral N? Much depends on the amount of available nitrogen and applying it as close to the maximum growth period as possible. In FRG, there are official regulations which prohibit spreading slurry between mid October and February. This has many pros and cons and the farmers try to empty their tanks by October which is worse still. Applications in late winter would be better for minimising nitrogen losses.

J H Williams
    I agree with your comments. This is why we are pursuing research on these applications, discouraging autumn application with the recommendation to store until winter. The efficiencies in the paper are based on total nitrogen content.

EFFECT OF A LONG TERM SLUDGE DISPOSAL
ON THE SOIL ORGANIC MATTER CHARACTERISTICS

M. LINERES, C. JUSTE, J. TAUZIN and A. GOMEZ

I.N.R.A., Station d'Agronomie, Centre de Recherches de Bordeaux (France)

Summary

The long term effet (8 years) of an anaerobic partially dried sewage sludge (10 t/ha/year or 100 t/ha/2 years in dry matter basis applications) on the organic status of a sandy soil cropped to continuous maize was compared to the effect of a similar farmyard manure application.
Organic status of soil (60 cm depth) has been evaluated using study of C and N soil transformation, biological activity (respirometric study), C.E.C. and water holding capacity of treated soils. Changes organic matter of sludge treated soils were also evaluated using C, N analysis and IR spectroscopy of extracted humic acids.
No increase in soil organic C occurs in plots as a result of 10 t/ha/year sewage sludge disposal. At the end of the experiment, 3 % of C added as sewage sludge was recovered whereas 49 % was recovered in case of farmyard manure disposal. In plots amended with heavy level of sewage sludge (100 t/ha/2 years) C recovery percentage was higher 29 %).
Nitrogen, added to soil as sewage sludge or farmyard manure, accumulated in the 0-60 cm layer andwwas more resistant to decomposition than carbon ; as a result, a strong decrease in C/N ratio of residual sewage sludge organic matter was observe (1,9 to 6,2 values).
The low capacity of sewage sludge disposal to increase soil organic carbon content was corroborated by the absence of sewage sludge effect on the C.E.C. and water holding capacity of treated soils.
Sewage sludge disposal brought however noticeable changes in soil organic matter composition as reflected by specific carbon biodegradability increase and N enrichment of extracted humic acids resulting from proteinaceous material addition.

INTRODUCTION

One of the main reasons why municipal sludge is promoted in agriculture is its organic content. The latter is regarded a priori as capable to maintain the stock of soil humus and thus to improve the behaviour of soils through its effect on their structural stability, cationic exchange capacity, water holding capacity and food-growing-potential.

This long term contribution assumes that this type of material is likely to leave a stable residue in the soil, as do the ligno-cellulosic residues usually included in organic fertilizers such as farmyard manure. Data from long term field experiment concerning the effects of continuous sludge applications on the composition of soil organic matter are extremely rare, due to the recent use of this type of wastes in agriculture.

Such informations are necessary, however, if the use of sewage sludges in agriculture is to be sensibly controlled. The purpose of this project is to provide some of the answers, with respect to environmental conditions (sandy soil, mid, damp Atlantic climate) and growing techniques (a single maize crop, massive fertilization, irrigation, weed control) which optimize the breakdown of organic matter. For that, on experimental plots conditioned over an eight-year period with either sludge or manure, the soil carbon and nitrogen balance and the impact of repeated applications on the cationic exchange capacity, water holding capacity, biological activity and composition of the humus fraction were studied.

MATERIAL AND METHODS

A/ Experimental design

The experiment is located in an INRA experimental farm, on an old clearing which had carried two stabilizing crops (maize in 1972 and oats in 1973).

The lay-out consists in a randomized block of four 18 m2 plots (3 x 6 m) each treatment being repeated five times.

The test started in 1974 and then bore eight successive maize crops (INRA 260 variety) until 1981.

The soil is gravelly, with a very coarse structure (74 % coarse sand), initially poor in organic matter and very acid. At the start of the experiment, a chemical analysis was done on the surface layer and a granulometric analysis of the surface and bottom layers (see Table 1).

The four experimental treatments are as follows :

C : control, mineral fertilization only ;

M : annual application of 10 t/ha farmyard manure (dry matter basis), giving a total of 80 t/ha in 1981 ;

S_1 : annual application of 10 t/ha sewage sludge, giving a total of 80 t/ha ;

S_2 : application of 100 t/ha sludge every two years, giving a total of 400 t/ha.

The annual applications of mineral fertilizers (N, P, K) for the C, M and S_1 treatments were adjusted to the same level, taking into account the inputs of the various organic amendements and supposing that the half of organic nitrogen they contain mineralize into one year.

The sludge came from a sewage plant in a mainly residential area : the sludge is biological anaerobically digested one, then floculated with a polyelectrolyte before being passed through a filter press. It had an medium heavy metal content.

B/ Organic balance

Before each incorporation, the organic matter was analysed in order to evaluate the total quantities of C and N incorporated each year into the soil from this source.

The balances were drawn up by comparing the quantities of C and N thus added with the quantities detected in the soil after eight years of spreading sludge.

In October 1981, soil samples were taken from the 0-20, 20-40 and 40-60 cm layers in each plot and the bulk density was measured. Once the percentage of gravel had been determined (0 > 2 mm), each air-dried sample was finely grinded and sifted to 315 µm. Carbon was determined by combustion at 1 050° C and coulometric analysis of the CO_2 released, and nitrogen by Kjeldahl's method.

C/ Cationic exchange capacity

The exchange capacity for the 0-20 and 20-40 cm layers is determined

by percolation with normal ammonium acetate ; it is expressed in milliequi-
valents per 100 g soil.

D/ Water holding capacity

This characteristic is measured in the upper 2 layers (0-20 and 20-40
cm). The samples, saturated by capillarity, are centrifuged to 1 000 g for
20 minutes. The results are expressed as a percentage of fine soil.

E/ Soil biological activity

The biological activity of the soil samples taken from the upper
layer (0-20 cm) was measured in the laboratory. The respirometric technique used consists in measuring quantities of carbon produced in the form
of CO_2 resulting from the microflora activity in the incubated soils. The
measuring vessels used are 250 ml erlenmeyer flasks with ground-glass
necks, sealed with washbottle stoppers with taps fitted to the inlet and
outlet tubes. At the determination moment, this rig made it possible to
ventilate the atmosphere of the vessel with a flow of gas (air or oxygen),
entraining the carbon dioxide produced by the soil to the measuring
cham-ber of an Eraly analyser, where CO_2 is determined by coulometric
analysis.

The experiment was corried out as follows :
- 50 g of air-dried soil, passed through a 2 mm sieven, from each of the 4
 plots in the block are placed in measuring flasks. After the soil has
been moistened by adding demineralized water, the flasks are sealed herme-
tically and placed in a bacteriological drying oven at 28° C.
- The quantity of carbon released in CO_2 form is measured after 1, 2, 4,
7, 9 and 11 days of incubation. After each measurement, the soils are
immediately put back in the incubator, the passage of air required for the
analysis being regarded as sufficient to renew the oxygen in the vessel
atmosphere.
- The results are plotted in a graph indicating, for each day and each mg
of carbon initially present in the sample, the average rate of carbon
produced (μg/h).

F/ Study of the soil humic fraction

The humic acids are extracted from the upper layer (0-20 cm) of the
C, M and S_2 plots.
- **Extraction of humic acids** - For each of the 5 plots, 75 g of soil are
added to 150 ml of 0,5 N sodium hydroxide and stirred for 4 hours. The suspension is centrifuged and then filtered. The filtrates from each plot
sample are collected, their pH being adjusted at 1 by adding several
quantities of HCl producing precipitation of humic acids over a 24 hour
period. The humic fraction is separated from the fulvic one by decanting,
centrifuging and dialysis. The final product is dried in the drying ovent
at 40° C and finely grinded.
- **Analytical measurements** - The carbon content of the humic acids is determined by combusting in oven and measure of CO_2 by coulometric analysis. Nitrogen is determined in a Technicon auto-analyser.
- **Infra-red spectrophotometry** - Potassium bromide pellets are used to produce the spectra, the organic products being added in a rate of 1 %. The
apparatus used is a Beckman Acculab 8 spectrophotometer.

3. RESULTS AND DISCUSSION

A/ Organic balance

The results of the analysis are expressed in concentrations and in
tons or kilograms per hectare for each layer, allowance being made for the
bulk density and weight of the fine soil.
- **Carbon** - The following conclusions can be drawn from the results in
Table 2 :

. carbon effect is only marked in the 0-20 cm layer : no significant deep removal of the organic matter applied is observed whatever its nature ;

. in the upper layer (0-20 cm) it was found that the carbon concentration resulting from the S_2 treatment is clearly greater than that from the other 3 treatments (significant at 0,1 % level). Among these, manure treatment (M) is the richest (significant at 1 % level). There is no significant variation, however, beetween the S_1 plots and the controls : the sludges applied at a rate of 10 t/ha/year dry matter do not therefore increase the final carbon content ;

. about 50 % of the organic matter brought by the manure remains after 8 years of annual application, whereas the residual amount left from the same dry matter weight of sludge represents only 3 % of the applications. By contrast, massive application of sludge leads to a much higher humic contribution (22 %).

In other words, the disappearance rate of the sludge incorporated at 10 t/ha dry matter is much higher than that of massive treatment, probably because, in the latter case, the soil biological purification capacity is saturated.

- **Nitrogen and C/N ratio** (see Table 3) - Down to 60 cm, the N content for the S_2 treatment method is significantly different at 0,1 % level from that for all other treatment methods, which means that only the nitrogen applied in the heavy rate of sludge had partly migrated to the lower layers. Just as for carbon, the persistence of nitrogen in strongly increased, therefore, when a heavy quantity of sludge is applied (S_2).

It was also found that much more nitrogen (incorporated from either sludge or manure) is conserved than carbon (the isohumic coefficient is higher and the disappearance rate therefore lower). As a result, the C/N ratio of the residual organic matter (see Table 4) is clearly lower than that for the initial organic matter, especially in the case of sludge.

Clearly, for either carbon or nitrogen, the disappearance rate can be regarded as a broad indication of the proportion mineralized during the year following incorporation, when the latter occurs every year. Nevertheless the identification of this average disappearance rate with an annual rate is more debatable when sludge is applied every 2 years.

B/ Cationic exchange capacity (C.E.C.)

It will be seen from Table 5 that cationic exchange capacity increases only in the case of treatments M and S_2, the latter being quite distinctive from the rst. By contrast, the low sludge application (S_1) does not affect the soil C.E.C.

For the whole experiment, a linear relationship is established between carbon content and cationic exchange capacity : C.E.C. (in me/100 g) = 0,30 C (per 1 000) - 0,78.

The curves for (a) just the control plots (y = 0,24 x -0,11) and (b) the control and manure plots (y = 0,25 x -0,25) have weaker slopes which hardly differ from each other. This leads to the conclusion that sewage sludges have a greater effect than manure on the the variation in C.E.C.. This preponderant effect can be attributable to the stable organic matter resulting from changes in the sludges applied since the beginning of the assay as well as to the sludges recently worked into the plot (1980) and still transforming.

C/ Water holding capacity (see Table 6)

Water holding capacity, which was initially very low, increases significantly only after high sludge applications (S_2) ; the manure treatment only tendes to have a favourable effect on this soil characteristic.

The following linear relationship between carbon content and water holding capacity is observed :

水 water holding capacity (%) = 0,32 x C (per 1 000) + 3,12
where r = 0,87 (significant at the 0,1 % level).

The looser correlations between carbon content and water holding capacity than those observed in the case of cationic exchange capacity are probably explained by manure having a lower specific effect on water holding capacity than on C.E.C.

D/ <u>Soil biological activity</u> (see Figure 1)

It was found that the biological activity of the plots conditioned with manure or sludge was greater than that of the control plots. The superiority of the S_1 treatment can be explained by the existence in these plots of a microflora activated at regular intervals by annual applications of sludge that was more fermentable than the original organic matter in the soil and also richer in fertilizing elements. Between treatments S_1 and C, therefore, there are differences in the organic matter composition and the behaviour of their respective microbial flora which are note revealed by simply determining total carbon.

E/ <u>Study of humic fraction of the soils</u>

- **Carbon and nitrogen** (see Table 7) - The large amount application of sludge principally incrases the nitrogen content of the humic acids, bringing about a significant decrease of their C/N ratios, a decrease comparable to that found in the residual organic matter of the higly conditioned plots (see Table 4).

- **Infrared spectra** (see Figures 2 and 3) - The humic acids extracted from plots conditioned with 100 t of sludge every two years are characterized by a shift in the OH band from 3 350 cm^{-1} to 3 300 cm^{-1} and by a peak at 1 640 cm^{-1} . These analytical and spectrometric results indicate the presence in the extracted compounds of protein material, a phenomenon wich certain authors, notably BOYD S.A. et al. (1980) have already reported.

4. CONCLUSION

The essential point to remember from the experiments just described is that, in the medium term, some types of sewage sludge are almost unable to enrich with stable carbon an sandy intensively cropped soil (irrigation, large scale use of mineral fertilizer) in a mild oceanic climate, despite applications far in excess (10 t/ha/yr dry matter) of those currently recommended in practice. From this point of wiew, even applied in caricatural quantities (100 t/ha/2 years dry matter), sludge proves a much less suitable way of increasing the soil stock of stable humus than farmyard manure applied in moderate rates.

According to Terry et al. (1979), GILMOUR J.T. and GILMOUR C.M. (1980) who showed with the help of laboratory experiment, that about 60 % of a sluge organic matter resists decomposition, a significant increase in the soil organic matter content should have been observed with lay-out, but this was not the case. It must be concluded that field conditions considerably enhance the biological brakdown of sludge, stable humus yield of which is ultimately quite low.

However, as regards quality, the suitability of sludge for enriching the soil humic fraction with protein products and the increase in the specific biodegradability of the carbon in the conditioned soil show that the pre-existing organic material is reworked, since a part is probably replaced by organic compounds deriving from the waste.

The current use of unlimed biological sludge with a moderate load of heavy metals may therefore, in the long terme, have a different effect on soil fetility to that of an organic ligno-cellulosic amendement such as manure.

REFERENCES

BOYD S.A., SOMMERS L.E., NELSON D.W. - Changes in humic acid fraction of
    soil resulting from sludge application. Sopil Sci. Soc. Amer. J.,
    1980, 44, 1179-1186.
GILMOUR J.T., GILMOUR C.M. - A simulation model for sludge decomposition
    in soil. J. Environ. Qual., 1980, 9, n°2, 194-199.
TERRY R.E., NELSON D.W., SOMMERS L.E. - Carbon cycling during sewage slud-
    ge decomposition in soils. Soil Sci. Soc. Amer. J., 1979, 43,
    494-499.

TABLE 1

Granulometric and chemical analysis of the soil

	cm		
	0-20	20-40	40-60
Gran. size (% of fine soil)			
- à 2 µ	4,7	4,5	5,2
- 2 µ à 20 µ	7,3	7,2	8,8
- 20 µ à 50 µ	5,4	5,1	6,2
- 50 µ à 200 µ	11,0	10,2	10,9
- 200 µ à 2 000 µ	71,6	73,1	68,9
Chemical analysis			
- % of fine soil			
. C	1,57		
. N	0,16		
C/N	14,0		
pH-water	5,3		
pH-KCl	4,6		
- ‰ of fine soil			
. assimilable P_2O_5 (Dyer)	0,249		
. Ca	0,579		
. K	0,053		
. Mg	0,020		

TABLE 2

Changes in carbon content after 8 years application of sludge or manure (average of 5 replications)

TREATMENT	UNIT	Distribution in the soil profile				Quant. C (kg/ha)		Isohumic coefficient	Disappearance rate
		0-20 cm	20-40 cm	40-60 cm	TOTAL	residual	applied		
CONTROL (C)	per thousand kg/ha	15,2 / 39 995	11,3 / 30 853	5,3 / 14 950	85 798				
MANURE (M)	per thousand kg/ha	18,1 / 48 628	12,1 / 33 640	6,2 / 16 860	99 128	13 330	27 040	49 %	51 %
SLUDGE 10 t/ha (S1)	per thousand kg/ha	15,2 / 40 633	11,1 / 30,813	5,1 / 15 046	86 492	694	23 178	3 %	97 %
SLUDGE 100 t/ha (2)	per thousand kg/ha	22,2 / 59 472	12,6 / 34,707	6,0 / 16 841	111 019	25 221	116 800	22 %	78 %

TABLE 3.

Changes in nitrogen content after 8 years application of sludge or manure (average of 5 replications)

TREATMENT	UNIT	Distribution in the soil profile				Quant. C (kg/ha)		Isohumic coefficient	Disappearance rate
		0-20 cm	20-40 cm	40-60 cm	TOTAL	residual	applied		
CONTROL (C)	per thousand	1,000	0,712	0,276					
	kg/ha	2 630	1 937	786	5 353				
MANURE (M)	per thousand	1,240	0,774	0,340				62 %	38 %
	kg/ha	3 328	2 158	962	6 448	1 095	1 773		
SLUDGE 10 t/ha (S1)	per thousand	1,082	0,704	0,288				17 %	83 %
	kg/ha	2 900	1 954	864	5 718	365	2 089		
SLUDGE 100 t/ha (2)	per thousand	2,170	0,930	0,386				36 %	64 %
	kg/ha	5 811	2 545	1 087	9 443	4 090	11 290		

TABLE 4

C/N ratio of the organic matter

| TREATMENTS | C/N organic matter | |
	applied	residual
Manure (M)	15,3	12,2
Sludge 10 t/ha (S1)	11,1	1,9
Sludge 100 t/ha (S2)	10,3	6,2

TABLE 5

Exchange capacity, me /100 g (average of 5 replications)

| | Depth (cm) | |
	0-20	20-40
Control (C)	3,6	2,5
Manure (M)	4,4	2,9
Sludge 10 t/ha (S1)	3,9	2,6
Sludge 100 t/ha (S2)	6,3	3,0

TABLE 6

Water holding capacity in % (average of 5 replications)

| | Depth (cm) | |
	0-20	20-40
Control (C)	8,0	7,1
Manure (M)	8,5	6,7
Sludge 10 t/ha (S1)	7,9	6,6
Sludge 100 t/ha (S2)	10,8	7,4

TABLE 7

C/N ratio of various humic acids extracted from the soils (Depth 0-20 cm)

TREATMENT	Carbon %	Nitrogen %	C/N	C/N of the residual organic matter
Control (C)	51,14	3,74	13,7	-
Manure (M)	55,76	3,92	14,2	12,2
Sludge 100 t/ha (S2)	54,51	5,06	10,8	6,2

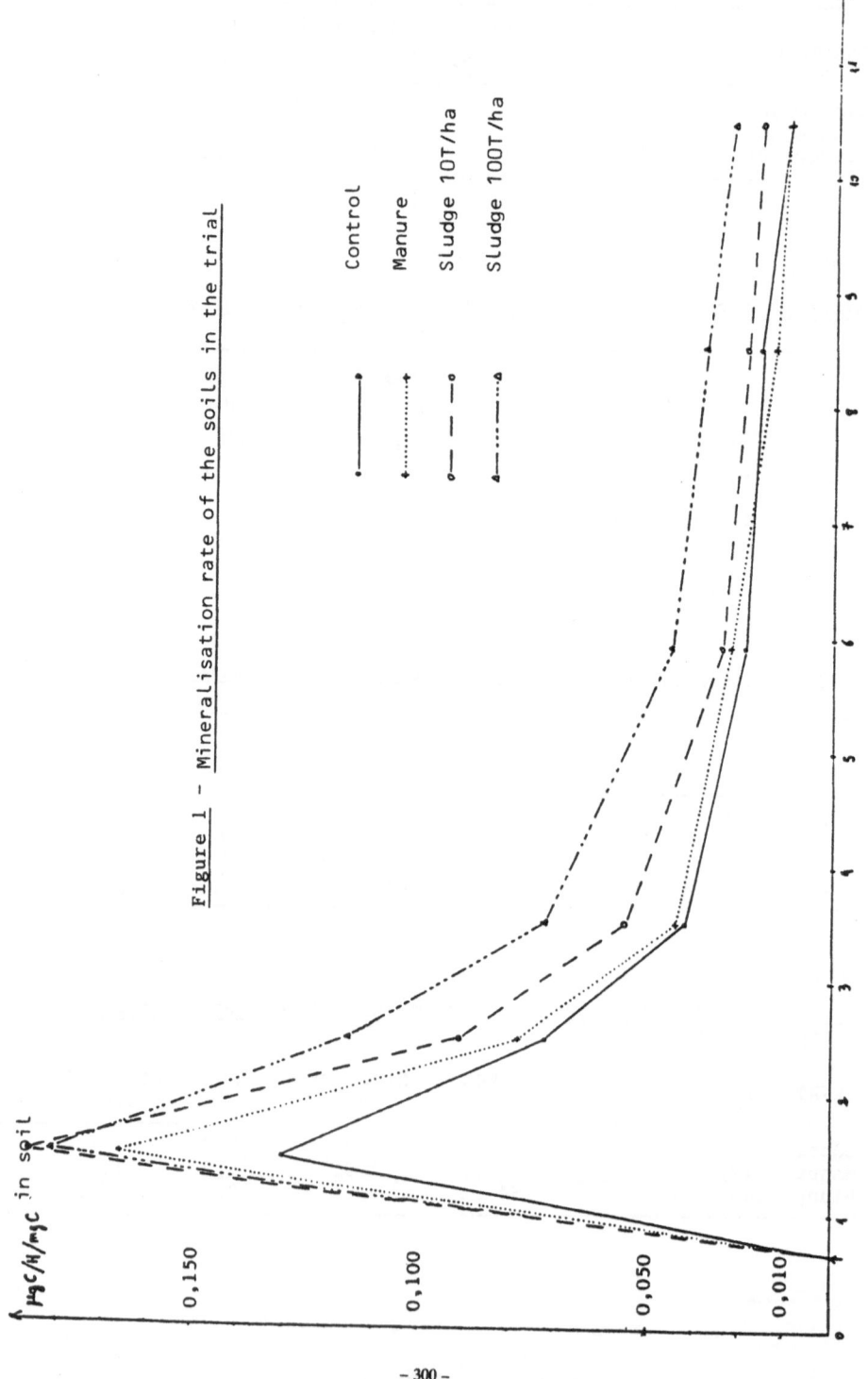

Figure 1 - Mineralisation rate of the soils in the trial

Control

Manure

Sludge 10T/ha

Sludge 100T/ha

µgC/H/mgC in soil

0,150

0,100

0,050

0,010

days

Figure 2

IR spectrum of humic acids extracted from the soil of the control plots (0-20 cm)

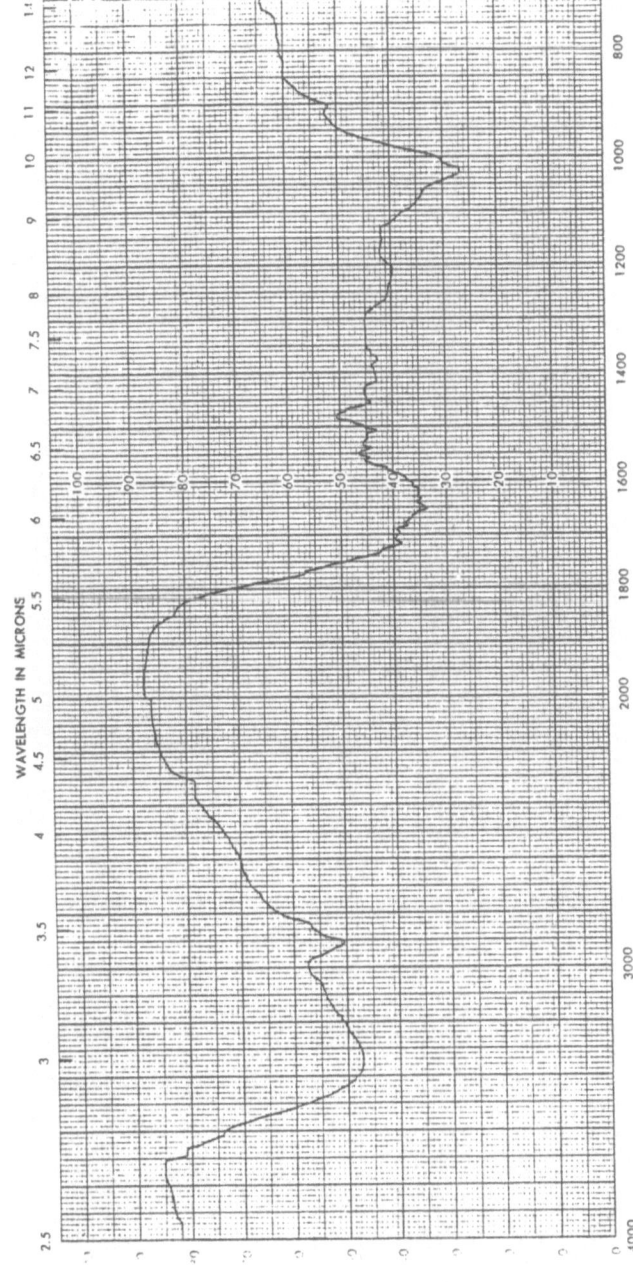

Figure 3

IR spectra of humic acids extracted from the soil of the S_2 plot

Upper horizon (0-20 cm)   40-60 cm horizon

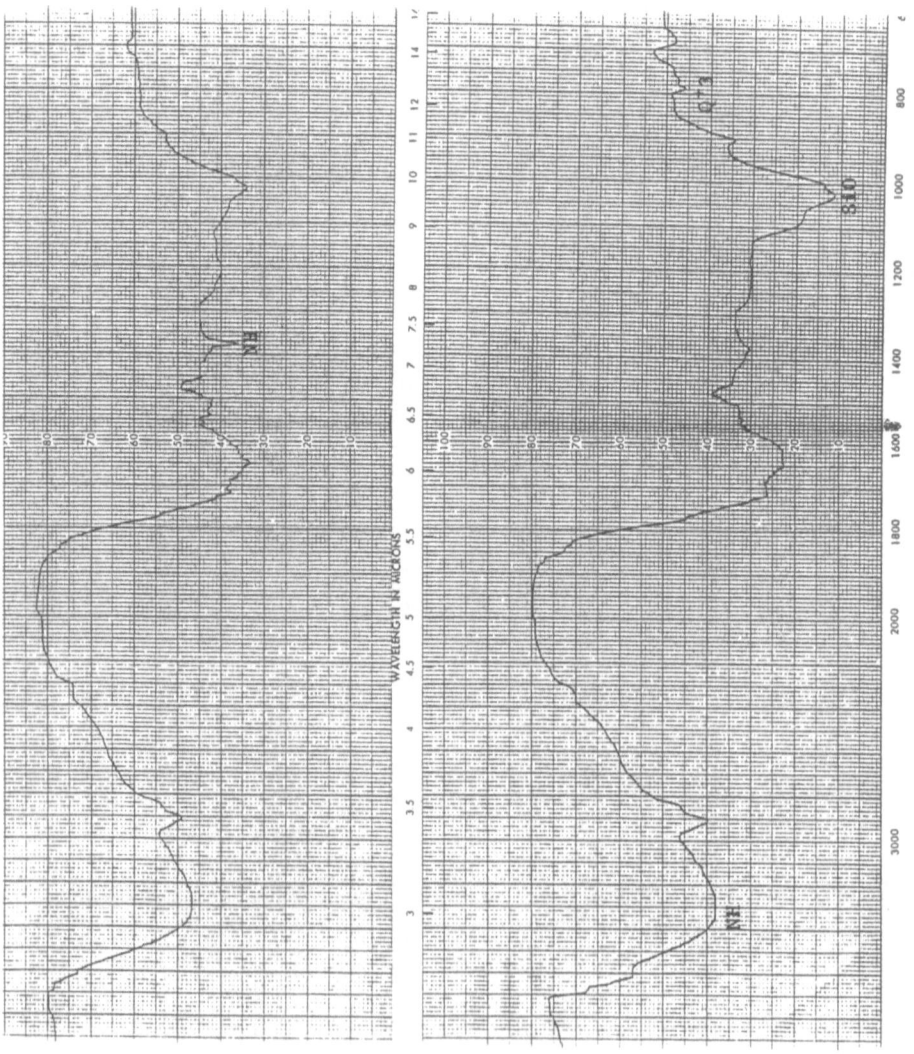

DISCUSSION

J E Hall
Whilst accepting that rates of organic matter decay vary between different soils and climates, this experiment suggests that farmyard manure has a longer term effect on soil conditions than digested sludge. This is contrary to commonly held views and other experiments which show that FYM decays rapidly in soil and the organic matter in digested sludge, because it is already humified, persists much longer. In such experiments a comparison on the basis of organic matter additions would be more informative than the approach chosen of equivalent dry solids additions. I would also question the validity of some of the calculations on residual C and N. Also you present only total nitrogen data, was mineral N measured? Nevertheless such experiments are important as there is relatively little data on the longer term effects of organic matter additions to soils.

M Lineres
Only total nitrogen was looked at.

C Juste
The experiment was set up because the industry was advocating the use of sludge on land to increase organic matter levels. A long term test was the best way to offer an answer to farmers wanting to know the humic value of sludge. Under this particular soil and climate, the stable humic contribution is virtually nil. These long term tests are useful as with FYM we found 50% of the carbon remaining in 0-60 cm. This agrees with the classical literature.

CONTAMINATION PROBLEMS IN RELATION TO LAND USE

G A FLEMING* and R D DAVIS**

* Plant Nutrition and Biochemistry Department, An Foras Taluntais,
Johnstown Castle Research Centre, Wexford, Ireland

** Environmental Impact Group, Water Research Centre, Henley Road,
Medmenham, P O Box 16, Marlow, Buckinghamshire SL7 2HD, UK

Summary

Use of sewage sludge on land has to be controlled to avoid
environmental problems from contaminants such as heavy metals which
usually occur in higher concentrations in sludge than in the soil to
which it is applied. The problems to be avoided are associated with
effects on crop growth and quality, animal nutrition and soil
fertility and contamination of ground and surface water. The
following general principles are suggested to minimise potential
problems. (1) Applications of sludge to the land should be
controlled in terms of quantity, timing and location; (2) The
receiving land should be managed according to good agricultural
practice; (3) Levels of contaminants in sludge and soil should be
monitored regularly to ensure compliance with limits; (4) Levels of
contaminants in the sludge should be controlled by appropriate limits
on industrial discharges to the sewer. The likelihood of
contamination problems arising depends on the way in which the land
receiving sludge is used and different land uses are discussed in
this context. Local conditions are important but a generalised
assessment suggests that horticulture, gardens and allotments are
high-risk outlets; arable land, grassland and forest land are
medium-risk outlets; whilst reclamation, green areas and viniculture
provide comparatively low-risk utilisation routes for sludge.

1. INTRODUCTION

Soil contamination or pollution is not new. Since the dawn of
history man's activities have resulted in the addition to soils of
unwanted materials of one form or another. Pollution from prehistoric
copper mining in Israel has, for instance, been revealed by archaeological
research (1). It is undeniable nevertheless that in the last one hundred
and fifty years or so, the rapid growth of cities and the development of
industry and agriculture have all conspired to accelerate the
contamination of soils. While soil contamination is primarily associated
with human activity, it is not exclusively so. Natural or geochemical
pollution resulting from the weathering of metal-rich rocks occurs in
sizeable land areas. Thus lead toxic soils resulting from the erosion of
a galena ore-body have been reported from Norway (2) and nickel-rich soils
in Scotland have developed from serpentine rock (3). In Somerset,
England, soils associated with severe animal health problems (and now
known to contain high levels of molybdenum) were recorded many years ago

(4). In Ireland, soils containing extremely high levels of both selenium and molybdenum occur (5). The source rock here is Carboniferous black shale. A disturbing aspect of soil pollution compared with air or water pollution is its long term nature. Atmospheric pollution ceases rapidly once the source has been eliminated while rivers appear to be cleansed within a year or so (1). Elements such as boron are relatively easily removed from soils by leaching but the retention of others e.g. copper, lead and zinc is virtually permanent (6). The application of sewage sludge to land while contributing much to the improvement of soil fertility has on the other hand frequently increased the heavy metal loading of soils.

The purpose of sewage treatment is to produce an effluent clean enough to be discharged safely to rivers and coastal waters. This is achieved partly by transferring the polluting load of raw sewage into sewage sludge which can then be separated out but subsequently has to be disposed of safely and economically. Sewage sludge contains organic matter, nitrogen and phosphorus of value to agriculture. Utilization on land is a cost-effective and environmentally acceptable method of disposal provided any problems due to odour nuisance, pathogen transmission and pollution by contaminants are avoided. This paper is concerned with the latter problem and how it relates to the way in which land receiving sludge is used. Background information about contamination problems is provided by Matthews (7) and the Report of a World Health Organisation Working Group (8).

2. GENERAL CONSIDERATIONS

Sewage sludge may contain a variety of different contaminants according to the nature of the catchment of the sewage-treatment works where it was produced. Vigilance is needed to anticipate and control unusual contaminants associated with particular industrial or commercial discharges but the ubiquitous contaminants which occur even in sludge of domestic origin are cadmium, copper, nickel, zinc and lead for which mandatory limits are proposed in the draft Directive on the Use of Sewage Sludge in Ariculture prepared by the European Commission (9). The potential contamination problems which have to be considered are effects on crop growth (phytotoxicity) and quality (food chain contamination), nutrition of farm animals and soil fertility (soil micro-organisms), and eutrophication of ground and surface water. All of these problems are less likely to occur if the application of sludge to land is controlled in terms of quantity timing and location. Thus sludge should be applied in accordance with crop requirements for nitrogen and phosphorus in quantities which take account of the need to avoid excessive build-up of contaminants in the soil. The timing of applications should take account of the growth cycle of the crop or future use of the land by livestock. Sludge should be spread so that contamination of surface water by runoff, for instance, is avoided. The operation as a whole is likely to be more successful if the land is well-managed both in terms of control of applications of sludge and all aspects of good farming practice. Levels of contaminants in sludge and sludge-treated soil should be monitored regularly to ensure compliance with limits. The toxicity of these metals depends principally on their concentration in soil not sludge (10) so contamination problems can be avoided by ensuring that soil limits are not exceeded by adjusting the rate of application of sludge according to its metal content. If the metal content of the sludge is comparatively high then effective control of rates of application must be assured which is not possible for all outlets. If the metal content of the sludge limits

its application to rates inadequate for crop requirements for nutrients then the farmer will not want the sludge and the operation is likely to be uneconomic. So control of contamination problems on disposal is made easier if the concentration of contaminants in sludge is kept as low as is realistically possible by appropriate limits on industrial discharges to the sewer.

So potential contamination problems can be minimised if:

1.   Application of sludge to the land is controlled in terms of quantity, timing and location to avoid contamination of soils, crops, animals and water courses.

2.   The receiving land is managed according to good agricultural practice.

3.   Levels of contaminants in sludge and sludge-treated soil are monitored regularly to ensure compliance with limits.

4.   Every reasonable effort is made to control levels of contaminants in the sludge by appropriate limits on industrial discharges to the sewer.

3. PARTICULAR LAND USES

Following on from these general principles, problems associated with different land uses are discussed below and an attempt is made to compare the relative hazard of the different utilisation routes.

3.1   Farmland

3.1.1 Arable - land used to grow a rotation principally of cereal and forage crops. Phytotoxicity problems and contamination of plant products for human and animal consumption depend largely on the availability for plant uptake of metals introduced to the soil in sludge. If the metals remain in the soil in insoluble or inert form they will be less hazardous than if they are readily solubilised in soil solution and available for crop uptake. Soil conditions influence availability as indicated in Fig.1 which presents results obtained in 1980 from the Royston and Cassington field trials, details of which were described at the Vienna Symposium by Davis and Stark (11). Concentrations of cadmium in wheat grain were lower on the calcareous soil containing free calcium carbonate with a pH value of about 8.0 compared with sandy loam and clay soils having a pH value of approximately 6.0. The precise effect of pH value on the availability of metals remains to be quantified further (12) but metal accumulations (with the exception of molybdenum for instance) are in general terms likely to be less hazardous in calcareous soils containing free calcium carbonate. Guidelines for sludge utilisation in the UK take this into account (13). The cation exchange capacity of the soil and its content of iron and manganese oxides also influence availability but for most metals (with the exception of cobalt for instance) the effect seems to be less marked than for soil pH value and again remains to be quantified. However, phytotoxic effects of cadmium on the yield of spinach and radish in pot trials at Johnstown Castle were found to be directly related to soil texture. Yields were more severely depressed in a light textured soil (6% clay) than in a medium textured one (15% clay) and the smallest yield depression occurred in a heavy textured soil (22% clay). Some metals such as lead and chromium appear to be almost completely unavailable in arable soils and therefore present little hazard.

The second important influence on the availability and hence hazard of metals in soils is the type of crop grown. Some crops achieve much higher concentrations in their tissues than others grown on the same soil. An example is lettuce which tends to achieve higher concentrations of cadmium in its tissues than do other crops. This effect is demonstrated in Fig.2 which presents results from the Royston field trial obtained in

1981. In Fig.2 crop concentrations of cadmium are shown on a dry weight basis. On a fresh weight basis or in terms of quantities eaten in the average diet, wheat grain and potato tubers assume much greater significance. Like lettuce, spinach and swiss chard (both members of the Chenopodiaceae) also show high foliar concentrations of cadmium and are often used as indicator crops in pot trials. Sensitivity in terms of tissue concentrations of metals is illustrated for 18 species of forage crop, classified according to botanical family on the basis of a glasshouse pot trial reported by Carlton-Smith and Davis (14). Leguminosae (clovers and lucerne) achieved comparatively high concentrations of molybdenum in their leaves whilst members of the Chenopodiaceae (mangold and sugar-beet) readily took up zinc. Highest concentrations of copper and nickel occurred in grasses but the Cruciferae (kale and turnip) and especially the cereals were less sensitive to metals in the soil. Use of arable land to grow cereals for animal feed purposes is a particularly effective way of reducing the transfer of metals such as cadmium from sludge-treated soil into the human food chain. According to one review, there would in these circumstances be no need for a cadmium soil limit to protect human health (15).

Metals limits for arable land receiving sludge are designed to provide long-term protection to the productive capacity of the soil and the quality of crops it produces. However, the incidence of contamination problems in arable farming systems using sewage sludge can be further minimised by maintenance of soil pH value at optimum levels for crop growth and by avoiding crops which are known to be sensitive to metals in the soil. Cultivation of forage crops or small grain crops for animal feed is a low-risk option particularly suitable for historically contaminated soils where soil metal levels may already exceed the maximum permissible concentrations set in current guidelines for sludge utilisation. Strict management along these lines can permit productive farming to continue on sacrificial land of this kind (16) which might otherwise be condemned to permanent dereliction.

In order to support crop growth and to comply with guidelines for sludge utilisation, the pH value of arable soils receiving sludge is usually maintained at a minimum of at least 5.5. In these circumstances it is thought that leaching of most metals is likely to be minimal. In relation to cadmium, Hansen and Tjell (17) concluded that high loadings of cadmium to soil do not lead to excessive contamination of groundwater. It should be noted that raising the soil pH value may actually increase the possibility of leaching for some elements such as molybdenum, selenium and chromium. Potential leaching of macronutrients especially nitrate presents more of a problem. Circumstances which would encourage leaching of nitrate include the application of heavy dressings of sludge in excess of crop nutrient requirements or the application of sludge at times of the year when crop uptake of nitrogen is likely to be minimal. Since sludge is produced continuously throughout the year it is often convenient for disposal authorities to spread sludge on the land at times when the application of inorganic fertilizer would not be contemplated. Working party 4 of COST 681 has defined the availability for crop uptake of the plant nutrients in sludge so that excessive applications can be avoided (18). As to unorthodox times of application, it appears that sludge nitrogen, being in the ammoniacal or organically bound form, is not leached as readily as fertiliser nitrogen (19). It is also reassuring that sludge is utilised on only perhaps 1% of agricultural land in Europe so that attention to conventional farm fertiliser practices and to the utilisation of animal slurries would be a more effective way of protecting

Figure 1. CADMIUM CONCENTRATIONS IN WHEAT GRAIN GROWN ON SOILS OF DIFFERENT TYPES TREATED WITH SEWAGE SLUDGE (for each soil, n = 26)

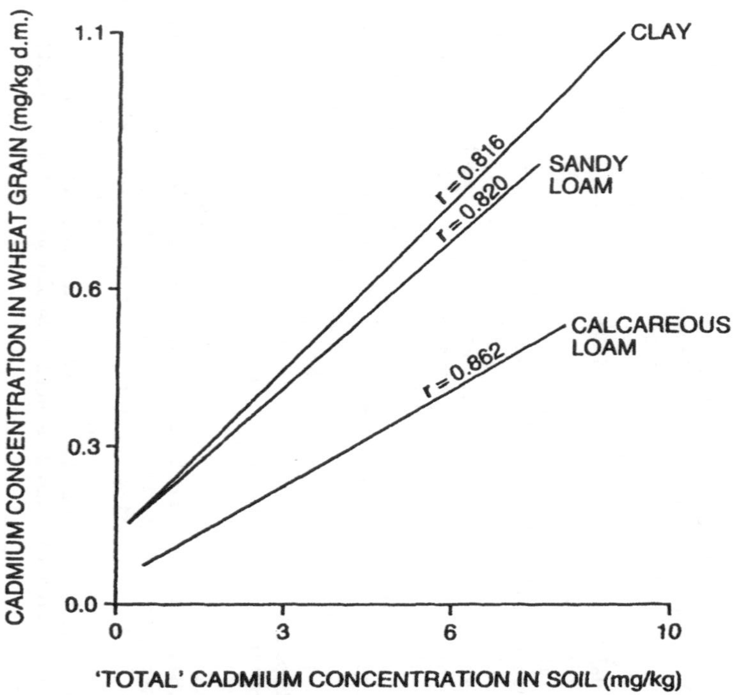

Figure 2. CADMIUM CONCENTRATIONS IN DIFFERENT TYPES OF CROP GROWN ON CALCAREOUS LOAM SOIL TREATED WITH SEWAGE SLUDGE (for each crop, n = 26)

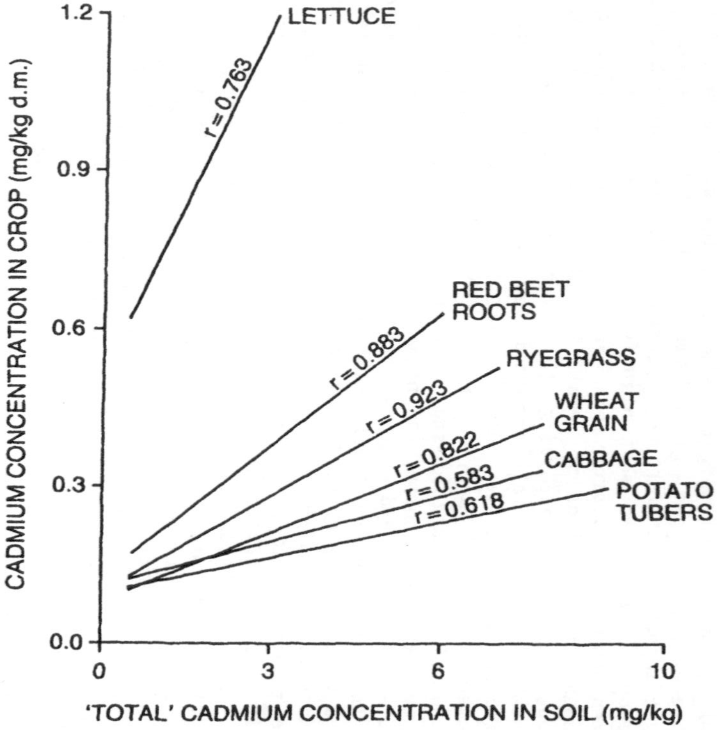

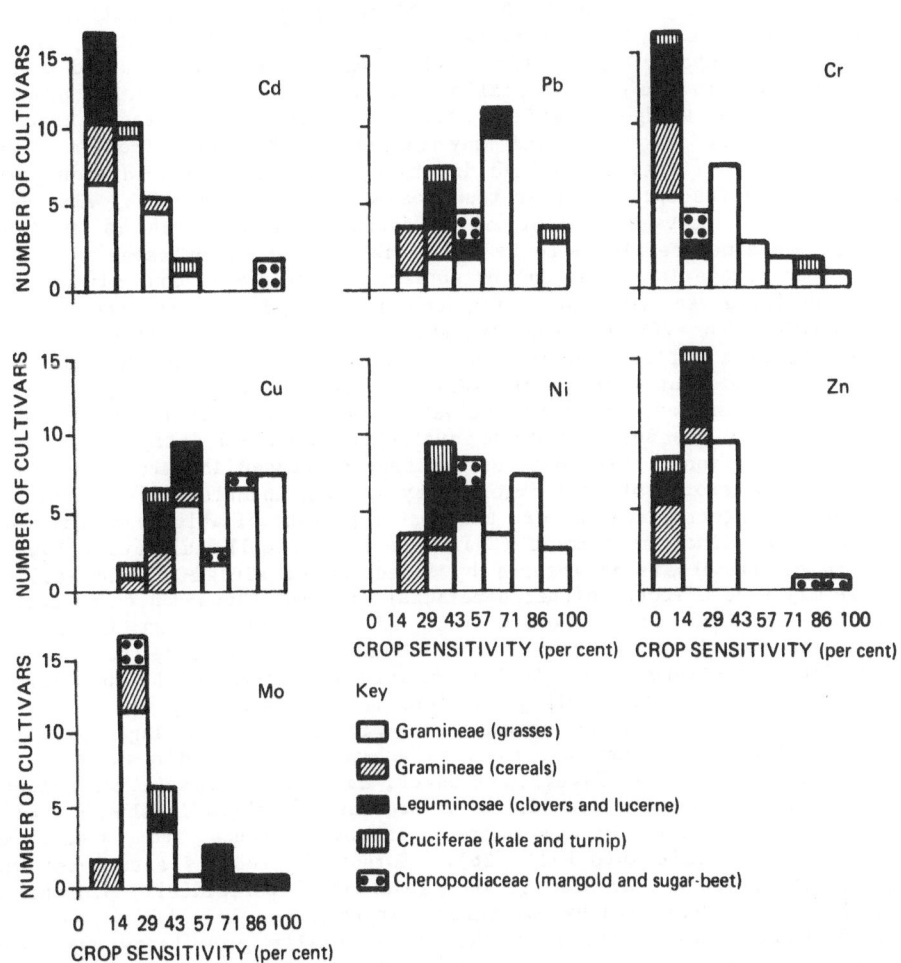

Figure 3 COMPARATIVE SENSITIVITY OF FORAGE CROPS TO METALS IN SLUDGE-TREATED SOIL

groundwater from pollution by nitrates. Guidelines for sludge utilisation should recommend that sludge is not spread near aquifers or water abstraction points or where pollution of surface water could occur following runoff.

The next paper (20) deals with long-term effects but recent studies in this area concerning effects on soil micro-organisms are of relevance to the use of sludge on arable land where maximum soil fertility is required to sustain crop production at profitable levels. Indications that metals from sludge may ultimately reduce soil microbial biomass and nitrogen-fixing capability (21) need to be further investigated to find out whether they are widespread and how they relate to crop production. Coppola's work (22) has shown that certain soil micro-organisms may be affected by remarkably low concentrations of cadmium.

3.1.2 Grassland. Provided that problems of pathogen transmission are avoided this outlet is in many ways very suitable for sludge. Grasses are not particularly susceptible to phytotoxic effects and respond well to the nutrients in sludge, especially to ammoniacal nitrogen. Grassland is accessible for sludge spreading throughout the year (provided that the ground is not too wet) and any excess of nitrogen from a winter application is likely to be held in the turf mat and utilised when growth commences in the spring rather than leached into groundwater. Also small increases in herbage content of essential elements such as selenium, copper and zinc are likely to be beneficial to grazing animals.

When sludge is spread on the surface of grassland, contaminants tend to accumulate in the upper few centimetres of the profile. Fig.4 illustrates how 76% of the increase in soil concentration of zinc resulting from surface applications of sludge remained in the upper 5 cm of soil. Ruminant animals inadvertently ingest soil as they graze and entry of contaminants into the animal by this route has implications for the health of the animal and the quality of meat and dairy products for human consumption. This question has been discussed in detail by Fleming (23,24). The amount of soil ingested by grazing animals is dependent on a variety of interrelated factors the most important of which are soil type, season and animal management (24). It is not well understood how the health of animals is influenced by ingestion of sludge-treated soil but there might be effects (either beneficial or deleterious) on trace element status depending on the composition of the sludge. An incident of fluorosis has been reported where cattle grazed pasture treated with fluoride-rich sludge (25). Such problems are exacerbated by poor farming practice such as over-stocking the land which breaks down the protective turf mat especially if the animals over-winter outside. Ingested metals are not thought to accumulate in dairy products or meat with the exception of offal (kidneys and liver). However, lipophilic organic contaminants which are not normally considered to present a problem in sludge used on land, many transfer by the direct ingestion route from sludge and sludge-treated soil into milk (26). Potential problems associated with direct ingestion are avoided if the grassland is periodically ploughed or if the sludge is applied by sub-surface injection.

3.1.3 Horticulture. Farms of this kind usually grow leafy vegetables known to be sensitive to metals in soil (Fig.2 and para 3.1.1 above), which are produced for human consumption with the minimum of processing. In order to maintain maximum productivity of the soil and take several crops each year it is often the practice to apply repeated heavy dressings of organic manure to horticultural land. On this basis, horticultural land would seem to present a comparatively high-risk outlet for sludge in terms of possible phytotoxicity and contamination of the food chain by

Figure 4. DISTRIBUTION OF ZINC THROUGH A SANDY LOAM
GRASSLAND SOIL PROFILE AFTER SURFACE
APPLICATIONS OF SLUDGE

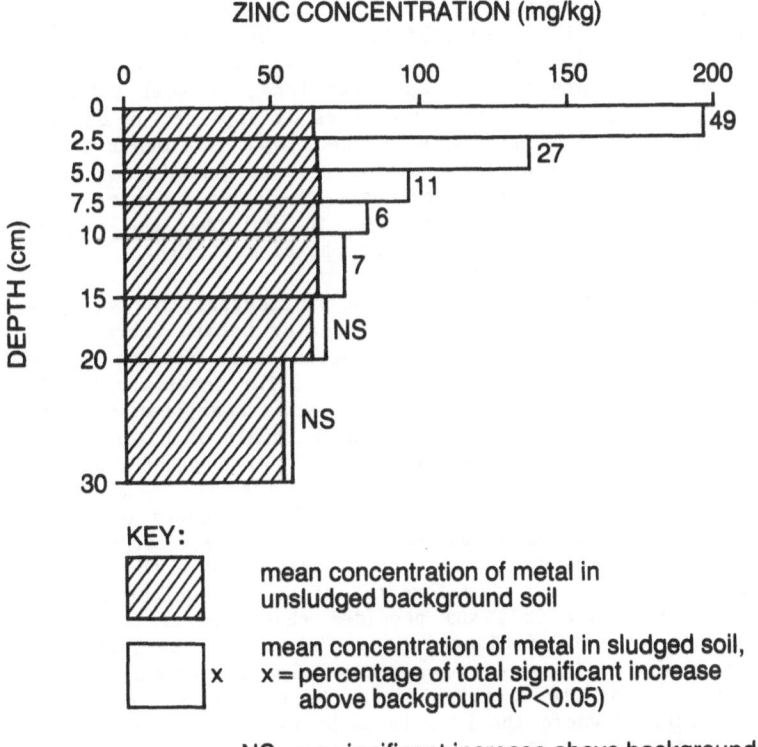

ZINC CONCENTRATION (mg/kg)

KEY:

mean concentration of metal in
unsludged background soil

x   mean concentration of metal in sludged soil,
x = percentage of total significant increase
above background (P<0.05)

NS = no significant increase above background

crop uptake and soil dust. It is probably suitable only for sludges or sludge compost of low metal content and requires careful management. Soil sampling on small horticultural units is likely to prove expensive and spreading equipment cannot be easily accommodated, leading to an uneconomic operation.

3.1.4 Viniculture. Vines are very deep-rooted and can tolerate high concentrations of copper at the soil surface since this builds up following repeated applications of fungicide based on Bordeaux mixture. Transfer of contaminants from sludge into wine for consumption seems unlikely. Vineyard soils tend to be poor in plant nutrients and organic matter and the use of dewatered sludge or sludge compost to establish young vine plants represent a novel outlet for sludge which could be exploited on a limited scale as economic considerations permit. Obviously, it may not be suitable for steeply sloping land where there is the likelihood of runoff.

3.2 Gardens and allotments

Potential contamination problems associated with gardens and allotments used for vegetable growing are similar to those described above for horticulture. Excessive applications of cadmium in particular are to be avoided because of the possibility that individuals or families might obtain the bulk of the vegetable part of their diet from allotments. Soil dust could contaminate these crops with lead, or soil could be ingested directly by children. Control over rates of application of sludge cannot be guaranteed. Use of sludge on gardens and allotments is therefore a comparatively high-risk outlet. Sludge used as a basis for composted soil conditioners for garden use obviously needs to be very low in metal content. In the UK it was recommended in 1981 that the supply of sewage sludge to the general public should be phased out (13).

3.3 Reclamation

Exploitation of natural resources particularly by the extraction of coal, minerals, clay, sand and gravel leaves behind barren land which is detrimental to the environment in aesthetic terms because it remains bare of vegetation and is therefore unsightly. The substrate is likely to be poorly structured and devoid of organic matter and plant nutrients and may be subject to erosion, acid runoff or high levels of toxic metals. Sewage sludge can supply the missing organic matter and plant nutrients so that vegetation can establish which will greatly improve the appearance of the site and help to prevent erosion. Lime will be needed to correct acidity but the high cation exchange capacity of sludge organic matter can contribute towards the detoxification of metals where these are a problem. Application of sludge can therefore transform the surface of derelict land into fertile top soil (27) and provided that logistical problems can be overcome, use in land reclamation can be a very satisfactory disposal route for sludge. In order to avoid problems, account must be taken of the proposed use of the restored land. With good management there should be no difficulties where the land is to be used for low grade agricultural purposes or for landscaping as is often the case. If housing is proposed, then the fact that the restored land will become gardens for vegetable growing has to be considered. Sludge is likely to be applied once only during land reclamation and a rate of application of the order of 100 tonne dry solids/hectare is necessary to restore soil fertility. Inevitably some contamination of ground and surface water may occur immediately after such an application. Similar problems occur during land reclamation with chemical fertilisers and such effects are usually negligible compared to the environmental problems present prior to reclamation (28). Indeed, leaching from chemical fertilisers with their high content of soluble nutrients may well be greater than from sludge. It has been suggested that the short-term disadvantage of groundwater contamination is far out-weighed by the long-term benefits of revegetation (29).

3.4 Green areas - amenity grasssland, golf courses, parks, playing fields. This outlet is closely related to land reclamation in that the after-use of reclaimed land will often be as green areas. The major advantage of this outlet is that the principal vegetation cover is grass. Grass is efficient in using nitrogen so that leaching of nitrate into groundwater should be minimal. Also green areas are not used for growing food chain crops. Like reclamation, use on green areas is a sensible outlet for sludge as summarised in one of Berglund's principles developed at the Uppsala Seminar of COST 681, "The use of sludge on green areas and for land reclamation should be encouraged" (30).

3.5 Forestry - timber (usually coniferous) production forest not broad leaf amenity woodland. Use of sludge on forest land is unlikely to become a major outlet for sludge because forests tend to be many kilometres distant from the conurbations where sludge is produced in quantity. Forests are usually developed on land which is unsuitable for agriculture because of low soil fertility, steep slopes or harsh conditions of climate. As with the reclamation of derelict land, it would seem that sludge should be suitable for improving the fertility of forest soils. The US Environmental Protection Agency reports that forest soils are in many ways well suited to sludge application (28). They have high rates of infiltration (which reduce runoff and ponding), large amounts of organic material (which immobilise metals from the sludge), and perennial root systems (which allow year-round application in mild climates). Although forest soils are frequently quite acid, research at the University of Washington has found no problems with metal leaching following sludge application (28). Also, there should be no food chain problems. Effects of sludge applications on tree growth appear to be equivocal. Work at the University of Washington has showed impressive improvements in coniferous tree growth following applications of sludge. In the UK two experiments are in progress in pine forests, one in Southern England and one in Scotland. Preliminary indications from the trial in Southern England with liquid digested sludge are that sludge is an effective source of nutrients for the trees (31). The trial in Scotland is with undigested sludge to which it is understood the trees are not responding so well. Experiments on the use of sludge in Mediterranean forests found a good initial response to sludge but three years after application growth rate tended to decrease compared with trees growing on control, untreated plots (32). More research is needed to evaluate the extent of any phytotoxic effects due to metals or disruption of mycorrhizal associations due to changes in nutrient balance or oxygen depletion in the surface soil. However, the primary environmental concern associated with the use of sludge in forests relates to contamination of surface water by runoff or groundwater by leaching, particularly of nitrate. Forests are often important catchment areas for potable water supplies. It is thought that these problems are avoidable by careful site selection to avoid runoff and applying the sludge at rates compatible with tree requirements for nitrogen. Whilst forest land remains a very minor outlet for sludge it is unlikely to contribute significantly to groundwater contamination. In some countries, for example the Netherlands, sludge application in coniferous forests is not recommended (33).

4. CONCLUSIONS

This paper has been concerned entirely with contamination problems and has inevitably treated the subject of sludge utilisation on land in a negative way. It must be appreciated that sludge utilisation on land is an environmentally acceptable method of disposal and is usually the principal outlet for countries practising sewage-treatment on a large-scale. Potential contamination problems can be minimised if:

1. Application of sludge to the land is controlled in terms of quantity, timing and location to avoid contamination of soils, crops, animals and water courses.

2. The receiving land is managed according to good agricultural practice.

3. Levels of contaminants in sludge and sludge-treated soil are monitored regularly to ensure compliance with limits.

4. Every reasonable effort is made to control levels of contaminants in the sludge by appropriate limits on industrial discharges to the sewer.

Much depends on local circumstances but Table 1 presents a general assessment of hazard associated with the various options for sludge utilisation on land. A natural extension would be to extend this approach to other disposal routes for sludge (landfill, incineration, sea) to help in planning the best practicable environmental option for sludge disposal from particular sewage-treatment works.

TABLE 1 GENERALISED ASSESSMENT OF HAZARD ASSOCIATED WITH SLUDGE UTILISATION ROUTES

Route	\multicolumn{4}{c}{Potential Contamination Problem}	Overall Hazard Rating
	Food Chain	Phyto-Toxicity	Zootoxicity (by direct ingestion)	Surface or Ground-water	
Farmland:					
- arable	2	2	2	1	2
- grassland	1*	2	3	1	2
- horticulture	3	3	2	2	3
- viniculture	1	1	1	2	1
Gardens and allotments	3	3	2	1	3
Reclamation	1	1	1	2	1
Green areas	1	2	1	1	1
Forestry	1	2	1	3	2

* except organics in milk   1 = low;   2 = medium;   3 = high

ACKNOWLEDGEMENTS

Results shown in Figures 1, 2 and 4 are from experiments at the Water Research Centre funded by the UK Department of the Environment.

REFERENCES

(1) DAVIES, B.E. (1980). Trace element pollution pp.287-351. In "Applied Soil Trace Elements". Ed. B.E. Davies. John Wiley & Sons N.Y..

(2) LAG, J., HVATUM, O.O. and BOLIKEN, B. (1969). An occurrence of naturally lead-poisoned soil at Kastad near Gjovik, Norway. Norges Geologiske Undersokelse No.266. 141-159.

(3) MITCHELL, R.L. (1964). Trace Elements. In "Chemistry of the Soil". Ed. F.E. Bear, Reinhold, N.Y. 320-368.

(4) GIMINGHAM, C.T. (1914). The scouring lands of Somerset and Warwickshire. J.Agric.Sci. Camb.6: 328-336.

(5) FLEMING, G.A. (1982). Geochemical Pollution - some effects on the selenium and molybdenum contents of crops. Commission of the European Communities. Environmental effects of organic and inorganic contaminants in sewage sludge. Ed. R.D. Davis, G. Hucker and P. L'Hermite. D. Reidel Publ. Co. Holland. 277-232

(6) PURVES, D. (1972). Consequences of trace element contamination of soils. Envir.Pollut. 3: 17-24.

(7) MATTHEWS, P.J. Control of metal application rates from sewage sludge utilisation in agriculture. CRC Critical Reviews in Environmental Control 1984, 14, 199-250.

(8) DEAN, R.B. and SUESS, M.J. eds. The risk to health of chemicals in sewage sludge applied to land. Report of a WHO Working Group. Waste Management and Research 1985, 3, 251-278.

(9) COMMISSION OF THE EUROPEAN COMMUNITIES. Proposal for a Council Directive on the Use of Sewage Sludge in Agriculture. Official Journal of the European Communities 14 6 84 No C 154/6-15.

(10) DAVIS, R.D. (1984). Crop uptake of metals from sludge-treated soil and its implications for soil fertility and the human diet pp.349-357, as reference 22.

(11) DAVIS, R.D. and STARK, J.H. Effects of sewage sludge on the heavy metal content of soils and crops: field trials at Cassington and Royston. pp 687-698 in Characterisation, Treatment and Use of Sludge ed. P. L'Hermite and H. Ott. Published for the Commission of the European Communities by D Reidel, Dordrecht, 1981.

(12) HANSEN, J.A. and TJELL, J.C. (1984). Agricultural use of sludge and compost: environmental contamination in context pp.77-101 in Sixth European Sewage and Refuse Symposium EWPCA-ISWA, Munich, 1984.

(13) DEPARTMENT OF THE ENVIRONMENT/NATIONAL WATER COUNCIL. Report of the Sub-Committee on the Disposal of Sewage Sludge to Land. DOE, London, 1981.

(14) CARLTON-SMITH, C.H. and DAVIS, R.D. Comparative uptake of heavy metals by forage crops grown on sludge-treated soil. pp 393-396 in Heavy Metals in the Environment, Heidelberg Vol.1, CEP Consultants, Edinburgh 1983.

(15) RYAN, J.A., PAHREN, H.R. and LUCAS, J.B. Controlling cadmium in the human foodchain: a review and rationale based on health effects. Environmental Research 1982, 28, 251-302.

(16) RUNDLE, H., CALCROFT, M. and HOLT, C. Agricultural disposal of sludges on a historic sludge disposal site. Water Pollution Control 1982, 81, 619-632.

(17) HANSEN, J.A. and TJELL, J.C. Sludge application to land - overview of the cadmium problem. pp 91-112 in Environmental Effects of Organic and Inorganic Contaminants in Sewage Sludge ed. R.D. Davis et al. Published for the Commission of the European Communities by D. Reidel, Dordrecht. 1983.

(18) COMMISION OF THE EUROPEAN COMMUNITIES. Nitrogen and phosphorus value of sewage sludges. G. Catroux et al. ed. SL/82/82, XII/ENV/35/82. Brussels, 1982.

(19) COMMISSION OF THE EUROPEAN COMUNITIES. Proceedings of a Seminar of WP4 and WP5 held at Ultana University, Uppsala, on The Utilisation of Sewage Sludge and Long-term Effects of Contaminants ed. S. Berglund et al. D Reidel, Dordrecht, 1983.

(20) SAUERBECK D.R. Long-term effects of contaminants. Session IV Paper 6, this Symposium.

(21) BROOKES, P.C. and McGRATH, S.P. Effects of metal toxicity on the size of the soil microbial biomass. Journal of Soil Science 1984, 35. 341-346.

(22) COPPOLA, S. Effects of sludge on soil microbial processes. Paper presented to a WP5 Seminar on Factors influencing Sludge Utilisation Practices in Europe, Liebefeld, May 7-10 1985. To be publised for the European Commission by Applied Science Publishers, Barking, 1986.

(23) FLEMING, G.A. Implications of soil ingestion by grazing animals. As reference 14.

(24) FLEMING, G.A. (1986). Implications of soil ingestion following sludge spreading on grassland. COST 681 Scientific Report WP5 Section. Commission of the European Communities (in preparation).

(25) WATER RESEARCH CENTRE. Agricultural and environmental aspects of fluorides in sewage sludge. Notes on Water Research No.21 WRc, 1979.

(26) DAVIS, R.D., HOWELL, K., OAKE, R.J. and WILCOX, P. Significance of organic contaminants in sewage sludges used on agricultural land pp 73-79 in Environmental Contamination, Conference held at Imperial College London July 1984. CEP Consultants, Edinburgh.

(27) HALL, J.E. and VIGERUST, E. The use of sewage sludge in restoring disturbed and derelict land pp 91-102, as reference 11.

(28) UNITED STATES ENVIRONMENTAL PROTECTION AGENCY. Use and disposal of municipal wastewater sludge. USEPA Office of Research Management, Cincinnati, 1984.

(29) ENVIRONMENT CANADA. Manual for Land application of treated municipal wastewater and sludge. Manual EPS 6-EP-84-1 Environmental Protection Programs Directorate, Ottawa, 1984.

(30) BERGLUND, S. Principles for sludge utilisation on land. Annex VI p.291 in Concerted Action Treatment and Use of Sewage Sludge COST 681 TER. Final report of the Community - Cost Concertation Committee. 11. Scientific Report SL/94/83. Commission of the European Communities, 1984.

(31) BINNS, W.O., DAVIS, R.D. and MUGLESTON, A.G. Preliminary results of an experiment on the use of sewage sludge as a phosphate fertiliser in a coniferous forest. pp.318-326 in Processing and Use of Sewage Sludge eds. P. L´Hermite and H. Ott. Published for the Commission of the European Communities by D Reidel, Dordrecht, 1984.

(32) THOMANN, C.H. Experimental study on the use of urban sewage sludge on Mediterranean forests pp61-78, as reference 11.

(33) VAN DEN BERG, J. H_2O 1978, 11. 482-488, cited in reference 20.

DISCUSSION

H Kuntze

You have shown us the strong correlation between pH, texture and heavy metal uptake. Large areas of Northern Europe are sandy and of low pH. Looking at your graphs liming could be a solution but the pH of soils are limited for other reasons. We found that iron content and quality, particularly for acid sandy soils, gives a good method for controlling metal additions.

G Fleming

There is a problem with liming as on peat land, grass grows best at pH 5.5. Iron and manganese are powerful scavengers of heavy metals. MnO_2 can be used for fixing cobalt 60 and iron oxide has an affinity for molybdenum but what are the costs of this approach?

M Sbaraglia

If you look at germination tests, I think you see if you dilute sludge, you get a better yield. This is because the salt load in the sludge is too high. So when considering metal effects, salt effects on germination must also be taken into account. When using germination tests it is important to state the level of salt otherwise, there is little point looking at heavy metal effects. Did you take account of the salt content?

R D Davis

In the case of the cadmium trial, levels were too low to cause an osmotic effect and there was no effect on germination. We should have mentioned that there may be occasions when the salinity may cause problems with some horticultural crops.

LONG-TERM EFFECTS OF CONTAMINANTS

D. R. SAUERBECK and P. STYPEREK
Institute of Plant Nutrition and Soil Science
Federal Research Center of Agriculture Braunschweig-Völkenrode (FRG)*)

Abstract

This paper summarizes the results from 15 European field experiments with sewage sludge concerning the long-term accumulation of 6 heavy metals in the treated soils, the availability of these sludge-derived contaminants to different plants, and the possible change in their availability after long-term residence in different soils.
According to this information about the transfer from field soils to plants, the data confirm that the German threshold values for most heavy metals in soils except cadmium are reasonably safe. As far as the long-term significance of these toxic substances is concerned, the following main conclusions have been drawn.

1. The long-term effect of contaminants cannot be judged on the basis of a few years or decades only but requires the scope of centuries to be fully realized.

2. Any incorporation of persistent contaminants, even if it is very gradual, will in the long term result in a gradual deterioration of soils.

3. Due to this fact, we should not only think in terms of what soils can stand now but instead aim at reducing the input of such contaminants by all possible means.

4. Accordingly, the long-term use of sludges is not a question of soils on which such wastes would be safely directed but of tracing and closing all possible sources of contaminants without compromise.

1. INTRODUCTION

Having the task to discuss "long-term effects of contaminants", the referee naturally wonders what the concept "long-term effects" actually means. There are three main aspects of contaminants in sludges and soils.

(1) the long-term accumulation of contaminants in soils which were treated repeatedly with polluted sludge

(2) the availability to plants of soil contaminants resulting from long-term sludge applications

(3) the change in availability of such contaminants after their long-term residence in soils.

---

*) Additional institutions cooperating in this project are listed at the end of this text.

2. LONG-TERM ACCUMULATION OF CONTAMINANTS IN SOILS FROM POLLUTED SLUDGE

For reasons of time these considerations have to be restricted to the most persistent group of contaminants which are the heavy metals. With regard to the first question of long-term accumulation the answer does not require specific experimentation but just to calculate the time period required to enrich a soil with these elements up to a certain extent (Table 1). Taking the threshold heavy metal values for German sludges and soils, and the maximum sludge application allowed per unit of time, we arrive at theoretical accumulation periods varying between 110 and 250 years. Since many sewage sludges have now much lower contents, the calculated accumulation periods increase correspondingly. On the other hand, there is a good chance for at least the cadmium threshold value in soil to be reduced soon from 3 to 2 ppm, and this would consequently decrease the accumulation period by a factor of about one third.

- Table 1 -

Threshold heavy metal contents in German sludges and soils and the corresponding accumulation periods

element	sewage sludge threshold ppm	sewage sludge loading g/ha·a	soil threshold ppm	soil average content ppm	enrichment period (years)
Cd	20	33	3	0.2	250
Cr	1200	2000	100	26	110
Cu	1200	2000	100	11	130
Hg	25	42	2	0.1	140
Ni	200	333	50	30	180
Pb	1200	2000	100	22	120
Zn	3000	5000	300	50	150

5 t sewage sludge d. m. in 3 years; 3 kg heavy metal/ha = 1 ppm

This means that the time spans allowed for safety are not very large, although for some elements like Cr and Pb the maximum permissible soil contents may be unnecessarily low. German authorities are going to reconsider the present sludge guidelines by 1988, and they will probably not accept the risk of sludge treated soils becoming further enriched with heavy metals in the future. The ongoing drop in the actual heavy metal content of current sludges since the present guidelines were issued in Germany gives rise to the hope that the threshold values can be lowered accordingly but still maintain a chance of sewage sludge being recycled into agricultural soils.

3. AVAILABILITY TO PLANTS OF CONTAMINANTS RESULTING FROM LONG-TERM SLUDGE APPLICATION

As far as the second question about the actual availability of contaminants derived from long-term sludge applications is concerned, Europe has a number of field experiments from which corresponding conclusions can be drawn (Table 2). However, the problem of evaluating and comparing such experiments is, that they were not originally designed for such inter-comparisons. The cumulative amounts of sludge applied and the extents of soil contamination resulting therefrom differ considerably as also do the time spans elapsed since the last treatment, the duration of the experiment and the crops grown on these plots.

- Table 2 -

Long-term field experiments with sewage sludge rendering information on heavy metal loading of soils and plants

Location		Start	Material used	Appl. rate tdm/ha		Soil type	pH	% C	CEC
Bonn	Bn1	1958	liq.sludge	5-	10/2.year	uL	6.5	0.9	14
	Bn2	1958	sl.compost	2-	12/2.year	uL	5.6	1.2	12
	Bn3	1972	liq.sludge	4.5-	9/2.year	lS	6.6	1.1	14
Braun-schweig	Bs1	1971	liq./filt.sl.	5-	15/year	ulS	5.5	1.6	9.3
	Bs2	1971	liq.sludge	5-	15/year	ulS	6.7	0.9	8.4
Bremen	Br1	1968	liq.sludge		20/year	peat	4.2	28.9	234
	Br2	1972	liq.sludge	12-	20/year	peat	4.0	16.1	190
Gießen	Gi1	1972	sl.compost	120-	200/3.year	L	6.6	1.1	11.6
	Gi2	1979	dehydr.sl.	4.5-	9/year	L	6.2	0.9	10.5
	Gi3	1969	liq.sludge	2.5-	5/year	L	5.9	0.8	12.4
	Gi4	1969	liq.sludge	2.5-	5/year	L	5.4	5.9	17.8
München	Mü1	1980	sludge cake	100-2000/single		sL	6.6	7.9	
	Mü2	1976	liq./filt.sl.	100-	690/single	lS	5.8	2.3	16
	Mü3	1978	liq.sludge	5-	10/year	sL	6.3	1.4	
	Mü4	1979	liq.sludge		2.5/year	lS	6.4	1.7	
Bordeaux	Bx1	1974	dehydr.sl.		10/year	S	5.5	2.2	
	Bx2	1974	dehydr.sl.		100/2.year	S	5.5	2.2	
Stevenage	Ste	1976	air dry sl.	35-	280/single	cL	8.0	1.1	

It is only by chance that certain crops have been grown at more than 2 or 3 places, and in some cases the number of comparable data sets is even lower. Up to now 15 German field experiments from 5 different research stations have been evaluated, of which only a summarizing review can be given here. As the analytical soil data and the treatments specified in Table 2 indicate, these experiments do not only represent a selection of different soils but also all kinds of possible sludge treatments varying from small regular applications up to large single doses at a certain time.

The following figures refer to these different field experiments and are all of the same design in order to enable a summarizing representation of the results to be made. They are arranged in the sequence of elements and crop plants, each showing the soil contents of the respective location in the lower part, and above the corresponding contents of the plant parts grown on these plots. In two cases we have also received corresponding data from comparable experiments carried out by colleagues abroad, which were incorporated in the corresponding evaluations for maize and herbage.

The abbreviations and numbers on the left of these graphs refer to the experiments listed in Table 2. The corresponding symbols represent the variants of the individual experiments, and where the same crop was grown several times, the degree of blackening in these symbols indicates the consecutive experimental years. As far as possible the results of each plot and year were represented, but in some cases these data were just too close together in order to be individually shown. Hence, the overall number of data pairs is also indicated numerically for each experiment, making clear that the visually distinguishable row of symbols only represents the actual range of metal contents but not always all single values concerned. The common scale for the metal contents in both soils and plants is shown at the bottom, but in some cases additional scales were required for the most highly polluted soil plots and plants. The arrows at this normal scale indicate the German threshold value for sludge treated soils.

Starting with the element Pb it can be concluded from almost all the existing results that not only soil contents lower than the German threshold value of 100 ppm but also much higher Pb concentrations did not result in an appreciable Pb uptake by the plants. This is not only true for both summer and winter wheat as well as for summer barley, for oats and for corn (Figure 1), but also for most dicotyledoneous plants.

Grains contained even less Pb than straw, and beet roots less than the leaves, but part of the Pb in the leaves from such row crops may actually be due to an external contamination with polluted soil (Figure 2).

A somewhat unexpected result were the relatively high Pb contents of potato leaves in one of the Munich experiments, but this exception was not confirmed by the results from the other experiments with high soil Pb pollution, and the tubers remained low in Pb anyway (Figure 3).

Similar to Pb were the general findings for Cr, against which most grain crops like wheat and barley, but also the corn discriminated most effectively (Figure 4).

Only in one case for oats there was a somewhat higher Cr recovery in the grain as compared with the straw, but the second plot with less organic matter near Munich did not reflect this in the same way (Figure 5).

Sugar beets also showed rather low Cr uptake into both leaves and beet roots, whereas potato leaves, as shown for Pb, seem to accumulate Cr somewhat more (Figure 6).

For the trace metal Cu its uptake by small grain crops from soils containing less than 100 ppm was not significantly increased above the control, but if soil contents exceed about 100 ppm; at least oat and corn grains may also in some cases reach about 20 ppm (Figure 7).

Sugar beet roots turned out not to have much lower Cu-contents than the corresponding leaves, so the 100 ppm threshold value may in fact be the acceptable maximum at least for more sandy soils (Figure 8).

As shown before for Pb and Cr, potato leaves also accumulated relatively large quantities of Cu, which at least for the soil plots exceeding 100 ppm may already have reached the phytotoxic level (Figure 9).

Coming to the Ni, this elements also hardly moved into summer or winter wheat grains, but accumulated to some extent on a high-carbon soil in barley and in both oat straw and grains (Figure 10).

On the other hand, even extreme Ni contents in a coarse French sand soil near Bordeaux (1) were not reflected by corresponding Ni contents in the corn grains (Figure 11).

A somewhat more pronounced Ni-accumulation was found in some sludge-treated sugar beet experiments, which suggest that the threshold value of 50 ppm for non-geological Ni in soils should not be exceeded (Figure 12).

A much more complex picture appears when Cd or Zn contents of soils and plants are compared, because the Cd guide values for plants are low and the availability in soil of this element is relatively high. Cereal grains always contain less Cd than the straw, but the difference is not so pronounced, and in soils near or above 2 ppm a grain content of about 1 ppm was frequently found. Even worse looks the situation on soils having 10 or more ppm Cd, although even at 40 - 50 ppm soil Cd, 3 ppm Cd was not much exceeded in summer or winter wheat grains and straw (Figure 13).

Barley, however, although only grown so far on the sludge plots near Munich translocated somewhat less Cd into its grains. The same seems to be true for corn as an effective discriminator against Cd in the grains (Figure 14), whereas oats did not differ from wheat as far as the grain Cd contents were concerned.

If 1 ppm Cd in sugar beet leaves are not to be exceeded, there is little doubt that soil contents should also be maintained below 2 ppm because 5 - 45 ppm Cd in the near-Munich soil already induced 3 - 18 ppm in the beet leaves. Even the beet roots in this case contained up to 3 ppm Cd, whereas in all other soils about 0.6 ppm in the sugar beets was not exceeded (Figure 15). A similar uptake pattern was found in the experiments near Munich for potatoes, but so far no comparison can be made with other soils except some relatively low Cd plots near Bonn.

Apart from much higher Zn levels, the transfer ratios from soils to plants for Zn were quite similar to what has been said for Cd. Hence, it may be sufficient to state here that in cereals and in corn even 400 - 1800 ppm soil Zn did not raise the Zn contents in the straw much above 300 ppm, which for grain crops is not yet phytotoxic (Figure 16).

Sugar beet leaves, on the other hand showed Zn contents in about the same range as that in the soil. This explains practical observations of Zn toxicity in sugar beet crops on highly polluted soils (Figure 17).

Not much Zn data exists for potatoes on sludged soil plots yet, but in two experiments near Munich critically high contents were recorded in the potato leaves grown on the highly Zn polluted plots (Figure 18).

Pulling all these field observations together, it can be safely concluded that Pb and Cr in such sludge-treated soils are no serious problem for plant growth or plant metal uptake, unless the soil contents are extraordinarily high. The present German threshold values of 100 ppm in soil (Table 1) are extremely cautious and do not justify stringent restrictions of soil use even if they are exceeded several times.

For Cu the 100 ppm threshold value has turned out to be fairly correct, but may already be too high for some crops on light sandy soils.

The Ni threshold value of 50 ppm guarantees non-toxic Ni contents in all crop plants tested, and needs to be taken seriously only if this soil Ni is pollution-derived and not of geological origin.

Regarding the safe upper Zn limit, some dicot plants contained this element in similar concentrations to those found in the soil in which they grew. Hence the soil threshold value of 300 ppm cannot be increased, but according to present knowledge there is also no pressing need to lower it.

The present Cd threshold value of 3 ppm was shown to result in Cd contents of some staple foods which cannot be tolerated in larger areas. Hence, in order to be on the safe side, a lowering to not more than 2 ppm Cd seems to be required at least in soils used for food production.

As far as the corresponding threshold values for grassland soils are concerned, only a few long-term experiments with higher degrees of metal pollution have been evaluated so far. Since grasses are known to absorb less trace metals than most dicotyledoneous plants do, the danger from Pb for animals is relatively small unless they take up air-borne Pb or pollu- ted soil along with their food. The Pb uptake by herbage itself, however, was relatively low even on acid peat soils from Bremen exceeding the German threshold value considerably (Figure 19). The same low uptake by grassland plants was also found in all cases for Cr.

The significance of the Cu contents in herbage depends on the kind of animals which eat it, but if 20 ppm is considered safe in the feed, more than 100 ppm in the soil cannot generally be allowed (Figure 20). Ni was obviously not problem in these 6 grassland experiments, because in no case was the threshold Ni value strongly surpassed.

Cd contents of much more than 1 ppm in ther herbage only occurred on a most strongly Cd polluted soil near Munich, but otherwise neither in peat nor in mineral soils Cd contents up to 5 ppm gave rise to critical Cd levels in the plants (Figure 21). Hence, probably no need exists to reconsider the 3 ppm threshold value here.

About the same as for Cd is also true for Zn. As already mentioned for cereals, even between 1500 and 1800 ppm Zn in soil did not harm the grass, although this does not imply that dicots in the sward were not negatively influenced. The 300 ppm threshold value, however, is certainly not too high.

4. AVAILABILITY OF CONTAMINANTS AFTER THEIR LONG-TERM RESIDENCE IN SOIL

Our last question which still remains to be answered in this context was whether there is a possible change in the availability of contaminants after their long-term residence in soils. In order to answer this, one requires experiments where former sludge treatments have resulted in a considerable degree of contamination, which after ceasing these treatments did not increase any more. Only a few experimental sites meet these two requirements, and little precise information about them is so far known.

The data of 2 such experiments have been considered here in order to tackle this question for grassland conditions. The results from Stevenage (2) for the Smallford site showed a considerable drop in Cd uptake from the first to the following year, but no consistent decrease in Cd availability later on. A similar tendency can be guessed for the behaviour of Zn, but the contamination with this element was not very high (Figure 22).

However, if this is compared with a similar experiment at Baumannshof near Munich (3), a less pronounced decrease in the Cd or Zn uptake was observed there. In order to explain this it needs to be mentioned that in this latter experiment the sludge application took place 2 years earlier, so that the possible initial drop during the first year was perhaps just not seen. Nevertheless the fact remains that in later years not much change was observed any more. Regarding the first year drop it should also be realized that a newly sown grass sward is very different from what develops later on, which was most obvious at the Smallford site where oats had been mixed with the ryegrass at the first cut.

Ryegrass because of its low heavy metal uptake is in fact not particularly suited as an indicator crop for the actual heavy metal availability. A comparison of the same two experiments for the elements Ni and Cu confirmed this as far as the contents in herbage were concerned, and did not confirm any decrease in availability as a function of time except perhaps for Ni (Figure 23).

In the experiment at Baumannshof near Munich not only herbage but also a number of other crops have been grown. The Figures 24 - 26 showing some of the results are somewhat difficult to understand because they were originally made for another purpose. However, if the squares representing the heavy metals in plants from the most highly polluted plots are compared, they indicate at least a tendency for the cereals of decreasing contents during the later years. However, this tendency was not free from exceptions especially for the grains, which is not surprising in view of the concomitant side-effects of the various sludge treatments made (Figure 24).

These contradictions were even more frequent for sugar beets, indicating that such a tendential drop in heavy metal availability as a function of time cannot be generalized (Figure 25).

As a matter of fact, the heavy metal accumulating crop spinach even exhibited the opposite, i. e. an average increase in heavy metal content during four consecutive years after the sludge application by a factor of almost two (Figure 26).

5. CONCLUSIONS

This actually means that we cannot expect the persistent heavy metals to become sufficiently immobilized later on in contaminated soils. Pending more reliable information from elsewhere, the final conclusion from the present findings may therefore be summarized as follows:

1. The long-term effect of contaminants cannot be judged on the basis of a few years or decades only but requires the scope of centuries to be fully realized.

2. Any incorporation of persistent contaminants, even if it is very gradual, will in the long term result in a gradual deterioration of soils.

3. Due to this fact, we should not only think in terms of what soils can stand now but instead aim at reducing the input of such contaminants by all possible means.

4. Accordingly, the long-term use of sludges is not a question of soils on which such wastes would be safely directed but of tracing and closing all possible sources of contaminants without compromise.

ACKNOWLEDGEMENTS

This paper summarizes experimental results from 18 long-term sludge experiments which have been carried out by the following institutions:

Bn 1 - 3: Institute of Agricultural Chemistry, University of Bonn, FRG (Prof. Dr. H. Kick) (4)

Bs 1 - 2: Institute of Plant Nutrition and Soil Science, Federal Research Center of Agriculture Braunschweig, FRG (Prof. Dr. D. Sauerbeck) (5)

Br 1 - 2: Institute of Soil Technologie, Bremen, FRG (Prof. Dr. H. Kuntze) (6)

Gi 1 - 4: Institute of Crop Science, University of Gießen, FRG (Prof. Dr. E. v. Boguslawski) (7)

Mü 1 - 4: Bavarian Institute of Soils and Agronomy, München, FRG (Dr. Th. Diez) (3)

Bx 1 - 2: Station d'Agronomie, I.N.R.A., Pont-de-la-Maye, France (Dr. C. Juste) (1)

St: Water Research Centre Stevenage, Great Britain (Dr. R. D. Davis) (2)

The cooperation by all these colleagues and the financial support by the Federal Office of the Environment (Umweltbundesamt) is gratefully acknowledged.

REFERENCES

(1) JUSTE, C. and SOLDA, P. (1984). Factors influencing heavy metal availability in field experiments with sewage sludges. In: Chemical methods for assessing bio-available metals in sludges and soils (Eds.: R. Leschber, R. D. Davis, P. L'Hermite), Elsevier Applied Sci. Publ., London - New York.

(2) COKER, E. G., DAVIS, R. E., HALL, J. E. and CARLTON-SMITH, C. H. (1982). Field experiments on the use of consolidated sewage sludge for land reclamation: effects on crop yield and composition and soil conditions, 1976 - 1981. Technical Report TR 183, Water Research Centre, Stevenage, Great Britain.

(3) DIEZ, Th. (1983). Einfluß der Klärschlammbehandlung und der Beschlammungsintensität auf den Schwermetalltransfer Boden - Pflanze. Landwirtsch. Forsch. Sh. 39, 213 - 223.

(4) KICK, H. and POLETSCHNY, H. (1982). Schwermetallgehalt im Boden und in verschiedenen Gemüsearten nach langjähriger Anwendung von Klärschlamm. Landwirtsch. Forsch. Sh. 38, 205 - 215.

(5) SAUERBECK, D. and STYPEREK, P. (1984). Evaluation of chemical methods for assessing the Cd and Zn availability from different soils and sources. In: Chemical methods for assessing bio-available metals in sludges and soils (Eds.: R. Leschber, R. D. Davis, P. L'Hermite), Elsevier Applied Sci. Publ., London - New York.

(6) KUNTZE, H. (1984). Sewage sludge on peatlands. In: Utilisation of sewage sludge on land: rates of application and long-term effects of metals (Eds.: S. Berglund, R. D. Davis and P. L'Hermite), D. Reidel Publ. Comp., Dordrecht - Boston - London.

(7) SCHAAF, H. and VON BOGUSLAWSKI, E. (1983). Schwermetallanreicherung in Böden und Pflanzen bei langjähriger Anwendung von Klärschlamm. Landwirtsch. Forsch. Sh. 39, 224 - 237.

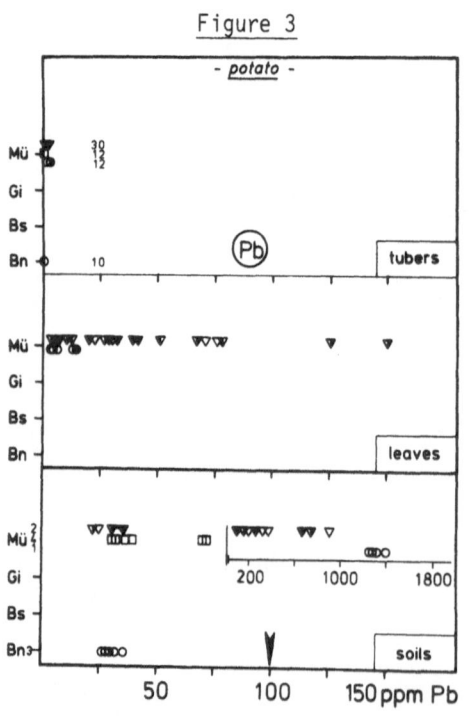

Figure 1

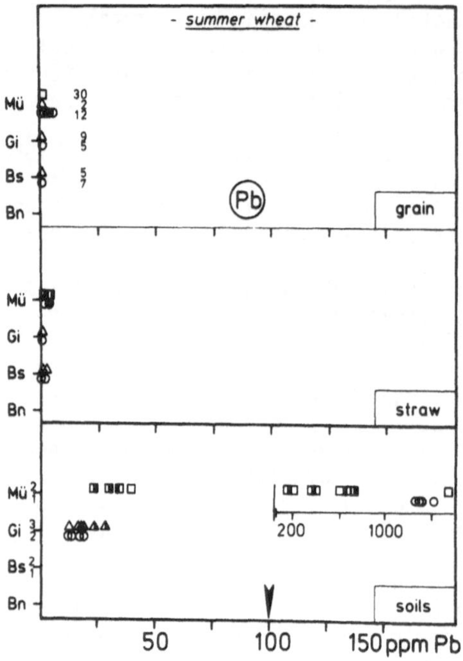

- *summer wheat* -

Figure 2

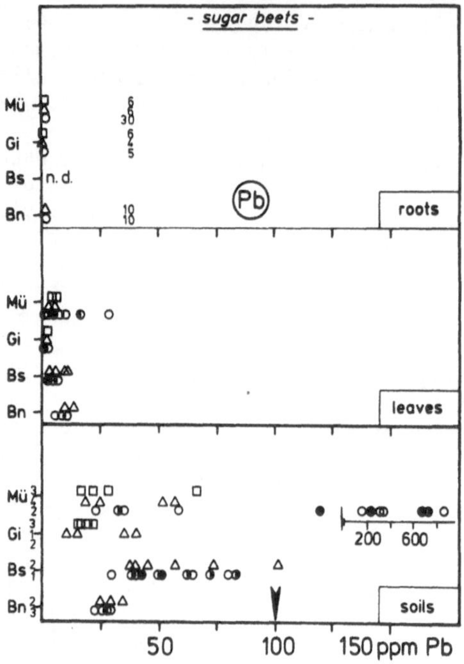

- *sugar beets* -

Figure 3

- *potato* -

Figure 1 - 3:

Lead contents in soils and crops of some long-term sewage sludge experiments

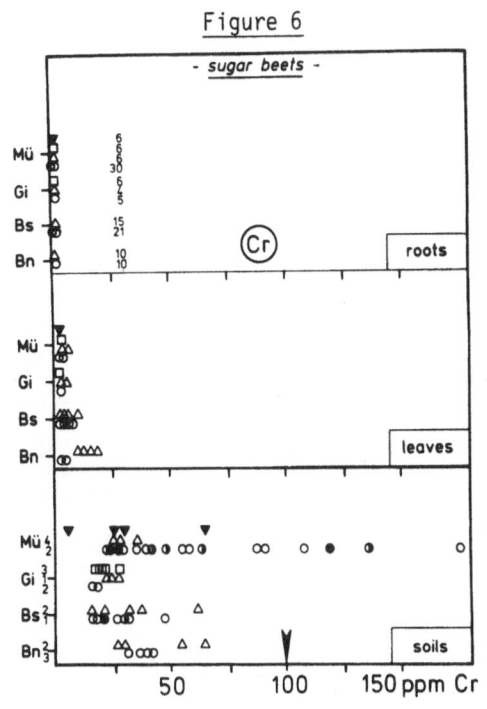

Figure 4

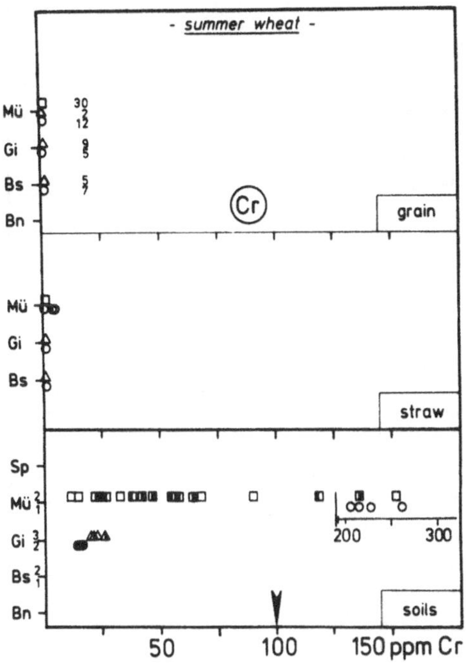

- summer wheat -

Mü
Gi
Bs
Bn

(Cr)   grain

Mü
Gi
Bs

straw

Sp
Mü
Gi
Bs
Bn

soils

Figure 5

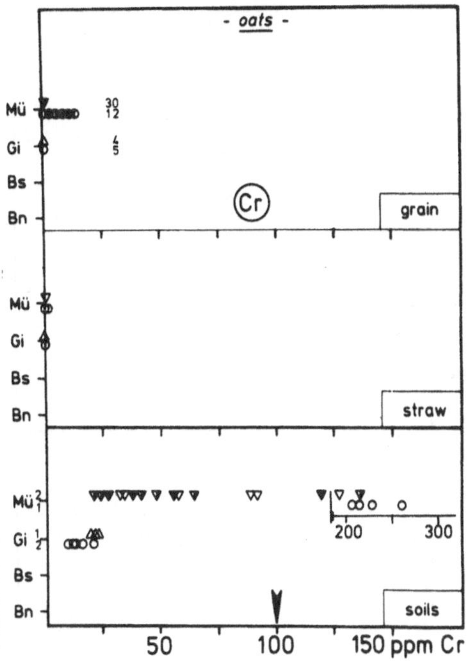

- oats -

Mü
Gi
Bs
Bn

(Cr)   grain

Mü
Gi
Bs
Bn

straw

Mü
Gi
Bs
Bn

soils

Figure 6

- sugar beets -

Mü
Gi
Bs
Bn

(Cr)   roots

Mü
Gi
Bs
Bn

leaves

Mü
Gi
Bs
Bn

soils

Figure 4 - 6:

Chromium contents in soils and crops of some long-term sewage sludge experiments

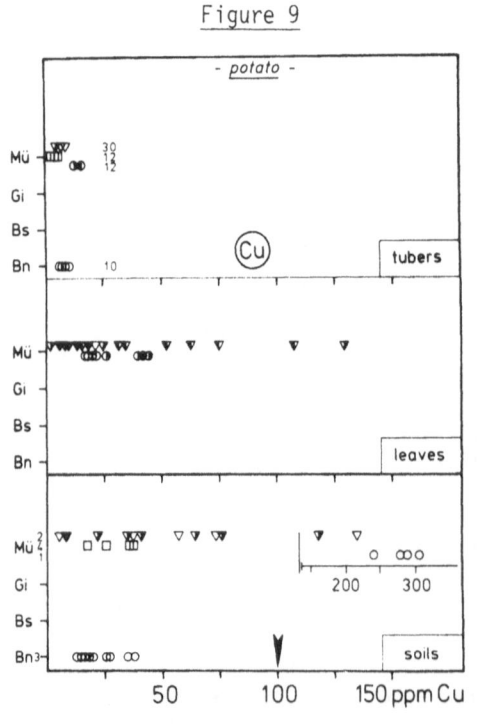

Figure 7

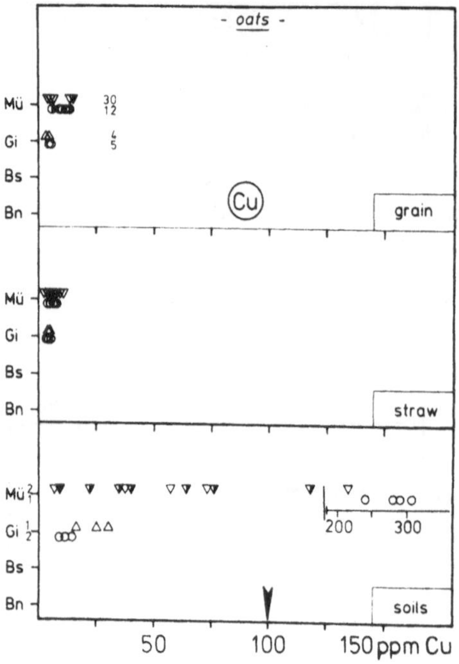

- oats -
Mü
Gi
Bs
Bn
30
12
4
5
Cu
grain
straw
soils
200
300
50
100
150 ppm Cu

Figure 8

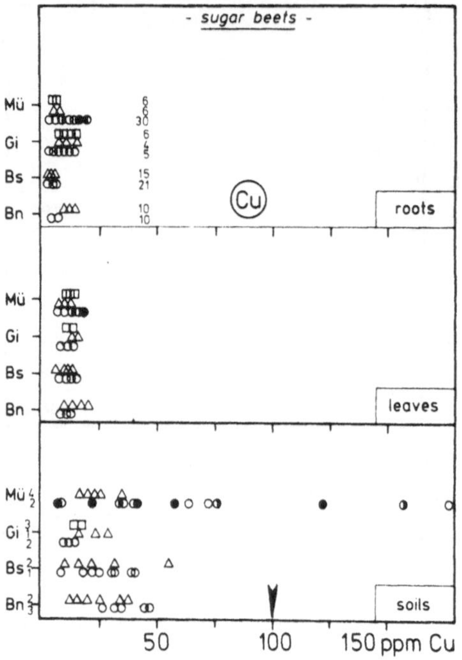

- sugar beets -
Mü
Gi
Bs
Bn
6
6
30
6
4
5
15
21
10
10
Cu
roots
leaves
soils
50
100
150 ppm Cu

Figure 9

- potato -
Mü
Gi
Bs
Bn
30
12
12
12
10
Cu
tubers
leaves
soils
200
300
50
100
150 ppm Cu

Figure 7 - 9:

Copper contents in soils and crops of some long-term sewage sludge experiments

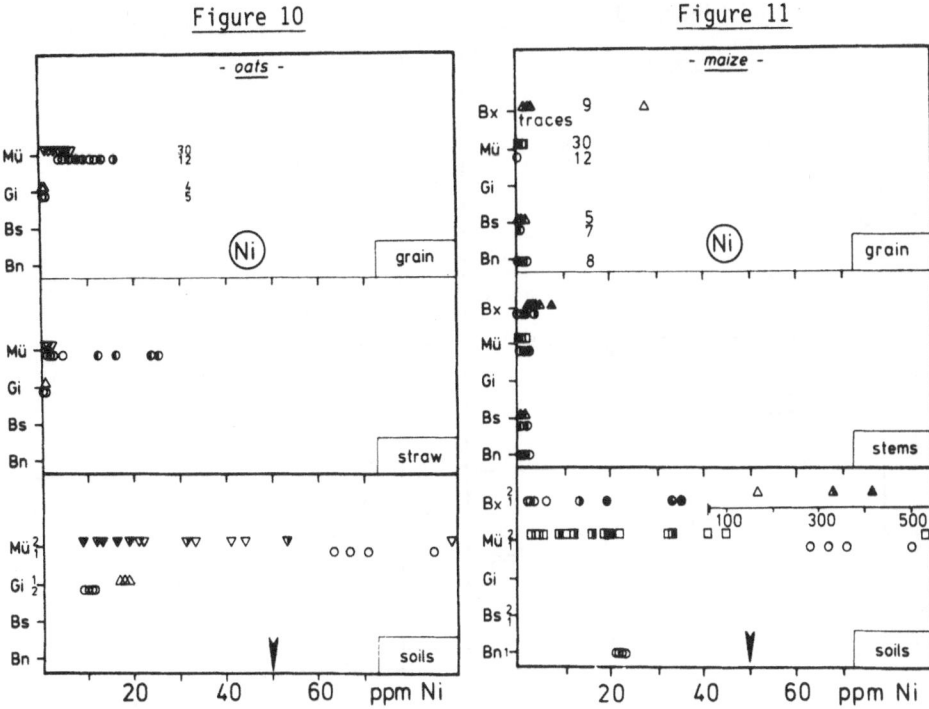

Figure 10

Figure 11

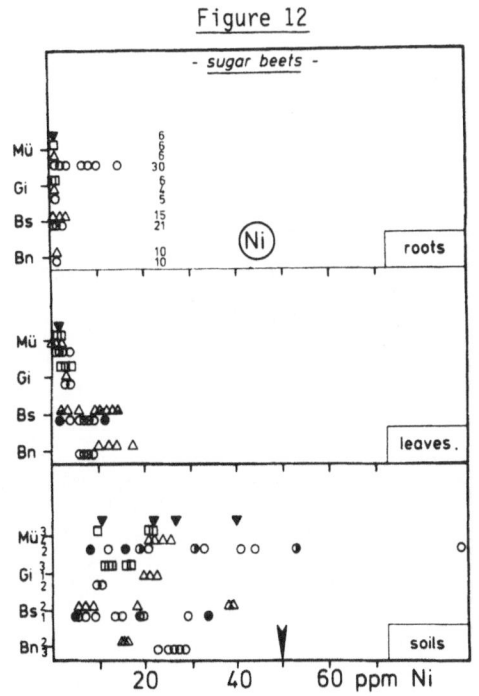

Figure 12

Figure 10 - 12:

Nickel contents in soils and crops of some long-term sewage sludge experiments

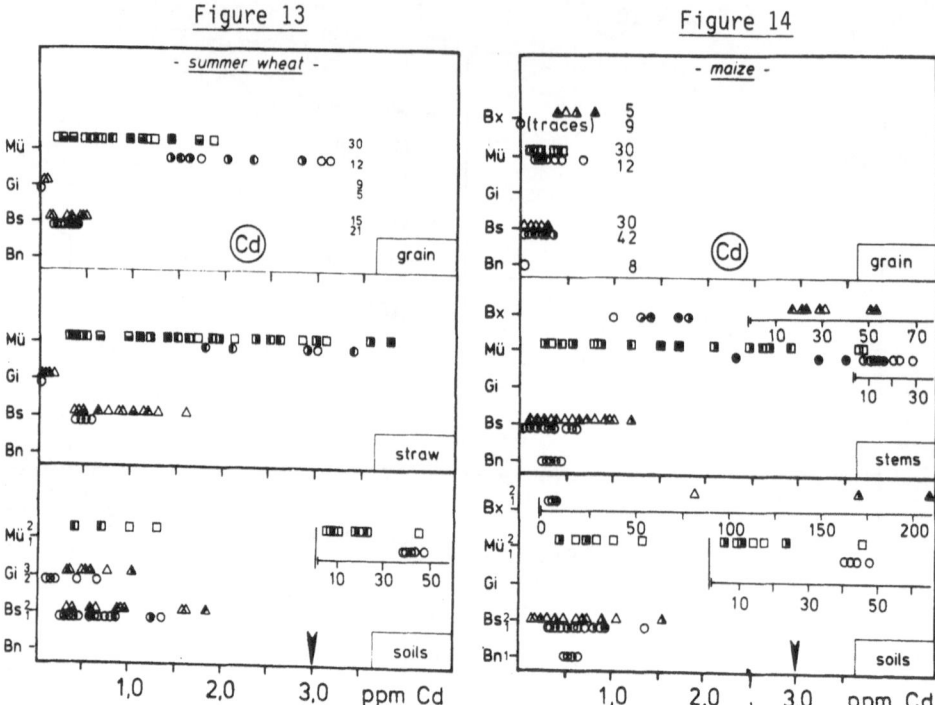

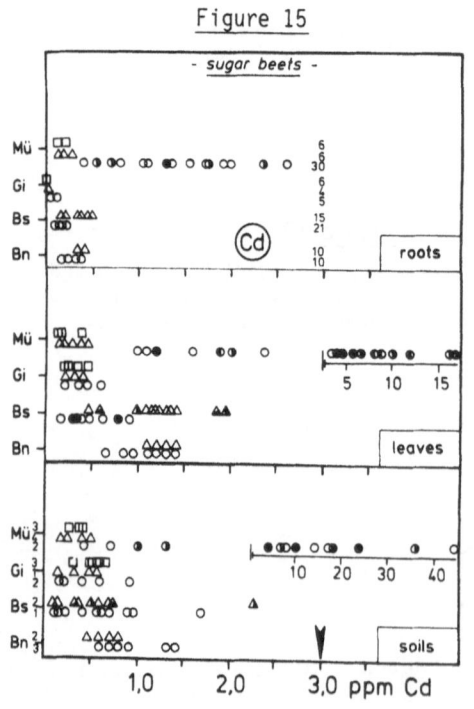

Figure 13 - 15:

Cadmium contents in soils and crops of some long-term sewage sludge experiments

Figure 16

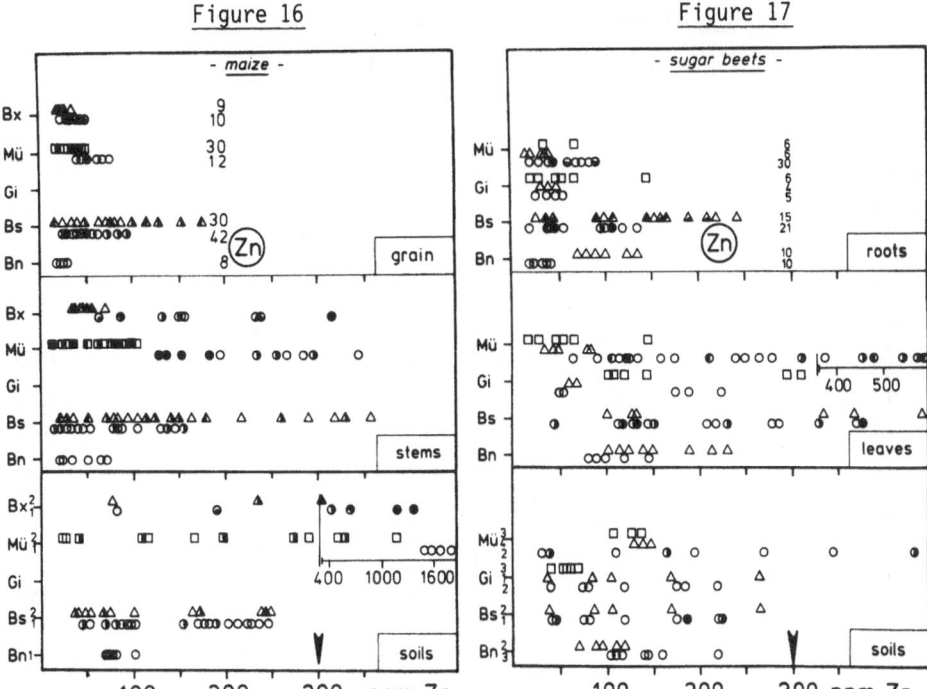

Figure 17

- maize -

Bx   9
      10
Mü   30
      12
Gi
Bs   30 Zn   grain
Bn   42
      8

Bx
Mü
Gi
Bs   stems
Bn

Bx₁²
Mü₁²   400 1000 1600
Gi
Bs₁²
Bn₁   soils

100   200   300   ppm Zn

- sugar beets -

Mü   6
      30
Gi   9
      5
Bs   15 Zn   roots
      21
Bn   10
      10

Mü   400 500
Gi
Bs
Bn   leaves

Mü₂³
Gi₂³
Bs₁²
Bn₃²   soils

100   200   300 ppm Zn

Figure 18

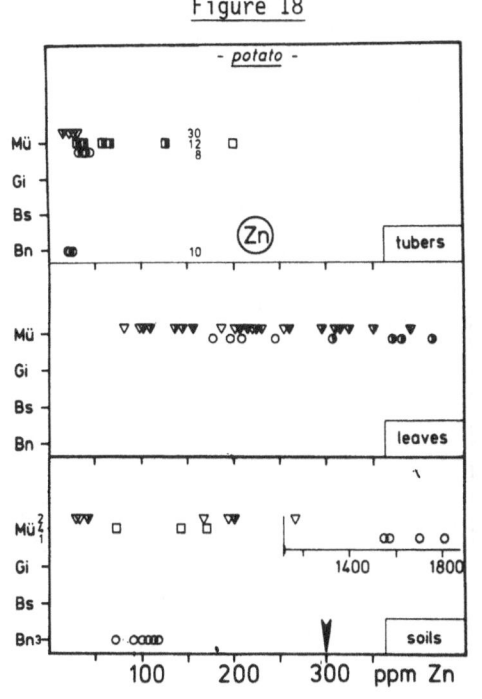

- potato -

Mü   30
      12
      8
Gi
Bs
Bn   10   Zn   tubers

Mü
Gi
Bs
Bn   leaves

Mü₁²   1400 1800
Gi
Bs
Bn₃   soils

100   200   300   ppm Zn

Figure 16 - 18:

Zinc contents in soils and crops of some long-term sewage sludge experiments

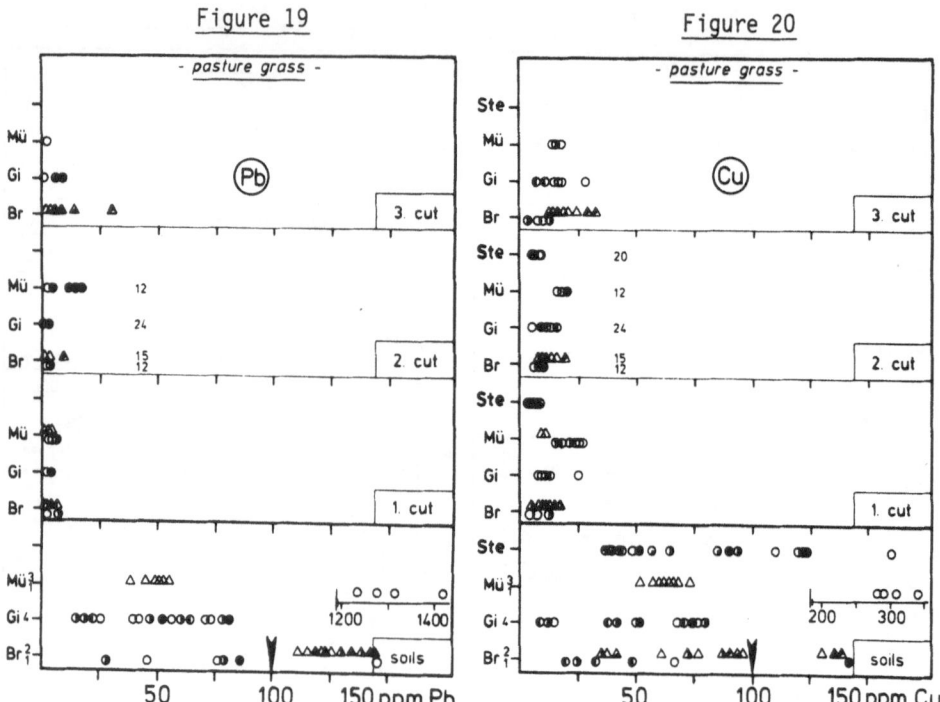

Figure 19

Figure 20

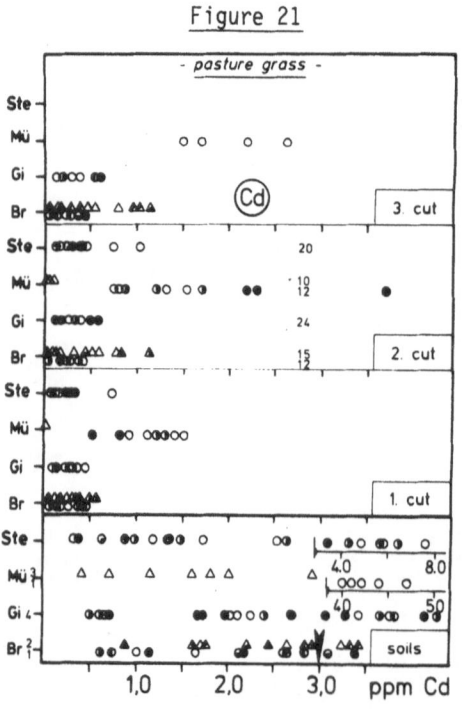

Figure 21

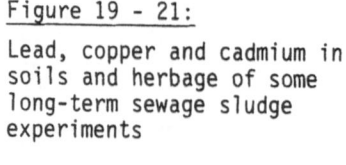

Figure 19 - 21:

Lead, copper and cadmium in soils and herbage of some long-term sewage sludge experiments

- 332 -

Figure 22

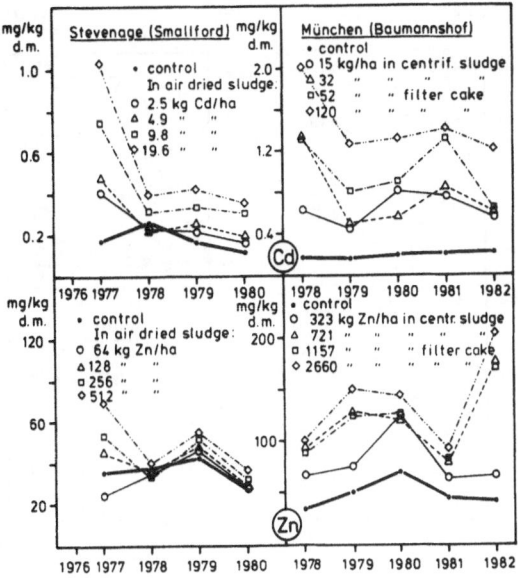

Figure 23

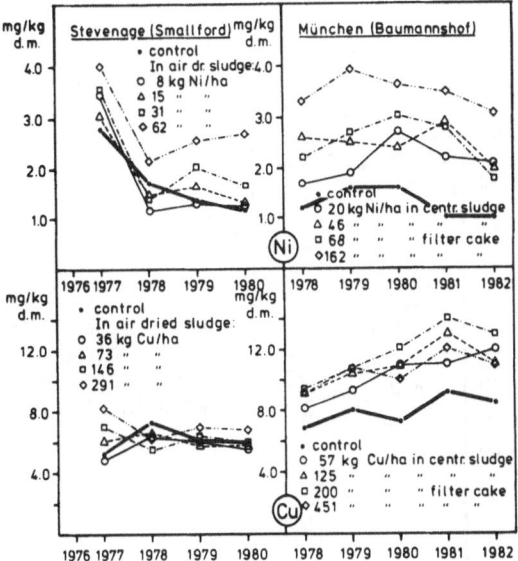

Figure 22 - 23:

Change in heavy metal contents of herbage
(2. cut) during the years following sludge
application in 1976

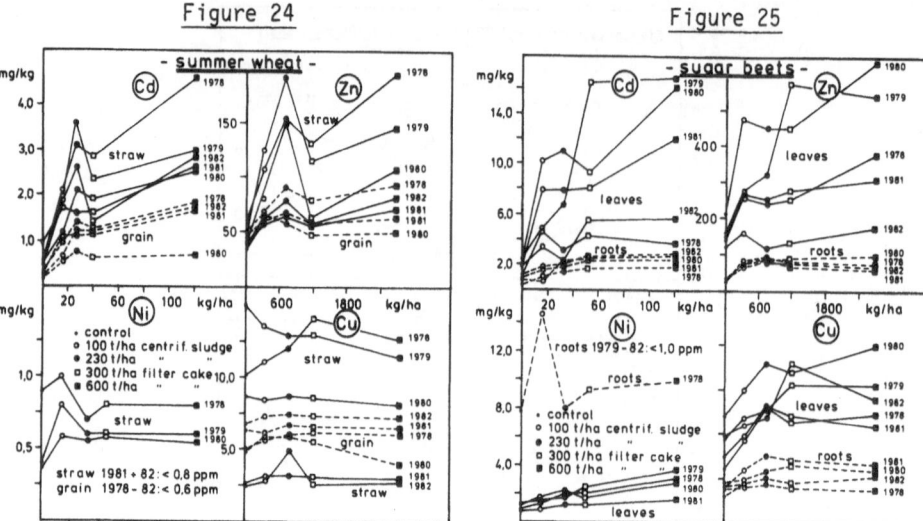

Figure 24

Figure 25

Figure 26

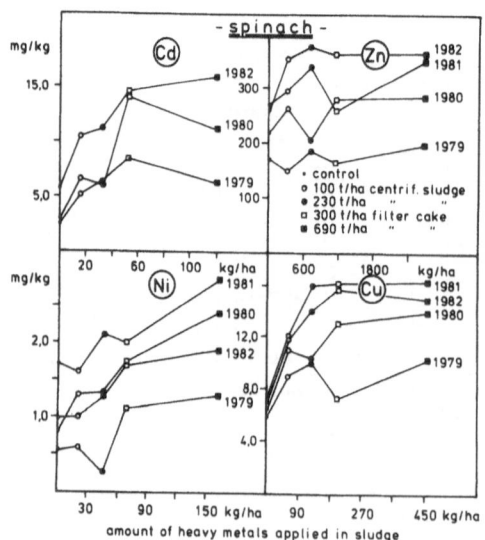

Figure 24 - 26

Heavy metal contents in crops
on a field soil at München-
Baumannshof treated in 1976
with increasing amounts of sludge

DISCUSSION

J Taradellas

We have heard some interesting results on the transfer of heavy metals to plants but what about the species of the metal. It is important to know total metal, species and route especially in sludge, soil and plant in relation to their impact.

G Fleming

The whole question of the speciation in which metals are combined is being worked on but more is needed. The absorption of metals results in the form of the metal is a truism as chromatography has shown for plants and animals. With selenium efficiency in animals, we are really talking about the enzyme; we know the active form is relative to ferric oxidase.

How valuable are total metal contents in soils and in plant? I see a great expansion in effort in the future in metal speciation. Total metals have given a lot of information but now is a bit too simplistic.

D Sauerbeck

I am not so optimistic. I have worked on speciation and reported this to the Working Party 5 meeting at Stevenage. These are empirical fractions and bear no direct relation to uptake and availability. I agree with Dr Fleming that the state of art is rather poor with soils and sludges however there has been some good work on river sludges and sediments but the connection with availability studies has not yet been achieved.

We have already discussed availability tests. With zinc, cadmium and to a lesser extent, nickel there is no question that the soluble fraction gives the most reliable indication of availability. In the UK, soil solution but we suggest neutral salts which are nicely related to uptake and availability. We need to take longterm trials where plant and soil contents are well known and test these samples with neutral salt extraction. Cadmium, zinc, nickel and to some extent copper are most critical and their availabilities should be predictable from such an extraction procedure. With EC collaboration, we could get sufficient data pairs to see if the neutral salt approach is valid.

R D Davis

Soil properties determine speciation with soil acidity posing a difficult problem with extractant. Dr Sauerbeck said that 1 ppm in wheat grain could be regarded as an acceptable limit. In figure 1 in the paper by Drs Fleming and Davis showing the cadmium concentration in wheat and soil, 1 ppm cadmium in wheat is reached at 8 ppm in the soil. So the limit of 3.5 ppm in soil is quite safe and does not support FRG lowering their soil cadmium limit to 2 ppm.

? (I)

What about the importance of transformations in the soil particularly with chromium.

D Sauerbeck

Hexavalent chromium is very toxic to plants but it is strongly fixed and unavailable. I do not know of any mechanism which can reverse this normal soil process.

EFFECT ON A LONG TERM SLUDGE DISPOSAL ON CADMIUM AND NICKEL TOXICITY TO A CONTINUOUS MAIZE CROP

C. JUSTE and P. SOLDA
Station d'Agronomie, I.N.R.A.
33140, Pont de la Maye, France

Summary

Anaerobically digested sewage sludge dewatered by the Porteous heat process was applied from 1976 to 1980 to an acidic coarse sandy soil, on which a continuous maize crop was grown from 1976 to 1984.
Tests on the sludge revealed very high cadmium and nickel contents, as the treatment plant received the effluent discharged by a battery manufacturer.
Treatments applied involved conditioning test plots with sludge (at doses of 10 tonnes D.M/hectare/year and 100 tonnes D.M./hectare/2 years) while control plots received only mineral fertilization. The field weight of the maize grain was determined, as were the Cd, Ni and Zn contents of the 6th leaf, sampled at the 12th leaf fully emerged stage.
Sludge application significantly decreased grain yield (-10 % and - 18 % respectively for the low and high rates of sludge application). However, in 1980 yield was drastically affected (-50 %) in the plots treated with high sludge dosages : purpling of seedling leaves, resembling signs of severe phosphorus deficiency, was observed in this case.
Cadmium uptake was greatly enhanced by the addition of sewage sludge : contents of up to 80 ppm were observed in some years. However in 1980 Cd uptake was much reduced. The Cd content of maize originating from the control plots increased over time, indicating appreciable cadmium pollution of the control plots.
Nickel uptake was also increased by the application of sludge, though to a lesser extent than Cd; in 1980 the Ni content of the 6th leaf was drastically increased.
The pattern of zinc uptake was similar to that for Cd, but a very large annual variation in the concentration of Zn in plants was observed during the experiment, uncorrelated with sludge applications.
Cd, and to a lesser extent Zn, should be seen as heavy metals which are highly available to the maize plant. Ni availability, on the other hand, was considerably lower.
No clear relationship between grain yields and heavy metal content in the 6th leaf was observed in the course of the experiment except in 1980; in that year a marked decrease in grain yield was associated with a significant increase in Ni uptake, accompanied by a decrease in the uptake of cadmium.
This data could be interpreted as evidence that Ni toxicity towards maize explains the marked drop in yield in 1980. The data for that year shows a probable mutual influence between the accumulation of Cd and that of Zn. The low soil temperature prevailing in the first stage of maize growth in 1980 provides an explanation for these

metals' behaviour : Cd uptake may have been reduced by the low soil temperature in the rooting zone whereas Ni uptake was less affected, resulting in a large decrease in maize yield.
More research is needed to assess the antagonisms involved in heavy metal uptake and the influence of translocation and temperature on this physiological process.

The heavy metal content of sewage sludge is the main factor which still restricts the general recycling of this type of urban waste for use on farming soils.
A great deal of research has been carried out to evaluate the effects of sludge spreading on the build-up in soils and plants and the behaviour of heavy metals. It must be admitted that very few such investigations have revealed a toxic effect whose precise cause is the heavy metal content of sludge, despite an often marked build-up in the soil and in plants. From this, one could draw the conclusion that the risk of heavy metal toxicity to plants arising from the spreading of sludge is a secondary risk, the main one being that of contamination of the food chain.
However, there are cases where the spreading of certain types of sludge does lead to a reduction in crop yields. The aim of this study is to describe one such case in which it was possible to ascribe the phytotoxicity of the sludge to its abnormally high cadmium and nickel contents.

MATERIALS AND METHODOLOGY

1. Design of experiment
The experiment was conducted in the open and was set up on an I.N.R.A. experimental farm near Bordeaux. It was carried out on a coarse sandy soil (70 % coarse sand) which was found on pretesting to be acidic (pH = 5.5.) with a low organic content (2.2 %).
It was set up in 1976 on land long cleared for cultivation on which crops had been rotated since 1972 (maize-oats-maize). The test crop was an irrigated maize monoculture, using a single variety from 1976 to 1984 : INRA 260. The mineral fertilizers applied, weed control practice and soil cultivation were modelled strictly on current farming practice in the sector under consideration. Only the grain was removed, crop residues (mostly stems and leaves) being ploughed in every year at the end of winter.

2. Treatments
Treatment A : control;
Treatment B : sludge (10 tonnes dry matter/hectare/year) spread from 1976 to 1980 inclusive;
Treatment C : sludge (100 tonnes/hectare/2 years) spread in 1976, 1978 and 1980.
The addition of sludge was deliberately halted in 1980 so as to monitor its after-effects.
Mineral fertilization was adjusted to the same level (200 N - 200 P2 05 - 200 K2 0) in the plots for treatments A and B, assuming that the nitrogen in the sludge was available for 50 % the year in which it was applied. Each plot was replicated five times in a randomized complete box design. The areas treated with sludge and fertilizer were 3 m x 6 m.

3. Sludge used

The sludge came from a sewage treatment plant in the Bordeaux conurbation, which treated an equivalent population of 300 000. It had been anaerobically digested and dewatered using heat (by the Porteous process). A battery factory upstream of the works, which discharged its effluent into it, caused heavy cadmium and nickel pollution of this sludge.

4. Measurements carried out

In addition to measuring crop yield, an analysis was made of the 6th leaf, which was removed after the 12th leaf had completely unfurled. After dry mineralization, the concentration of Cd, Ni and Zn in the mineralizate (hydrochloric acid) was measured by atomic absorption flame photometry. The zinc was measured because of its relative abundance in the sludge used and the fact that it is often observed to behave similarly to cadmium.

The results were subjected to multifactorial analysis, which enabled, among other things, the multiannual effect to be taken into account. The level of significance was determined using the Bonferoni test.

RESULTS

(a) sludge composition

Table 1 sets out average contents measured over the five years of application and the range of values obtained. Cd, Ni and Zn were found to be the most abundant metals in the sludge. The cumulative quantities added to the soil thus reached very high levels in the case of the higher sludge dosage (Table 2) and these values were confirmed by the extraction of exceptionally high quantities of Cd and Ni with aqua regia.

(b) Crop performance – yield

Table 3 shows that, for the period considered (nine years, 1976 to 1984) sludge application had a depressive effect (fall in yield of the order of - 10 % for the lower sludge dosage and - 18 % for the higher dosage). Altogether, this depressive effect was quite small in view of the enormous quantities of reputedly very toxic metals (Cd and Ni) which were added to the soil through the application of sludge.
However, the yearly results (Figure 1) indicate that the depressive effect the last time that the sludge was spread (1980) was far greater, amounting to - 50 % in the case of the higher sludge dosage. It was also in this year that the most spectacular poisoning symptoms occurred, at the first stage of development of the plant. These consisted of intense reddening of the leaves and were very similar to those observed in cases of acute phosphorus deficiency. These symptoms then disappeared and were replaced by chlorosis between the veins, affecting the whole plant. In addition to these symptoms, the plants in the sludge-treated plots were distinguished by being shorter and more slender in appearance.

(c) Metal content of the 6th leaf
Cadmium

Table 4 shows clearly that this metal was strongly accumulated in the 6th leaf, although there was a tendency for it to level out for the highest dosage of sludge applied. It also shows that the leaves of plants from the control plots contained a much higher quantity of cadmium than that usually found in plants from non-contaminated soils (0.5. ppm on average). Figure 2, representing concentration trends over the years, shows a progressive increase in cadmium content in plants originating from control plots : it thus seems very likely that there was increasing pollution of the control plots by the metal, no doubt owing to soil cultivation practices or erosion, possibly even to the transfer by lateral migration of Cd in the form of ions or complexes.

Figure 2 also shows that plant cadmium content dropped considerably in 1980. In that year an exceptional drop in yield was recorded for the sludge-treated plots. Despite the halting of the application of organic waste after 1980, plant cadmium contents continued to be very high.

Nickel (Table 5)

The application of sludge resulted in a highly significant increase in nickel concentration in the plant. However, this increase was lower than that observed for cadmium.

Figure 3 shows that in 1980, when there was a very marked drop in yield, leaf nickel content reached its maximum value (approx. 18 ppm for the high rate of sewage application. Unlike the situation observed for cadmium, nickel concentrations in the control plants did not show a progressive increase over time : it would thus appear that the control plots were not polluted with this metal.

Zinc (Table 6)

Concentrations of this metal were monitored because many research results had indicated certain parallels between the behaviour of this metal and that of cadmium. It was found that zinc, the content of which in the sludge was of a similar order of magnitude to that of nickel, accumulated much more in the plant than the later. Figure 4 shows that the effect for each year cannot be related to additions of sludge, as is shown by the fluctuations of the concentration in plants from the control plots.

(d) Availability indices of metals

This index was calculated (Table 7) for both sludge application rates by considering the increase in concentration of a given metal in the plant and the total quantity of metals added in the sludge for each treatment.

The value of this index reveals that cadmium is most easily absorbed by the plant, zinc less so by a factor of 2, while nickel is distinguished by a much lower bio-availability. These results confirm the relative similarity of behaviour of Cd and Zn in terms of bio-availability. The very low availability of nickel may be associated with the lower solubility of this metal, phenomena of competition between metals for absorption or with the self-restriction by the plant of absorption and translocation of a metal whose high phytotoxicity is well established.

(e) Relationship between plant yield and metal concentration

In 1980, as indicated above, there were simultaneously a very large fall in yields, a very appreciable increase in plant nickel content and a no less marked drop in Cd concentration, which reached, in the case of the highest sludge rate of application the lowest value observed for the whole period of the experiment.

This observation would seem to indicate that it is above all nickel which has a phytotoxic effect, whereas cadmium may in some years reach high concentrations (approaching 80 ppm) without apparently causing too much damage to the plant.

The exceptional uptake of Ni in 1980 raises a question, as the quantities added in the form of sludge in 1980 were of the same order of magnitude as those applied in 1976 and 1978. One possible hypothesis is that this is due to the lower ambient temperatures during establishment of the crop in 1980; Figure 5 does show that average soil temperatures measured at a depth of 10 cm during 10 day periods in May and June were lower in 1980 than in 1976 and 1978. It is possible that under these conditions absorption of Cd could have been reduced. As a result Ni uptake was enhanced because likely there is competition for plant uptake between the two metals.

CONCLUSIONS

A marked increase in concentrations of Cd and Ni originating from heavily polluted sludge in a sandy soil resulted in a very significant decrease in yields, even after the end of sludge application.

Detailed examination of the results tends to show that toxicity to maize is more related to the soil content of nickel than that of cadmium, which is much more easily absorbed by the plant; in the case of maize, phytotoxicity does not therefore constitute an effective barrier to the translocation of cadmium into the food chain.

Supplementary studies would appear to be necessary to investigate the antagonisms which appear to exist between heavy metals regarding absorption and transfer and into the conditions which may increase such competition.

Table 1 : Average composition of sludge samples tested and range of values

		Average	Range
% D.M.	(Carbon	16.78	15.89 – 17.66
	(Nitrogen	1.26	0.89 – 1.42
	(P_2O_5	4.22	3.16 – 4.75
	(Ca	10.59	7.17 – 14.45
	(Mg	0.29	0.16 – 0.40
	(K	0.12	0.08 – 0.15
	(Fe	3.65	2.76 – 4.15
ppm D.M.	(Cu	488	248 – 663
	(Mn	699	405 – 984
	(Zn	3 066	2 300 – 4 718
	(Cd	1 830	910 – 2 672
	(Cr	219	127 – 365
	(Ni	4 071	2 920 – 5 000
	(Pb	722	578 – 875

Table 2 : Cumulative amounts of metals added to the soil by applying high
dosages of sludge : period 1976 – 1980 (kg/ha);
quantities extracted with aqua regia (ppm).

Metal	Cumulative amounts of metal 1976 – 1980 (kg/ha)	Metal extracted with aqua regia (ppm)
Fe	11 354	3 470
Mn	234	81
Cu	170	46
Zn	976	158
Cd	641	110
Ni	1 337	269
Pb	231	49
Cr	64	18

Table 3 : Effect of adding sludge on average yields from 1976 to 1984

	Cumulative amounts (t/ha) of sludge		
	0	50	300
Grain at 0 % moisture content (kg/ha)	8 544	7 733	7 122
0 t/ha of sludge	–	v.s.	v.s.
50 t/ha of sludge	v.s.	–	v.s.

v.s. = difference greater than the 0.1 % significance
       level

Table 4 : Cadmium content of the 6th leaf

	Cumulative amounts (t/ha) of sludge		
	0	50	300
Cd (ppm / D.M.)	6.9	35.6	54.6
0 t/ha of sludge	–	v.s.	v.s.
50 t/ha of sludge	v.s.	–	v.s.

v.s. = difference greater than the 0.1 % significance
       level

Table 5 : Nickel content of the 6th leaf

	Cumulative amounts (t/ha) of sludge		
	0	50	300
Ni (ppm/D.M.)	1.3	3.2	9.0
0 t/ha of sludge	–	v.s.	v.s.
50 t/ha of sludge	v.s.	–	v.s.

v.s. = difference greater than the 0.1 % significance
       level

Table 6 : Zinc content of the 6th leaf

	Cumulative amounts (t/ha) of sludge		
	0	50	300
Zn (ppm / D.M.)	24.6	46.0	56.7
0 t/ha of sludge	-	v.s.	v.s.
50 t/ha of sludge	v.s.	-	v.s.

v.s. = difference greater than the 0.1 % significance level

Table 7 : Availability index of metals

Cumulative amounts sludge spread (t/ha)	Mean increase in concentrations compared with the control (ppm) (a)		Cumulative amounts of metal added to soil (kg/ha) (b)		Availability index (c)	
	50	300	50	300	50	300
Cd	28.7	47.7	107	641	26.8	7.4
Ni	1.9	7.7	223	1 337	0.9	0.6
Zn	21.4	32.1	163	976	13.1	3.3

$$(c) = \frac{(a)}{(b)} \times 100$$

Figure 1

EFFECT OF ADDING SLUDGE ON THE MAIZE GRAIN YIELD

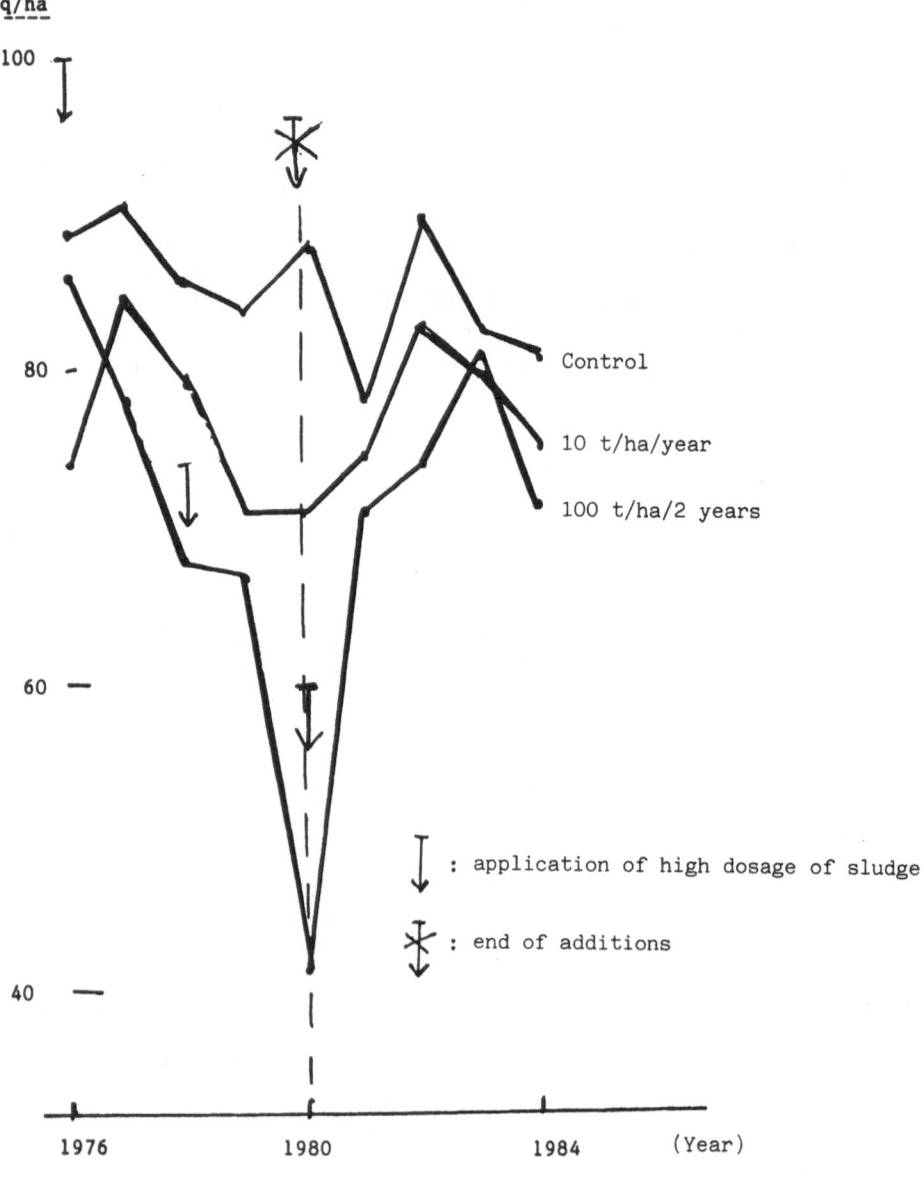

Figure 2
EFFECT OF ADDING SLUDGE ON THE CADMIUM CONTENT OF THE 6th LEAF

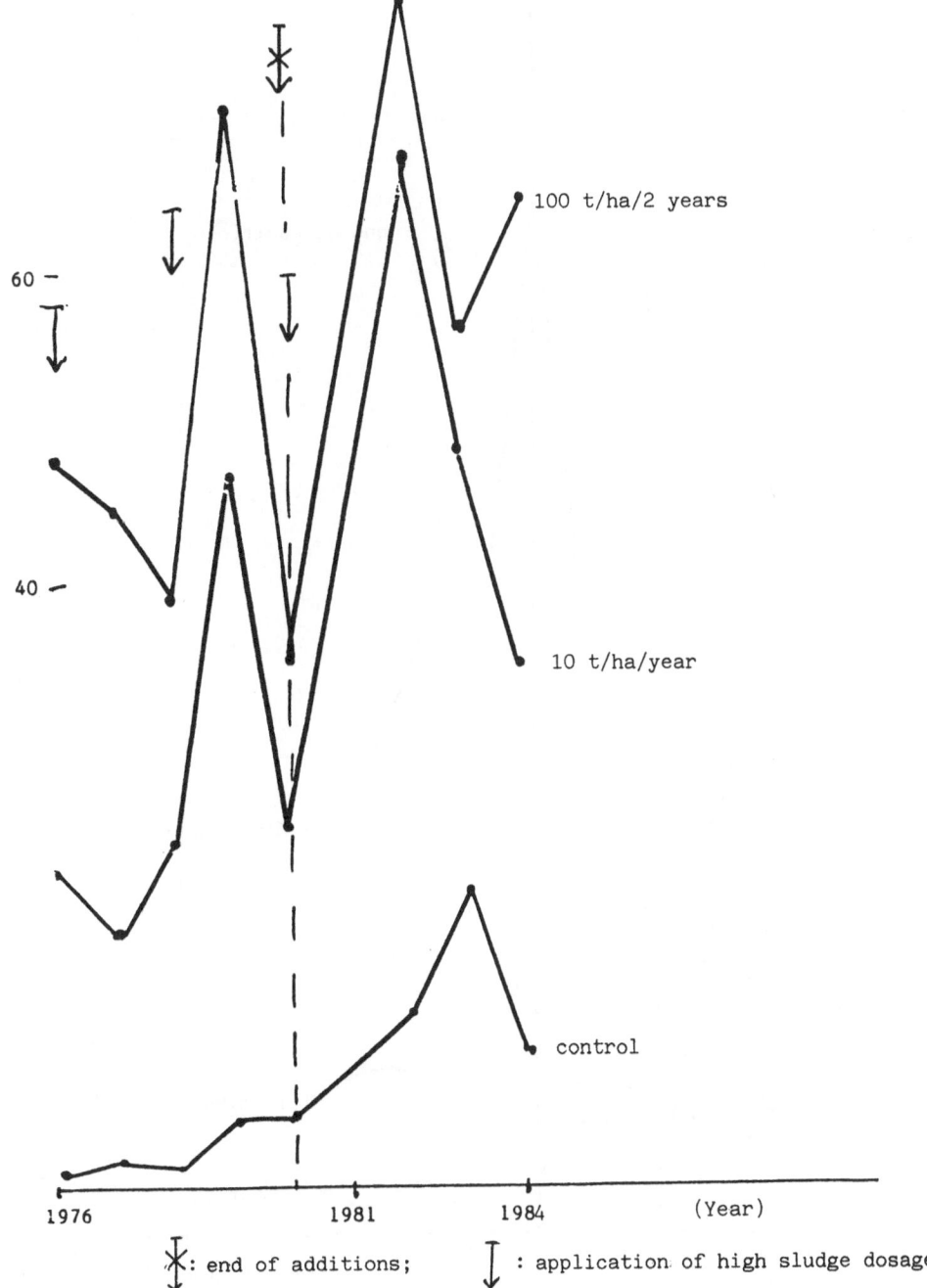

ppm (D.M.)

100 t/ha/2 years

10 t/ha/year

control

1976   1981   1984   (Year)

✳: end of additions;   ⌡ : application of high sludge dosage

Figure 3

EFFECT OF ADDING SLUDGE ON THE NICKEL CONTENT OF THE 6th LEAF

ppm (D.M.)

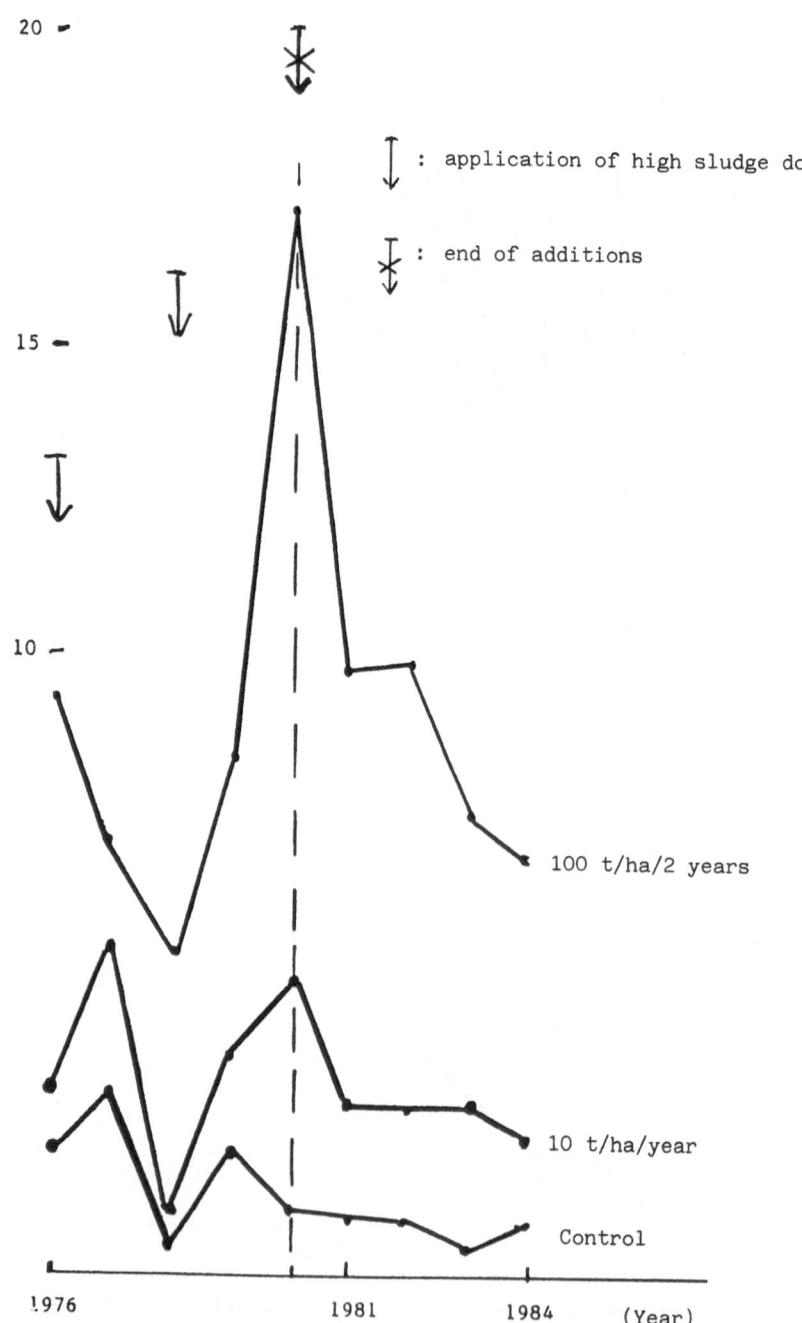

$\downarrow$: application of high sludge dosage

$\underset{\downarrow}{\ast}$: end of additions

100 t/ha/2 years

10 t/ha/year

Control

1976   1981   1984   (Year)

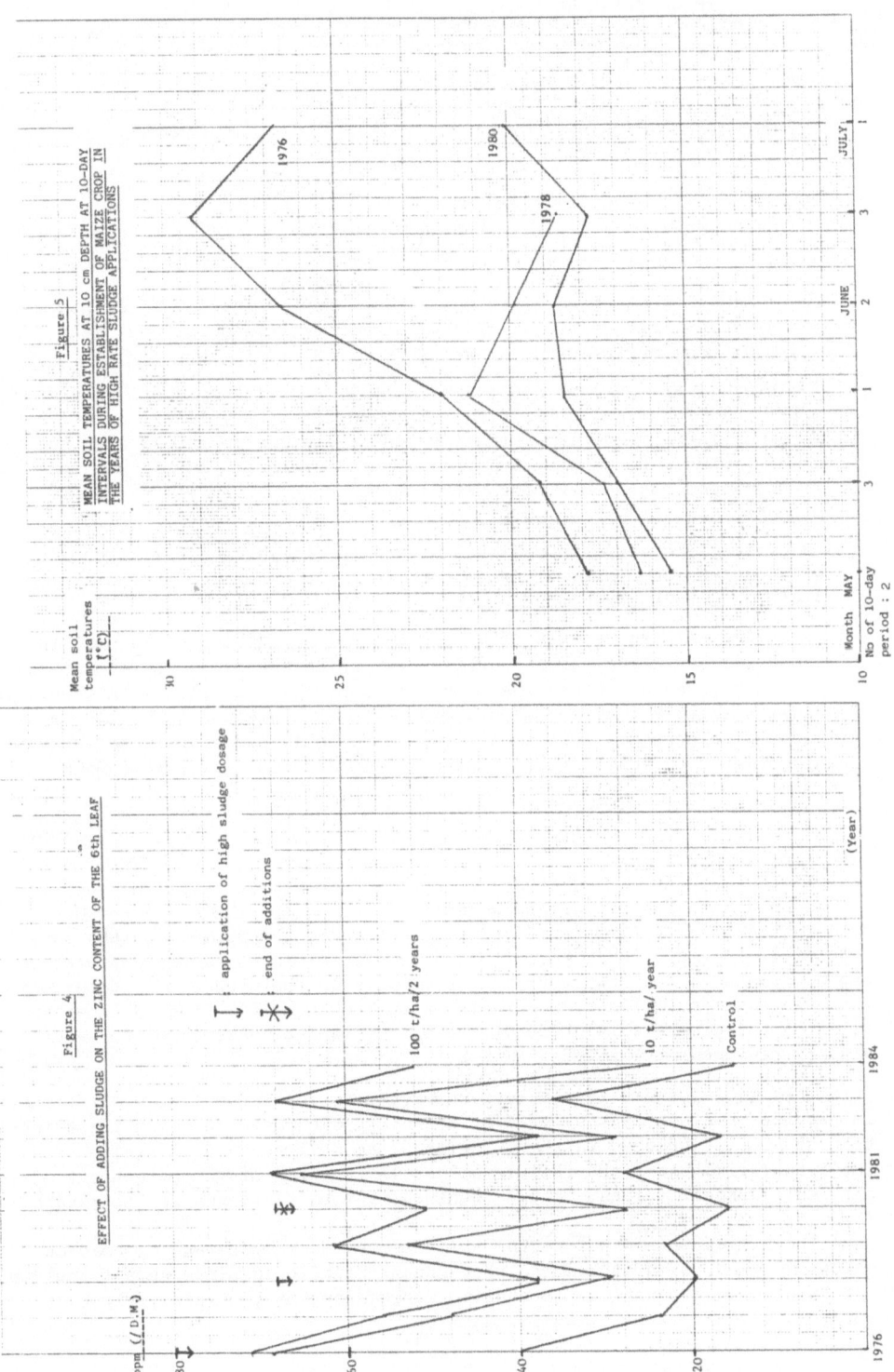

Figure 5

MEAN SOIL TEMPERATURES AT 10 cm DEPTH AT 10-DAY INTERVALS DURING ESTABLISHMENT OF MAIZE CROP IN THE YEARS OF HIGH RATE SLUDGE APPLICATIONS

Figure 4

EFFECT OF ADDING SLUDGE ON THE ZINC CONTENT OF THE 6th LEAF

Mean soil temperatures [°C]

application of high sludge dosage

end of additions

100 t/ha/2 years

10 t/ha/year

Control

(Year)

ppm (/ D.M.)

Month MAY   JUNE   JULY

No of 10-day period : 2

TRACE METAL REGULATIONS FOR SLUDGE UTILIZATION IN AGRICULTURE; A CRITICAL REVIEW.

J.C. TJELL
Department of Environmental Engineering
Technical University of Denmark, DK-2800 Lyngby, Denmark

Summary

It is argued that three different approaches 1) historical, 2) pragmatic, and 3) scientific have been adopted when formulating national regulations on sludge utilization in agriculture.

The historical approach recognized metal toxicities to crops (phytotoxicity), and regulations limiting metal additions to soils to avoid this were developed. The pragmatic approach recognized phytotoxicity, but also current sludge quality and the need for disposal, and regulations were developed based on sludge metal concentrations and useful sludge application rates.

The scientific approach involves human health risk analysis, and regulations can be developed to protect the quality of animal and human foodstuff, and thus the human health.

It is further argued that metal concentrations in sludges in certain countries are diminishing considerably, which may clear the ground for stricter international guidelines acceptable to countries with sensitive environments. Guidelines should also take into account naturally occurring differences in soil conditions, especially pH as an important governing factor in the dietary transfer of metals from soils to humans.

INTRODUCTION

Utilization of sewage sludge in agriculture as a fertilizing material is regarded as a beneficial and economical outlet for the solid residues from waste water treatment. In order to safeguard the public health, many countries have in recent years adopted some form of regulatory measures to avoid unwanted side effects from contaminants in the sludges. A number of these national guidelines were reviewed by Webber et al. (1984), regarding rules on trace metals. Additional guidelines, and amendments to those previously reviewed are listed in Table 1.

The major outcome of the review was the impression of large differences in the national approaches, leading to remarkable differences in the selection of metals included in the guidelines, as well as large variations in the actual limits for the individual metals in terms of sludge concentrations, soil concentrations, annual loadings or total loadings of the sludged land.

The aim of this paper is to explore possible factors and approaches which might have been considered when the national guidelines were formulated, and to give an account of the ongoing discussions on the scientific basis for regulations, exemplified for Cadmium.

Table 1. Additional guidelines, and guideline modifications since the review by webber et al. 1984.

	Max.Conc. in sludges to land ppm in DS				Annual loading limit kg/ha,y					Total loading limits				
	USA N-E	Nether- lands	Scotland	Austria	USA N-E	Nether- lands	Scotland	Austria	Italy	US N-E kg/ha	Scotland Soil ppm	Scotland kg/ha	Italy Soil ppm	South Africa kg/ha
As		10	150			0.02	0.3				12	30		
Cd	25	5	20	10	0.4-0.9	0.01	0.04	0.038	0.04	2-4.5	1.6	4	3	1
Co														10
Cr	1000	500	2000	500	20-120	1.0	4	1.875	3.6	100-600	120	300		380
Cu	1000	600	1500	500	5-30	1.2	3	1.875	4.5	25-150	80	200	100	150
Hg	10	5	7.5	10		0.02	0.015	0.038	0.025		0.4	1		2
Mo			25						0.05		2	5		5
Ni	200	100	600	100	2-12	0.2	1.2	0.38	1.2	10-60	48	120	50	36
Pb	1000	500	1500	500	20-120	1.0	3	1.875	3	100-600	90	225	100	225
Se			40						0.08		2.4	6		2
Zn	2500	2000	2500	2000	6-50	4	5	7.5	9	50-300	150	375	300	870
Minimum pH					6.0		5.5		6.0					
Maximal sludge loading t/ha,y							2	2						

USA Northeast — Metal loadings are according to three soil classes. pH is measured in 1:1 soil: water paste. Sludge should be used as a fertilizer (no specific limits). Metal loadings are originally in pounds/acre. (Baker et al. 1985).

Netherlands — Loadings are for arable land. Loading to pasture land are one-half the values shown (revised April, 1984 for rules covering Cd and Hg).

Scotland — The total loadings are inclusive the indigenous soil contents. If low sludges are applied higher DS may be allowed. (Scottish Agricultural Colleges 1981).

Austria — Metal limits may be raised two fold if only one metal exceeds the limits. A maximum sludge loading of 5 t DS/ha,y is recommended. It is anticipated that it will take more than 100 years to reach the tolerable soil concentrations as used in FRG. (ÖWWV 1984).

Italy — The values are from a proposal by Genevini et al. (1983). The annual loadings are meant for soils with CEC >12 meq/ 100g, and should be halved for soils with a CEC 6-12 meq/100. Presumably soils, with lower CEC ought not receive sludge.

South Africa — The values are from a proposeal by Nell & Engelbrecht 1982. Only cumulative metal loading is limited. Sludge should be utilized as a fertilizer.

THE HISTORICAL APPROACH IN ESTABLISHING REGULATIONS

When the first guidelines for management of sludge utilization were formulated 10-15 years back, the scientific background for setting limiting values was rather scarce. However, toxicity of metals to plants (phytotoxicity) was recognized and consequently guidelines were formulated to avoid this unwanted side effect, often as a limitation on the maximum concentrations of mobile metals in soils to be reached after sludge use. The metals often referred to as phytoxic are Zn, Ni and Cu, but also Cd may become phytotoxic, as well as the non-metal B.

Over the years several modifications to the essentially phytotoxicity limitations have led to decreasing figures for tolerable contamination of soils. There is no clear indication of how the emphasis on the protection of the plant production was shifted to protection of the consumers of plant materials, which is the stated purpose of all current guidelines.

An illustrative example of this revision proces may be taken from Germany (FRG), see Table 2. The tolerable soil concentrations have decreased considerably over the years, especially for Pb and Cd. The guideline for sludge application (Klärschlammverordnung 1982) states an intention of protecting the food chain, and one of the regulations is soil concentration limits to fulfil this objective.

A possible clue to the present german tolerable soil concentrations may be the wish to accomodate most german soils in the tolerable range. The concentrations of trace metals in soils in Nordrhein-Westfalen were reported by Kick et al. (1979), and by Vetter et al. (1983) for Baden-Württemberg. In both areas a small but significant part of the soil samples analyzed exceeded the present tolerable soil concentration limits for the FRG (largest for Ni 14.9%).

Table 2. Development in tolerable soil concentrations (in mg/kg) in the Federal Republic of Germany.

Metal	1974*	1977*	1982**
Cd	50	5	3
Cu		100	100
Hg	5	5	2
Ni		50	50
Pb	2000	100	100
Zn	500	300	300

* Referred by Claussen 1979
** Klärschlammverordnung 1982

But whatever the reason for the continued enforcement in the FRG of the values in Table 2, it may not be safe to tolerate in the future that for instance the Cd content of all german soils may increase to 10 times higher level (3 ppm), than at present (~0.3 ppm, Augustin et al. 1984), presumably rising the human dietary intake of that metal to well over the tolerable level.

Included in the historical approach is also international copying of limiting values. It may be noted from the tables produced by Webber et al. 1984, that the limits to soil concentrations for the FRG for several metals appear rather identical in guidelines issued in France, Austria and Switzerland and also seems to have been the basis for the proposal from the Commission of the European Communities (CEC 1982).

THE PRAGMATIC APPROACH IN ESTABLISHING REGULATIONS

The stage following the historical approach in establishing guidelines may be called the pragmatic approach. Besides recognizing the need to protect soils from becoming phytotoxic, the pragmatic approach recognizes the current sludge qualities and the wish or need for disposal of sludges in agriculture as a fertilizer material containing Nitrogen and Phosphorous. Most current guidelines appear to be of the pragmatic type, which may be illustrated by considering the current national limitations for Cadmium.

In Figure 1 is plotted for various countries the allowable annual loadings of Cd against the sludge concentrations of Cd reported shortly before the guidelines were issued. The maximum allowable sludge concentrations are also shown. In the figure is further shown the annual Cd loadings following sludge applications of 2, 5 or 10 t DS/ha,y with the indicated sludge Cd concentrations. These doses cover the normal range of sludge utilization practice. The 10 t/ha,y corresponds to a total-N of 300-500 kg/ha which is considered a useful dose when sludge is meant as a N-fertilizer, and the 2 t/ha,y corresponds to a total P of 15-30 kg/ha,y which should be sufficient to meet the P requirements of the crops, but not of N.

Looking at Figure 1 it is obvious that the maximum sludge Cd concentrations often are chosen to 2-3 times the known mean or median sludge concentrations. In defining the maximum annual Cd loading the countries appear to have used a presumed optimal sludge dosage of 5-10 t/ha,y, when N fertilization is considered feasible, and the then recent sludge concentrations of Cd known. For countries with a probable policy of using sludge as a P fertilizer (D, DK, N, S and SF) the chosen sludge dose appears to be 2-5 t/ha,y.

It should be mentioned that the regulations for the US (out of scale in the figure) fits into this pattern, even when stricter regulations are enforced in 1987, (500 g Cd/ha,y and a max. sludge conc. of 25 ppm Cd for vegetable crops). The CEC regulations do not either deviate from this pattern except that the values approach the highest values for the member states.

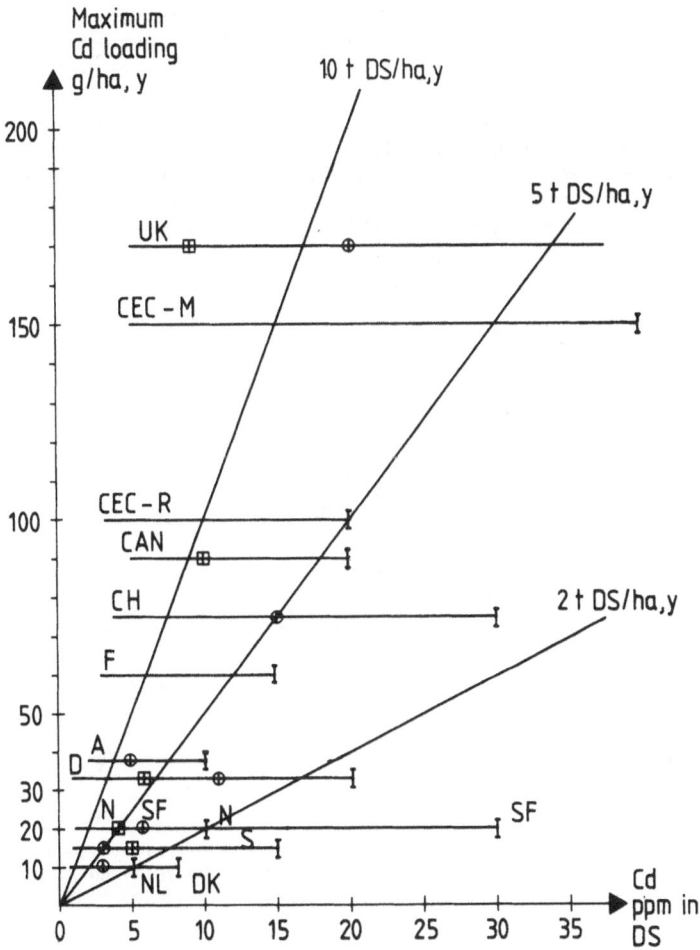

Figure 1. Compilation of national guidelines on Cadmium for allowable sludge concentrations ($\dashv$) and maximum annual loading of agricultural land receiving sludge. For each country is further shown reported sludge concentrations prior to the issue of the guidelines, as mean ($\oplus$), or median ($\boxplus$) values. The three lines indicate the annual Cd loading following sludge dosing of 2-5-10 t/DS,ha,y. The data are from Webber et al. 1984 and Table 1.

Although most of the national regulations are said to protect the productivity of agricultural soils, as well as protect the consumers of the agricultural products, it is doubtful if any of the regulations actually are based on calculations to fulfil the said objectives. It appears more likely that the regulations are based on rather pragmatic considerations on what was practical and possible, at the time of formulating the regulations, taking into account politics and economics of sludge disposal.

Although the pragmatic approach may not be founded on clear principles it does work for a continuous improvement of the situation, simply by putting pressure on reductions in the industrial discharges of metals to the sewers. In Figure 2 and 3 are shown two examples of steadily decreasing metal concentrations in sludge from larger municipal waste water treatment plants, with a considerable influx of trade waste. Especially for the problematic metal Cadmium it has been possible in both cases to reach the rather low concentration of 7 ppm in DS. The concentration levels for the four metals shown are not necessarily the ultimate which can be obtained. In the Danish case the lead content is still high, caused by a large battery manufacturing company, which still decrease its discharges despite increased production. When Pb in gasoline finally is faded out, the Pb concentrations in sludges may fall much further.

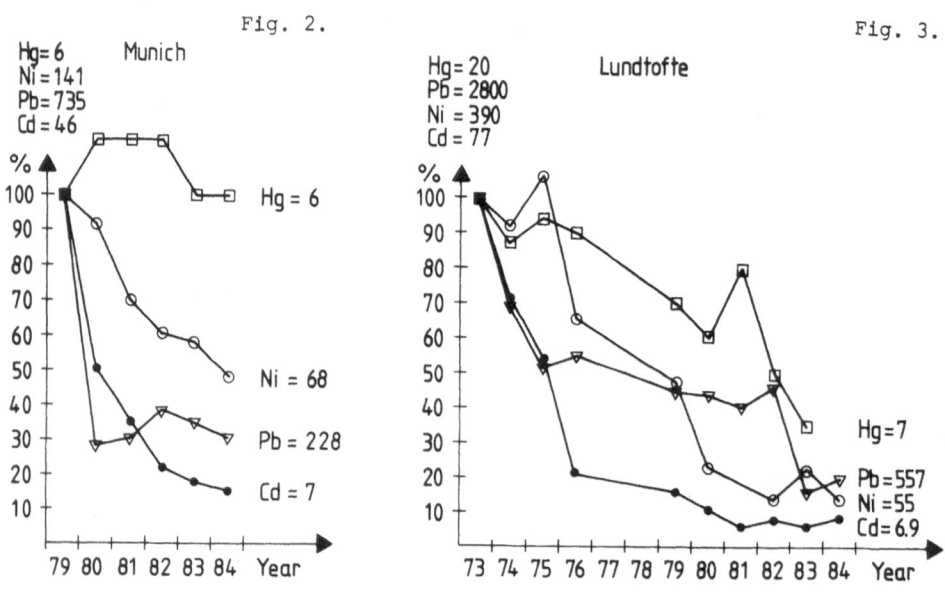

Fig. 2.

Fig. 3.

Fig. 2. Development in concentrations (in ppm in DS) of four metals in sludge from Munich (FRG) sewage treatment plant producing 35.000 t/ year of sludge dry matter. Eichinger 1985.

Fig. 3. Development in concentrations (in ppm in DS) of four metals in sludge from Lundtofte sewage treatment plant (outside Copenhagen, Denmark) producing 5000 t/year of sludge dry matter. (Values obtained from Lyngby municipality).

THE SCIENTIFIC APPROACH IN ESTABLISHING REGULATIONS

The scientific approach in examining the effects of metals after sludge application to agricultural land is timewise relatively short, but has resulted in a great volume of work, with great many details being studied. With the object of using the wealth of scientific envidence in establishing well founded regulations for use of sludge it is logical to follow a line of reasoning as used by e.g. Ryan et al. 1982:

1. Establish relationships between soil-plant-animal concentrations of individual metals, under various circumstances, like soil conditions (pH, texture, temperature, water regime, etc.) species of plants, and nutritional availability to animals.
2. Define the various transport routes for metals in the food chains relevant to sensitive species (mostly the human species) and individuals.
3. Define the dose-effect of individual metals to sensitive species and individuals.
4. On this basis (hopefully) define acceptable doses of metals for sensitive species and individuals.
5. Define the section of the populations to be protected under the regulation.
6. Calculate on these assumptions the acceptable loadings of metals to land, and/or define ameliorating conditions (e.g. soil pH).

Table 3. Approximate metal concentrations in purely domestic sludges, and suggested limits for metal loadings to agricultural land not giving rise to unwanted long term side effects, see text for details.

Metal	Approximate concentration in domestic sludge ppm in DS		Suggested Loading rate at 3 t TS/ha,y g/ha,y
As	5	++	50
Cd	3	+++	10
Cr(III)	50	+	1500
Co	5	++	50
Cu	200	++	1800
Hg	2	++	20
Mn	500	+	15000
Mo	5	++	50
Ni	25	++	250
Pb	100	++	1000
Se	2	++	20
Sn	10	+	300
Zn	700	++	6500
B	25	++	250
F	200	++	2000

Provided an annual application rate of <3 t DS/ha, then for each metal:

+ 10 x higher concentrations may be tolerated.
++ 3 x higher concentrations may be tolerated.
+++ Higher concentrations should not be tolerated

With this type of reasoning process for a long row of metallic contaminants in sludges, it may be possible to define under which circumstances
there should not be any long term unwanted side effects of utilization of
sewage sludges in agriculture. In Table 3 is shown the approximate metal
concentration levels in purely domestic sludges together with an assessment
of possible allowable deviations from these concentration levels, provided
the sludge is used for P-fertilization at application rates up to 3 t DS/ha,y.
The table is based on evidence compiled by Chaney 1982 (the "soil-plant barrier" concept), CAST 1976, Webber 1984, and a selection of data on sludge
concentrations, Sjöqvist 1984, Skov 1982, DOE 1983, and Augustin 1984.

There may still be some additional metals which are not examined to
such an extent that they may be deemed harmless, but until we know more
about metals as Ag, Be, Ce, La, Sb, Te, Tl and V it is most likely that
they are not able to cross the "soil-plant barrier".

THE SCIENTIFIC APPROACH APPLIED TO CADMIUM

It is well known that cadmium is considered to be the most troublesome
metal often limiting sludge application in agriculture, and certainly the
most disputed of the contaminants in sludges. Digging a little deeper into
the current scientific approach on this metal may cast light on the problems.

1.   Based on extensive reviews by CAST 1981, Logan & Chaney 1983, Ryan et
al. 1982, Page 1981 and Davis & Coker 1980 it may presently be the most likely assumption that the plant uptake of cadmium from soil is not declining
with time, is largely proportional to the cadmium concentration in a soil,
and that the other most influential soil parameter is pH, with an approximate increase in Cd uptake of 2-3 x for each pH unit decrease. A lot of anomalies from these three rules of thumb have been found (Logan & Chancy 1983),
but for a crop producing region at large this seems to hold. A special case
to note it the very low plant availability of Cd at higher concentrations,
in calcareous soils, as described by Davis 1984, and with a possible explanation by Christensen & Tjell 1984.

The transfer of cadmium to animals are not as well studied, but it is
most likely that a moderate increase in intake with fodder raises the organ
concentrations proportionally (Logan & Chaney 1983).

2.   For humans, dietary studies in industrialized countries, have revealed
that the present intakes of Cd often is around 25 and range from <20 to
>40 µg/P,d and that a rough sharing between food classes is as shown in Tab-
le 4.

Table 4. Approximate amounts of Cd in human food classes in
industrialized countries.

Food class	% Cd	Intake µg/P,d
Vegetables (& fruit)	15	4
Potatoes	20	5
Cereals	25	6
Animalia	20	5
Other	20	5
Total	100	25

Note: The food class "other" contains items like beverage,
Water, fish etc. which do not obtain Cadmium from
Soil.

3 + 4. The provisional tolerable weekly intake of Cd (400-500 µg/P,w) with food defined by WHO/FAO 1972, seems still valid, although suggestions for raising the figure are numerous. (e.g. Taylor 1984, Epstein & Alpert 1983, Logan and Chaney 1983). A recent discussion of the ptwi for Cd is presented by Dean and Suess 1985, with a suggestion for a reevaluation of the figure.

5.   The definition of the section of the population to be covered under a safe regulation is not really a scientific problem. An individual (smoker) consuming copious amounts of lettuce, wild champignons and kidney pie may take in excessive amounts of cadmium, whatever regulations be adopted. But it may be argued that the higher the average human food intake is, the more individuals are exceeding their tolerance. This type of argument is not easy to quantify in a scientific way. But a short resumé may be useful to show some of the key problems in that process.
    A method for deciding on the future intake of cadmium is the so called scenario, where various assumptions are used in predicting the impact of raising the soil concentrations of Cadmium, with or without sludge application to land.
    An average hypothetical situation may be defined as in Table 5 (based on scenarios by Ryan et al. 1982, Naylor and Loehr 1981, Davis et al. 1983, Hansen and Tjell 1983 etc.).
    The assumed transfer factor soil-human for Cd is not easy to calculate with great accuracy, but taking literature informations on daily intakes and soil concentrations the magnitude is as shown in Table 6. The selected value is 4, although it may be justified to use a lower figure. All figures in Table 5 are proportional to this transfer factor.

Table 5. Average of various scenarios predicting increases in human dietary intakes of Cadmium following a 1 kg/ha addition of the metal to the soils producing the listed food items. See text.

| Food items | Increase in dietary Cd µg/P,d at soil pH | | | |
	Calcareous	+ 1 pH	Neutral	- 1 pH
All food	3	16	40	100
Vegetable + potatoes	1	7	18	45
Cereals	1	5	12	30
Vegetables	0.5	3	8	20

Assumptions:

$$\frac{\Delta \text{ Cd plant in \%}}{\Delta \text{ Cd soil in \%}} = 1 \qquad \frac{\text{Cd plant, pH}}{\text{Cd plant, pH + 1}} = 2.5$$

Transfer factor soil - human:

$$\frac{\text{Ave. soil derived Cd intake}}{\text{Ave. top soil content}} = \frac{20 \text{ µg/P,d}}{0.5 \text{ kg/ha}} = 4 \frac{\text{cm}^2}{\text{P,d}}$$

Table 6. A compilation of recorded human food intakes and average soil concentrations of cadmium, and the corresponding transfer factors for the metal from soil - human.

Country	Cd intake µg/P,d	Cd in soil conc. ppm	Cd in soil content kg/ha	Transfer factor cm^2/P,d
US Northeast[a]	32	0.19	0.57	5.6
W-Germany[b]	35	0.25	0.75	4.7
Denmark[c]	25	0.22	0.66	3.8
Sweden[d]	18	0.22	0.66	2.7
Canada[e]	52	0.3	0.9	5.8
Finland[f]	13	0.22	0.66	2.0

Ref.: a Naylor & Loehr 1981   Luce 1985
      b Weigert et al. 1984   Augustin et al. 1984
      c Hansen & Tjell 1983
      d Kjellström et al. 1978   Andersson 1977
      e Ryan et al. 1982   McKeague & Wolynetz 1979
      f Varo & Koivistoinen 1980   Sippola 1985

The scenario summary in Table 5 points to that predictions of the increase in dietary Cadmium intake following addition of 1 kg Cd/ha, may range from 0.5 to 100 µg/P,d depending on assumptions on soil parameters and food classes assumed to be affected. Relating the content of Table 5 to the actual situations in various regions of the world will inevitably produce highly differing opinions on the level of tolerable soil contamination with Cadmium.

A point to note is the high impact of vegetables and potatoes alone. This may have special significance to sludge utilization schemes near larger cities. Although not always the case, a high proportion of vegetables and some potatoes are grown near the cities, and thus a higher recycling of contaminants back to the population may occur, which is not anticipated in the scenarios. The same possibility exists of excessive exposure to individual families from utilization of sludge in private gardens.

6. Now the key question is raised mostly to the health authorities: Which end of the scenario to base limiting values for Cd addition to land, also in sludge. And this is where the dispute gets heated, as the health authorities mostly opt for the high impact end to be the most realistic leading to strict regulations (as in the Nordic countries and Scotland). The moderating effect of a limited available sludge amount which can be disposed in high dosage rates on limited land areas, and thus limiting the impact (Ryan et al. 1982), may not be very obvious in densely populated countries. And with increasing interest in the fertilizing value of the sludge, especially the phosphorous value, this possibility may not be optimal either.

Reason for opting for the low impact end of the scenarios are not all that clear, but a reasonable guess is according to the pragmatic approach, that the situation is not yet mature for too tight regulations, as they might jeopardize ongoing sludge operations, or that the general soil pollution status in some localized areas would not comply with tighter regulations issued for sludge application to land.

The especially low impact of rising soil contents of Cadmium on calcareous soils have led to the least strict regulations for sludge on land (eq. Davis 1983, DOE 1981), which is quite justifiable if UK soils are predominantly calcareous.

This is approximately the situation at present, leading to the great disparity among countries on which limiting values for Cd to choose.

It is difficult yet to say that the scientific approach in establishing guidelines is much better than the two previous ones. But if there is a need of unification of soil pollution regulations covering metals, it is probably the only possibility for getting the differing opinions to meet.

IS THERE A WAY TO MORE UNIFORM AND JUSTIFIABLE REGULATIONS?

It is quite obvious that the utilization of low metal sludges in agriculture is of no major concern in respect to associated risks to the public. As suggested in Table 3 for most metals even 3-10 times higher metal concentrations than found in domestic sludges may be tolerated for moderate sludge application rates. How restictive to ongoing sludge utilization these limitations may be is shown in Table 7 for sludges from FRG and UK. It is clearly Cadmium and to a lesser extent lead and Nickel which are the troublesome metals according to this rather practical approach. On the other hand it is also clear that sludge concentrations are on the way down, probably as a result of the pragmatic regulations imposed on sludges and because of public awareness in general. This point was illustrated in Figures 2 and 3, and in Table 8 is shown the improvements in the situation with time for Cadmium in three countries. Although not exactly compatible (and maybe not entirely correctly read from figures in the publications) the trends for UK and Sweden are clear. It is possible to reduce sludge concentrations considerably through efforts in the industries and waste water authorities.

But the tables also show that considerable quantities of sludges (especially from larger towns) are still not satisfactorily low in metal contents to be used as fertilizers. It is a philosophical question if the farming community, and indeed the whole community, should accept cheep fertilization and disposal of heavily contaminated sludges, with all the possible risks involved.

Table 7. The amount of sludge (UK), or number of sludge samples (FRG) in two surveys not complying with the maximum loadings listed for individual metals in Table 3 (at 3 t DS/ha,y).

Metal	UK DOE 1980	FRG Riegler 1981
	%	%
Cd	63	60
Cr	20	10
Cu	20	10
Hg	–	8
Ni	35	25
Pb	40	20
Zn	15	25

Table 8. Approximate classification (in %) of sludges in three countries
according to the Cadmium content.

Country Conc. group	UK			FRG[c] ~1980	SWEDEN[c]			
	1977[a]	1980[a]	1980[b]		1969	1973	1976	1981
0-5 ppm	13	23	37	40	20	30	40	85
0-10 ppm	37	45	60	74	70	70	70	98

a Surveys in 1977 and 1980 of the amount of sludge disposed of to agricultural land and landfills.

b Survey in 1980 of the amount of sludge disposed of to agricultural
land alone.

c Percent of number of sludges analyzed

References: DOE 1981 & 1983   Riegler 1981
                                     Sjöqvist 1984 & 1978

Regarding the prospects for future unified sludge utilization regulations concerning trace metals three options may be mentioned:

1. A rather elaborate system of rules covering the various nature given differences in soil conditions (e.g. in Europe). There will have to be provision for variations in pH covering acid sandy soils and calcareous
   clays. Even then the risk assessment (scenarios) will be viewed upon with
   some suspicion, as long as the rules may have to accomodate heavily contaminated sludges.
2. Adopt stricter rules (e.g. Table 3) and a limit on soil pH with a period
   of grace for either bringing down the sludge concentrations, or to halt
   the most risky sludge utilization schemes.
3. Leave it entirely to the national authorities. It is anyhow their own
   public health situation which is most at stake. Countries contemplating
   on sludge utilization regulation (new or amendments) are able to chose
   between a whole array of possibilities.

IS THERE A NEED FOR UNIFORM REGULATION?

In the review by Webber et al. (1984) it was pointed out that international trade in agricultural products is large and increasing, and in order to
ensure uniform quality of food chain products amongst countries, it is useful to standardize allowable loadings of contaminants to agricultural land.
It may be argued (e.g. Matthews 1984) that sludge is only a minor contribu-
tor of contaminants to land, and as such not important enough to warrant efforts to harmonize the national regulations. The tricky point is however
that regulation of contaminant loadings to land with sludge often appears to
be the first serious national attempt to control the quality of soil and
hence agricultural products, and thus may serve as a model for later regulations of the much larger contaminant loadings from the atmosphere and through
the use of chemical fertilizers, as pointed out by several authors (e.g.
Hansen & Tjell 1983).

It is difficult to argue that sludge regulations ought to be internationally standardized, for the sake of protecting consumers from unsafe imported
agricultural produce due to improper use of sludge. But the eventual adoption
of sound international rules or guidelines for use of sludge may open up a
fruitful discussion on the necessity of controlling the mentioned much larger loadings.

The question of international harmonization of rules for the sake of
promoting equal economic competiveness may not be very important for this
particular subject, not either in the EEC. Although disposal of sludge seems
costly, each citizen may annually pay less than 2 £ extra for even the most
expensive sludge disposal method, mechanical dewatering and sanitary landfill, compared to agricultural disposal, thus averting rather more costly
industrial measures to curb excessive metal discharges to the sewers.

REFERENCES

(1)   ANDERSSON, A. (1977). Heavy metals in Swedish soils: on their reten-
      tion, distribution and amounts. Swedish J. Agric. Res., Vol. 7, 7-20.
(2)   AUGUSTIN, K., FISCHER, D., HOLZ, A., KUMM, A. and LANGER, H. (1984).
      Daten zur Umwelt 1984. Umweltbundesamt, Berlin.
(3)   BAKER, D., BOULDIN, D.R., ELLIOT, H.A. and MILLER, J.R. (1985). Recom-
      mendations for guidelines. Ch. 1 in: Criteria and Recommendations for
      land application of sludges in the Northeast. The Pennsylvania Univer-
      sity, University Park, Pennsylvania, U.S.A., Bulletin 851.
(4)   CAST (1976). Council for Agricultural Science and Technology, AMES.
      Application of Sewage Sludge to Cropland: Appraisal of Potential Haz-
      ards of the Heavy Metals to Plants and Animals. U.S. Department of
      Commerce. NTIS, Report PB-264 015.
(5)   CAST (1981). Council for Agricultural Science and Technology. Effects
      of Sewage Sludge on the Cadmium and Zinc Content of Crops. U.S. Envi-
      ronmental Protection Agency, EPA 600/8-81-003.
(6)   CEC (1982). Proposal for a Council Directive on the Use of Sewage
      Sludge in Agriculture. Official Journal of the European Communities.
      No. C264/3-C264/7, 8. October 1982.
(7)   CHANEY, R.L. (1982). Fate of toxic substances in sludge applied to
      cropland. In: Proceedings of the International Symposium on Land
      Application of Sewage Sludge, October 13-15, Tokyo.
(8)   CHRISTENSEN, T.H. and TJELL, J.C. (1984). Interpretation of experimen-
      tal results on cadmium crop uptake from sewage sludge amended soil.
      In: Processing and Use of Sewage Sludge, pp. 358-369. D. Reidel Publ.
      Co., Dordrecht.
(9)   CLAUSSEN, T. (1979). Schwermetalle und andere toxische Spurenelemente
      in der Nahrungskette der Menschen. Berichte über Landwirtschaft, Vol.
      57, 105-117.
(10) DAVIS, R.D. and COKER, E.G. (1980). Cadmium in Agriculture, with Spe-
      cial Reference to the Utilization of Sewage Sludge on Land. Water
      Research Centre, Stevenage, Herts, UK. WRC Technical Report TR 139.
(11) DAVIS, R.D., STARK, J.H. and CARLTON-SMITH, C.H. (1983). Cadmium in
      sludge treated soil in relation to potential human dietary intake of
      cadmium. In: Environmental Effects of Organic and Inorganic Contami-
      nants in Sewage Sludge, pp. 137-146. D. Reidel Publ. Co., Dordrecht.
(12) DAVIS, R.D. (1984). Crop uptake of metals (cadmium, lead, mercury,
      copper, mickel, zinc and chromium) from sludge-treated soil and its
      implications for soil fertility and for the human diet. In: Processing
      and Use of Sewage Sludge, pp. 349-357. D. Reidel Publ. Co., Dordrecht.

(13) DEAN,R.B. and SUESS, M.J. (1985). The risk to health of chemicals in sewage sludge applied to land. Waste Management & Research, Vol. 3, 251-78.

(14) DOE (1981). Report of the Sub-Committee on the Disposal of Sewage Sludge to Land. Department of the Environment, London. Standing Technical Committee Reports Number 20.

(15) DOE (1983). Sewage Sludge Survey 1980 Data. Department of the Environment, London.

(16) EICHINGER, J. (1985). Massnahmen zur Reduktion des schwermetallgehaltes im Klärschlamm am Beispiel der Stadt München. Technische Universität Wien. Wiener Mitteilungen - Wasser - Abwasser - Gewässer, Band 58, H1-H12.

(17) EPSTEIN, E. and ALPERT, J.E. (1983). Cadmium: Crop Uptake and Human Health Risks Resulting from Sludge Application. In: Developments in Ecology & Environmental Quality, Vol. II, pp. 563-75. Balaban Int. Science Services, Rehovot.

(18) FURRER, O.J., CANDINAS, T. and LISCHER, P. (1982). Schwermetallgehalt der Klärschlämme in der Schweiz. Landwirtschaftliche Forschung, Sonderheft 39, 309-316.

(19) GENEVINI, P.L., VISMARA, R. and MEZZANOTTE, V. (1983). Utilizzo agricolo dei fanghi di depurazione. Ingegneria Ambientale, Vol. 12, No. 9, 1-133.

(20) HANSEN, J.AA. and TJELL, J.C. (1983). Sludge application to land. Overview. Cadmium. In: Environmental Effects of Organic and Inorganic Contaminants in Sewage Sludge, pp. 91-112. D. Reidel Publ.Co., Dordrecht.

(21) KICK, H., BURGER, H. and SOMMER, K. (1980). Gesamthalte an Pb, Zn, Sn, As, Cd, Hg, Cu, Ni, Cr und Co in Landwirtschaftlich und gärtnerisch genutzten Böden Nordrhein-Westfalens. Landwirtschaftliche Forschung, Vol. 33, 12-21.

(22) KJELLSTRÖM, T., BORG, K. and LIND, B. (1977). Heavy metals in Swedish soils: on their retention, distribution and amounts. Swedish J. Agric. Res., Vol. 7, 7-22.

(23) Klärschlammverordnung AbfklärV 25.6 (1982). Bundesgesetzblatt, Teil I, 734-9. Bonn: Bundesdruckerei.

(24) LOGAN, T.J., and CHANEY, R.L. (1983). Utilization of municipal wastewater and sludge on land - metals. In: Proceedings of the 1983 Workshop on Utilization of Municipal Wastewater and Sludge on Land, pp. 235-323. University of California, Riverside, U.S.A.

(25) LUCE, H.D. (1985). Soil Resources of the Northeast. Ch. 2 in: Criteria and Recommendations for Land Application of Sludges in the Northeast. The Pennsylvania University, University Park, Pennsylvania, U.S.A. Bulletin 851.

(26) MCKEAGUE, J.A. and WOLYNETZ, M.S. (1980): Background levels of minor elements in some Canadian soils. Geoderma, Vol. 24, 299-307.

(27) MATTHEWS, P.J. (1984). Control of metal application rates from sewage sludge utilization in agriculture. CRC Critical Reviews in Environmental Control. Vol. 14, No. 3, 199-250.

(28) NELL, J.H. and ENGELBRECHT, J.F.P. (1982) Guidelines for sewage disposal to land. In: Symposium on Water Research: Selected In-Depth Studies. CSIR Conference Centre, Pretoria, 18. November.

(29) OMOE (1977). Ontario Ministry of the Environment. Plant Operating Summary, Water Pollution Control Projects, Toronto, Ontario.

(30) PAGE, A.L. (1981). Cadmium in soils and its accumulation by food crops. In: Heavy Metals in the Environment, International Conference, Amsterdam, pp. 206-13. CEP Consultants Ltd., Edinburgh, UK.

(31) RIEGLER, G., ECKHARDT, H. and KARGES, J. (1981). Inhaltstoffe und Beseitigung von Klärschlamm in der Bundesrepublik Deutschland. Der Technischen Hochschule, Darmstadt, Inst. f. Wasserversorgung, Abwasserbeseitigung und Raumplanung, Wasser- und Abfallwirtschaft Forschungsbericht Nr. 103 03 10B.

(32) RYAN, J.A., PAHREN, H.R. and LUCAS, J.B. (1982). Controlling Cadmium in the Human Food Chain: A Review and Rationale Based on Health Effects. Environmental Research, Vol. 28, 251-302.

(33) THE SCOTTISH AGRICULTURAL COLLEGES (1981). Disposal of sewage sludge on agricultural land. Edinburgh, UK. Publication No. 76.

(34) SIPPOLA, J. (1985). Personal communication, for later publication.

(35) SJÖQVIST, T. (1978). Tungmetaller i slam från svenska reningsverk 1968-1977. Statens Lantbrukskemiska Laboratorium, Uppsala.

(36) SJÖQVIST, T. (1984). Trends in heavy metal, PCB and DDT contents of sludges in Sweden. In: Utilisation of Sewage Sludge on Land: Rates of Application and Long-Term Effects of Metals, pp. 194-7. D. Reidel Publ.Co., Dordrecht.

(37) SKOV, T. (1982). Koncentrationen af metaller i spildevandsslam er på vej ned. Stads- og Havneingeniøren, Vol. 73, 375-80.

(38) TAYLOR, D. (1984). Cadmium - A Case of Mistaken Identity? Marine Pollution Bulletin, Vol. 15, 168-70.

(39) VARO, P. and KOIVISTOINEN, P. (1980). Mineral element composition of Finnish food. XII General Discussion and Nutritional Evaluation. Acta Agric. Scand. Suppl. Vol. 22, 165-71.

(40) VETTER, H., KOWALEWSKY, H.-H. and SÄLE, M. (1983). Datenerhebung über die Cadmiumbelastung von Böden und Pflanzen in der Bundesrepublik Deutschland im Auftrag des Bundesministerium für Ernährung, Landwirtschaft und Forsten. VDLUFA.

(41) WEBBER, M.D. (1984). Land utilization of sewage sludge - a discussion paper. Research Branch, Agriculture Canada.

(42) WEBBER, M.D., KLOKE, A. and TJELL, J.C. (1984). A review of current sludge use guidelines for the control of heavy metal contamination in soils. In: Processing and Use of Sewage Sludge, pp. 371-85. D. Reidel Publ.Co., Dordrecht.

(43) WEIGERT, P., MÜLLER, J., KLEIN, H., ZUFELDE, K.P. and HILLEBRAND, J. (1984). Arsen, Blei, Cadmium und Quecksilber in und auf Lebensmitteln. ZEBS Hefte 1. Bundesgesundheitsamtes, Berlin.

(44) WHO/FAO (1972). Joint FAO/WHO Expert Committee on Food Additives. Evaluation of certain food additives and the contaminants mercury, lead, and cadmium. Sixteenth report. FAO Nutrition Meetings Report Series No. 51. WHO Technical Report Series No. 505.

(45) ÖWWV (1984). Landwirtschaftliche Verwertung von Klärschlämmen - Empfehlungen für Betreiber von Abwasserreinigungsanlagen. Österr. Wasserwirtschaftsverband, Wien, Regelblatt 17.

Leaching resulting from land application of sewage sludge

Germination tests for the determination of sludges agricultural value

Utilization of sewage sludge as fertilizer in energy plantations on peatland

The use of composted agricultural waste as peat substitute in horticulture

Labscale simulation of isman-cotton batch phase anaerobic digestion for prediction of plant design and performance

Solid phase anaerobic digestion of farm wastes mixed with sewage sludge

Aerobic stabilization in the solid state of partially dewatered sewage sludge

Environmental impact assessment of agricultural use of sewage sludge

Analytical characterization of biological sewage sludges especially about some organic compounds

Sludge composting and utilization in West Switzerland

Productivity and quality of cereal crops grown on sludge-treated soils

Biological evaluation of sludge phytotoxicity

The fate and behaviour of selected heavy metals during pyrolysis of sewage sludge

Latest test results in sludge dewatering with the CHP-filter press

A pot experience with a high level copper sewage sludge

Chemical properties of sewage sludges produced in the piedmontese area (Italy)

Autothermic sludge incineration

Modification of heavy metals solubility in soil treated with sewage sludge

Methods of treating pig slurry to increase the volumes which can be used on crops

Chemical composition of sludges from sewage treatment systems in the Emilia Romagna region

Application of $CaCl_2$-extraction for assessment of cadmium and zinc mobility in a wastewater-polluted soil

Anaerobic contact digestion of biochemical sludge (from simultaneous precipitation) - results of a semi full-scale study

The alternative "earthworm" in the organic wastes recycle

Slurry meter instructions

Composting of sewage sludge containing polyelectrolytes

An alternative application for sewage sludge : black earth

Toxic organic substances in sewage sludges : case study of transfer between soil and plant

Varietal tolerance in cereals to metal contamination in a sewage treated soil

Effect of sample storage on the extraction of metals from raw, activated and digested sludges

INTRODUCTION TO THE POSTER SESSION

The Chairman introduced the session and said that a total of 33 posters had been exhibited. He said that it was interesting to note the numbers of posters relevant to the activities of the respective Working Parties. Thus there were some 10 Posters each of special interest to Working Party 1 (Processing), Working Party 4 (Agricultural Effects) and to Working Party 5 (Environmental Aspects) respectively. However, there were only 3 Posters for Working Party 2 (Chemical Aspects) and none for Working Party 3 (biological Aspects).

LEACHING RESULTING FROM LAND APPLICATION OF SEWAGE SLUDGE[1]

M. MELKAS, M. MELANEN, A. JAAKKOLA, M. AHTIAINEN *

Summary

Sewage sludge (16 - 20 t ha^{-1} DM) was applied on a
clay soil and movement of nutrients, heavy metals,
organic matter and (indicator) micro-organisms was
monitored in a five-year experiment. Two of the years
were hydrologically exceptional: weather conditions
had a strong effect on leaching. Following results
were obtained: 1) mostly 80 - 90 per cent of total
leaching was through subdrainage; 2) no harmful
leaching of heavy metals or micro-organisms through
subdrainage or surface was found; 3) the treatments
had the most profound effect on nitrogen leaching
and on conductivity of runoff waters; effect duration
was over two years for nitrogen; 4) application of
sludge on snow immediately increased NH_4 concentrations
of surface and subdrainage waters; in the long run,
applications on snow and on nonfrozen soil yielded
equal nitrogen leaching.

1. INTRODUCTION

The amount of municipal sewage sludge has grown
rapidly in Finland and it was over 100 000 tons DM (dry
matter) in 1980. About half of the sludge is at present
disposed of for agricultural purposes.

In Finland sewage sludge should not be spread more
than 20 t ha^{-1} DM once in every five years. If applied
more often, the maximum dose is 4 t ha^{-1} DM a year. Cadmium
content may limit the use of sludge: the highest acceptable
amount is 100 g ha^{-1} Cd once in five years or 20 g a year.
Particularly liquid sludge should be mixed into soil
immediately after application. Dewatered sludge is also
recommended to be spread on nonfrozen soil and mixed into
it within one day. A protection zone of 20 - 50 metres
with no sludge application is recommended by watercourses.

1 This paper is based on the original paper by Melanen et al.
  (1985).

Spreading is not allowed in the vicinity of drinking-water wells or important groundwater aquifers (Latostenmaa 1976 (1), Lääkintöhallitus 1977 (2), Vesihallitus 1979 a) (5).

A test field was constructed in 1978 in Liperi (Siika-salmi Agricultural School: 62°32'N, 29°22'E) in eastern Finland for studying movement of the constituents of sewage sludge resulting from land application. Uptake by crops was also studied in the project which was carried out by the National Board of Waters, the North Karelia Water District Office, the Agricultural Research Centre and the local municipalities. A detailed description of the study is given in Melanen et al. (1985).

2. MATERIALS AND METHODS

The soil type of the Liperi field is clay. In the top layer, 0.1 - 0.6 metres, the clay fraction is over 60 per cent. The average slope of the field is 0.5 - 1 per mille. Typically, just 0 - 10 per cent of precipitation percolates to ground-water storage in soils of this kind.

The test field was arranged in the form of a latin square with 16 test plots, about 600 m^2 each (Fig. 1).

The year 1979 was a calibration year without fertilization (the cultivated crop was a mixture of peas and oats).

The treatments were given in 1980. Dewatered municipal sewage sludge (9 - 12 % DM) was spread on snow in the end of March (treatment b) and on nonfrozen soil in May (treatment c). Limed sludge (pH 8.4) was also spread in May (treatment d). The amount of sludge was 16 - 20 t ha^{-1} DM. The reference plots were not fertilized (treatment a). Each treatment had four replicates. After spreading of sludge on nonfrozen soil it was immediately mixed into the soil.

The amounts of the three major nutrients spread in the different treatments in 1980 were as follows (in kg ha^{-1}):

Treatment	N	P	K
a	-	-	-
b	1 300 - 1 400	500 - 600	130 - 150
c	750 - 950	400 - 500	80 - 100
d	450 - 550	450 - 500	25 - 30

The heavy metal contents of the sludges were close to averages (Vesihallitus 1979b).

In 1981, no fertilization was given. In 1982, all plots were fertilized with NPK fertilizer (16 - 7 - 13) (350 kg ha^{-1}). The cultivated crop in 1980 - 1982 was barley.

A large number of constituents - nutrients and heavy metals (N_{tot}, N_{NH4}, N_{NO3}, P_{tot}, P_{PO4}, K, Ca, Mg, S_{SO4}, Na, Cl, Fe, Mn, Cu, Cr, Ni, Cd, Hg, Pb), conductivity, pH, organic matter (COD_{Mn}), suspended solids (SS) and indicator

micro-organisms (coliform bacteria and enterococci) - were
analyzed on the subdrainage and surface waters.

The Mann-Whitney rank-sum test (Malik and Mullen 1973)(3)
was used to test the hypothesis that two samples come from
populations with the same distribution. The latinsquare
procedure of the analysis of variance (Cochran and Cox 1957)
was applied to test the effect of the treatments.

3. RESULTS AND DISCUSSION

Hydrological conditions and soil effects

Two of the years - 1981 and 1982 - were hydrologically
exceptional: the 1981 mean runoff has an average return
period of about 30 years; the 1982 spring runoff is exceeded
only once in 20 - 30 years on an average.

Subdrainage runoff caused 75 - 90 per cent of total
runoff. Most of both surface and subdrainage runoff came
during the snowmelting period.

After the application, in 1980, in all sludge treatments
(b, c, d) a higher content of soil phosphorus, iron and zinc
and a slightly higher content of potassium, manganese and
copper was observed than in the reference treatment (a).
A difference could also be found in nickel, cadmium, lead
and conductivity in treatments b (sludge on snow) and c
(sludge on nonfrozen soil) compared to treatment a. In
1981, higher contents still persisted in the sludge
treatments as regards phosphorus and zinc (and iron). The
contents had however decreased since 1980. In 1982 the
difference in soil phosphorus still existed. (Heavy metals
were not measured in 1982.) The content of soil ammonium
and nitrate nitrogen was measured in September 1981. The
content was raised in treatments with unlimed sludges (b and
c) in all soil layers down to the depth of one metre.

Quality of and leaching with subdrainage and surface waters

Two seasonal peaks were generally observed in
concentrations: one in the snow-melting period and another
in the autumn.

On the order of 70 - 90 per cent of nitrogen was in
nitrate form. Treatment effects could also be seen in
the annual total leaching rates (Table I). The effect of
the limed sludge was smaller than that of the other sludges.

Raised concentrations of nitrogen could be seen in
subdrainage water still in autumn 1982. In autumn 1983
differences no more existed.

Raised contents of ammonium nitrogen were observed in
both subdrainage and surface waters immediately after
spreading of sludge on snow (treatment b) in 1980: the
monthly flowweighted mean concentration was 1.7 mg l^{-1} in the
subdrainage water in April and 1.0 - 1.2 mg l^{-1} in the

surface water in April-May. The concentrations in the other
treatments were below 0.1 mg l^{-1}.

On the order of 30 - 50 per cent of phosphorus was in
phosphate form. The increase in the total phosphorus
content in 1980 - 1981 and 1981 - 1982 was obviously mostly
caused by increased runoff. The phosphorus content was
generally higher in the surface water than in subdrainage
water. Only a small increase could be observed in the
flow-weighted mean concentrations of potassium after the
treatments and this growth is, obviously, mostly explained
by runoff. The small increase in organic matter content
(COD_{Mn}) was obviously also mostly explained by increased
runoff.

The median concentrations of heavy metals remained at
the detection limit of analysis (0.003 mg l^{-1} Cr, 0.0001
mg l^{-1} Cd, 0.0001 mg l^{-1} Hg, 0.001 mg l^{-1} Pb). Land
application with normal sludge, performed according to the
present guidelines, thus does not seem to cause any harmful
runoff of heavy metals.

A reliable evaluation of bacteria runoff was not
possible because of too few samples. Fecal coliforms
seemed, however, to disappear in a short time after the
application.

The effect of the treatments was also visible in the
annual leaching rates (Table I). On an average, 90 per
cent of total nitrogen, almost 80 per cent of total
phosphorus and over 80 per cent of potassium came through
subdrainage. The exceptionally rainy years had a
considerable effect on leaching rates.

The runoff rates had a great influence on leaching
(Table I, Fig. 2). The leaching through subdrainage
in 1980 - 1981 was exceptionally high because of exceptional
runoff and large amount of nitrogen in soil storage.

CONCLUSIONS

The various analyses on the effects of the different
sludge treatments can be summarized as the following main
findings:

1) The treatments had the most profound effect on
    nitrogen leaching and on the conductivity of runoff
    waters.
2) The duration of treatment effect was over two years
    for nitrogen.
3) The application of limed sludge on nonfrozen soil
    caused, in general, the lowest leaching among the
    three studied sludge treatments.
4) Application of sludge on snow raised the content
    of ammonium nitrogen in runoff waters for some period.
    In the long run, applications on snow and on nonfrozen
    soil yielded equal nitrogen leaching.

5)   No harmful leaching of heavy metals or bacteria was
     observed.

     The smaller amount of nitrogen in the limed sludge
may have corresponded to plant needs better than the
amount in the other sludges. The amount of nitrogen has,
however, been adequate and the time of application
favourable because the crop yield was as good as in the
other treatments.
     The observations on constituent accumulation in soil
support the leaching results.

REFERENCES

(1) LATOSTENMAA, H. 1976. Jätevesilietteen hyödyntämisen
     perusteet (Principles of the use of sewage sludge -
     Finnish original with English summary). YVY-tutkimus
     21. Yhdyskuntien vesi- ja ympäristöprojekti.
     Helsinki. 86 p.

(2) LÄÄKINTÖHALLITUS (NATIONAL BOARD OF HEALTH). 1977.
     Ohjeet terveydellisten haittojen estämiseksi jätevesi-
     lietettä hyödynnettäessä (Instructions for prevention
     of health hazards in the use of sewage sludge -
     Finnish original). Yleiskirje n:o 1637. 13 p.

(3) MALIK, H. & MULLEN, K.   1973. A first course in
     probability. Reading. p. 296-297.

(4) MELANEN, M., JAAKKOLA, A., MELKAS, M., AHTIAINEN, M. &
     MATINVESI, J. 1985. Leaching resulting from land
     application of sewage sludge and slurry. In press.
     Publications of the Water Research Institute.
     National Board of Waters, Finland. Helsinki.

(5) VESIHALLITUS (NATIONAL BOARD OF WATERS). 1979a.
     Vesiensuojelunäkökohdat yhdyskuntien viemärilaitos-
     lietteen sijoittamisen valvonnassa (Water protection
     aspects in the supervision of the use of sewage
     sludge - Finnish original). Valvontaohje 41. 8 p.

*   Marjatta Melkas, Matti Melanen, Water Research
     Institute, National Board of Waters, Finland, P. O.
     Box 250, SF-00101 Helsinki, Finland.

     Antti Jaakkola, Department of Agricultural Chemistry,
     University of Helsinki, Finland

     Marketta Ahtiainen, North Karelia Water District
     Office, P.O.Box 69, 80101 Joensuu, Finland

Table I Annual subdrainage leaching (four replicates per treatment) (Melanen et al. 1985).

Hydrological year	Treatment	Leaching (kg km^{-2}a^{-1})				
		N$_{tot}$	N$_{NO_3}$	P$_{tot}$	P$_{PO_4}$	K
1978 – 1979	a	1 600	1 300	12	3.5	1 100
	b	1 800	1 600	13	4.0	1 400
	c	1 800	1 500	13	4.0	1 700
	d	2 000	1 800	13	4.3	1 300
1979 – 1980	a	1 200	930	11	4.6	1 200
	b	1 900	1 400	15	5.2	1 300
	c	1 900	1 700	12	5.4	1 700
	d	1 300	1 100	11	5.0	1 100
1980 – 1981	a	5 500	5 700	55	22	4 000
	b	16 000	16 000	83	41	4 800
	c	24 000	23 000	79	35	6 400
	d	7 900	6 600	83	40	4 000
1981 – 1982	a	1 700	1 400	54	15	3 900
	b	4 000	3 800	74	33	5 300
	c	4 900	4 600	82	29	6 400
	d	2 700	2 500	81	29	4 000
1982 – 1983	a	540	180	36	21	1 700
	b	720	370	30	15	1 800
	c	1 040	520	50	29	3 100
	d	650	230	44	28	1 700

1 From Nov. 1 to Oct. 31

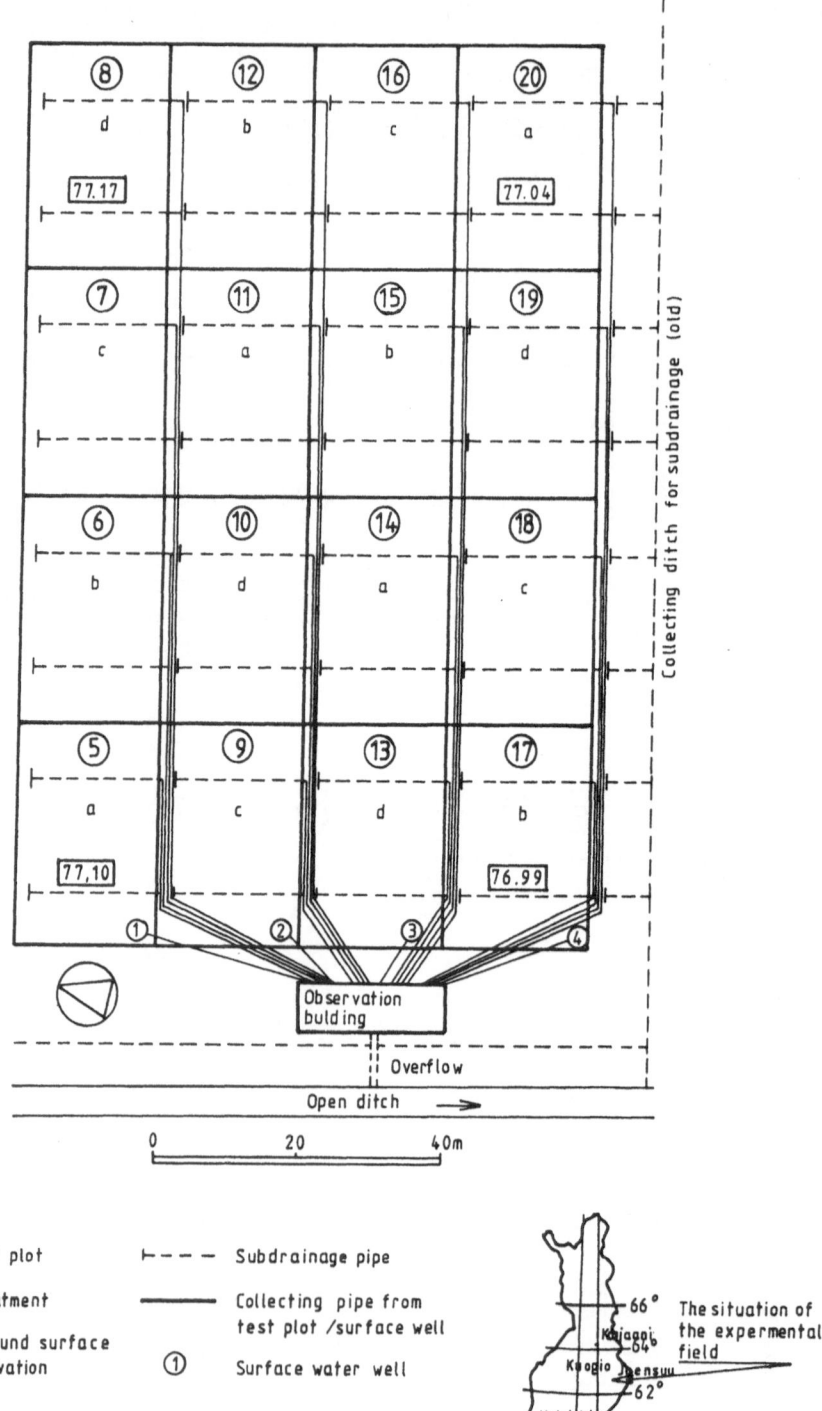

Fig. 1. THE EXPERIMENTAL FIELD OF LIPERI

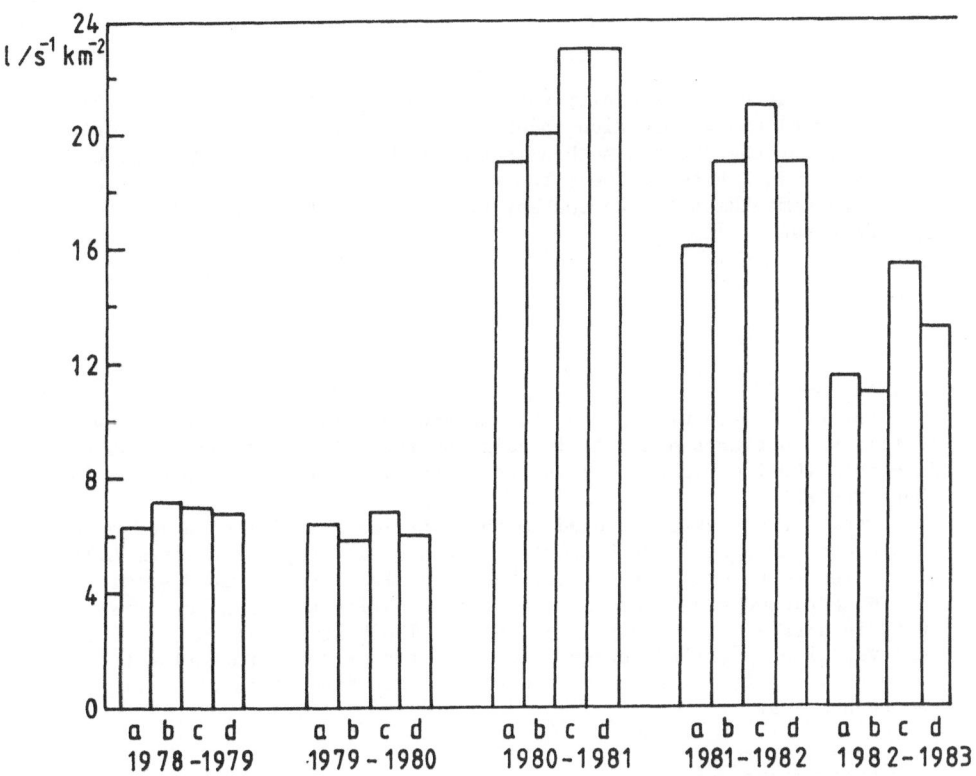

Fig.2 Mean subdrainage runoff measured in the hydrological
years studied (Melanen et al. 1985).   Hydrological
year = Nov. 1 – Oct. 31.

GERMINATION TESTS FOR THE DETERMINATION OF SLUDGES AGRICULTURAL VALUE

L. BARIDEAU
Groupe Valorisation des Boues
Faculté des Sciences Agronomiques
B-5800 GEMBLOUX

Summary

The relation between nitrogen availability for ray-grass (Lolium perenne) and germination index of cress (Lepidum sativum) as determined by the Zucconi method is examined. A logarithmic relation is established between these two variables. The possibility of measuring ammonium nitrogen content of sludges with a germination test is discussed.

1. INTRODUCTION

Pot trials are widely used to measure soil fertility, organic matters fertility, nutrients dynamics in soil and for a lot of other purposes. Ray-grass (Lolium perenne L.) is one of the very often used species in these tests.

Germination tests are used to characterize toxicity of a large range of products amongst which composts. Cress seeds (Lepidum sativum L.) have been selected by many authors for their sensitivity and rapid germination.

Discordance was observed between germination test and pot trial results in a recent work (Barideau, 1984). The best result in pots being achieved with a "toxic" sewage sludge, as far as the germination test was concerned. We try in this work to verify what kind of relation links the two techniques.

2. GERMINATION TESTS

2.1. Method

The method used for this test is the method proposed by Zucconi et al. (1984) adapted for liquid sludges : no pressure extraction is performed, the sludges are just homogenized in a mixer and the dilutions made at once. The same concentrations are used than those proposed in the Zucconi method : 3, 10, 30, 100 %. 2 ml of dilution and 10 cress (Lepidum sativum L.) are introduced in Petri dishes lined with a filter paper. The dishes are incubated during 24 h at 27° C in the dark.

The number of germinated seeds and the length of the roots are recorded at the end of the incubation period. The results are expressed as germination index :

$$GI : \frac{G}{Gc} \cdot \frac{L}{Lc} \cdot 100$$

with G : germinated seeds.

    Gc : germinated seeds in the control.

    L : total root length.

    Lc : total root length of the control.

2.2. Origins of sludges

Fourteen sludges of different origins have been compared in this germination test. Table 1 summarizes the characteristics of the sewage plants where they were sampled. Most of the plants treat sewage from rural areas or small towns. Philippeville, Eghezée, Fleurus are larger towns (7500-15000 inhabitants). Wasmuel is a very large sewage plant (420000 e.i.) and Baileux purifies the sewage of a cheese-dairy.

2.3. Results

Table 2 contains the results of the germination test. The first five sludges have a very favourable effect on germination and root length : germination index (GI) is well over 100, mainly because of root growth, stimulated by these sludges. At the other end of the table, the three last sludges depress germination and root growth. There is often no root growth at all, giving GI values of zero.

Nothing justifies these differences : sewage plants use very different purification and stabilisation processes and sewages are of very different natures. At first sight no satisfying explanation can be given to explain the phenomenon.

3. POT TRIALS

3.1. Method

Ten out of the fourteen sludges have been compared in a pot trial in order to measure the nitrogen they were able to provide to a ray-grass (Lolium perenne L.) culture.

The sludge volume containing 5 gr of dry matter was mixed with 1.5 kg of soil (loamy agricultural soil) and the all placed in 13 x 13 x 13 cm plastic containers. 1.5 gr of ray-grass was sown in each container. Sludges containing less than 3 % dry matter were not included in this trial, their volume containing 5 gr of dry matter being too large.

Pots were maintained at 20° C with a photoperiod of 12 hours. Ray-grass was harvested after six weeks of growth, dry matter production and nitrogen content were measured.

3.2. Results

The net amount of nitrogen exported by ray-grass can then be computed for each sludge as the difference between the total exported nitrogen and the nitrogen exported by the control (table 3). Statistical analysis (Newman and Keuls test) segregates several groups of sludges shown by identical letters in table 3. To simplify this, one may consider two great sludges groups : one of poor sludges containing 10-30 mg available nitrogen per 5 gr dry matter and another of rich sludges containing 70-120 mg nitrogen per 5 gr dry matter. The Fleurus sludge occupies an intermediary position with 40 mg N/5 gr dry matter.

4. RELATION BETWEEN THE TWO TESTS

A relation exists between the results of the germination and the pot trials, more precisely between the net nitrogen exportations (N exp.) and the GI of the 3 % and 10 % concentrations (GI 3 and GI 10). The relations can be best described by the following regression equations :

$$Ln (1 + GI\ 3) = -\ 0,0273\ N_{exp} + 5,203$$

with $r = -\ 0,82$ and $r^2 = 0,67$

and $Ln (1 + GI\ 10) = -\ 0,0424\ N_{exp} + 5,2203$

with $r = -\ 0,81$ and $r^2 = 0,66$

The correlation is high and the regression assumes two thirds of the variation.

These two equations show that Zucconi's germination index has some relation with the sludge nitrogen availability and thus with their agricultural value.

If the results obtained here were confirmed for other sludges, this germination test should be a useful tool for rapid sewage sludge characterization.

5. AN HYPOTHESIS

One may try to go somewhat further on and try to explain the observed correlation between the two tests. One serious hypothesis seems to be the présence of ammonium nitrogen in the sludges. This should explain all together the effect on ray-grass yields, the amounts of exported nitrogen and the decreasing function describing the relation between these variables and germination index, germination and root growth being inhibited by ammonium nitrogen.

To verify this, we performed a germination trial with ammonium nitrate, ammonium chloride and ammonium hydroxide solutions at various concentrations starting from $0,5\ NH_4^+$ g/l. This concentration can be found in a sludge with 5 % dry matter, 3 % total nitrogen, 1 % ammonium nitrogen.

Table 4 shows that the GI variations are widest with the ammonium hydroxide solutions. These variations are very similar to those observed with the sewage sludges. The solution pH was measured (table 5) and however it varies from 7.85 to 10.55 for the NH_4OH solutions, it does not seem to be involved.

Another germination test with dilutions of a 0.01 N solution of NaOH providing a large scale of pH showed that GI was not a function of pH in this conditions (table 6) and that consequently, its variations could be attributed to the presence of NH_4^+.

6. CONCLUSION

The germination test developped by Zucconi to measure compost maturity has shown to be a mean of agricultural value evaluation of sewage sludges. Ammonium nitrogen, rapidly mineralized in the soil, is largely responsible of this agricultural value. Our observations show that it is most probably at this factor that cress reacts when sewage sludges are used in this method. Zucconi's test should then be a biologic method for measuring ammonium nitrogen with reasonable accuracy.

7. BIBLIOGRAPHY

- BARIDEAU L., IMPENS R. : Sludge origins and nitrogen efficiency. EEC seminar, Pisa, 1984, to be published.
- ZUCCONI F., MONACO A., FORTE M. : Phytotoxins during the stabilization of organic matter. Composting of agricultural and other wastes. EEC Oxford seminar. J.K.R. Gasser ed., pp 73-86, 1984.

Sewage plants	Activated sludges	Bacterial beds	Biodisks	Aerobic stabilisation	Anaerobic digestion	No stabilisa- tion	Fe Cl₃ Precipitation	Dry matter content
Saive		x			x			-
Biercée	x							2,65
Moustier			x		x			8,15
Han s/ Lesse	x							12,0
Winenne			x			x		5,3
Philippeville	x				x			7,9
Rance			x		x			0,12
Wasmuel	x				x			3,7
Cerfontaine	x			x			x	6,17
Beaumont	x					x		0,12
Eghezée		x			x			2,0
Fleurus		x			x			5,3
Durbuy	x			x				3,83
Baileux	x					x		4,52

Table 1. : Sewage plant and sludges characteristics.

Sewage plant	Concentration			
	3	10	30	100
Saive	238,4	33,9	0,3	0
Biercée	234,2	221,3	82,0	0
Moustier	215,8	169,0	86,7	32,3
Han s/ Lesse	155,6	88,5	44,0	2,7
Winenne	151,0	110,4	41,8	0
Philippeville	88,3	69,2	38,4	0,8
Rance	82,0	22,6	5,5	25,0
Wasmuel	75,5	76,9	3,6	0
Cerfontaine	61,4	41,9	2,7	0
Beaumont	48,5	85,7	9,2	0
Eghezée	46,8	170,6	65,2	15,5
Fleurus	23,8	19,7	0	0,2
Durbuy	16,7	4,8	0,3	0
Baileux	9,5	0	0	0

Table 2. : Germination index of the fourteen sludges at
four different concentrations.

Sewage plant	Exported nitrogen	
	Total	Net
Moustier	48,1 b	16,7
Han s/ Lesse	41,0 a	9,5
Winenne	49,7 b	18,3
Philippeville	57,5 c	26,0
Wasmuel	100,2 e	68,7
Cerfontaine	44,0 b	12,5
Fleurus	72,5 d	41,1
Durbuy	110,9 f	79,4
Baileux	124,3 g	92,8
Control	31,5 a	-

Table 3. : Total and net exported
nitrogen (mg/pot).

NH_4^+ g/l	$NH_4 NO_3$	$NH_4 Cl$	$NH_4 OH$
0,5	73,9	92,8	2,2
0,15	78,9	87,6	63,0
0,05	60,2	155,5	95,8
0,015	71,3	153,5	135,7

Table 4. : Germination indexes for three
ammonium sources.

NH_4^+ g/l	$NH_4 NO_3$	$NH_4 Cl$	$NH_4 OH$
0,5	5,80	5,55	10,55
0,15	6,00	6,10	9,95
0,05	6,00	5,95	9,25
0,015	6,15	6,05	7,85

Table 5. : Ammonium sources solution pH.

Normality	pH	GI
0,005	11,7	123,1
0,002	10,7	177,0
0,001	7,8	157,0
0,0005	6,9	112,9
0,0002	6,1	147,1
0,0001	5,9	95,6

Table 6. : Relation between pH
and GI.

UTILIZATION OF SEWAGE SLUDGE AS FERTILIZER IN ENERGY PLANTATIONS ON PEATLAND

T. BRAMRYD
Department of Plant Ecology, University of Lund,
Ecology Building, Helgonavägen 5, S-223 62 Lund, SWEDEN

Summary

Sewage sludge was used as fertilizer in a plantation with
fastgrowing willow (Salix viminalis) in a former peat excava-
tion area. Experimental plots were fertilized with 3 and 5 kg
dw per m^2 of sewage sludge (20 % dry matter). The sludge was
mixed with the upper 30 cm peat layer. In a pilotstudy plastic
tubes were filled with peat, humus and mineral soil and were
fertilized with sewage sludge, peatashes, woodashes or commercial NPK fertilizer. The columns were irrigated daily during
14 days with 2x100 ml of simulated acidified rain (pH 4.0).
Sewage sludge fertilization increased the concentrations of
potassium, phosphorus and nitrogen in the peat soil and increased the productivity of the Salix seedlings compared with
control areas. The accumulation of Cd and Cr was slightly increased in the upper peat layers after sludge fertilization,
but the uptake in the vegetation was small. Copper showed both
the highest accumulation and the most significant uptake in
Salix biomass. This was explained by a relatively high concentration in the sludge. Sewage sludge increased the pH in the
peat soil from about 5.2 to about 7.7 in the upper peat layers.
This was about one pH unit more than after fertilization with
the· same amounts of peat- or woodashes or with calcium carbonate. The irrigation experiments showed that the soil substrate has a significant effect on the leaching rate. Metals
were effectively immobilized by complex bindings in low decomposed peat. The highest leaching rate was found in mineral
soil with low concentrations of particles with complex-forming
capacity as organic matter or clay particles.

1. INTRODUCTION

Sewage sludge from the treatment of domestic waste water
is rich in nutrients, especially nitrogen, but also phosphorus
and many trace elements. Due to high concentrations of heavy
metals in the sewage a large proportion of the sewage sludge
in the world has been landfilled or dumped in the oceans.
During the last years efforts have been made in many industrialized countries to separate metal-containing industrial effluents in order to reduce the concentrations of heavy metals
in the sewage. An increasing amount of sewage sludge is used
in agriculture where it is regarded as an alternative to commercial nitrogen fertilizers. In addition to nutrients sewage
sludge is a valuable source for humus substance in the soil
which improves the water- and nutrientholding capacity (1,2,3).

In some cities sewage sludge compost is used as top soil in parks, along roads a s o. In this case higher concentrations of heavy metals can be allowed than if the sludge is used in agriculture.

Investigations in eg the US (4,5) and in Sweden (1,2,3) with sewage sludge fertilization in pine- and spruce forests have reported positiv effects on the nutrient balance in soil and on tree productivity.

In order to produce biomass as a fuel large scale plantations of fast growing tree species have been discussed. In Scandinavia willow species, birch and aspen have been considered as the most suitable species. To achieve a fast assimilation and a high productivity it is essential to supply with fertilizers and sometimes water in relatively high amounts. Generally nitrogen, potassium and phosphorus are considered to be the most essential limiting factors for the productivity. This is especially the case in organic soils with normally low natural concentrations of phosphorus and potassium.

In large scale experiments in Sweden fertilization with ashes from wood and peat has resulted in increased productivity of willow (Salix sp) (6). Within this research program field experiments have also been started with sewage as a fertilizer in willow plantations. In field experiments and in pilot studies the effects of sewage sludge have been compared to those of wood- and peatashes. The aim of this study was to investigate the effects of sewage sludge on the nutrient status and biomass production. Accumulation and uptake of heavy metals were investigated and compared with the situation after fertilization with ashes. In irrigation experiments the leaching of nutrients and heavy metals from different soil types was compared.

2. MATERIAL AND METHODS
2.1 Experimental design
Experimental plots on peat soil, each with a size of 50 m^2, were fertilized with sewage sludge in the amounts of 3 and 5 kg dw per m^2. The research area was a former drained bog in the north of Scania, south Sweden, where peat had been excavated during World War II. About 1-2 m of the bottom peat layer was left. The upper 50 cm of the peat layer was milled and carried no herb vegetation. The research area is situated in the nemoral phytogeographical region and has an annual mean temperature of about +7°C. The relative humidity is about 100 mm (according to Tamm (7)) and a mean precipitation of about 500-600 mm.

The sewage sludge was obtained from the sewage works in Lund, a university city of about 70 000 inhabitants and with a mixed industrial and domestic waste water. The sludge contained residues from the first mechanical, the biological, and the third phosphorreducing processes of the sewage treatment. The sludge was dewatered to about 20 % dry matter. (Chemical composition, see table 1)

After fertilization the sludge was mixed within the upper 30 cm of the peat layer. The experimental plots were planted with seedlings of Salix viminalis. The experiments were started in spring before the growing season had started.

In a pilot study plastic tubes (50 cm high with a diameter of 10 cm) were filled with milled peat, with humus, and with mineral soil from a podsolic spruce forest soil. These columns were treated with sewage sludge, peatashes, woodashes or commercial N P K fertilizer. The fertilizers were mixed within the upper 10 cm of the soil column. The columns were irrigated during 14 days with 2x100 ml of a simulated acidified rain (destilled water acidified with sulphuric acid to pH 4.0). This volume corresponds to the normal precipitation during approximately 2-3 years.

	P	K	Ca	Mg	Cd	Cr	Cu	Mn	Ni	Pb	Zn
					(μg/g dw)						
Peat ashes	3780	12500	50	7.0	2.9	65	79	801	200	42	283
Wood ashes	5630	24080	300	10.0	21.2	26	52	6000	50	19	1130
Sewage sludge	28000	4300	27	6.4	9.9	260	1070	265	110	315	3510

Table 1. Concentrations of phosphorus, potassium, magnesium, and heavy metals in sewage sludge and in ashes from peat and woodchips.

2.2 Field sampling and chemical analyses

In the field experiments peat samples were taken before and one year after fertilization from different layers down to 50 cm. From each research area 2x20 subsamples were taken and then put together to two general samples from each area; one sample containing peat from the upper 0-30 cm layer and one from the layer 30-50 cm.

The peat samples were analyzed for pH, K, P and the heavy metals Cd, Cr, Cu, Mn, Ni, Pb, and Zn. Before the analyzation of metals and phosphorus the soil samples were dried and digested in concentrated nitric acid. The metal concentrations were determined with a Varian AA-475 or Varian Techtron AA6 atomic absorption spectrophotometer. The total concentration of phosphorus was analyzed colorimetrically according to Kitson and Mellon (8).

During late summer the height increment of the Salix viminalis species were measured and biomass samples were taken from the leaves and the stem for chemical analyzes. The biomass samples were treated similar to the peat samples and were analyzed for phosphorus, potassium and the heavy metals above.

The collected leaching water samples from the soil columns in the pilot study in the laboratory were evaporated to about 20 ml each and were then digested with concentrated nitric acid. The heavy metals Cd, Cr, Cu, Ni, Pb, and Zn were analyzed with the same technique as the peat and biomass samples above. pH was measured in the original leaching water.

3. RESULTS AND DISCUSSION

Fertilization with sewage sludge increased the concentrations of potassium and to some extent phosphorus in the peat soil (table 2). As K and P are limiting nutrients in most peatlands in Scandinavia the sludge treatment resulted in increased productivity of the Salix seedlings. Table 3 shows the length increment of the seedlings during the first vegetation period after fertilization compared to parallel experiments with peat- and woodashes. The table indicates that sewage sludge gives a better growth effect than ashes although P and K are added in both experiments. This is probably due to the supply of nitrogen through the sewage sludge. The root systems also developed faster for seedlings fertilized with sewage sludge, and therefore these were less sensitive to drought. The differenses in growth between areas fertilized with 3 resp 5 kg dw of sewage sludge per m^2 were relatively small.

Soil analysis

	pH_{H2O}	P	K	Cd	Cr	Cu	Mn	Ni	Pb	Zn
				(μg/g dw)						
Sewage sludge 3 kg dw/m^2										
0 - 30 cm	7.7	402	605	0.21	5.4	6.9	173	6.2	6.7	75
30 - 50 cm	6.2	585	154	0.05	2.1	8.6	155	3.2	4.1	18
Sewage sludge 5 kg dw/m^2										
0 - 30 cm	7.7	338	537	0.28	3.8	13.8	111	3.8	2.5	17
30 - 50 cm	7.9	232	169	0.04	3.0	12.3	136	4.6	0.5	26
Control										
0 - 30 cm	5.2	245	174	0.14	2.2	6.7	114	3.6	2.8	14
30 - 50 cm	5.6	219	284	0.09	0.8	3.1	84	3.2	4.4	25

Table 2. Concentrations of nutrients and heavy metals in soil samples from two different layers. Soil samples taken at the end of the first vegetation period after sewage sludge fertilization in a drained and excavated peatland. pH in water solution.

Average height increment (cm)

	year 1	year 2
Sewage sludge		
3 kg dw per m^2	80	110
5 kg dw per m^2	85	115
Peat ashes		
3 kg dw per m^2	70	100
Wood ashes		
3 kg dw per m^2	40	70
Control	30	60

Table 3. Average height increment (cm) of Salix during the first and second vegetation periods after sewage sludge fertilization. The Salix seedlings were planted immediately after fertilization.

The soil concentrations of cadmium in the sewage sludge fertilized areas were about twice as high as in control areas (table 2). The increase could only be noticed in the upper

0-30 cm. No significant difference could be seen between the two different doses of sewage sludge. A slight uptake of Cd in the Salix biomass was noticed (table 4).

The accumulation of chromium was slightly increased after fertilization with sewage sludge, but the uptake in the biomass was not influenced. Chromium very easily form stable complexes with organic matter and therefore has a very low mobility and rate of uptake (9).

	Plant biomass analysis						
	Cd	Cr	Cu	Mn	Ni	Pb	Zn
		(μg/g dw)					
Sewage sludge 3 kg dw/m^2							
stem	1.9	1.2	2.3	261	0.59	1.6	51
leaves	1.7	2.0	3.4	170	0.93	5.3	119
Sewage sludge 5 kg dw/m^2							
stem	1.3	1.5	7.4	365	0.64	0.9	137
leaves	2.1	3.3	9.3	340	1.15	6.4	355
Control							
stem	0.8	1.4	1.1	427	0.54	1.1	36
leaves	1.3	3.2	2.1	1319	1.20	6.5	101

Table 4. Concentrations of heavy metals in different fractions of Salix biomass. Samples taken during the first vegetation period after sewage sludge fertilization. Salix seedlings planted immediately after fertilization.

According to table 1 sewage sludge from the city of Lund contains 15-20 times as much copper as eg ashes from peat and wood. This is to a great extent due to the widespread use of copper tubes for drinkingwater in new houses, but also to industrial emissions. The twofold increase in soil concentrations of copper in the area fertilized with 5 kg dw of sewage sludge per m^2 was therefore expected (table 2).

The noticable increase of the Cu concentrations at the depth of 30-50 cm can be explained by the fact that copper like most other heavy metals have a relatively strong tendency to form complexes with organic matter. Although this will lead to some immobilization in the upper peat layers, copper bound to the soluble organic fraction of the sludge will penetrate to deeper soil layers. This tendency has also been shown by earlier investigations (eg 6). The uptake of copper in Salix biomass was about four times higher in the experiments with a higher dose of sludge (table 2).

The concentrations of manganese and nickel are only to a minor extent affected by the sewage sludge fertilization.

In mechanically managed research areas the contamination of lead is quite pronounced and no clear trends can be seen for this metal. In most treated areas the concentrations of lead is higher in the top-soil than in the 30-50 cm horizon. Lead form strong complexes with the peat material (10). These complexes have a high stability and the rate of leaching is very slow (11). This is especially the case at the relatively high pH values (around pH 7-8) prevailing after sewage sludge fertilization of these research plots. The pH in the reference areas is around 5.0-5.5 (table 2).

The concentrations of zinc increase slightly after sludge fertilization. In the area with a higher dose of sludge the uptake of zinc in the Salix biomass seems to be more pronounced (about three times higher) than in the area with 3 kg dw of sludge per m^2. This probably explains the lower concentrations in the peatsoil compared to the experiments with the lower dose. Zinc forms the weakest complex bindings among the heavy metals and is easily leached down to deeper layers of the soil.

The concentrations of most heavy metals and nutrients are higher in the green leaves with an active metabolism than in the stem tissues. The leaves are also more direct subject to deposition of airpollutants.

Compared to parallel experiments in the same peatland locality with ashes from peat and wood burning, the sewage sludge fertilization increased the pH value of the soil more than the same amounts of ashes. After fertilization with 3 kg dw of peat- or woodashes the pH increased approximately 1 pH unit less than after sewage sludge fertilization (6). This indicates that sewage sludge has a good buffering capacity. Similar effects have been recorded from sewage sludge experiments in pineforests on sand sediment soils (1,2,3).

The accumulation of Cd is more pronounced after fertilization with wood ashes than with sewage sludge (approximately 7-10 times higher concentrations (6). However the uptake in the plant biomass is of the same magnitude which indicates that the retentive capacity of the peat through the formation of complexes is rather high.

The accumulation and uptake of Cu is somewhat higher after sewage sludge fertilization than after fertilization with peat- or woodashes (6). This was expected as the sludge contains higher concentrations of Cu.

In pilotstudies plastic tubes filled with peat, humus (from a podzol), and mineral soil (from a podzol) were fertilized with sewage sludge, ashes or N P K. These soil columns were leached with simulated acidified rainwater with pH 4.0. The experiments showed that the soil substrate had a great influence on the leaching rate (table 5). The humus substrate was more susceptible for leaching than the peat. This is probably due to a higher content of soluble organic humus compounds which form complexes with the metals and carry them away in the leachates. Peat is rather low decomposed and the metals are more effectively immobilized by complex bindings (12). Relatively high leaching rates were also found for the mineral soil with low concentrations of clay particles or organic matter which could immobilize the metals through complexbinding.

The investigations show that there is a slight tendency that an addition of well-buffering fertilizers, as ashes or sewage sludge, sometimes decreases or sustains the leaching of heavy metals, in spite of the addition of heavy metals through these fertilizers. This is the case for eg Cu and Cr in the column with mineral soil. For these metals the complex binding processes are favoured by an increased pH which leads to an immobilization. As expected the sustaining effects on the leaching of Cd and Zn were smaller. These metals form weaker complexes than the other heavy metals (13). Fertilization with the commercial fertilizer N P K decreased the pH in all substrates and increased the leaching of most of the investigated heavy metals.

Fertilizer	pH	Cd	Cr	Cu	Ni	Zn
			(μg/g dw)			
Peat						
Sewage sludge	5.0	2.0	0.3	4.5	1.0	45.0
Peatashes	5.6	1.6	0.6	7.5	2.3	360.0
Woodashes	5.5	1.4	0.8	4.9	2.0	40.0
N P K	4.6	4.3	6.9	14.5	0.7	30.0
Control	4.8	1.8	0.7	5.5	1.0	10.0
Humus						
Sewage sludge	4.6	2.9	4.2	22.3	9.7	320.0
Peatashes	4.5	2.3	3.7	10.0	7.5	170.0
Woodashes	4.5	2.6	3.9	16.0	8.7	380.0
N P K	3.9	4.5	5.5	13.8	21.0	1168.0
Control	4.3	2.9	5.6	16.0	61.0	110.0
Mineral soil						
Sewage sludge	4.9	4.0	0.2	6.8	9.5	315.0
Peatashes	4.8	3.1	0.5	9.5	5.6	280.0
Woodashes	4.8	3.8	0.5	6.0	8.5	270.0
N P K	4.1	1.7	2.9	48.0	4.7	3225.0
Control	4.4	2.0	0.6	14.8	4.4	260.0

Table 5. Irrigation of fertilized soil columns of peat, humus (from a podzol) and mineral soil (from a podzol) with acidified water (pH 4.0). Soilcolumns fertilized with sewage sludge, ashes from peat or woodchips and commercial N P K fertilizer, pH_{H_2O} and concentrations of heavy metals in the collected leaching water.

REFERENCES

(1) BRAMRYD, T. (1976). Effekt av rötslamdeponering på kväve-mineraliseringen i en planterad granskog (The effects of digested sludge application on the nitrogen mineralization in a planted spruce forest, English abstract). Meddn. Avd. Ekol. Bot., Lunds Univ. 4:1. 1-38.
(2) BRAMRYD, T. (1980). Sewage sludge fertilization in pine forests - Ecological effects on soil and vegetation. In: Biogeochemistry of ancient and modern environments (Trudinger, Walter, Ralph, eds). Australian Academy of Science, Canberra. 405-412.
(3) BRAMRYD, T. (1981). Comparative studies of nitrogen mineralization in forest soils fertilized with fluid and dewatered sewage sludge. In: Characterization, treatment and use of sewage sludge (L´Hermite, Ott. eds). Commission of the European Communities, Reidel Publicity Company, Dordrecht. 475-483.

(4) HINESLY, T.D., BRAIDS, O.C., DICK, R.I., JONES, R.L. and
     MOLINA, J.-A.E. (1971). Agricultural benefits and envi-
     ronmental changes resulting from the use of digested
     sludge on field crops. Results of a Cooperative Project
     between the University of Illinois, the Metropolitan
     Sanitary District of Greater Chicago and U.S.E.P.A.,
     1-375.
(5) SOPPER, W.E. and KARDOS, L.T. (1973). Vegetation respon-
     ses to irrigation with treated municipal wastewater.
     In: Sopper, E. and Kardos, T. (eds) Recycling treated
     municipal wastewater and sludge through forest and crop-
     land, The Penn. State University Press, University Park.
     269-294.
(6) BRAMRYD, T. (1985). Torv- och vedaska som gödselmedel -
     effekter på produktion, näringsbalans och tungmetallupp-
     tag. (Peat- and woodashes as fertilizer - Effects on
     production, nutrient balance and uptake of heavy metals.
     In Swedish. English summary). National Swedish Environ-
     ment Protection Board Report. 84 pp (in press).
(7) TAMM, O. (1959). Studier över klimatets humiditet i
     Sverige. (Studies of the climatic humidity in Sweden).
     Bulletin of the Royal School of Forestry, Stockholm,
     Sweden, 32. 1-48.
(8) KITSON and MELLON. (1944). Ind. Eng. Chem. Anal. Ed. 16,
     379-383.
(9) BRAMRYD, T. (1983). Uptake of heavy metals in pineforest
     vegetation fertilized with sewage sludge. Proceeding
     Third International Symposium on Treatment and Use of
     Sewage Sludge, Brighton, UK, 26-29 Sept. 1983. Commission
     of the European Communities (EEC). 3 pp.
(10) OLSON, K.W. and SKOGERBOE, R.K. (1975). Identification
     of soil lead compounds from automotive sources. Environ-
     mental Science and Technology 9. 227-230.
(11) SICCAMA, T.G. and SMITH, W.H. (1978). Lead accumulation
     in a northern hardwood forest. Environmental Science and
     Technology 12. 593-594.
(12) BRAMRYD, T. and FRANSMAN, B. (1985). Utvärdering av äldre
     gödslings- och kalkningsförsök med torv- och vedaska i
     Finland och Sverige. (Evaluation of older experiments
     with peat- and woodash-fertilization in Finland and
     Sweden. In Swedish). National Swedish Environment Pro-
     tection Board Report 1991. 1-60.
(13) TYLER, G. (1978). Leaching rates of heavy metal ions in
     forest soil. Water, Air and Soil Pollution 9. 137-148.

THE USE OF COMPOSTED AGRICULTURAL WASTE AS PEAT SUBSTITUTE IN HORTICULTURE

Y. CHEN[1], Y. HADAR[2], and Y. INBAR[1]

[1]Dept. of Soil and Water Sciences, and [2]Dept. of Plant Pathology and Microbiology, The Hebrew University of Jerusalem

Summary

Composted agricultural wastes were tested as substitutes for peat in container media. The wastes used in this study, all composted prior to their use as media, were: (1) anaerobically fermented cattle manure; (2) separated cattle manure; and (3) grape marc. The physical and chemical properties of the composts were determined. The composts were tested as growth media for the production of vegetables' seedlings and for growing ornamentals up to an age of 8 months. Plant growth response was either equal or better as compared to peat. Mixtures of 1:1 of the composts with peat seemed to provide a recommended growth medium. Mechanisms for the improvement of growth in the compost are proposed.

1. INTRODUCTION

In the last decades the demand for peat as a substrate in horticulture had increased continuously while its availability is decreasing. A number of organic wastes such as bark, leaf mould, town refuse, sawdust, spent mushroom compost, treated animal excreta and many other organic materials were introduced as peat substitute in container media after proper composting (Bik, 1983; Cull, 1981; Lohr et al. 1984; Raviv et al. 1985; Verdonck, 1984). Both cattle manure and grape marc (the residue of wine processing) are agricultural wastes which are produced all over the world. By proper treatment these wastes can be converted into container substrates.

This paper summarizes some research conducted on the recycling of these agricultural wastes to container media.

2. COMPOST TYPES - DISCUSSION OF PROPERTIES

I. The use of composted slurry produced by methanogenic fermentation of cow manure (cabutz).

The anaerobic digestion of organic matter is a well known technology which is used to produce biogas from wastes, especially manures. The digested slurry is usually used for direct application in the field. We found that by sieving and leaching the digested slurry on a vibrating screen, two major products are obtained: (i) a fibrous material which resembles peat in its physical structure. This fraction is commercially named "cabutz", and may serve as a growth substrate. (ii) an effluent which is similar in its composition to liquid fertilizers. The cabutz has been tested for physical and chemical properties in our laboratory and was found to maintain high hydraulic conductivity and air capacity as well as an adequate water and nutrients retention. As a result of the high temperature in the digesters ($55\,°C$), plant disease problems were not observed in any of our experiments. The particle size of cabutz ranges from 1-5 mm, which is slightly larger than sphagnum peat moss from Finland. The bulk density is 0.08-0.12 g/cm and the porosity reaches 93-95%. Water and air capacity of the cabutz are 62% and 31-33%, respectively. Hydraulic conductivity at saturation is as high as 150 cm/h. Similar values were measured on peat moss. The chemical properties of cabutz are related to the leaching intensity and duration. The electrical conductivity ranges from 0.5-3.0 mmhos/cm or less, if required, the pH range is 7.0-7.6 and the nutrient contents resembles that of enriched sphagnum peat moss from Finland (Chen et al. 1984 a & b).

Cabutz was composted in windrows for 100 days. Temperatures rose to 55°C within a week. Chemical analysis of samples taken during composting revealed that the total contents of macroelements (N,P,K), minor elements (Fe, Mn, Cu, Zn) and ash increased about two-fold, and the C/N ratio decreased from 40 to 15. The pH dropped from 7.5 to 6.6, and the amount of soluble salts rose from 15 to 37 meq/100 g. Growth experiments were conducted under greenhouse conditions. Germination and seedling development of peppers, cucumbers and tomatoes were compared in growth media consisting of raw cabutz, composted cabutz and peat. The raw cabutz had a strong inhibitory effect on tomatoes. This inhibition was not removed by fertilization. For all plants, the composted cabutz appeared to be as good or superior as compared to peat. Fertilization with inorganic fertilizers or slurry effluent increased both plant dry weight and height (Inbar et al. 1985; Hadar et al. 1985). This compost can be used as a sole medium or as a component in a mixture at levels similar to those known for peat.

II. Separated cattle manure

The traditional way of treating cattle manure is to spread it on the soil as source of nutritional elements or as conditioner, either raw or after composting.

Huijsmans and Lindley (1984) investigated solid-liquid separation systems for pre-processing dairy cattle manure prior to anaerobic digestion. They suggested that the main product is the liquid fraction which may be used either for irrigation or for biogas production. In their system, the solids fraction is a by-product which can be applied to land as soil conditioner and if composted, as a mulch for nurseries and landscape use. Cull (1981) stated that animal fibers appear to be promising substrates but composting and distribution could present problems.

In our system, the cattle manure is separated to liquid and solids. The liquid is used after dilution as organic liquid fertilizer, while the fibrous solids are composted in windrows and later used as growth medium (Inbar et al. 1980). Composted slurry manure was successfully tested as peat substitute or peat complementary media for vegetables seedling production. Development of pepper, cucumber and tomato seedlings was faster in compost containing media in comparison to peat+vermiculite media.

III. Grape marc

In the wine producing industry, the marc which accumulates in relatively large quantities is usually considered a waste and its disposal seems expensive. Grape marc is not suitable for direct soil application since during its degradation materials that are highly incompatible with plant root systems are released. As a result, it was not previously considered to utilize grape marc as an organic fertilizer. Graefe (1980) and Streichsbier et al. (1982) investigated the biological process in which marc material was decomposed by microbes. They propose to use the composted grape marc as a high grade organic fertilizer, while recovering heat and CO_2 which are produced during the process.

Because of their high sugar content, grapes belong to the class of fruits which are highest in calories. Grape pressing residues which contain somewhat more than 50% moisture have a sufficient amount of easily degradable carbohydrates in solution to provide the micro-organisms with easily available nutrients. Thus, in 3-4 days temperatures higher than 50°C are reached during aerobic decay. If adequate supply of moisture and oxygen is available, skins and stalks can be converted after only a few weeks at the high decaying temperature prevailing in the compost pile to a fine friable humus. Only the grape seeds can stand the action of the micro-organisms and maintain their external structure and shape. Two to four years in a compost heap or in soil are required to achieve decomposition (Graefe, 1980) of the seeds.

In our study, grape marc which consisted of grape skins, seeds and stalks was composted in windrows for 6 months. The composted material was tested as growth media (Inbar et al. 1986). The composted grape marc was found to be a promising organic component in growth media. It was tested for several vegetable crops as well as for ornamentals such as carnations and Ficus.

3. CONCLUSIONS

Composts obtained from slurry produced by methanogenic fermentation of cattle manure, separated cattle manure and grape marc were tested for their performance as the organic component of growth media, or as a sole substrate. These composts exhibited some advantages over peat and commercial mixtures when vegetable seedlings were grown. The growth period of seedlings was shortened by 6-10 days. The reaons for the composts superiority over the control treatments are not fully understood. However, the following possible mechanisms are suggested:
1. The composts intially contain high nutrients level. In spite of the intensive leaching prior to planting, the nutrients, especially P and K, are slowly released into the solution during the growing period.
2. Water and air capacity are within the range of the "ideal substrate" (de Boodt and Verdonck, 1972).
3. The composts contain high levels of humic substances which are potential growth stimulators (Schnitzer and Poapst, 1967; Lee and Barlett, 1976).
4. High activity of the microbial population may suggest increased growth response due to rhizosphere micro-organisms.

The composts studied provide inexpensive, high quality peat substitute as well as a solution for environmental problems of waste disposal.

4. ACKNOWLEDGMENTS

This research was supported by a grant from the National Council for Research and Development, Israel and the European Economic Community.

REFERENCES

(1) BIK, A.R. (1983). Substrates in floriculture. Proc. XXI Intern. Hortic. Congr. 1982, Hamburg, II:811-822.
(2) de BOODT, M. and VERDONCK, O. (1972). The physical properties of the substrates used in horticulture. Acta Hort. 26:37-44.
(3) CHEN, Y., INBAR, Y., RAVIV, M. and DOVRAT, A. (1984a). The use of slurry produced by methanogenic fermentation of cow manure as a peat substitute in horticulture - physical and chemical properties. Acta Hort. 150:553-561.
(4) CHEN, Y., INBAR, Y. and RAVIV, M. (1984b). Slurry produced by methanogenic fermentation of cow manure as a peat substitute in horticulture. Proc. 2nd Intern. Symp. Peat in Agriculture and Horticulture, pp. 297-317, Bet-Dagan, Israel.
(5) CULL, D.C. (1981). Alternatives to peat as container media. Organic resources in the U.K. Acta Hort. 126:69-81.
(6) GRAEFE, (1980). Methods and apparatus for preparing high grade fertilizer. United States Patent 4,211,545, July 8, 1980.
(7) HADAR, Y., INBAR, Y. and CHEN, Y. (1985). Effect of compost maturity on tomato seedling growth. Scientia Hortic. 27 (in press).
(8) HUIJSMANS, J. and LINDLEY, J.A. (1984). Evaluation of a solid-liquid separator. Transactions of the ASAE:1854-1858.
(9) INBAR, Y., CHEN, Y. and HADAR, Y. (1985). The use of composted slurry produced by methanogenic fermentation of cow manure as a growth media. Acta Hort. 172:75-82.
(10) INBAR, Y., CHEN, Y. and HADAR, Y. (1986). The use of composted separated cattle manure and grape marc as peat substitute in horticulture. Acta Hort. Intl. Symp. on Plant Substrates and Growing Techniques, As, Norway (in press).
(11) LEE, Y.S., and BARTLETT, R.J., 1976. Stimulation of plant growth by humic substances. J. Soil Sci. Soc. Am. 40:876-879.
(12) LOHR, V.I., O'BRIEN, R.G., and COFFEY, D.L. (1984). Spent mushroom compost in soilless media and its effects on the yield and quality of transplants. J. Amer. Soc. Hort. Sci. 109:693-697.
(13) RAVIV, M., CHEN, Y. and INBAR, Y. (1985). Peat and peat substitutes as growth media for container-growth plants. In: Chen, Y. and Avnimelech Y. (eds.): The Role of Organic Matter in Modern Agriculture. Martinus Nijhof/Dr. W. Junk Publ., The Hague (in press).
(14) SCHNITZER, M., and POAPST, P.A. (1967). Effect of soil humic compounds on root initiation. Nature (Lond.) 213:548-599.
(15) STREICHSBIER, F., MESSNER, K., WESSELEY, M. and ROHR, M. (1982). The microbiological aspects of grape marc humification. European J. Appl. Microbiol. Biotechnol. 14:182-186.
(16) VERDONCK, O. (1984). Reviewing and evaluation of new materials used as substrates. Acta Hort. 150:467-473.

LABSCALE SIMULATION OF ISMAN-COTTON BATCH PHASE ANAEROBIC

DIGESTION FOR PREDICTION OF PLANT DESIGN AND PERFORMANCE

Pierre AMMANN [1], Armand COTTON [2], Nicolas MAIRE [3],

1) Ecole Polytechnique Fédérale de Lausanne
   Génie biologique, CH - 1015 LAUSANNE

2) A. COTTON S.A. - CH - 1253 VANDŒUVRES-Genève

3) ACEPSA - CH - 1041 OULENS

SUMMARY

Process design parameters and environmental conditions are of main importance during digestion of mixed wastes, but full scale optimisation is tedious and expensive.

The aim of this work is to try to simulate different experimental conditions (farm wastes to sludge ratio, different C : N : P, pre-aeration time, etc.) in small labscale units in order to collect rapidly and simply a lot of informations and to evaluate to which extent such simulations can be used for prediction of plan design and performance.

In this preliminary study we chose the C/N ratio as the only variable biological paramter in conjunction with two scales for the laboratory apparatus used for these simulations.

The C/N ratio is of fundamental importance in biological waste treatment, especially when these wastes contain lignocellulosic materials, because the carbon contained in the wastes is not readily bioavailable. This means that the biological C/N ratio is lower than the measured C/N ratio of the substrate.

The raw material used are horse manure and digested sewage sludge. In a first step, the horse manure was air-dried and milled to lower heterogeneity (smaller than 2 mm) of the material. The raw material was rehumidified with water to obtain a substrate with a C/N ratio of 80 and with different volumes of sludge for C/N ratios of about 60 and 30.

After incubation at 35° during 12 hours to restore microbial life, the mixtures were incubated 4 days at 55° to simulate aerobic prefermentation. After the completion of the first step, the CO_2 production and the ATP-biomass were measured in replicates. The remaining samples were then incubated anaerobically at 35°. The evolution of pressure and gas quality (H_2, CO_2, CH_4) were monitored for estimation of biogas production and quality. At the end of the experiments (10 weeks), the ATP-biomass and substrate quality were measured.

The results of the "gram-scale" experiments were compared with those obtained in a similar way with "kilogram-scale" and full scale runs.

MATERIAL AND METHODS

The horse manure treated in the lab-scale reactors (fig. 2) and full scale units (fig. 3) is of the same origin (farm), but not necessarily sampled at the same time. The same remark can be made for digested sludge (different dry matter contents).

Aerobic prefermentation in the gram-scale experiments has been performed in 25 ml glass beekers, perforated at the bottom for aeration.

The labscale simulation procedure has to be as simple as possible. We tested aerobic pretreatment at room temperature (25ºC)and at 55º C (thermophile pretreatment) in order to see if we obtain comparable results or conclusions with regard to the biological evolution of the fermented material. Only the samples incubated at 55º C have been digested anaerobically.

Aerobic prefermentation for the kilogram-scale experiment has been performed in "home made" stainless steel anaerobic jars of a volume of 3 l, under a pressure of 3 bars, with daily air renewal. These jars are equipped with a pressure gauge and a septum port for gas sampling. The roughly cut material (< 30 mm) was placed in a polyethylene basket.

The main difference between the full scale and labscale procedure (fig. 1) lies in the fact that in reality, liquid of a previous digestion run is added as an innoculum to the aerobically pretreated manure, while this has not been the case in the simulations.

For the anaerobic digestion step, the pretreated manure of the gram-scale experiments has been transferred in 50 ml incubation flasks and sealed with a thick black septum.

The initial air atmosphere has been replaced by nitrogen at a pressure of 1 bar (atmospheric).

The pressure monitoring in the 50 ml incubation flasks has been achieved by measurement through the septa with a manometer equipped with a syringe needle (fig. 2). The dead volume of the manometer assembly is 3.6 ml.

ATP-biomass has been measured with a bioluminometer (SKAN, Basel) and the luciferine/luciverase system.

CO_2 production has been measured by trapping in KOH and titration. Gas composition has been measured with a gas chromatograph equipped with a thermal conductivity detector (TCD) (VARIAN Vista 6000/401) and a Carbosieve SII (2mm x 2m) glass column (SUPELCO).

RESULTS

1. PREFERMENTATION

1.1. Lowering the C/N ratio of horse manure with digested sewage sludge gives higher biological activity (CO_2 production) and ATP-biomass (fig.4A-B). These values are higher at 55º C than at 25º C, but the evolution in time and the relation between the different C/N ratios are comparable.

1.2. The respiratory activity at C/N ration = 30 is higher than at C/N = 80, but the time evolution of this parameter is quite similar and consistent with a stabilization process. On the other hand, at C/N = 60, the evolution of the activity is on the opposite sens (fig. 4A).

1.3. The degree of mineralization is highest at C/N = 30 and quite similar for C/N = 60 and 80 (fig. 4C).

2. ANAEROBIC DIGESTION

The mean gas production and composition observed in the laboratory for C/N = 30 and 60 are similar, but the reproducibility is much better at C/N = 30. The mean gas production at C/N = 80 has been very low and gas quality

poor (fig. 6 - 9).
The lag phase of the methane production is quite short at C/N = 30, a little bit longer at C/N = 60 and long at C/N = 80.

3. FINAL PRODUCTS
The ATP-biomass and respiratory activities are significantly lower for the final products than during prefermentation. For lab- and fullscale we observed at C/N = 60 the same abnormal evolution of the activity during pre-fermentation and in the final products, (fig. 4 - 5).

4. SCALE COMPARISION
The specific gas production rates expressed in m^3/ton DM.day are lower than in the lab than in full-scale, but comparable at C/N = 30 and 60, (fig. 10). This comparison is not so good when the specific gas production is expressed in m^3 gas/m^3 reactor·day.
The comparision for the 50 ml run at C/N = 80 is bad. The runs in the kilogram-scale with a particle size 30 mm are not comparable with the full scale runs.

CONCLUSIONS
At C/N ratio = 80 (no sludge added) the biological evolution is consis-tent, but productivity and the degree of mineralization are much lower in the simulation than in fullscale and the gas quality is poor (differences in the procedure). At C/N = 60 we observed an abnormal biological behaviour during prefermentation and in the final product. This is true for all scales. The reproducibility was bad, but the gas production very good.
C/N = 30 seems actually to be an optimum for this type of solid phase process, even if productivity is lower than at C/N = 60, because the over-all process is more stable. The biological parameters are very consistent with a good stabilization. The biogas production is rapid and very repro-ducible. The final products show the best equilibrium for agricultural use.
The results show that if labscale units are fed with raw material with a particle size lower than 2 mm, it is possible to have results which are comparable with those obtained in full scale. Further improve-ment of the simulation procedure and measuring technics is ncesssary for design purposes.
The simulation is good with regard to the biological parameters, like respiratory activity, ATP-biomass and gas composition.
The main advantage if labscale simulation is an economical one :
. The eleven experiments described here lasted approx. 100 days and could be performed exactly in the same environment and with the same substrate quality.
. 11 full-scale runs should last more than 3 years with different environmental conditions and substrate qualities.

Further research will be made in order to :
- Improve simulation procedure
- Test other substrates
- Test time and temperature of pre-fermentation
- Test thermophile digestion
- Clear up the abnormal behaviour of C/N = 60

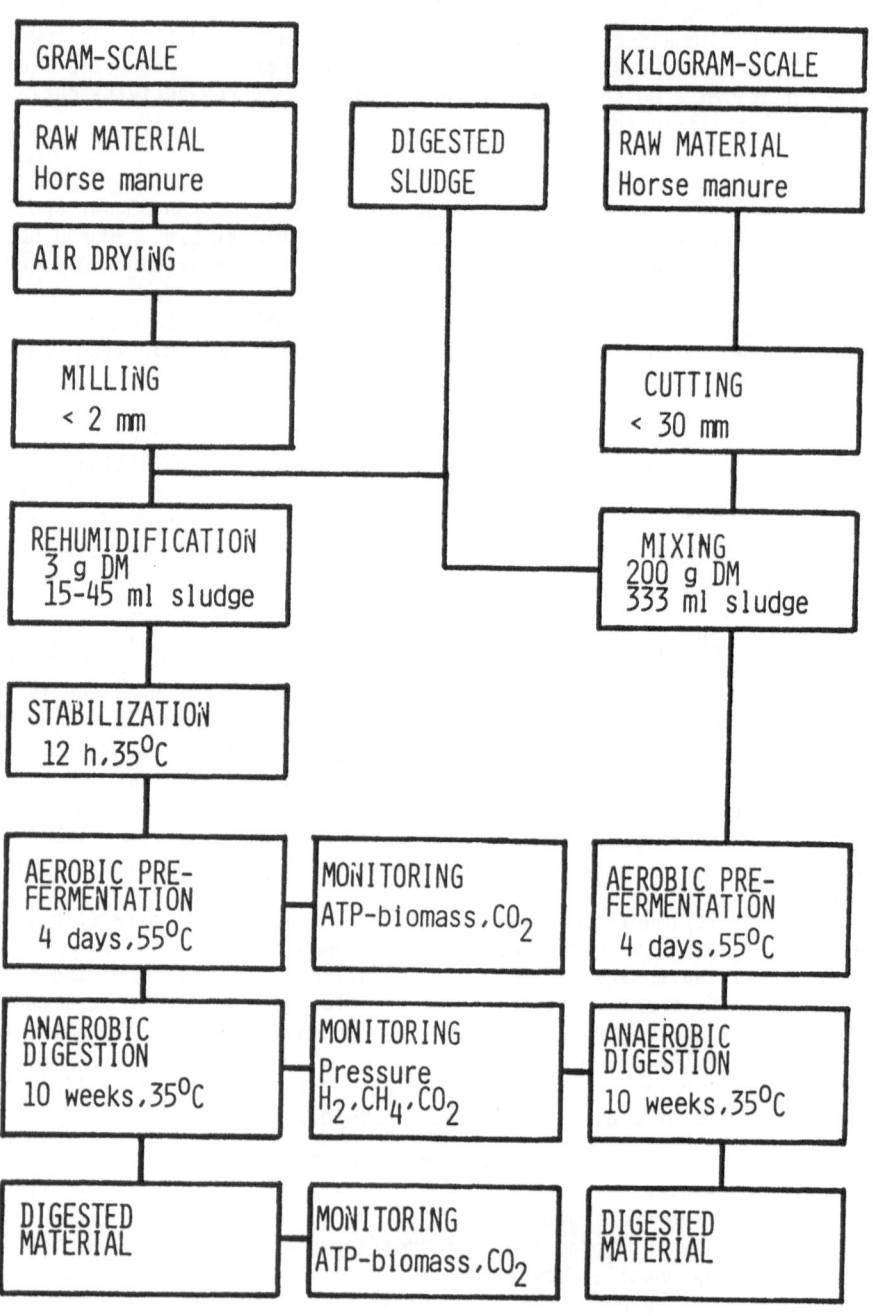

SIMULATION PROCEDURE

GRAM-SCALE

RAW MATERIAL
Horse manure

AIR DRYING

MILLING
< 2 mm

REHUMIDIFICATION
3 g DM
15-45 ml sludge

STABILIZATION
12 h, 35°C

AEROBIC PRE-
FERMENTATION
4 days, 55°C

ANAEROBIC
DIGESTION
10 weeks, 35°C

DIGESTED
MATERIAL

DIGESTED
SLUDGE

MONITORING
ATP-biomass, CO_2

MONITORING
Pressure
H_2, CH_4, CO_2

MONITORING
ATP-biomass, CO_2

KILOGRAM-SCALE

RAW MATERIAL
Horse manure

CUTTING
< 30 mm

MIXING
200 g DM
333 ml sludge

AEROBIC PRE-
FERMENTATION
4 days, 55°C

ANAEROBIC
DIGESTION
10 weeks, 35°C

DIGESTED
MATERIAL

FIGURE 1 : SIMULATION PROCEDURE

FIGURE 2 :
LABSCALE REACTORS OF
3 LITERS AND 50 ML

FIGURE 3 : FULLSCALE UNITS OF 4.5 M^3

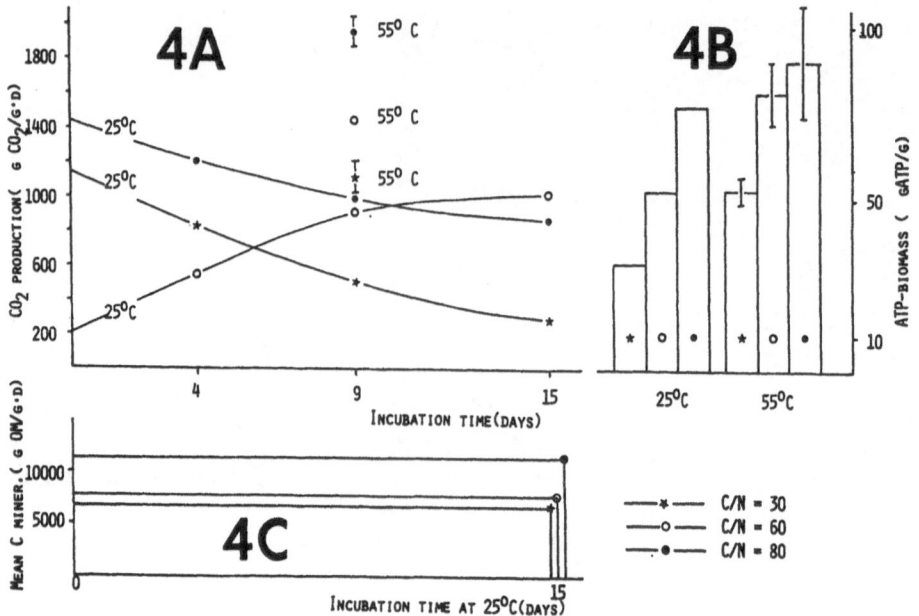

FIGURE 4 : PREFERMENTATION   A) RESPIRATORY ACTYVITY
                             B) ATP-BIOMASS
                             C) OVERALL CARBON MINERALIZATION

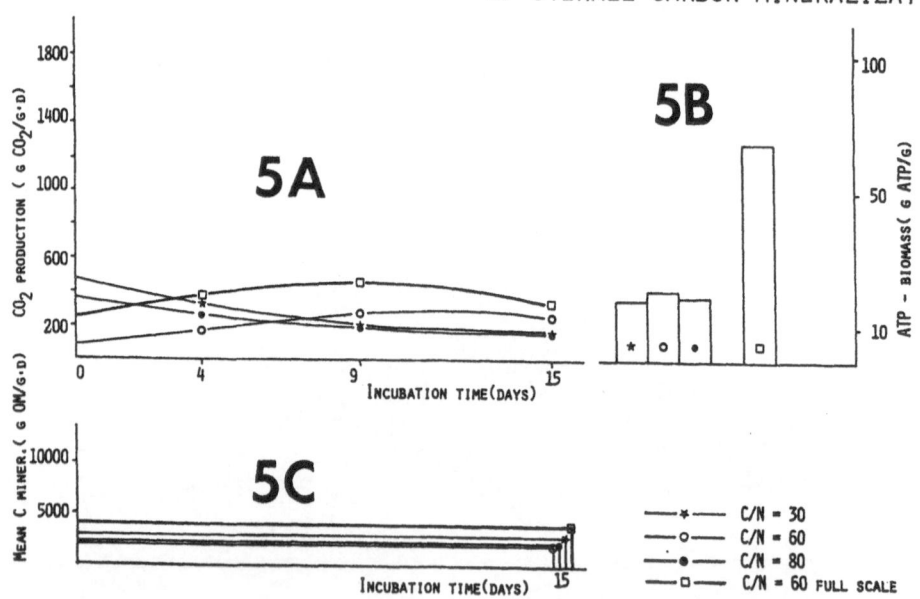

FIGURE 5 : ANAEROBIC DIGESTION   A) RESPIRATORY ACTIVITY
                                 B) ATP-BIOMASS
                                 C) OVERALL C MINERALIZATION

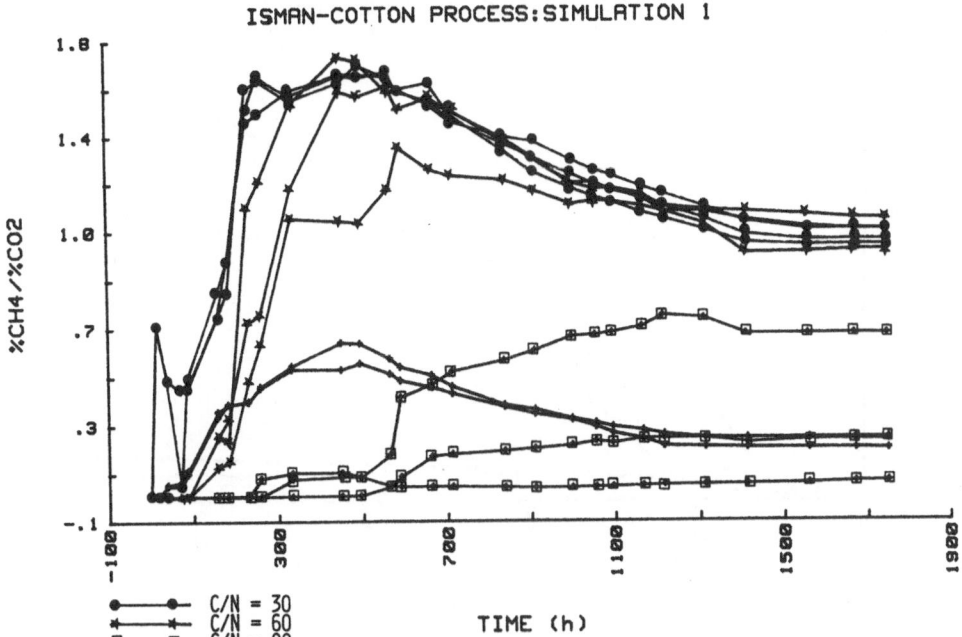

ISMAN-COTTON PROCESS:SIMULATION 1
%CH4/%CO2
1.8
1.4
1.0
.7
.3
-.1
-100
300
700
1100
1500
1900
TIME (h)
C/N = 30
C/N = 60
C/N = 80
C/N = 60*(KILOGRMA-SCALE)

FIGURE 6 : TIME EVOLUTION OF THE CH_4/CO_2 RATIO IN THE LABSCALE UNITS

ISMAN-COTTON PROCESS:SIMULATION 1
%CH4
56
45
33
22
10
-1
-100
300
700
1100
1500
1900
TIME (h)
C/N = 30
C/N = 60
C/N = 80
C/N = 60*(KILOGRAM-SCALE)

FIGURE 7 : TIME EVOLUTION OF THE % CH_4 IN THE LABSCALE UNITS

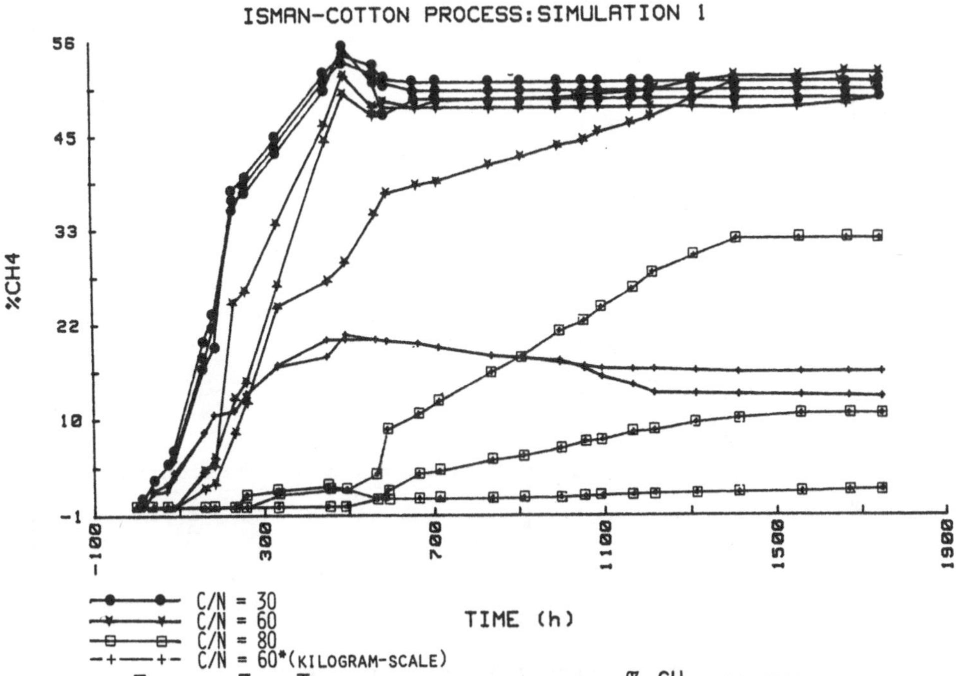

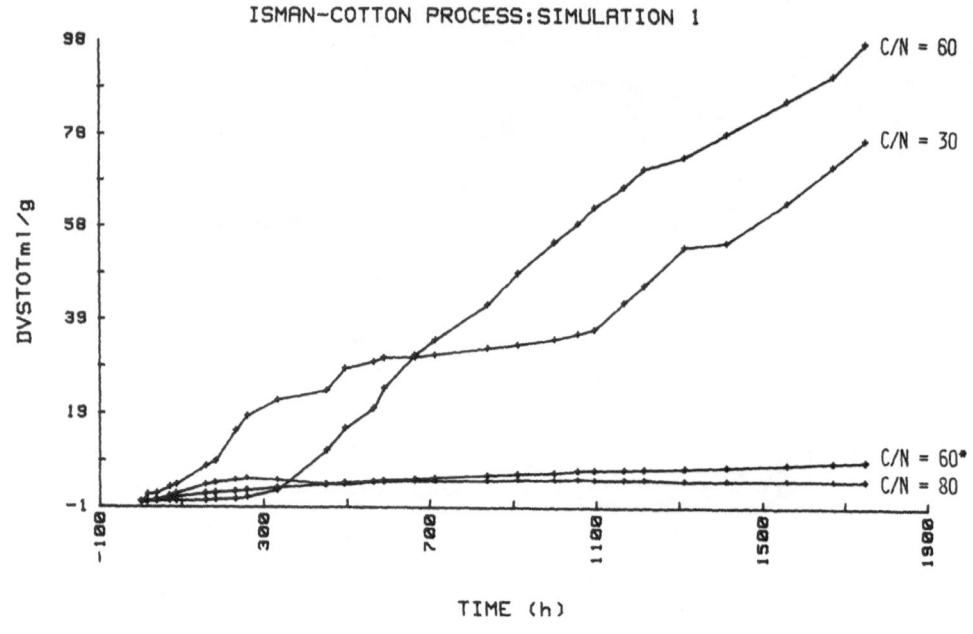

ISMAN–COTTON PROCESS:SIMULATION 1

C/N = 60* IN KILOGRAM-SCALE

FIGURE 8 : CUMMULATIVE SPECIFIC GAS PRODUCTION RATES
IN THE LABSCALE UNITS

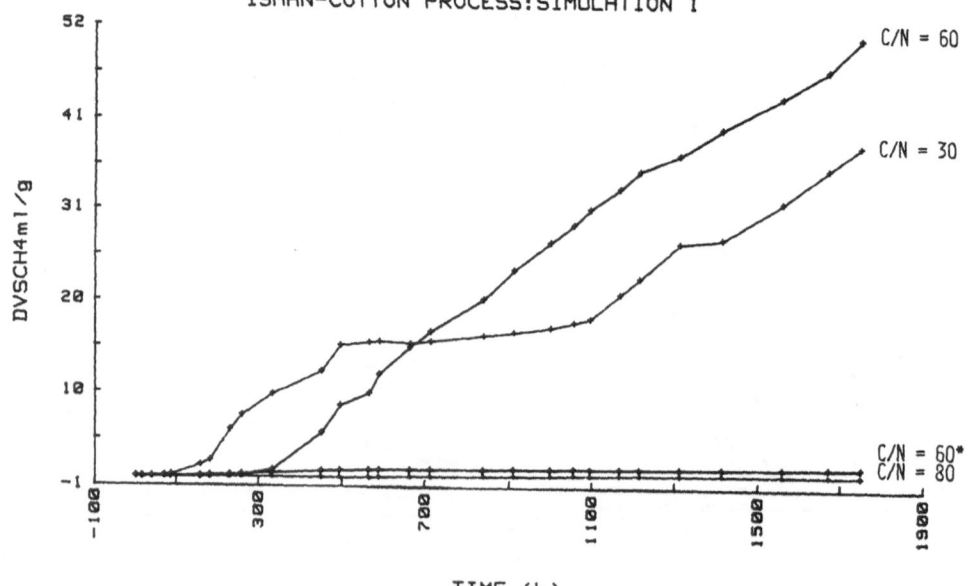

ISMAN–COTTON PROCESS:SIMULATION 1

C/N = 60* IN KILOGRAM-SCALE

FIGURE 9 : CUMMULATIVE SPECIFIC CH_4 PRODUCTION RATES
IN THE LABSCALE UNITS

FIGURE 10

COMPARISON OF THE DIFFERENT SCALES

Scale		50 ml		3 l		4.5 m^3	
Specific gas production		$\dfrac{m^3}{tDM \cdot d}$	$\dfrac{m^3}{m^3 \cdot d}$	$\dfrac{m^3}{tDM \cdot d}$	$\dfrac{m^3}{m^3 \cdot d}$	$\dfrac{m^3}{tDM \cdot d}$	$\dfrac{m^3}{m^3 \cdot d}$
C/N ratio	30	5.6	0.34	-	-	10-17	1.48
	60	8.2	0.49	0.14	9.4	9,5-12	1.06
	80	0,02	-	-	-	8-11	0.97

SOLIDE PHASE ANAEROBIC DIGESTION OF FARM WASTES MIXED WITH SEWAGE SLUDGE

Armand COTTON
Biogaz-Bureau d'études techniques
et réalisations, A. Cotton S.A.
CH - 1253 Vandoeuvres

Following Professor Isman's research work, dating back to the 40's, we created a COMPOST/METHANE GAS pilot-plant, of the BATCH/SOLID type, which we have further improved, then patented.

1. DESCRIPTION OF THE PILOT PLANT

The prototype consists of the following elements :

2 digesters (1) of 4,5 m3 each. Each digester includes 1 removable door (2) and a cover (3). The liquids pour into a pit (4) by settling. The transfer of the liquids is made by means of the pump (5) which pours them either into the pit (1), or into the liquids stock tank (6).

A fan (7) injects air under the solid substrate. The biogas is tapped in the cover on position (8). The sewage sludge or liquid manure is introduced into the pit by the pipe (9). The handling of the solid substrate is done with the help of a frontal pitchfork tractor (10).

The digesters are kept warm by auto-consumption of the produced methane gas (about 25 %).

2. STAGES UNDERGONE BY THE SUBSTRATE

The tests concentrated on various vegetables (straw), but mainly on horse manure, alone and in addition with sewage sludge. The A, B and C stages take place in the digester, the D stage out of it (compost pile).

A - Filtration

The straw acts as a filter; it holds back the organic matter in suspension within the sludge, thus enriching itself with nitrogenous material. Besides, the straw becomes impregnated and swollen with liquid.

B - Hygienization

With forced aeration, the solid substrate warms up to reach a temperature of about 70° C and even more.

C - Methane gas producing stage

The contents of the digester are immersed with the liquid from the preceeding fermentation - stocked in (6). The liquid is still warm and contains all methanogenous micro-organisms (inoculum). The temperature of the mixture reaches 35/37° C. The cover is set in place, the methane gas production begins and raises progressively.
The methane gas producing stage goes on for about 4 weeks.

D - Composting

When the production becomes too low, the liquid is drained into the stock tank (6), the solids are taken out and piled up. The substrate warms up again to reach a level of 45/50° C and cools down progressively.
After 4 weeks, a good compost is obtained.

3. ENERGY PRODUCTION

The methane gas production varies in time. It begins right after the airtight closing of the digester. It shows a peak after 1 week, then lowers slowly. Some substrates have produced during 75 days or more, but the production becomes lower. During the first 3 or 4 days, the methane gas is poor, with a CO_2 content of more than 60 %. Then the CO_2 averages 40 %. The average production with various substrates during 28 days shows :

cattle manure	1,06 m3 methane gas/day/m3 digester
horse manure	0,97 "
horse manure + pig liquid manure	1,08 "
horse manure + grass	1,19 "
horse manure + sewage sludge	1,48 "

With the introduction of sewage sludge, the production is higher because of a better-balanced C/N ratio.

Usually, a production plant includes 3 digesters or more. The total production of the digesters is thus continuous or slightly serrated.

4. TRANSFORMATION OF THE SUBSTRATE

Action of the hygienization stage (B)

This aerobic stage has two consequences :

- reduction of the percentage of dry matter (DM). With horse manure, the percentage goes from 50 to 20 % in 4 days, because it is swollen with liquid.

- reduction of the C/N ratio by oxydation of the carbon. However, 15 days are needed in order to reduce the C/N ratio of horse manure from 80 to 30. This too long a time is a waste and can be avoided by adding sewage sludge, rich in nitrogen.

Action of the methane gas producing stage (C)

Usually, the percentage of DM goes from 20 to 12 % after 4 weeks, rather constantly,.whatever the substrate may be.

The C/N ratio by the end of the process will depend on the ratio it started with. For horse manure, it will be located between 35 and 45; for horse manure enriched with sewage sludge, it will reach 20 to 25.

Action of the composting stage (D)

The final aerobic stage, after methane gas production, completes the maturation of the compost. A horse manure without adjunction of sewage sludge evolved in the following manner, after 30 days :

	before	after
C/N ratio	44	24
totN (% of DM)	1,24	2,12
P_2O_5	0,72	1,34
K_2O	3,32	4,30

5. HORSE MANURE ENRICHED WITH SEWAGE SLUDGE

Analyses, leaded after the methane gas production stage, show that the mixture is ready to be used as a mature compost.

	after the methane gas production stage
C/N ratio	22
totN (% of DM)	2,26
P_2O_5	1,43
K_2O	2,96

6. CONCLUSIONS

Horse manure, as well as other farm wastes, processed in the COMPOST/ METHANE GAS production plant can provide an excellent compost. It has to undergo 4 working stages (filtration, hygienization, methane gas production, composting) during about 2 months.

With the initial adjunction of sewage sludge, which improves the C/N ratio, the same quality is reached in 3 stages only (filtration, hygienization, methane gas production), within 1 month only. Furthermore, from the ecological point of view, it is advantageous to turn liquid sewage sludge, which drain into the soil, into a solid compost which remains on the ground.

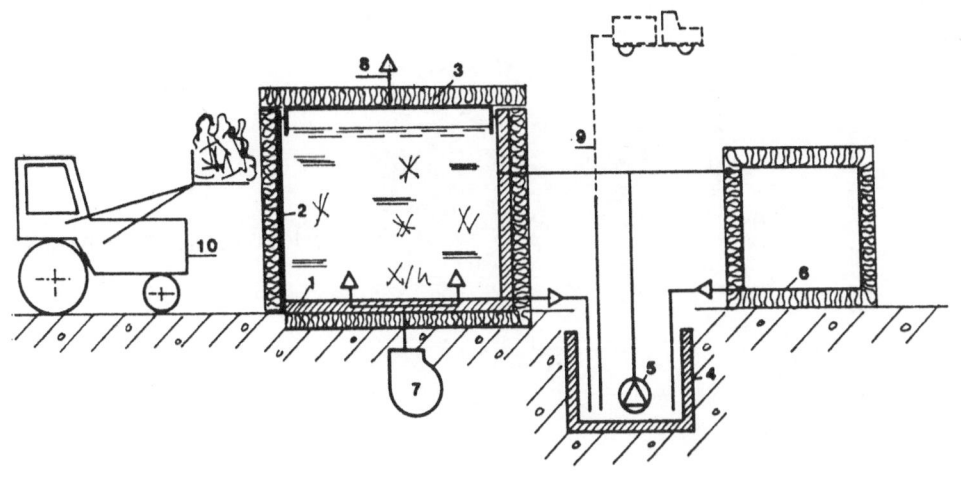

FIGURE 1 : FLOWSHEET OF ISMAN-COTTON PROCESS

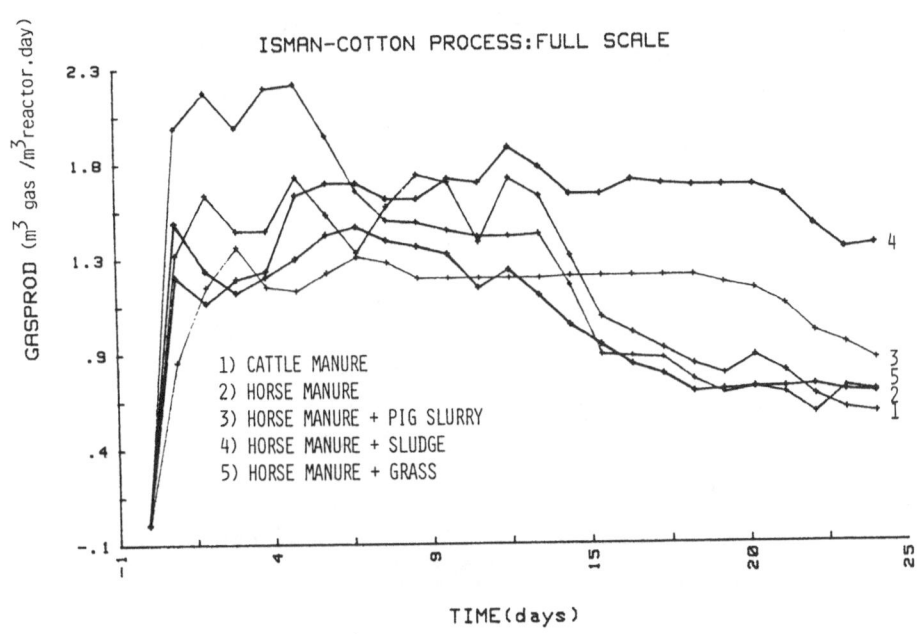

FIGURE 2 : SPECIFIC GASPRODUCTIONS IN M³ GAS/M³ REACTOR.DAY
FOR DIFFERENT SUBSTRATES

A) Loading the digester

B) Anaerobic digestion

C) End of run

Figure 3

AEROBIC STABILIZATION IN THE SOLID STATE OF PARTIALLY DEWATERED SEWAGE SLUDGE

S. COPPOLA, F. VILLANI
Istituto di Microbiologia agraria e
Stazione di Microbiologia industriale
F. ROMANO
Istituto di Meccanica agraria
Università degli Studi di Napoli
I-80055 Portici, Italy

Summary

A method of aerobic stabilization in the solid state of partially dewatered sewage sludge in the form of filter-cake pieces has been pointed out. The process includes a brief treatment (11 hours long) with pure oxygen in a closed system followed by ripening through the conventional windrow system. The resulting product is hygienically safe and richer in nitrogen in comparison with other composted sewage sludges.

Sewage sludge can be aerobically stabilized in the solid state by means of composting in mixture with other organic solid waste materials: sawdust, wood chips, domestic refuse, tree barks a.s.o. Synthetic and inert bulking agents have also been successfully assayed (Coppola et al., 1983). This poster deals with experiments of stabilization of partially dewatered raw sludge in the absence of bulking agents, as a treatment able to produce a solid and hygienically safe material suitable to agricultural utilization. With composting, stabilization is achieved as the effect of microbial activities promoted by appropriate aeration. The transformation of organic matter is typically coupled in this process to self-heating of the mass up to temperature values absolutely effective to kill all pathogenic micro-organisms. A similar treatment of partially dewatered sewage sludge in the form of filter-cake (72÷73 per cent of moisture content) does not produce sufficient temperature ascent because the law C/N ratio and the high water content of the material. Depletion of the carbonaceous energy supply and C/N ratios <10 are reported by Sikora and Sowers (1983) as responsible for the cooling of compost piles. Water content thermodinamically affects the composting process, where energy should be available both for heating and water evaporation, according to Haugh (1983). In addition, excessive moistu re prevents from aerobic conditions slowing oxygen supply to the reacting mass.

The experimental adiabatic equipment outlined in figure 1 has been utilized to define operative conditions necessary to obtain intense bio-oxidative reactions and self-heating of the sludge. The Dewar vat contained 2 Kg of pieces of filter cake resulting from press-filtration of raw sewage sludge from the urban waste water treatment plant of Torre del Greco-Villa Inglese (Naples) previously characterized (Coppola et al., 1983). Sludge moisture was 72÷73%, organic matter 60÷61%, Carbon 34÷35%, Nitrogen 4.3÷4.4 per cent.

Experiments have shown that flowing of pure oxygen through sludge quickly promotes temperature ascent up to 63°C. A similar result can be achieved by forcing into the closed system a high flow of air yielding the internal pressure to >0.025 atm, probably able to increase the oxygen solubility in the sludge water, according to the Henry law. Figure 2(A) showes that the best result is produced by an oxygen flow of 0.5 l.min^{-1}. Temperature monitoring presents a very interesting flexion around 45°C, the critical value corresponding to the upper limit for mesophilic microorgani sms. Anyway ascents can be fitted to straight-lines, whose equations are reported in table 1 for an easy comparison of their slope.

Table 1. Temperature ascent by self-heating up to the highest value, of sludge treated with different flowes of air or oxygen. Temperatures (°C) are considered as a linear function of time (h).

Air, 0.2 l.min^{-1}	°C=0.692 h + 23.591	r = 0.958
Oxygen, 0.2 l.min^{-1}	°C=2.204 h + 26.838	r = 0.950
Air, 0.5 l.min^{-1}	°C=2.099 h + 29.901	r = 0.955
Oxygen, 0.5 l.min^{-1}	°C=2.607 h + 32.154	r = 0.954

In figure 2(B) the cumulative release of CO_2 is reported as g of C.Kg^{-1} of sludge dry matter. The highest oxygen flow rate seems responsible for the highest CO_2 production, but mineralization of sludge organic matter is not correlated to temperature. On the other hand it is well established that lower temperatures allow a larger number and variety of microorganisms to be active (Finstein et al., 1982). Considering $C-CO_2$ released per liter of oxygen supplied during the treatment, the time-course of organic carbon mineralization appeared as reported in figures 3 and 4 showing two characteristic steps when temperature ascent caused the mesophiles-thermo philes succession and a regular trend following the treatment with 0.2 l min^{-1} of air, unable to increase significantly the temperature of the mass.

Therefore the process results interesting mainly as a sanitizing treatment. This objective can be achieved the more quickly the larger the reacting surface (Figure 5). Then the sludge is discharged from the closed system and ripened by conventional windrow system. Figure 6 showes the decrease of moisture content and of the water activity (a$_w$) during this last stage of the transformation. At the end the product showed 10^2 C.F.U. of

Enterobacteriaceae per gram, about 8 per cent of water content, 55 per cent of organic matter and 3.5 per cent of total nitrogen.

Experiments at pilot-plant scale and further microbiological investigations are still in progress.

REFERENCES

(1) COPPOLA, S., DUMONTET, S. and P. MARINO (1983). Composting raw sewage sludge in mixture with organic or inert bulking agents. International Conference on "Composting of solid wastes and slurries", Department of Civil Engeenering, University of Leeds (U.K.), 28-30 september.

(2) FINSTEIN, M.S., MILLER, F.C., STROM, P.F., MAC GREGOR, S.T. and K.M. PSARIANOS (1982). Composting ecosystem management for waste treatment. Biotechnology, 1, 347-352.

(3) HAUG, R.T. (1983). Thermodinamic and kinetic constraints in compost system design. In: "Biological reclamation and land utilization of urban wastes" (F.Zucconi, M.de Bertoldi and S.Coppola Eds.), Proc. of the Int.Symposium held in Naples, October 11-14.

(4) SIKORA, L.J. and M.A. SOWERS (1983). Factors affecting the composting process. International Conference on "Composting of solid wastes and slurries", Department of Civil Engeenering, University of Leeds (U.K.) 28-30 september.

AKNOWLEDGMENTS

This research was supported by Consiglio Nazionale delle Ricerche, Rome.

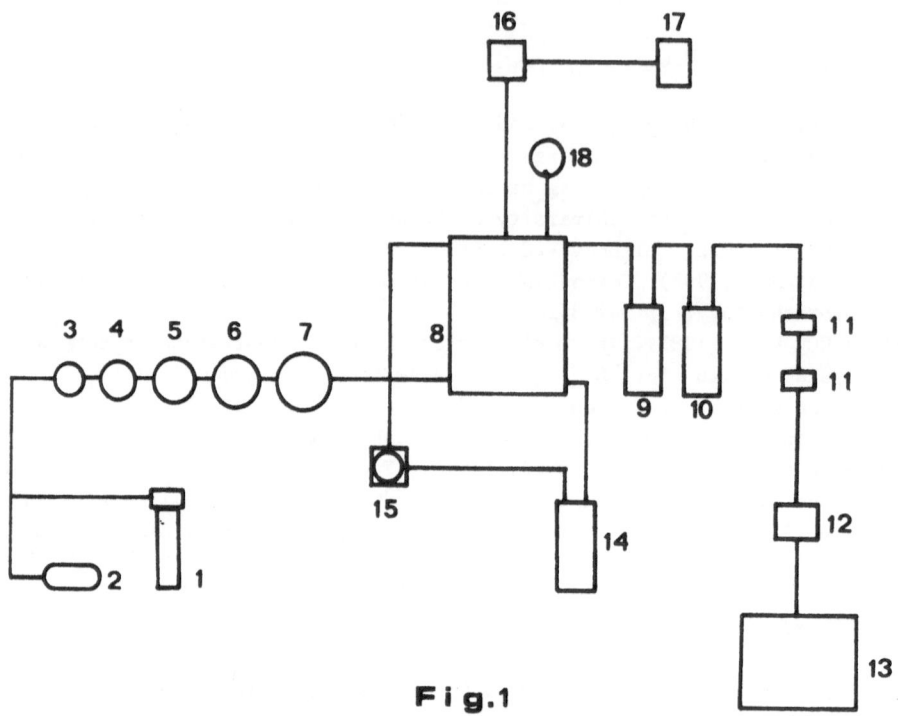

Fig.1

1: Oxygen; 2: Air compressor; 3: Filter; 4: Pressure regulator;
5: Flow regulator; 6: Flow-meter; 7: Flow-couter; 8: Dewar vat;
9: Condenser; 10: Acid trap; 11: Drier; 12: Oxygen analyzer; 13: Base
trap; 14: Condensate trap; 15: Suction-pressure pump; 16: Electronic
thermometer; 17: Temperature recorder; 18: Manometer.

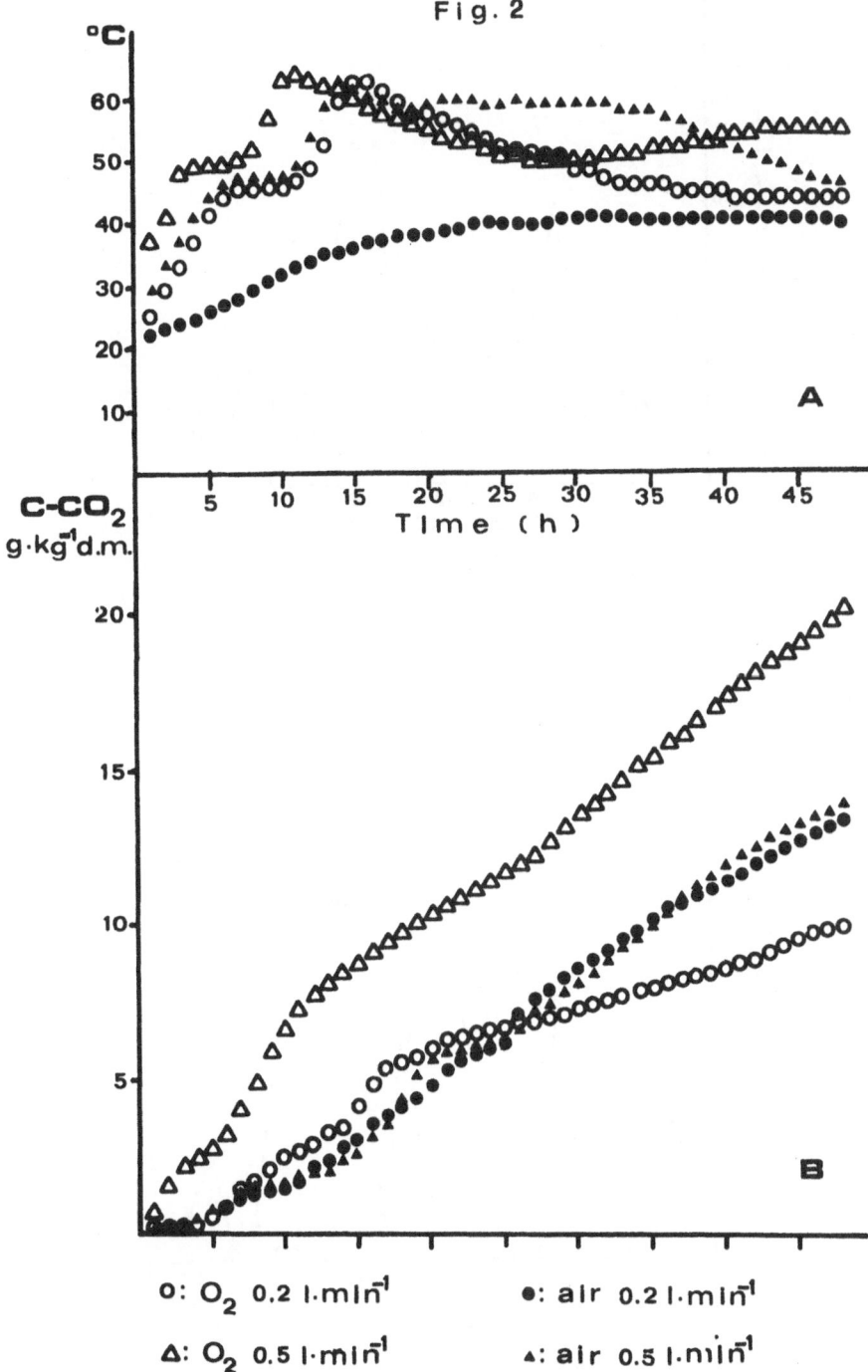

Fig. 2

o: O$_2$ 0.2 l·min^{-1} •: air 0.2 l·min^{-1}

△: O$_2$ 0.5 l·min^{-1} ▲: air 0.5 l·min^{-1}

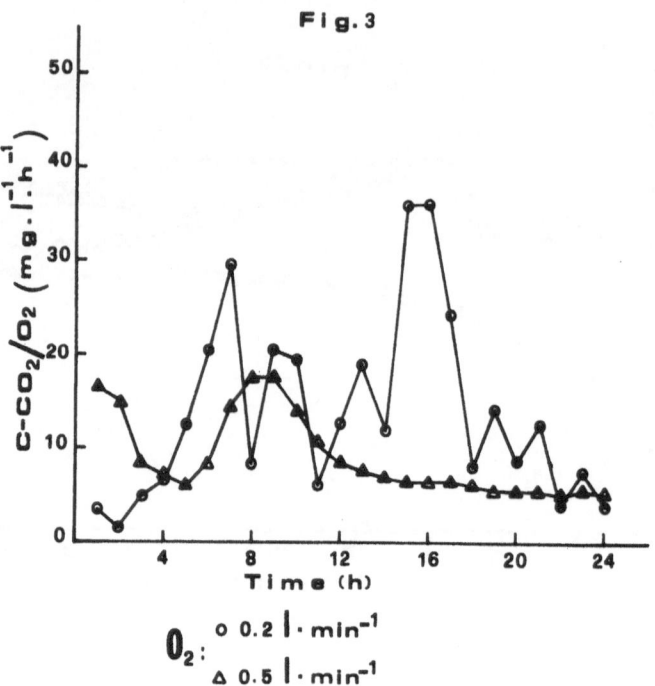

Fig. 3

$O_2:$ o 0.2 $l \cdot min^{-1}$
      Δ 0.5 $l \cdot min^{-1}$

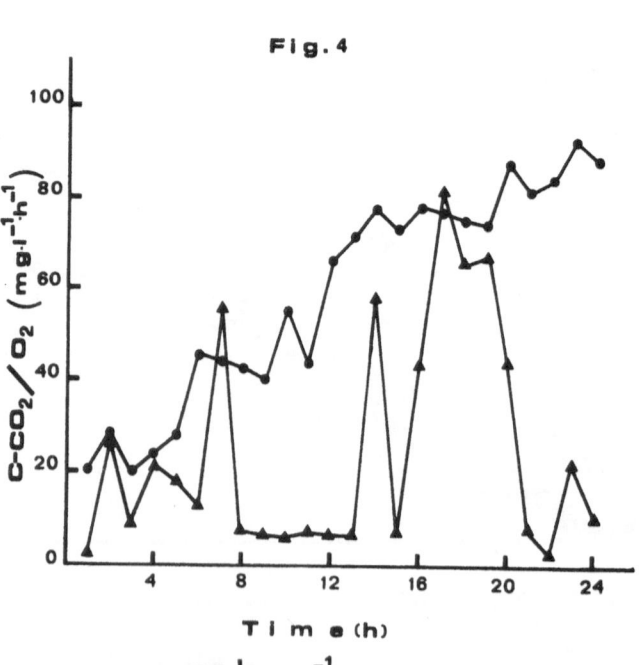

Fig. 4

Air: • 0.2 $l \cdot min^{-1}$
     ▲ 0.5 $l \cdot min^{-1}$

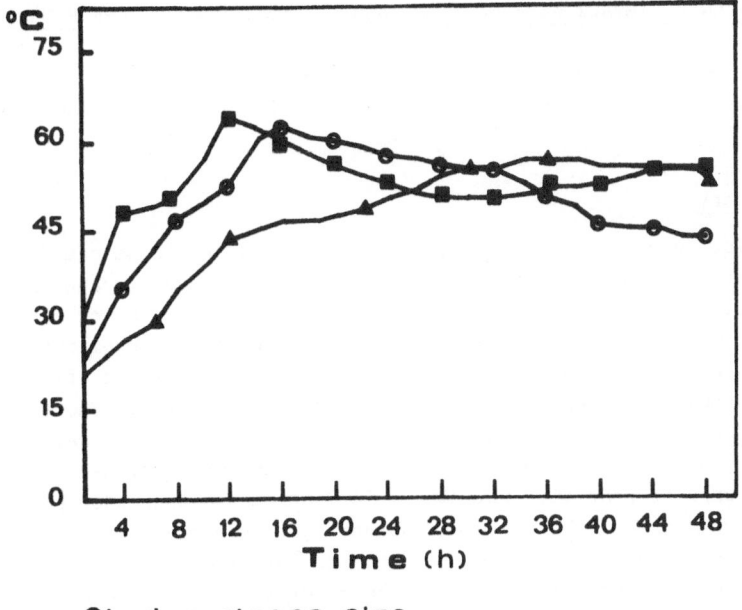

Sludge pieces size
■-∅ < 8.5 mm   o-∅ 8.5 ÷ 11 mm   ▲-∅ 45 ÷ 110 mm

Fig. 5

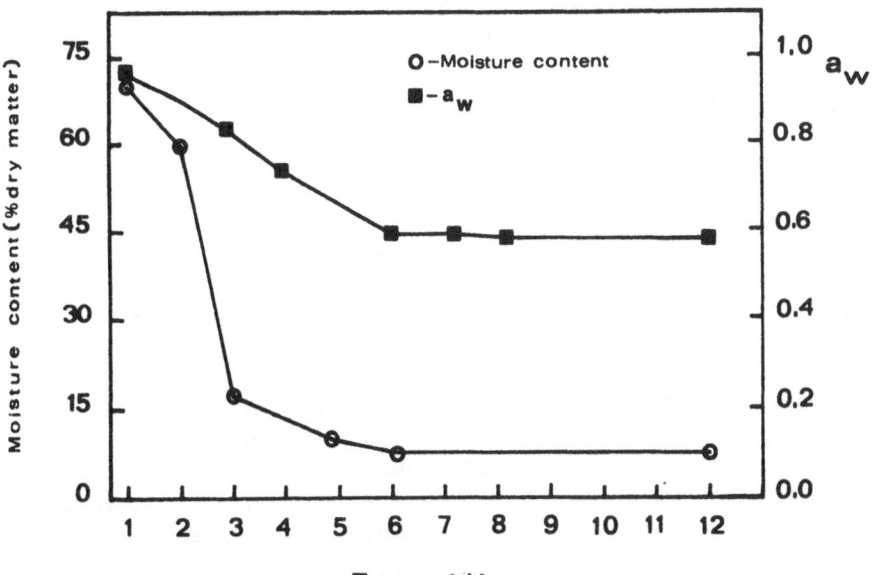

Fig. 6

ENVIRONMENTAL IMPACT ASSESSMENT OF AGRICULTURAL USE OF SEWAGE SLUDGE

E. DE FRAJA FRANGIPANE, R. VISMARA, V. MEZZANOTTE
Istituto di Ingegneria Sanitaria del Politecnico di Milano

Summary

Agricultural use of sewage sludge emerges as an economic and convenient disposal practice for many aspects (reduced technological treatment cycle, resource recycling, etc.) but can induce remarkable environmental problems, yet widely documented, involving soil and water (surface and groundwater) pollution, negative effects on crops and animals, and on human health. To evaluate the environmental impact, and thus the feasibility of such disposal system, a scheme is proposed, based on the use of matrices, which can provide general, preliminary indications on the major problems arising from the different situations, from the environmental and economic point of view, and on the relative tools to minimize the environmental impact. Such scheme will then have to be integrated by specific matrices, calibrated according to the local situation.

1. INTRODUCTION

The environmental impact study of agricultural use of sewage sludge is needed as a tool to evaluate the ecological feasibility of such disposal system.

The pollutants removed from the aqueous phase, though concentrated in sludge in a relatively small volume, still constituite a great amount of undesired by-products, to be disposed in a hygienically safe and ecologically correct way.

Even if agricultural use appears firstly as simple and convenient it can't be looked at as the only disposal solution, for both ecological-environmental reasons and for economical and perspective problems. Moreover, it is not suitable for all situations (1) and must be considered to integrate other disposal ways, such as landfilling or incineration, rather than to substitute then. Actually, fertilizing demand commonly concentrates in twice a year, thus allowing agricultural use of sewage sludge for no more than 4 months per year.

The reasons for such kind of disposal must be searched in its low cost, so as in the possibility to recycle agronomic valuable resources, due to the organic matter and fertilizing element (especially nitrogen and phosphorus) content of sewage sludge. Thus agricultural use appears not only as a disposal system, but also as a resource recycling way.

On the other hand, just being a sewage treatment by-product, sewage sludge can carry potentially hazardous elements and compounds, so as pathogenic organisms and micro-organisms reducing its agronomic value and inducing a certain number of risks, which must be evaluated both in general terms of frequency and for the single case.

2. AGRONOMIC FEATURES OF SEWAGE SLUDGE AND EFFECTS ON THE AGRARIAN ENVIRONMENT

Due to the great variety of sewage sludge to be disposed, from a planning point of view agricultural use is practically restricted to domestic sludge, though it is difficult to demonstrate its better quality with respect to the whole of industrial sludges, and to exclude that a domestic treatment plant receives also some industrial sewage. Sludge composition is greatly variable, due to the fluctuating of both sewage composition and of the plant efficiency.

Anyway, the concentrations of various elements can be indicatively grouped in frequency classes as reported in Tab. 1. Sewage sludge carries an important amount of organic matter and its fertilizing element contribution must be taken into account, especially for nitrogen and phosphorus, even if they're prevalently present in not readily available chemical forms. Potassium contribution is, on the contrary, pratically negligible.

A feature observed for nearly the whole of sewage sludges is their high metal content. These, actually, are removed from sewage with a variable but anyway considerable efficiency (Tab. 2), and thus concentrate in sludge.

Depending on their concentration and form, metals can have, in case of agricultural use of sewage sludge, consequences of various kind and importance on the environment.

Toxic effects on plants can actually appear, so as quality modifications in the plant tissues which can result, in their turn, in bio-accumulation along the food chain. Though less frequent, some risks can emerge for groundwater pollution, due to leaching, and for surface water, due to run-off.

An excess and long-lasting metal enrichment of soil, even if, in particular cases, may not show clear negative effects, will anyway result in the pollution of soil, which will thus be altered in its chemical and biological functions (1).

Sludge may also carry a certain number of pathogenic organisms or micro-organisms, and this fact limits, in practice, agricultural use to suitable sites and to particular application techniques (6, 8). Toxic and/or persistent molecules can also be present but, generally, do not cause any particular problem.

With respect to agricultural use, sludge treatments have also a great importance. A good stabilization is needed, in fact, to reduce bad odours, and further treatments do not only determine sludge physical properties, but pass through the addition of chemicals as well,

and lead anyway to chemical modifications of sludge. As to the normal
ly used conditioning chemicals, polyelectrolytes do not pose any limita
tion to sludge agricultural use, while for aluminium an immobilizing
action has been detected on phosphorus, which consequently decreases
its plants availability. In case of limed sludge, a further criteria
must be applied, consisting in the pH evaluation in the chosen soil:
a high calcium application rate affects soil pH, and, consequently,
the availability of various elements. At high pH values the solubili
ty (and thus the toxicity) of metals decreases, but the plant availa
bility of phosphorus decreases too.

A similarly negative effect on phosphorus availability can be
attributed to iron salts in acid soils, where sludge application should
anyway be avoided, with respect to the higher mobility of metals.

3. ENVIRONMENTAL IMPACT ASSESSMENT

The weighted evaluation of the differently important environmental impacts, both desired and undesired, has been done in a wide stu
dy, presently in press, of which a short synthesis is reported here,
in the usual terms of environmental impact matrices. Such matrices aim
chiefly at general evaluations about the importance of the various effects, and thus represent methodological tools to extend environmental
impact studies to specific situations. Environmental impact assessment
will then obviously need to be integrated by an economic appraisement.
In Tab. 3 the impact matrix is reported, in which the main impact sectors can be individuated (soil, crops, water, animals, public health),
so as the operative conditions, defined by a series of parameters, grou
ped to describe:
- soil characters (1);
- land use features (2);
- sludge characters (3);
- characters of the sewage treatment plant form which sludge derives
  (4);
- sludge application system (5);
- sludge application period (6).

As to the features of soil (1) and of the treatment plant (4), at
least two alternatives are possible for each parameter (e:g.: C.E.C.
6-12 meq/100 g, or C.E.C. $\geq$ 12 meq/100 g), while for the other groups
(2, 3, 5, 6) one line excludes the others.

For each impact sector different indicators have been chosen, to
which a value included between 0 and 3 has been assigned.

0 does not necessarily mean no effect, but rather effects compara
ble to those deriving from the conventional agronomic practice.

Actually, the various impact sectors are strictly related among
them, since the effects on plants, on waters, on animals and on man can
be mostly considered to be indirect, and deriving from the effects on
soil. In practice then the environmental impact assessment of sludge
agricultural use must take into account desired and undesired effects,

the first including fertilizing element and organic matter enrichment, and the latter being relative to most of the other influences of sludge on the various environmental indicators.

Column totals represent the general importance of the various impacts, meaning the relative frequency for the various indicators are affected in the whole of different situations.

Line totals, from which the values corresponding to desired effects (first two columns), have been deducted, are indices of the risks to which the various impact sectors are subjected, in connection to the single characters of the specific operative situation. The most favourable situation will thus be the one for which summing up the whole of line totals the lowest value will be obtained.

Looking at the column totals, the main, or more frequent, problem can be individuated in the metal enrichment of soil, carrying alterations in the product quality, drops in crop yield and bioaccumulation along the food chain. Relatively high values of column totals also emerge for bad odours and for alterations in soil biological activities, also connected, to a certain extent, to the metal enrichment of soil. However, effects showing a lower column total can be important too. This can easily be explained taking into account the fact that, for instance, water or animal impact will only be observed when water or animal are present, and will thus be characterized by a low column total, related to their limited frequency. A similar trend emerges with respect to public health and hygienic conditions for operators, getting greater importance with increasing direct contact between man and sludge.

Considering separately biological and chemical effects of agricultural use of sewage sludge, the first should include the diffusion of diseases (among men, in general, among the operators and among animals) and of bad odours, and the latter the enrichment of soil with substances undesired or, anyway, quantitatively exceeding the right or tolerable amount, with all the previously listed consequences some of such substances can be subjected to a more or less rapid metabolism in soil, while others (e.g. metals) are not bio-reacting. The preminent importance of soil metal inputs must be practically evaluated, not only in frequency, but also in duration terms. Actually while in case of biological pollution the consequences will last for a relative short time, in the order of hundreds of days, the effects of chemical pollution will last much longer (some decades) in the absence of any possible recovery intervention.

4. QUANTIFIABLE COSTS AND BENEFITS

With respect to other disposal ways, agricultural use of sewage sludge entails costs and benefits which are not entirely quantifiable a priori. Actually, while on the one hand the relative incidence of costs for storage, transport, treatment and control can be evaluated, on the other hand the economic results of being able to dispose at low costs a certain amount of sludge, or of other factors, strictly connected to the specific characters of the disposed sludge and to the practical situation can difficultly be foreseen.

Anyway, though sludge can have its intrinsic value, due to its organic matter and fertilizing element content, the farmer must be allowed to get it with no cost as shows the consolidated example of other european countries. Often, it can even be convenient for the plant management to take also the transport expenses upon itself.

From the plant management point of view, a comparison is needed between the costs of the technological sludge treatment cycle plus final disposal (generally landfilling) and those of the reduced technological cycle (excluding, or instance, artificial dewatering) plus agricultural land disposal.

From such comparison, within the various possible hypotheses, agricultural use of liquid sludge emerges as more convenient than landfilling, leaving other factors, such as the economic value of sludge, out of consideration. We can also observe that, when the covered transport distances are shorter than 20 km (there and back), the cost for dewatering equals that for transport, so that, in the end, liquid sludge should be preferred for agricultural use.

As to the different local and operative conditions, anyway, the various costs can change, taking into account that the application of precautionary measures to minimize environmental impact is reflected in increased costs. Tab. 4 has been complied according to the same criteria followed for the environmental impact matrix (Tab. 3). Line totals, calculated summing up costs, off the savings, allow to compare the different situations defined by the whole of corresponding items. In Tab. 5 too, the inherent alternative must be chosen among the proposed ones. The item "competition towards animal slurry (surrounding production centres)" can be either considered or simply left out, according to the case.

Column total, calculated summing up the values corresponding to the chosen items, are representative of the incidence of the various costs and benefits.

It is obviously question of purely indicative values, giving anyway a quite reliable idea of reality. Though it is evident that various local and/or engineering problems, not appraisable in general terms, can affects the distribution of costs, the scheme can anyway be proposed as an evaluation tool, with the needed adjustments related to the specific situation.

5. MEASURES TO MINIMIZE ENVIRONMENTAL IMPACT

For a correct approach to the problem of agricultural use of sewage sludge the double-dealing need emerges to evaluate previously the incidence of the various impacts in the specific situation we're referring to and, consequently, to minimize them keeping the situation controlled during the whole disposal period.

The first question can be dealt with by an accurate survey on the climatic, hygienic, agronomic and territorial features (area, slope,

hydrogeological characters, proximity to surface waters and built-up a-
reas), jointly to sludge and soil analyses and phytoxicity tests (6, 7).
Anyway, aiming at minimizing the environmental impact, we can group the
constraints to be respected as belonging to the following main categories: use prohibitions, use criteria and modality, tools for management and control.

Use prohibitions

According to what has been pointed out by a previous study carried out by the Authors (1), agricultural use of sewage sludge must be
prevented:
- on soils subject to hydrogeological constraint, and anyway at distan
  ces below 300 m from the nearest well;
- for fresh (not digested) sludge, both liquid and dewatered;
- for sludge from hospitals or nursing-homes;
- at distances below 300 m from built-up areas, farmhouses excepted;
- on vegetable crops;
- on vegetative phase crops;
- for sludge with undesired substances exceeding the fixed concentrations, and on soils not coming into the estabilished quality standards;
- for sludge coming from plants receiving loads of not domestic assimi
  lable sewage in quantities over 40% (cooling waters excepted), unless
  particular cases, needing anyway to be controlled and authorized;
- on pasture lands during animal grazing;
- on pasture lands where animals will enter before 60 days have passed
  from land application, or where mowings will be effected within the
  same period;
- on water soaked, or snow-covered, or frozen soils;
- in public used parks and gardens, if people can enter the sludge trea
  ted areas;
- on rice-fields;
- anyway and on any crop at application rates over 200 m^3/ha day, unless more restrictive limits depending on soil and sludge characters;
- for liquid sludge on soil having a slope over 5%;
- at distances below 300 m from lakes and 50 m from rivers;
- for liquid sludge on forest lands and in the protected oasis;
- in greenhouses and tunnels.

Use criteria and modality

A substantial criteria for any impact minimizing strategy consists
in determining the exact bonds of the future actions. In the case of
agricultural use of sewage sludge, it is thus needed to establish, first of all, how long such practice will be carried out, what kind of
sludge will be used, if, and to what degree, soil can bear pollution.

The following bareers will thus be posed: a quality standard for
sludge, a quality standard for soil, and a maximum lasting for the di
sposal period.

Aiming at the respect of the quality standard for soil, sludge ap plication rates will be functions of the soil characters and of the disposal period, and will anyway be inversely proportional to the undesired substance concentrations in sludge.

Sludge treatment, then, affects the importance and kind of environ mental impact, and a remarkable weight can be attributed to the modality adopted for sludge application. While a certain range of possibilities can be admitted for the kind of sludge to be used, in relation to practical and economic facilities, the modality for sludge application are generally restricted to the immediate earthing up, or, in case of liquid sludge, to the sub-surface injection.

Tools for management and control

There is evidence for the need to put the indicated countraints into operation by an efficient structure for management and control, allowed to use all the possible means to carry out its task. Without such a structure, the efficacy of both use prohibitions and use criteria and modality will drop considerably, thus undermining the whole operation. Its tasks will include the control on the disposal modality and authorization, the management of a cadastre for agricultural usable sludges and of experimental plots.

The public administration will then have to produce periodical reports on the agricultural use of sewage sludge, based on data collocted by the structure for control and management, which will thus be verified in their turn. To evaluate the efficacy of the various measures to minimize the environmental impact of agricultural use of sewage sludge a matrix has been set up (Tab. 5), according to the same criteria followed for Tab. 3 and Tab. 4, attributing the maximum value (10) to the most efficient measures and the minimum value (0) in the contrary cases. Apart from the possible alternatives about the kind of sludge to be used and of the management (public or private) the indicated measures make up a single whole of the interventions held to be the most efficient.

The sector where environmental impact can be minimized at best is that inherent public health. The highest values for line totals, corresponding to the greatest efficacy, can be observed, in the or der, for the following items: sludge composted with carbon residues at low metal content, - sludge application rate as a function of soil pH, C.E.C. and quality standard, of sludge quality standard and of the lasting disposal period, - obligatoriness of periodical reports, affecting a greater number of environmental impact sectors. Anyway, the importance of measures characterized by a lower line total must not be undervalued, because the impact minimization may be determinant even if concerning only one sector. Such efficacy evaluation must anyway take into account the incidence of the various impacts in the specific considered situation, and must be integrated by and compared to the costs that each measures entails.

The weight values assigned in the matrix have an essentially general meaning and admit adjustments according to the pratical reality.

From a merely economic point of view, we must remind that all use prohibitions result, at first, in increasing costs for storage, transport
and treatment. Such increase can hardly be quantified a priori, but in
case agricultural use would emerge as not suitable, the whole of prohibitions would induce a remarkable rise of the disposal costs, due to
the need for alternative sludge destinations (landfilling, incineration,
etc.) not considered in the present work.

LITERATURE
(1) - P.L. Genevini, R. Vismara, V. Mezzanotte (1983): Utilizzo agricolo dei fanghi di depurazione. Ingegneria Ambientale, 12, 9, 1-133
(2) - A. Garbarino, P.L. Genevini (1985): Ipotesi di un metodo chimico-
-biologico per la valutazione del possibile impiego agricolo dei
fanghi di depurazione. Atti del Convegno: Trattamento e Smaltimen
to delle Acque Reflue e dei Fanghi; Nuovi Aspetti Tecnologici e
Normativi. Milano, 30-31 gennaio 1985. In Stampa
(3) - V. Mezzanotte, R. Vismara (1984): Problemi igienici legati allo
smaltimento di fanghi in agricoltura. Ingegneria Sanitaria, 32,
2, 24-31
(4) - EPA (1977): Trace metal removal by wastewater treatment. Techno
logy Transfer , gennaio 1977
(5) - Commissione delle Comunità Europee (1982): Proposta di direttiva
del Consiglio concernente l'utilizzazione in agricoltura dei fanghi residuati dai processi di depurazione. Gazzetta Ufficiale del
le Comunità Europee 264/3 - 264/7

Tab. 1 - Composition of digested domestic sludge (data on dry weight ba sis) (1).

Parameter		Minimum	Maximum	More frequent concentration classes
Org. C.		20	44.4	20÷30 (67%); 30 (33%)
Org. matter	(%)	35	73	40÷60 (54%); 60÷70 (30%)
Tot. N	(%)	0.22	21	3÷4 (27%); 2÷3 (22%)
Org. N	(%)	2.59	4.81	3÷4 (44%)
NH$_4$-N	(%)	0.012	4.3	0.01÷1 (54%)
Tot. P	(%)	traces	15	0÷1 (38%); 1÷2 (30%); 2÷3 (25%)
K	(%)	traces	70	0÷2 (57%); 2÷4 (19%); 4 (23%)
Ca	(%)	0.8	27	0÷2 (20%); 2÷4 (31%); 4-6 (22%)
Mg	(‰)	0.11	16	0÷5 (51%); 5÷10 (34%)
Na	(‰)	0.5	13.9	0÷5 (75%); 5÷10 (20%)
Al	(%)	0.2	16	0÷1 (64%)
Fe	(%)	0.34	18	0÷10 (40%); 10÷20 (38%)
As	(ppm)	1	1,800	
B	(ppm)	0.2	1,000	0÷50 (35%); 50÷100 (25%)
Cd	(ppm)	traces	3,000	0÷10 (41%); 10÷20 (18%); 20÷30 (10%)
Cr	(ppm)	traces	40,615	0÷100 (32%); 100÷200 (11%); >1,000 (25%)
Mn	(ppm)	20	4,021	0÷200 (23%); 200÷400 (23%); >1,000 (25%)
Mo	(ppm)	0	1,000	
Ni	(ppm)	traces	14,150	0÷50 (45%); 50÷100 (16%)
Pb	(ppm)	10	26,000	100÷200 (16%); 0÷100 (15%); >500 (40%)
Zn	(ppm)	58	50,000	0÷1,000 (27%); 1,000÷2,000 (22%); 2,000÷3,000 (24%)
Cu	(ppm)	16	22,682	0÷200 (22%); 200÷400 (19%)
Co	(ppm)	0.6	1,565	0÷10 (71%)
Hg	(ppm)	0,1	61	0÷5 (41%); 5÷10 (32%)

Tab. 2 - Metal removal by activated sludge treatment (4).

Metals	Concentration in raw sewage (mg/l) Range	Mean values	Removal after activated sludge treatment (%)
Cd	0.008 - 0.142	0.02	20 - 45
Cr	0.020 - 0.700	0.05	40 - 80
Cu	0.020 - 3.360	0.10	0 - 70
Hg	0.0002 - 0.044	0.0013	20 - 75
Ni	0.0020 - 8.80	0.10	15 - 40
Pb	0.050 - 1.27	0.20	50 - 90
Zn	0.030 - 8.31	0.18	35 - 80

Tab. 3 - Environmental Impact Matrix for agricultural use of sewage sludge. A value included between 0 and 3 has been assigned to the various indicators, according to the impact importance with respect to the operative conditions, defined, in their turn, by various items chosen among the line listed ones.

| | | IMPACT SECTORS | | | | | | | | | | | | | | | |
| | | Effects on soil | | | | Effects on crops | | Effects on waters | | Effects on animals | | Effects on human health | | | | | |
		Organic matter enrichment	Fertilizing element enrichment	Metal enrichment	Alterations in soil biologic activities	Drop in crop productivity	Alterations in crop quality	Groundwater pollution	Surface water pollution	Uptake of toxic elements or compounds	Diffusion of infectious diseases	Metal bio-accumulation	Organic pollutant bio-accumulation	Diffusion of infectious diseases	Bad odours	Negative effects on the operator hygiene	T O T A L
Soil characters (1)	C.E.C. 6-12 meq/100	2	2	2	2	3	3	2	0	3	0	2	0	0	1	0	14
	C.E.C. > 12 meq/100	2	2	2	1	1	2	1	0	2	0	1	0	0	1	0	7
	pH < 7	2	2	2	3	3	3	2	0	3	0	2	0	0	1	0	15
	pH > 7	2	2	2	1	1	2	1	0	2	0	1	0	0	1	0	7
	Slope < 5%	2	2	2	1	2	2	1	0	0	0	0	0	0	1	0	5
	Slope > 5%	2	2	2	1	2	2	1	2	0	0	0	0	0	1	0	7
	Distance from built-up areas < 300 m	2	2	2	1	2	2	1	0	0	0	0	0	0	3	0	7
	Distance from built-up areas > 300 m	2	2	2	1	2	2	1	0	0	0	0	0	0	1	0	5
	Distance from wells < 300 m	2	2	2	1	2	2	2	0	0	0	1	0	0	1	0	7
	Distance from wells > 300 m	2	2	2	1	2	2	1	0	0	0	0	0	0	1	0	5
	Distance from rivers < 50 m and from lakes < 300 m	2	2	2	1	2	2	1	3	0	0	0	0	0	1	0	8
	Distance from rivers > 50 m and from lakes > 300 m	2	2	2	1	2	2	1	0	0	0	0	0	0	1	0	5
Land use features (2)	Public parks and gardens	2	2	2	1	0	0	1	0	0	0	0	0	3	3	0	6
	Horticulture	2	2	2	1	2	3	1	0	0	0	3	1	3	1	2	15
	Pasture lands	2	2	2	1	2	3	1	0	3	3	3	1	2	1	1	19
	Grain cereal culture	2	2	2	1	1	2	1	0	0	0	0	0	0	1	1	4
	Forage culture	2	2	2	1	1	2	1	0	2	1	2	0	0	1	0	9
	Floriculture	2	2	2	1	1	0	1	0	0	0	0	0	0	3	2	6
	Protected cultures (green houses and tunnels)	2	2	2	1	2	3	1	0	0	0	3	0	3	3	3	7
	Forest	2	2	2	1	1	0	1	0	0	0	0	0	0	1	0	2
	Rice-field	2	2	2	1	2	3	2	1	0	0	2	0	0	2	1	12
Sludge characters (3)	Liquid sludge	2	2	2	1	2	2	1	1	0	1	2	0	1	2	0	11
	Dewatered sludge	2	1	3	2	2	3	1	0	0	1	2	0	1	1	0	13
	Dried sludge	2	1	2	1	2	2	1	0	0	0	2	0	1	0	0	8
	Hygienized sludge	2	1	2	1	2	2	1	0	0	0	2	0	0	2	0	9
	Composted sludge	2	1	2	1	2	2	1	0	0	0	1	0	0	0	0	6
Plant characters (4)	Served population: < 5,000 inhabitants	2	2	2	1	1	1	1	0	0	0	1	0	0	1	0	4
	Served population: 5,000-20,000 inhabitants	2	2	2	1	2	2	1	0	1	0	1	0	0	1	0	7
	Served population: > 20,000 inhabitants	2	2	3	2	3	3	1	0	2	0	2	1	0	1	0	14
	Industrial wastewaters < 15%	2	2	2	1	2	2	1	0	1	0	1	0	0	1	0	7
	Industrial wastewaters	2	2	3	2	3	3	1	0	1	0	2	1	0	1	0	14
Applic. system (5)	Surface application	2	1	2	1	2	2	1	1	3	1	2	1	1	3	1	18
	Sub-surface injection or immediate burial	2	2	2	1	2	2	1	0	0	0	2	0	0	1	0	7
Applic. period (6)	Before seeding or inter-rows (weeded cultures)	2	2	2	1	2	2	1	0	0	0	2	0	0	1	0	7
	During vegetative phase	2	2	2	1	3	3	1	0	3	2	2	1	1	1	0	16
T O T A L		70	65	73	41	66	74	39	8	26	9	44	5	16	46	10	=

Tab. 4 - Comparison between costs and benefits of the agricultural use of sewage sludge. On line, a value from 0 to 10 has been assigned to each item, referred to the various costs and benefits, except for treatment costs, for which a wider range has been used, from 0 to 30, because of their high variability. The maximum value (10 or 30) corresponds to the maximum cost or saving. On each line, for costs non-dependent from the operative condition the minimum value in the whole column has been attributed. Totals have been obtained summing up the costs, off the benefits.

			COSTS						BENEFITS			TOTAL
			Transport costs	Distribution costs	Treatment costs	Burocratic costs	Monitoring costs	Storage costs	Saving due to the possibility to dispose a greater amount of sludge at law cost	Saving on fertilizing units	Saving on organic matter	
LOCAL AND OPERATIVE CONDITIONS	Features of the plant area	Available agricultural land at distances > 15 km	8	0	0	2	4	7	4	3	4	10
		Needed extension for the whole produced sludge available within 15 km	4	0	0	2	4	4	8	7	8	-8
		Needed extension for more than twice the produced sludge within 15 km	4	0	0	2	4	2	10	3	4	-5
		Competition towards animal slurry (surrounding production centres)	9	0	0	2	4	8	4	0	0	19
	Kind of sludge	Liquid sludge	10	0	0	2	4	9	10	7	8	0
		Dewatered sludge	8	4	4	2	4	6	6	3	8	11
		Dried sludge	4	0	30	2	4	4	8	2	8	26
		Composted sludge	6	4	15	2	4	4	5	4	8	15
		Hygienized sludge	10	0	15	2	2	9	10	6	8	14
	Features of the plant and of the treated sewage	Served population: < 5,000 inhabitants	8	0	0	2	4	5	10	7	8	-4
		Served population: 5,000÷20,000 inhabitants	7	0	0	2	5	6	9	6	7	-2
		Served population: > 20,000 inhabitants	7	4	8	2	6	6	5	3	4	21
		Industrial wastewaters < 15%	4	0	0	2	4	2	10	7	8	-13
		Industrial wastewaters > 15%	5	0	0	4	6	2	5	3	4	5
	Applic. system	Surface application	4	0	0	2	10	2	10	5	8	-5
		Sludge earthing-up or injection	4	5	0	2	4	2	10	7	8	-8

Tab. 5 – Matrix to evaluate the efficacy of the different possible measures to minimize environmental impact in the different sectors. Column partial totals have been calculated only for use prohibitions; for the other groups of measures the totals must be calculated case by case according to the chosen alternative. As to use criteria and modality, the possible alternatives refer to the kind of sludge to be used, while the other measures must be adopted jointly; for management and control too, once the kind of management (public or private) has been chosen, the adoption of the whole of other measures is recommended.

MEASURES TO MINIMIZE ENVIRONMENTAL IMPACT	Effects on soils		Effects on crops		Effects on waters		Effects on animals		Effects on human health					TOTAL
	Metal enrichment	Alterations in soil biologic activities	Drop in crop productivity	Alterations in crop quality	Groundwater pollution	Surface water pollution	Diffusion of infectious diseases	Uptake of toxic elements or compounds	Metal bio-accumulation	Organic pollutant bio-accumulation Cumulation	Diffusion of infectious diseases	Bad odours	Negative effects on the operator hygienic conditions	
Use Prohibitions														
On soils subject to hydrogeological constraint	0	0	0	0	10	0	0	0	2	1	1	0	0	14
At distance < 300 m from the nearest well	0	0	0	0	10	0	0	0	2	1	1	0	0	14
For not digested sludge, both liquid and dewatered	0	8	5	0	3	0	4	3	1	1	3	8	4	40
For sludge from hospitals or nursing-homes	0	2	0	0	0	0	5	2	0	0	8	0	8	25
On vegetable crops	0	0	7	10	0	0	0	0	10	1	10	0	8	46
On vegetative phase crops	0	0	4	4	0	0	4	2	3	1	8	0	5	31
At distance <300 m from built-up areas	0	0	0	0	0	0	0	0	0	0	0	5	0	5
On pasture lands where animals will enter before 60 dd from sludge application	0	0	0	0	0	0	8	7	2	2	3	0	3	25
On water soaked, or snow-covered, or frozen soils	0	2	1	0	4	6	0	0	1	0	0	2	0	15
On public parks and gardens	0	0	0	0	0	0	1	0	0	0	7	8	3	19
At distance < 300 m from lakes and < 50 m from rivers	0	0	0	0	0	10	0	0	0	0	0	0	0	10
In application rates over 200 m^3/ha day (unless more restrictive limits)	1.	8	3	2	1	7	0	2	2	2	2	4	3	34
For liquid sludge on soil having a slope > 5%	0	0	0	0	0	5	0	0	0	0	0	0	0	5
On rice-field	0	2	0	0	4	6	0	0	1	0	0	2	0	15
In greenhouses and tunnels	0	0	0	5	0	0	0	0	0	0	5	0	10	20
PARTIAL TOTAL	8	21	21	26	38	32	28	19	26	9	49	29	44	==
Use Criteria and Modality														
Established disposal period	5	3	3	5	7	0	2	5	7	1	1	0	1	40
Liquid sludge	0	0	0	0	0	0	0	0	0	0	0	0	0	0
Dried sludge	0	0	0	0	3	4	10	0	0	0	10	9	4	40
Sludge composted with low metal content residues	8	6	5	6	5	5	9	3	5	0	9	9	4	74
Sludge composted with municipal solid wastes	0	4	3	4	3	3	9	2	4	0	9	9	4	54
Hygienized sludge	0	0	0	0	0	0	10	0	0	0	10	0	10	30
Sludge application rate as a function of soil pH, C.E.C. and quality standard, of sludge quality standard and of the duration of disposal period	10	10	10	10	10	3	1	2	10	2	0	0	0	68
Earthing-up or sub-surface injection of sludge	0	0	0	0	0	10	8	7	0	2	8	6	3	42
Sludge quality standard	7	4	6	7	4	5	0	7	7	1	0	0	0	48
Soil quality standard	9	5	8	9	7	1	0	6	10	0	0	0	0	55
Management and Control Tools														
Centralized systematic control on the disposal modality	8	4	2	2	5	5	6	3	8	2	6	2	3	56
Need for obligatory public authorizations	8	3	2	4	3	3	4	2	8	2	4	1	1	45
Cadastre of agricultural usable sludges	4	2	1	2	1	2	2	2	4	1	2	1	1	25
Control and management of experimental plats	5	2	7	7	7	1	1	3	5	2	1	1	0	42
Compulsory periodical reports by the Public Administrat.	8	4	5	7	8	5	4	4	8	2	4	2	1	62
Public management of the disposal practice	8	4	2	2	4	2	6	2	8	1	6	2	3	50
Private management of the disposal practice	4	2	1	1	1	1	3	1	4	1	3	2	2	26

ANALYTICAL CHARACTERIZATION OF BIOLOGICAL SEWAGE SLUDGES

EXPECIALLY ABOUT SOME ORGANIC COMPOUNDS

P.L. Genevini (*) and M.C. Negri (**)

(*) Istituto di Chimica Agraria, Università di Milano
(**) Ecodeco s.p.a., Giussago - Pavia

Summary

On a lot of 57 biological sewage sludges coming from civil and industrial
plants, the following parameters have been researched: total organic carbon, animal and vegetal fats and oils, mineral oils, surfactants, chlorinated solvents, aromatic solvents, chlorinated pesticides, phenols, aldehydes. All the alalysed sludges had previously passed the biological assay
and had a eavy metal content below the prescripted limits (EEC proposal).
At the doses prescribed in Lombardy Region, all the detected compounds sup
ply to soils seems roughly negligible.

1. INTRODUCTION

Since a long time most authors have found in mineral fraction of sewage
sludges and particularly in heavy metals, the most affecting factor of
land disposal of sewage sludges.
Former studies of the authors pointed out the not negligible importance of
the composition of the organic fraction: as a matter of fact these one can
give, like heavy metals, phitotoxcity problems or more general problems
connected with an agricultural recovery of such by-products (1).
Problems involved in heavy metals differ from those related with organic
compounds not only for a different behaviour concerning soil and crops me-
thabolism, but also for well known analitycal difficulties. These difficul
ties are underlined by the poor number of studies concerning organic compounds compared with and extremely large number of works on heavy metals
vehiculated by sewage sludges.
Also, most laws give quality limits in land disposal of sewage sludges re
lated only to heavy metal concentration. Rules adopted in Lombardy Region
differ from other countries' adding to heavy metals determination the research of some organic compounds and a more general evaluation of the sludge with a biological assay on crops.
On a lot of 57 sewage sludges coming from civil and industrial plants, the
following parameters have been researched: total organic carbon, animal
and vegetal fats and oils, mineral oils, surfactants (MBAS), chlorinated
and aromatic solvents, chlorinated pesticides, phenols and aldehydes. All
the analysed sludges had previously passed the biological assay and had a
heavy metal content below the prescripted limits (EEC proposal).

2. MATERIAL and METHODS

Parameters have been detected following the most usual laboratory methods going from usual analytical techniques to the use of mass spectrometry.

3. RESULTS

3.1. Total Organic Carbon (% d.w.)

average	median	s.d.	min.	max.
26	27	12	4	56

A certain variability is shown, owing to the different origin and treatment of tested materials. However average and median values are very close and agree with literature data (2). It is noticeable that these values are comparable at a rough estimate to those of a typical FYM and support the opportunity of an agricultural recovery of sewage sludges.

3.2. Animal and Vegetal Fats and Oils (% d.w.)

average	median	s.d.	min.	max.
5.3	2.1	9.4	tr	58.5

The values are largely variable. A rilevant difference is revealed between average and median values.
Toxicity of those compounds has already been studied by the authors (3): temporary problems were evidenced caused by the supply of 100 q/ha of triglycerides. Those problems, mostly due to a nitrogen immobilization with consequent yield decrease were easily bypassed by an adequate balancing of C/N ratio. In the same study authors found that the most outstanding fatty acids in sludges were: oleic, lauric and palmitic.

3.3. Mineral Oils (% d.w.)

average	median	s.d.	min.	max.
2.3	2.2	3.3	0.02	21.5

Mineral oils presence in sewage sludges shows a certain coincidence of average and median values, settled on 2 %.
Considering the different origin of samples and the normal appearance of the values distribution, it is likely that the main source of mineral oils in such sludges can be ascribed to the use of lubricants in plant management.

3.4. Surfactants (MBAS) (ppm d.w.)

average	median	s.d.	min.	max.
219	19	1098	tr	8337

Literature shows that some depressing activity concerning crops and soil

biological activities is ascribed to surfactants compounds, expecially a-
nionic (4, 5, 6). Critical values on crops is estimated in 1,000 ppm and
on soils in 30 q/ha. In the worst case of the highest registered value,
calculating a supply to the soil of 3 t/ha.year of dried sludge, the critical values are however very far to be reached.

3.5. Chlorinated and Aromatic Solvents, Chlorinated Pesticides (ppm f.w.)

	average	median	s.d.	min.	max.
Chlor. Solv.	0.12	0.10	0.05	0.08	0.35
Arom. Solv.	11.70	2.00	28.20	2.00	160.00
Chlor. Pest.	.... below 0.002 ....

Data show a general preminence of aromatic solvents on chlorinated solvents. However the concentration values are unassuming, even if considered
on a dry matter basis.
Chlorinated pesticides, researched with a global screening, always showed
values below 2 ppb. In few cases chlorinated compounds peaks were detected
but these were unlikely to be related with pesticides compounds; further
informations will income after a deeper outstanding study of the processes
from whom the involved sludges were originated.

3.6. Phenols (ppm d.w.)

average	median	s.d.	min.	max.
214	78	357	tr	1908

A consistent difference between average and median values is evidenced.
Fifty percent of samples shows a phenol concentration below 78 ppm, that
would involve a supply of about 300 g/ha of phenols to soil.
This amount wouldn't seem to cause any environmental problem: no problems
would arise either in the use of the sludge containing the maximum detected amount (about 6 kg/ha of phenols with the prescripted 3 t/ha.year of
dried sludge (7)).

3.7. Aldehydes (ppm d.w.)

average	median	s.d.	min.	max.
761	203	1460	tr	6272

The rilevant variability in obtained data can be ascribed to the production of aldehydes in natural methabolic processes, mostly depending on the
sludge stabilization operations.
For this, and because of the biodegradability of such compounds, authors
consider this parameter not necessary to be detected in a sewage sludge,
even if required by the Regional rules.

4. CONCLUSIONS

This survey, based on a sufficiently representative number of samples, shows a very low presence of the detected organic compounds, with the exception of animal and vegetal fats and oils and mineral oils.
On this subject, it is important to specify thet environmental problems would arise only related to mineral oils; however literature references are scarce and contradictory.
At the doses prescripted in Lombardy Region, other compounds supply to soils seems roughly negligible.
It is important to underline that suche values concern only sludges that have previously passed a standadized biological assay.
Being this correspondence more proved it would be possible to bypass the expensive and difficult research of the organic compounds detected in this survey, giving further attention to the more exhaustive biological assay which globally can verify the impact of sludge on agricultural environment

REFERENCES

(1)   P.L. Genevini, M.C. Negri, A. Garbarino (1985): Prime risultanze di un'indagine in Lombardia sulla disponibilità e caratterizzazione analitica di fanghi e reflui potenzialmente utilizzabili in agricoltura. Proc. "Trattamento e Smaltimento delle Acque Reflue e dei Fanghi" Milano, 30-31 Gennaio. in press

(2)   P.L. Genevini, R. Vismara, V. Mezzanotte (1983): Utilizzo agricolo di fanghi da depurazione. Ing. Ambientale, 12,9

(3)   P. Zaccheo, P.L. Genevini, C. Cassinis (1984): Studio della tossicità della frazione lipidica in reflui industriali. Proc. XIII Int. Simp. di AGROCHIMICA, Pisa, 8-11 Ottobre. in press

(4)   A. Luzzati (1981): The effect of detergents on some plant species. II Laboratory and field tests on oats, red clover, alfalfa and peas. In "Decomposition of toxic and non toxic organic compounds in soils" Ann Arbor Sci. Ed., 379-387

(5)   A. Luzzati (1981): The effects of surfactants on plants. III Field experiments with potatoes. Idibem, 389-394

(6)   M.V. Vandoni, L.G. Federico (1981): Synthetic detergents behaviour in agricultural soils. Idibem, 395-406

(7)   L.G. Dolgova, V.N. Kuchma (1981): The ability of the soil to decompose phenol under industrial pollution conditions. Ididem, 169-174

SLUDGE COMPOSTING AND UTILIZATION IN WEST SWITZERLAND

Interdisciplinary research conducted by the "Institut de Génie Rural", E.P.F.L. (Prof. P. REGAMEY as initiator and G. HUBERT as coordinator)

Research partners:
- Ecole Polytechnique Fédérale de Lausanne (E.P.F.L.), Lausanne (Instituts de Génie Rural et de Génie de l' Environnement)
- Station Fédérale de Recherches Agronomiques de Changins (RAC) Nyon
- Service Intercommunal SIEG, Vevey-Montreux.

1. INTRODUCTION
     Studies on the treatment, agricultural use and disposal of sewage sludge have been in progress in French-speaking Switzerland since 1980. Up to now they have been carried out in the catchment area of 113 water treatment plants, mostly in the Canton of Vaud, treating effluent from a population of 275 000. The studies are divided into four parts:
a) An agronomic part determines direct agricultural markets for liquid and dewatered sludges according to composition, environmental constraints and agricultural demand. This is assessed after an indepth information campaign and a survey of interested parties. The quantities of sludge which could be usefully used on each farm are calculated on a basis of the phosphate manure requirement.
b) A "technology" and "transportation" part proposes procedures and equipment best adapted to deal with the particular situations of developing direct agricultural outlets, of transforming the sludge into compost, or of sludge disposal (incineration, landfill).
c) An "economic" part is concerned with the costs of sludge treatment and delivery. It also takes into account possible revenue from the sale of compost.
d) A final report synthesizes these three parts and makes precise recommandations.
Sludge composting with a carbon-based bulking agent has often appeared as a possible future solution for recycling sludge which cannot be directly used in agriculture. From the technical point of view, several firms could provide reliable composting facilities; however, this process is only attractive economically if the compost produced meets the precise needs of the agricultural market and can be sold.
     It is thus essential that the advantages of the product be demonstrated before a decision is made to build the facilities.
     In 1984-85, intersiciplinary research was undertaken, bringing together the various french-speaking swiss organizations listed above.
     In paralled with a bibliographical study (G.HUBERT and E. RUEGG), summing up current knowledge of risks in the use of sludge compost and of its performances in practice, compost was prepared at the SIEG to serve as the base material for different trials: the use of compost as an organic soil conditioner to reduce erosion (J.P. VIANI and P. BARIL) and as a substrate in floriculture (J.P. RYSER and D. PIVOT).

2. COMPOST PREPARATION

G. HUBERT, Institut de Génie Rural (Hydrologie et aménagements) E.P.F.L.

The manufacture of compost consisting of digested sewage sludge and a carbon-based bulking agent (sawdust or crushed conifer bark) was carried out at the SIEG, Vevey-Montreux, sewage sludge treatment plant at Roche. The composting took place in two 24m^3 cells with aeration from below, according to the method recommended by M.S. FINSTEIN (temperature control by forced ventilation to ensure that 55-60°C, the range of maximum microbial activity, is not exceeded).

It was only possible to control the process manually. Continuous or intermittent forced aeration could be brought about by blowing or by suction. Preparation of the bulking agents and mixtures, control of the operating parameters (temperature and aeration) and turning over of the compost were carried out by the SIEG, (D. KRATZER and M. CHAMMARTIN). The total composting period was six months, the compost being turned over five times during this period to ensure complete disinfection.

During the composting period, samples were taken:
- for routine analysis: pH, dry matter, organic matter by loss on ignition, total and water soluble nitrogen
- for specific analysis: measurements of adenosine triphosphate (ATP) and biological activity (CO_2 given off) by J.P. DUBOIS; of specific respiratory activity (O_2 consumption), polysaccharides and volatile fatty acids by P. AMMANN; of organic micropollutants (polychlorinated biphenils by P. DIERCXSENS and 4-nonylphenol by Miss M. TROCME).
- for growth tests by S. PACHE.

The initial sludge /sawdust and sludge/ bark mixtures had the following characteristics: moisture 62.8 and 60.2 (% fresh weight) respectively; wet bulk weight 0.73 and 0.72 t/m^3; volume of free air space 32.5 and 32.9 %; thermodynamic balance (weight of water/weight of biodegradable organic matter) 12.3 and 10.5.

In both cases the overall percentage degradation of organic matter in the mixture was 19 %. If the sludge is assigned a degradation coefficient of 40 %, then the rates of degradation of the sawdust and the bark were 11.3 and 15.5 % respectively. Water losses were 40 and 25 % respectively and total nitrogen losses 18 and 0 %.

The following points emerge from examination of the temperature curves:
- It is difficult not to exceed 55-60°C by switching between intermittent and continuous forced ventilation without automatic temperature control of the aeration system.
- There is a temperature gradient in the direction of aeration, the coldest area being on the air inlet side (at the bottom of the cell in the case of blowing, at the top where suction is used). Alternating between blowing and suction does not appear to be sufficient to create even temperature distribution throughout the cells. To achieve this, and hence complete disinfection, turning over the compost is essential.

3. CHANGES WITH TIME OF CHEMICAL AND BIOLOGICAL PARAMETERS

J.P. DUBOIS, Institut de Génie Rural (Pédologie) E.P.F.L.

Changes observed in the values of chemical parameters (pH, watersoluble nitrogen) and biological ones (CO_2 given off, ATP) show that there are three phases in the composting process.
- The first, lasting a week, is characterized by the rapid multiplication of microorganisms resulting in considerable quantities of CO_2 being given off and an increase in ATP biomass.
- During the second phase (10-80th day) the system passes through a transitional state. Respiratory activity decreases as a result of the exhaustion of readily biodegradable matter while at the same time the steady increase in the biomass shows that the microbial spectrum is being reestablished. The drop in activity also reflects a reduction in microbial synthesis and explains the accumulation of nitrate in the system and the concomitant decreases in pH. Instability at this stage, which is more marked in bark-based compost, is attributed to temperature variations in the heap. These variations can be quite significant (45-55°C for sawdust and 20-70°C for bark). The latter type of compost also has a lower content of water-soluble NH_4 because of the influence of lignoproteinic complexes which retard the biodegradation of the organic nitrogen and thus reduce the production of nitrate. This "buffer" effect explains the relative stability of the pH.
- In the final phase (from 80th day onwards) all the parameters are in a steady state.

4. ESTIMATION OF MICROBIAL ACTIVITY

P. AMMANN, Institut de Génie de l'Environnement (Génie Biologique) E.P.F.L.

Measurements of specific respiratory activity (O_2 consumption) on compost samples at different stages of maturity have shown that microbial activity drops considerably after six weeks for both types of compost. There is little change in polysaccharide content whether extracted by water or acid: hence, there is little readily available organic matter. Concentrations of volatile organic acids were not measurable.

5. FATE OF POLYCHLORINATED BIPHENYLS (PCBs)

P.DIERCXSENS, Institut de Génie de l'Environnement (Ecologie des polluants) E.P.F.L.

Nearly all the PCBs found in the compost originated from the sludge (0.9 ppm/DM as against 0.007 ppm for sawdust and 0.010 ppm for bark). No degradation of PCBs was observed during composting. On the contrary, a relative increase was observed owing to the partial decomposition of the organic matter.

This contradicts the results of work by Hunter et al. (0 to 78 % decomposition of Aroclor 1016 and 1254 in a laboratory composting test).

6. FATE OF 4-NONYLPHENOL (4-NP)

Miss M. TROCME, Institut de Génie de l'Environnement (Ecologie des Polluants) E.P.F.L.

Nonylphenol is a toxic non-ionic surfactant which accumulates to high concentrations in sewage sludge. In Switzerland, it originates mainly from the degradation of nonylphenol polyethoxylate used in detergents.

The 4-NP content (the most common nonylphenol compound) in the sludge/sawdust mixture described above was monitered over a nine-months composting period using gas-phase chromatography. The concentration in the starting mixture was 418 ppm/DM while that in the sludge varied from 850-1100 ppm. After two months of composting, the concentration fell to 143 ppm and after nine months to 3 ppm.

A change in the chromatogram pattern was also observed. The relative size of some peaks increased while that of others decreased, resulting in a completely different pattern after nine months. 4-NP is a mixture of isomers with alkyl chains branched to varying extents. The change in pattern could be related to the fact that the more the alkyl chain is branched the more difficult it is to decompose.

At the high temperatures reached during composting, volatilization of 4-NP may be significant, mainly owing to the codistillation effect. Volatilization of 4-NP was monitored during incubation of mature compost doped to a concentration of 600 ppm/DM at 55°C and 65 % moisture. No significant volatilization was observed. These results show that the almost complete disappearance of 4-NP during the composting of sludge is due to degradation.

7. BIOLOGICAL TESTS AND ANALYSES OF COMPOSTS USED AS SUBSTRATES

S. PACHE, Sol-Conseil, Nyon

Sludge/sawdust and sludge/bark composts were assessed at different stages of maturity using a growth test for barley in pots (pure substrates) and by the modified Standford and De Ment test (measuring root development in an aqueous extract of the substrate). The growth tests did not show any phytotoxicity. The yield of dry matter was higher from the sludge/sawdust compost than from the sludge/bark compost. Leaf analysis reveals nutritive unbalance for the sludge/sawdust compost (high nitrogen and low potassium). On the other hand, the sludge/bark compost show an equilibrium close to that for the control.

Chemical analysis of the water extract from the sludge/sawdust compost showed high nitrogen level and salinity. During the Standford and De Ment tests this extract gave the best results after composting for 140 days, whereas it showed symptoms of phytotoxycity at the start of composting.

8. EFFECT OF SLUDGE/SAWDUST COMPOST ON THE ERODIBILITY OF SANDY LOAM: LABORATORY AND FIELD TRIALS

J.P. VIANI and P. BARIL, Institut de Génie Rural (Hydrologie et Aménagements) E.P.F.L.

The effects of a sludge/sawdust compost on the erodibility of a leached brown pseudogley soil from the Vaud plateau were measured using two experimental apparatuses:
- a laboratory test bench consisting of a tank (6 x 2 x 0.6 m) divided into two compartments, and a rainfall simulator.
- two plots of land (2 x 18m) subject to natural rainfall (St-Cierges/Vaud area).

For each of the two plots (bare soil and soil treated with compost), soil losses were quantified and characterized analytically; changes in the physical and hydrodynamic properties of the remaining soil were monitored.

Laboratory simulations were effected, varying the amount of compost, slope, rainfall intensity and rain sequences. The addition of compost was accompanied by a reduction in mineral soil losses varying from 40 to 60 %.

In the field, 16 rainstorms recorded over seven months led to soil losses of 1.8 kg DM/m^2 for the bare plot and 0.8 kg DM/m^2 for the treated plot.

As for the remaining soil, the application of compost was accompanied by an increase in its structural stability, total porosity, permeability and water retention capacity. These results, taken together, serve to explain the reduction of erosion by splashing and run-off.

During these tests, the compost had a beneficial effect, reducing erosion. It would be desirable to carry out tests on a field scale over long periods. Such tests should support these initial results and enable the amount of compost and the period and manner of its incorporation into the soil to be determined.

9. TEST ON THE USE OF COMPOST AS A SUBSTRATE IN FLORICULTURE

J.P. RYSER and D. PIVOT, Station Fédérale de Recherches Agronomiques de Changins, Nyon

Two tests on the use of sludge/sawdust or bark composts for growing pot gerberas were carried out simultaneously at Changins and Conthey. The aim of these tests was:
- To measure the fertilizing capacity of the composts and the required mineral fertilizer supplementation (J.P. RYSER)
  The sludge/sawdust substrates were too rich in fertilizers (nitrogen and salinity) and had to be diluted with peat in order to fit the plant requirements. On the other hand, the sludge/bark substrates had low fertility levels, and presented a good plant growth when supplemented with adequate amounts of mineral fertilizers.
- To improve the water retention characteristics through addition of peat or rockwool-Grodan® "green" (20 or 40 % by vol.)(D. PIVOT).
  The experimental mixtures were checked against a standard of peat and perlite (80:20 %). Mixtures of sludge/sawdust or bark and peat (60:40 %) or rockwool (80:20 %) proved more satisfactory than the control for plant growth. The raw composts give uneven results due to the difficulty of proper watering (permeability index is much too high).

PRODUCTIVITY AND QUALITY OF CEREAL CROPS
GROWN ON SLUDGE-TREATED SOILS

M. CONSIGLIO, R. BARBERIS

Istituto per le Piante da Legno e l'Ambiente , Torino (Italy)

G. PICCONE, G. DE LUCA

Istituto di Chimica Agraria, Università di Torino (Italy)

A. TROMBETTA

Assessorato Ambiente ed Energia, Regione Piemonte, Torino (Italy)

Summary

A series of studies concerning the agricultural use of sewage sludges produced in the Piedmont region was carried out.In this context the application effect of increasing amounts of a sludge with a high concentration of heavy metals on a wheat crop with or without the addition of a mineral fertiliser was assessed.The grain production data offered no evidence of a toxic effect, though a negative interaction between sludge and fertiliser was noted. After the first year of the experiment, it was found that the grain content of Cd, Cu, Ni, Pb and Zn was not markedly increased by the application of greater amounts of sludge. The latter displayed a significant correlation with Cd and Ni content only. An interaction between sludge and mineral fertilisation can be postulated in the case of Cd.

1. INTRODUCTION

The Piedmont Regional Government, the Faculty of Agriculture, Turin University, and IPLA (Istituto per le Piante da Legno e l'Ambiente) have set up a long-term applied research project designed to investigate the possibility of using sewage sludges in agriculture. During the first three years of this project, several experiments have been carried out <u>inter alià</u> on the fertilisation of cereals (wheat, maize, rice), fruit crops, vegetables and flowers (1,2) to obtain information on the following points:
- the productivity of crops treated with sludges and the possible interaction between sludge and ordinary mineral fertilisers application;
- changes in crop quality,with particular reference to the absorption of heavy metals;
- long-term changes in the physical, chemical and biological features of the main Piedmontese soil types following treatment with sewage sludges.

This paper presents preliminary results observed during a multi-year full-field fertilizing of wheat crops growing on a highly fertile alluvial soil (3) near a large Piedmontese wastewater treatment plant.

2. MATERIAL AND METHODS

2.1. Soil

The soil on which these first agronomic tests have been carried out is located in the municipality of Settimo Torinese. It is a typical udifluent on the left bank of the Po river and extends over middle-recent alluvial deposits consisting of gravel with sand and clay lenses that are occasionally terraced. The mean depth is around 1.5 m. The surface horizons (0-40 cm) have a free sand texture (USDA) and are sufficiently structured with small-sized aggregates. Their biological activity is good. The mean chemical and chemico-physical values are:

Texture (SISS)
Clay 4.48%, fine silt 14.37%, coarse silt 6.24%, fine sand 62.84%, coarse sand 12.07%.

Chemistry
pH (H_2O) 7.3, pH (KCl) 6.6, organic matter 2.69%, org. C 1.56 %, total N 0.158 %, C/N ratio 9.9, available P_2O_5 24 ppm, available K_2O 43 ppm.

Exchange capacity (meq/100 g)
C.E.C. 10.2 ; Ca 7.8, Mg 1.6, K 0.09, Na 0 20, B.S. 95%.

Extractable elements (Lakanen)(ppm)
Al 26, Cd 0 32, Cr 0.42, Cu 10, Fe 160, Mn 106, Ni 22, Pb 10, Zn 12.

The soil displays a sublkaline reaction and is well endowed with organic matter and humus, whereas it has little assimilable P and K. Its CEC is average to low. The BS is high, mainly due to the presence of CA. Extractable metals are poorly represented, with the exception of Cu, Ni,and Mn.

2.2. Sludge

Sewage sludge with the following composition was applied :
pH 7.5, soluble salts 94.7 meq/100, moisture 74.4%, ash 41.3; organic matter 58.7, total C %, total N 3.44 %, C/N ratio 9.5, P 1.25 %, K 0,21 %, As 0.8 ppm, B 66 ppm, Cd 6.38 ppm, Cr 1044 ppm, Cu 694, Hg 6.1 ppm, Ni 267 ppm, Pb 109 ppm, Se not detected, Zn 3280 ppm.

It was obtained from the Collegno-Grugliasco-Rivoli joint municipal treatment plant in the province of Turin. Owing to its composition, it was thought that the crops would be polluted by components released from the sludge and hence to plants.

Cd, Cu, Ni and Zn were the heavy metals that appeared to offer the greatest risk, both because they were present at high concentrations and also because they are readily absorbed by plants. Pb was also regarded as potentially dangerous on account of its high concentration, even though it is poorly transported into plants.

3. EXPERIMENTAL PROCEDURES

What follows refers to the production of grain in the first two years, and its quality (in terms of heavy metal content) for the first year only, since the experiment is still in progress.

Plots measuring 50 m were studied. The factors examined were: sludge and mineral fertilisation (sludge 50 and 100 tonnes/hectare (wet basis);mineral fertiliser added or not). Three trials were run, giving a factorial pattern of (3 x 2) x 3 (see table 1).

The fertilised plots received 130 kg of N per ha in the form of a ternary complex and calcium nitrate. The sludge was applied on ploughed soil with a manure spreader and turned in with a disc harrow.

4. RESULTS AND COMMENTS

First and second-year crop production is illustrated in Table 2 and the heavy-metal content (first year only) in Table 3.

Field observations were also made during the growing seasons. A difference in plant height was noted in both years between fertilised and non-fertilised soils only when 0 or 50 tonnes/ha of sludge were applied. When 100 tonnes/ha were accompanied by the application of mineral fertiliser (treatment n° 6), laying occurred. This was probably caused by high amounts of nitrogen introduced through the combined application of sludge and the mineral fertiliser.

5. DISCUSSION

It is clear from table 2 that in the first year the plots treated with sludge alone (n° 3 and n° 5) are comparable with the control plot (n° 2). Fifty tonnes of sludge per hectare provide 340 units of total nitrogen per hectare. This can be compared with the 130 units given by mineral fertilisation, assuming 50% mineralisation. It can also be seen that massive application of nitrogen, whether in organic form (plot 5) or in organic and mineral form (plots 4 and 6), does not result in a further increase in production. There is, indeed, a slight fall in plot 4.

Statistical analysis shows that a significant negative interaction (p 0.05) exists between the amount of sludge and the presence of mineral fertiliser. Over-heavy introduction of nitrogen leads to growth imbalance by encouraging stem and leaf development. This in turn leads to the risk of laying, whereas grain production is unchanged. All these impressions were confirmed during the second year. As expected, production fell in the non-fertilised control due to soil impoverishment. The negative interaction between sludge and mineral fertiliser was again significant. The main "sludge dose" and fertilisation effects, on the the other hand, were not significant.

Apart from the risk of flattening and depressed production owing to an excess of nitrogen, no adverse effects associated with the use of sludge have been observed in the first two years.

It is evident, however, that application of 100 tonnes of sludge per hectare is equivalent to application of an ordinary mineral fertiliser (see plots 2 and 5). The quality data are set out in Table 3. Since wheat is a plant whose grain is used as food, no account was taken of possible direct pollution due to contact with sludge particles. Consideration was therefore confined to indirect absorption through the soil.

The data obtained are in line with those in the literature (4). Statistical analysis led to the following results: there is a highly significant (p = 0.01) correlation between fertilisers and Ni, and between the presence of sludge and Cd and Ni concentrations; there is a significant interaction between mineral fertilisation and the presence of sludge on the case of Cd only.

As far the sludge doses are concerned, there was a significant (p = 0.05) relation with regard to Cu (dose 1 vs dose 2), Ni (Dose 0 vs dose 2) and Cd (dose 0 vs dose 1 and dose 2).

6. CONCLUSIONS

At this stage of the research, the conclusions can be drawn:
- potential toxicity of the sludge used, with its high heavy metal concentrations, has no adverse effect on the wheat at any stage of its growth, even when applied in an amount cooresponding to 20 tonnes of dry matter per hectare
- a limit to the application of sludge can be seen in the amount of nitrogen introduced into the soil. Too much nitrogen, in fact, is followed by growth imbalance and the risk of laying.
- after one year's treatment with sludge there was no dangerous accumulation of heavy metals in the grain, though an increase in Cd and Ni absorption was observed.

As the research programme continues particular attention will also be directed to the accumulation of heavy metals in the soil.

Table 1 – Results (t/ha) concerning grain production in the first and second year; the results are the average of three tests and refer to 15 % moisture content.

Treatment	Sludge t/ha	Fertilizer	First year	Second year
1	0	absence	3,92	2,85
2	0	presence	4,54	4,44
3	500	absence	4,50	3,01
4	500	presence	3,86	3,39
5	1000	absence	4,50	4,34
6	1000	presence	4,57	3,57

Table 2 - Results(t/ha) concerning grain production in the first and second year; the results are the average of three tests and refer to 15% moisture content.

===

Treatment	First year	s.d.*	Second year	s.d.*
1	3.92	0.33	2.85	0.29
2	4.54	0.11	4.44	0.14
3	4.50	0.44	3.01	0.35
4	3.86	0.56	3.39	0.02
5	4.50	0.12	4.34	0.16
6	4.57	0.34	3.57	0.42

* s.d. = Standard deviation

Table 3 - Analytical data concerning heavy metal content in grain produced during the first year of the experimentation; the results are the average of three tests expressed as ppm.

===

	Cu	s.d.	Zn	s.d.	Ni	s.d.	Cd	s.d.	Pb	s.d.
1	4.57	0.55	115	15.1	1.03	0.23	0.09	0.006	0.73	0.058
2	4.47	0.03	115	27.5	1.73	0.40	0.11	0.030	0.40	0.26
3	4.73	0.21	109	18.2	1.30	0.26	0.19	0.010	0.60	0.17
4	4.70	0.36	136	9.45	2.50	0.46	0.19	0.006	0.67	0.31
5	4.27	0.115	134	12.12	1.97	0.42	0.23	0.025	0.70	0.10
6	4.80	0.36	130	2.52	2.57	0.32	0.18	0.006	0.87	0.15
$\bar{x}$	4.49	0.35	123	17.32	1.85	0.65	0.16	0.05	0.66	0.22

REFERENCES
(1) NAPPI, P., BARBERIS, R., CONSIGLIO, M., JODICE, R. and TROMBETTA, A.(1985).Biological evaluation of sludge phytotoxicity. Symposium on Processing and Use of Organic Sludge and Liquid Agricultural Wastes, Roma.
(2) PICCONE, G., SAPETTI, C., BIASIOL, B., DE LUCA, G. and AJMONE MARSAN, F. (1985). Chemical properties of sewage sludges in the Piedmontese area (ITALY). Symposium on Processing and Use of Organic Sludge and Liquid Agricultural Wastes, E.C. Roma.
(3) REGIONE PIEMONTE, IPLA (1982). La capacità d'uso dei suoli del Piemonte ai fini agricoli e forestali. Published by Edizioni l'Equipe, Turin.
(4) GENEVINI, P.L., VISMARA, R., MEZZANOTTE, V. (1983). Utilizzo agricolo dei fanghi di depurazione. Ingegneria Ambientale, Milano.

BIOLOGICAL EVALUATION OF SLUDGE PHYTOTOXICITY

P. NAPPI, R. BARBERIS, M. CONSIGLIO, R. JODICE
Istituto per le Piante da Legno e l'Ambiente - Torino (Italy)
A. TROMBETTA
Assessorato all'Ambiente ed Energia, Regione Piemonte - Torino
(Italy)

ABSTRACT
An account is given of two years' work on the biological
determination of sludge phytotoxicity using three methods:
phytotoxicity of sludge with respect to the germinative capacity
of Lepidium sativum seeds; phytotoxicity of sludge with respect
to the root development of a pregerminated test essence (Lupinus
albus) grown in a liquid medium; phytoxicity of sludge with
regard to the growth and development of L. sativum and Lolium
italicum raised in pots as closely as possible to ordinary
cultivation condition.

1. INTRODUCTION

As part of a long-term research programme conducted by several
public instituts in Piedmont (I)(1-2), the need has emerged to
evaluate and perfect analytical methods to determine the possible
phytotoxicity of sludges.

It must be stressed that toxicity is a variable function in the
stabilisation process of organic matter. Different levels of toxicity
are present and require different methods for their determination.

The most commonly used lab test is the Lepidium test devised by
Zucconi et al. (3). This evaluates the influence of sludge on
germination and the early root growth of seeds. It can thus be
regarded as a test of instantaneous toxicity, subject to change each
time the substrate and the process are altered.

This analysis, however, is not in itself enough for overall
evaluation of processes as complex as the biodegradation and
stabilisation of organic matter. Next, an assessment must be made of
the effect of sludge on another critical point in plant development,
namely root growth. For this purpose, the Lupinus test is used.

The results given by these two tests have been checked and
supplemented by testing in pots. In this case, the results, even
though tey are not immediate and mediated by the substrates with
which the sludge is mixed, are more complete, since the sludge is
regarded in its entirety and not just as an extract. In all tests,

but especially those in pots, account must also be taken of the fertility factors introduced by the nutrient elements contained in the sludge itself.

This paper describes two years' work on evaluation of these methods.

2. MATERIALS

Sludges were obtained from wastewater treatment plant using aerobic digestion (at Bra and Canelli) and anaerobic digestion (at Collegno and Pianezza). Their chemistry is illustrated in Table 1.

To determine the efficiency and correspondence of the assays used, and as further basis for comparison, parallel tests were run with three organic fertilizers obtained by composting poplar bark se-wage sludges vith vine stalks for 5 months, and with poplar bark for 9 months.

Table 1 - Chemical characteristics of sludges utilised in biological toxicity assays.

	pH	moisture %	ash %	N %	C/N	P %	K %	B ppm	Cd ppm	Cr ppm	Cu ppm	Hg ppm	Ni ppm	Pb ppm	Zn ppm
Collegno	7.5	74.4	41.3	3.44	9.5	1.25	0.21	66	6.38	1044	694	6.1	267	109	3280
Pianezza	7.7	68.3	32.3	4.11	9.2	0.65	0.20	89	1.02	3450	1020	2.6	700	538	5750
Canelli	7.5	77.6	37.0	5.15	6.8	0.53	0.38	52	0.30	58	248	1.1	54	212	1050
Bra	7.5	80.8	30.8	4.57	8.3	0.91	0.44	59	0.54	148	370	2.7	75	367	1727

3. RESULTS AND DISCUSSION

3.1. Germination assay with Lepidium sativum

Germination and root growth of L. sativum seeds at various extract concentrations are evaluated in Petri dishes and compared with a control consisting of distilled water. The influence of temperature, time and light on germination were considered during the experiment.

The extract was obtained by adding water to the sludge up to 90% moisture content. After 12 h the sample was centrifuged for 15' at 6000 rpm and filtered with a sterilising membrane using a pressurised filtering apparatus.

The toxicity of the filtrates was shown in the case of anaerobic sludges from the Collegno plant. (Fig. 1)

The anaerobic sludge from the Pianezza plant, on the other hand, was highly toxic at high concentrations, whereas its toxicity was low at greater dilutions. Lastly, the aerobic sludge from Bra was not toxic or barely toxic, even at high concentrations.

As far as the filtrates of organic fertilisers are concerned, the response of the biological assay is essentially positive, since their germination indexes are high even at high concentrations and in the sludge as such(Fig.2)In addition, organic fertilisers 1 and 3 (from sludge with poplar bark and sludge with vine stalk respectively), display higher values than the control at high dilutions. It may thus be supposed that they contain physiologically active substances that stimulate germination and root growth.

3.2. Lupinus albus growth assay

The L. albus test can be regarded as midway between the toxicity test in dishes and that in pots. Briefly, seeds germinated in distilled water and with similar root lengths are placed in test tubes containing sludge extracts at different dilutions and compared with a distilled water control. The Collegno anaerobic sludge was used for this assay.

Table 2 - Total weight, root weight and total length of Lupinus albus tested in different concentrations; data are the average of three tests.

===

Sludge extract concentration (%)	Total weight (g)	Root system weight (g)	Total length (cm)
100	0,91	0,23	5,43
50	1,33	0,45	7,00
30	1,33	0,50	8,03
10	1,25	0,43	8,23
3	1,42	0,48	10,53
Control(water)	1,48	0,63	11,00

As can be seen, Collegno sludge as such is highly toxic. When brought into contact with the extract, the Lupin root necrotises and dies. At intermediate concentrations (50% to 10%), there is no substantial difference with regard to weight and total length. The roots have difficulty in growing, but get the critical stage. At the lowest concentration (3%), indeed, the results observed with the extract are virtually the same as in the control. The roots easily withstand the transplantation crisis and develop considerably.

3.3. Growth assay in pots

The experiments carried out were designed to devise a suitable method for evaluating the toxicity of sludges. Attention was given to:

- the identification of an effective "control";
- the choice of materials for the preparation of a basic substrate to which different amounts of sludge could be added;
- the choice of sludge amounts in the light of two considerations: detection of the amount at which toxicity first appears; recognition of the fact that sludge also contributes fertility in the form of organic matter and nutrient elements.

Various materials were tested singularly or in combination to serve as the control: agricultural soil, farmyard manure, peat, river sand and perlite. It was found that a mixture of acid peat with sand and a suitable mineral fertiliser made the best control. These materials are easily obtained and ensure efficient repeatability compared with soil and manure. Furthermore, the assay can be checked, since no considerable variability factors are included: river sand has a very low amount of organic matter and nutrient substances, while peat has few nutrient elements and its organic matter is difficult to degrade.

The control mixture was therefore used as the basic substrate for addition of the sludge.

As to amounts of sludge employed, experience showed that 50-100 g/dry sludge per litre of substrate gave rise to toxicity. To enable the fertility contributed by the sludge to be evaluated, therefore, amounts of 12.5 and 25 g/litre were chosen. These are close to the levels used in agriculture (50-70 tonnes/ha). As a further reference for evaluating the response of the assay, the organic fertilisers described earlier were also employed.

The following method was then elaborated: sludge as such was added to the substrate in the proportion of 12.5, 25, 50 and 100 g dry sludge per litre of substrate and mixed in. The acid reaction of the peat was previously buffered with lime to pH 6.5 - 7.4. The mixtures were placed in 2-litre pots containing a 1 cm layer of expanded clay for drainage. To ensure that germination would take place and make it possible to evaluate the toxic effect on plant growth only, the mixture was covered with a 1 cm layer of river sand. For each test plant the number of seeds sown per pot was sufficient to ensure the germination of at least 100 seeds. The seeds were then covered with a 1 cm layer of perlite.

Data on the production of Lolium italicum and Lepidium sativum are shown in Table 3 and Fig. 3-4.

Table 3 - Lolium italicum and Lepidium sativum production (g/pot) obtained in growth tests in pot. Data are the average of three tests.

Typology	Dose (g/pot)	Lolium italicum production First cut (g dry matter/pot) average	s.d.	Second cut (g dry matter/pot) average	s.d.	Lepidium sativum production (g dry matter/pot) average	s.d.
	0	1,81	0,09	1,58	0,12	18,36	2,34
	25	2,48	0,43	4,12	0,60	24,46	1,59
Collegno	50	1,73	0,04	4,15	0,42	9,48	2,14
sludge	100	1,44	0,18	4,17	0,43	4,68	1,92
	200	0,57	0,25	3,09	1,14	0,76	0,21
	0	1,81	0,09	1,58	0,12	18,36	2,34
Pianezza	25	1,80	0,12	3,44	0,27	12,84	2,29
sludge	50	1,36	0,12	3,51	0,94	6,19	0,53
	100	0,43	0,11	1,94	0,72	0,79	0,32
	200	0,09	0,05	0,03	0,03	0,71	0,17
Compost A as whole		2,47	0,18	2,78	0,47	43,05	1,28
Compost B as whole		2,79	0,09	7,13	1,36	37,47	2,19

Notes :
Compost A = obtained from sludge and poplar bark
Compost B = obtained from sludge and vine stalks
s.d. = Standard deviation

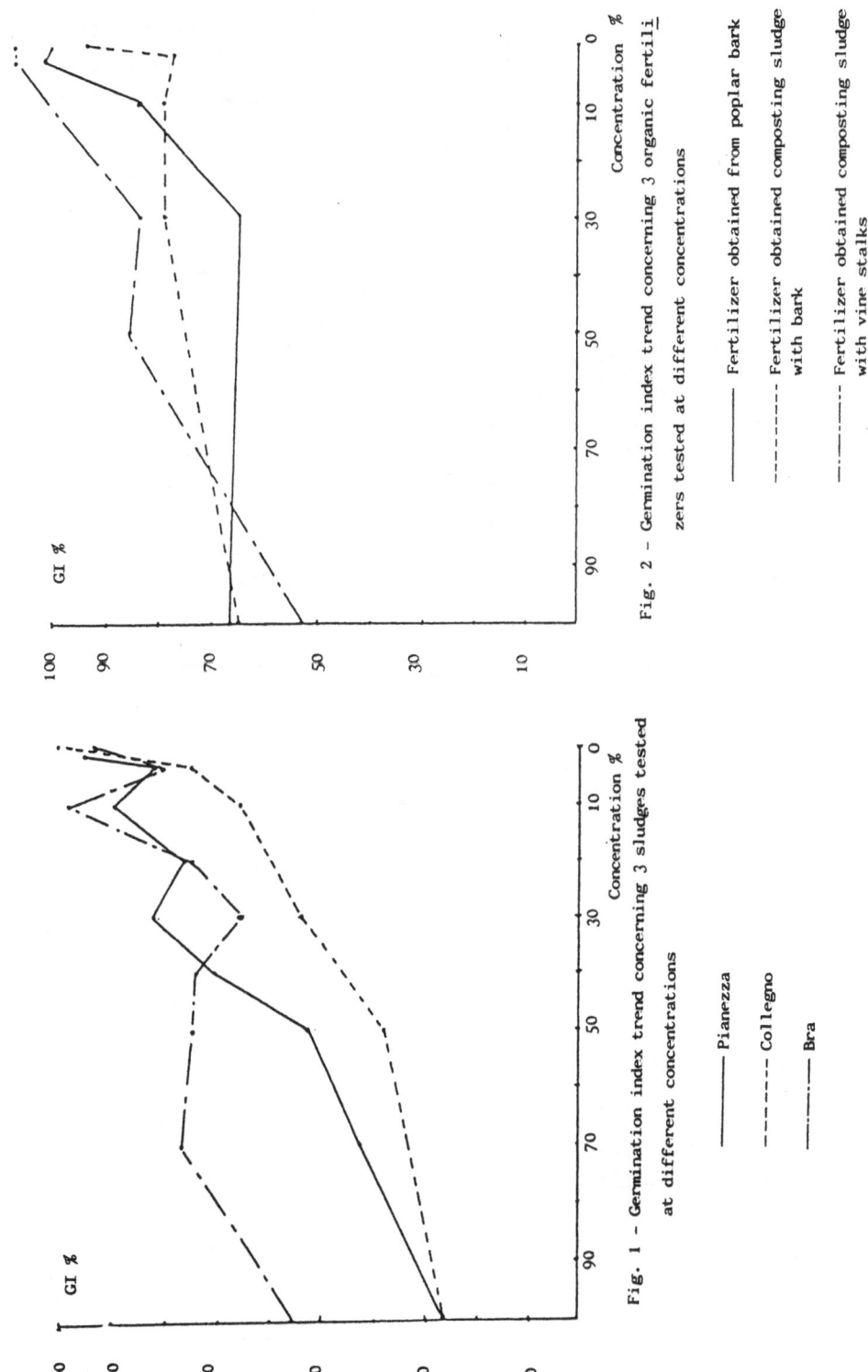

Fig. 1 - Germination index trend concerning 3 sludges tested at different concentrations

Pianezza
Collegno
Bra

Fig. 2 - Germination index trend concerning 3 organic fertili zers tested at different concentrations

Fertilizer obtained from poplar bark

Fertilizer obtained composting sludge with bark

Fertilizer obtained composting sludge with vine stalks

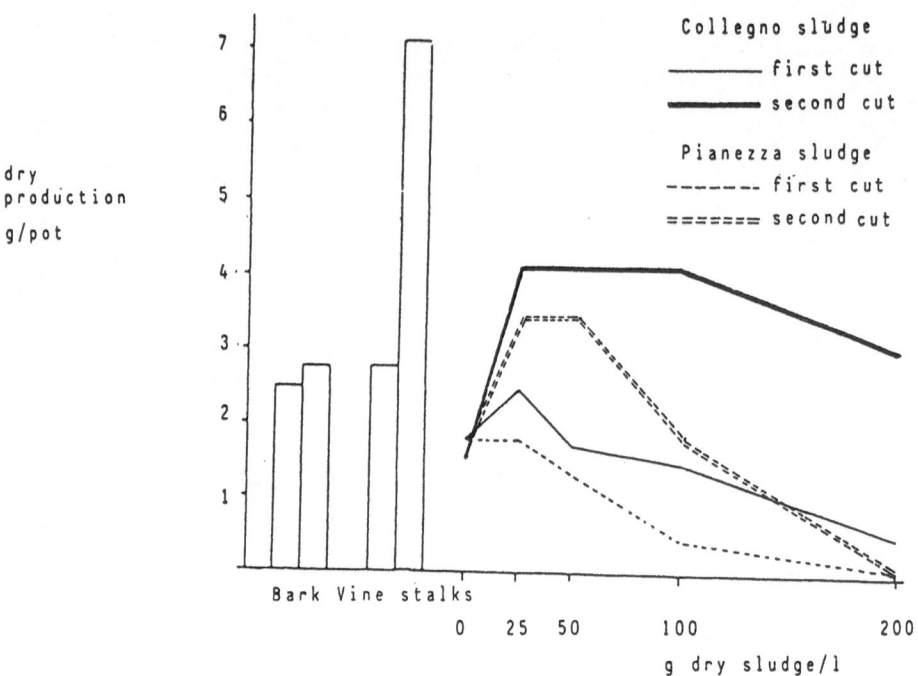

Fig. 3 – Lolium italicum first and second cut production trend

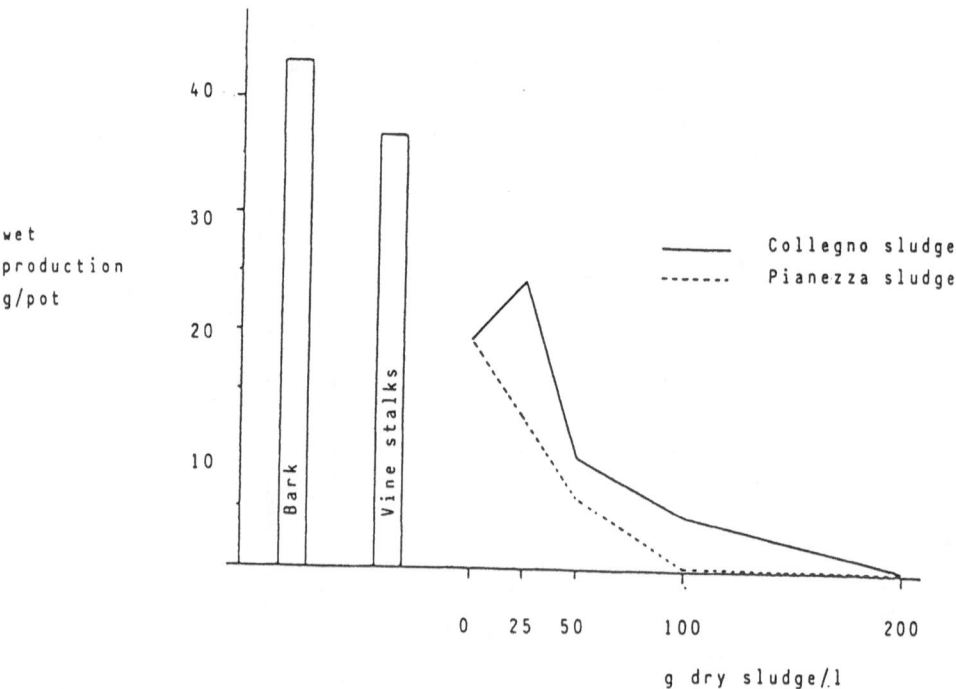

Fig. 4 – Lepidium sativum production trend

Higher sludge doses brought about marked toxicity, with a distinct fall in the production of biomass for both sludges and for both plants.Toxicity appeared at the commencement of growth (5-6 days after germination), when the roots came into contact with the substrate, and took the form of yellowing and root tip necrosis. This was more marked as the amount of sludge increased. At 200 g of sludge per pot, the seedlings withered almost completely. This was perticularly true in the case of L. sativum, which is more sensitive than Lolium.

The difficulties encountered by the plants during the early growth stages seem to be less apparent as the assays proceed, though they are reflected in the size of the final crop. This improvement may be the result of adaptation to the substrate, or a falling off of toxicity owing to chemical and physical changes in the substrate itself. These phenomena are particularly evident in the case of Lolium, where the second-cut biomass is much greater than that of the control, even when larger amounts of sludge are applied. It would thus seem that sludge exerts a temporary toxicity that appears in the early growth stage and the first harvest, whereas in the second stage changes in the substrate are such that fertility factors due to nutrient elements (especially nitrogen) in the sludge gain the upper hand. Not even temporary toxicity, however, was noted when smaller amounts of sludge were used. The production of both Lolium and Lepidium, in fact, was always higher than those responsible for toxicity.

Particular importance may also be attached to the results obtained with two composts obtained from sludge mixed with vine stalks and poplar bark. Their productivity was always higher than that of the control both with Lolium and Lepidium.

REFERENCES

(1) PICCONE, G., SAPETTI, C., BIASIOL, B., DE LUCA, G. and AJMONE MARSAN, F.A. (1985). Chemical properties of sewage sludges produced in the Piedmontese area (ITALY). Symposium on Processing and Use of Organic Sludge and Liquid Agricutural Wastes, E.C. Roma.

(2) CONSIGLIO, M., BARBERIS, R., PICCONE, G., DE LUCA, G. and TROMBETTA, A. Productivity and quality of cereal crops grown on sludge-treated soils. Symposium on Processing and Use of Organic Sludge and Liquid Agricultural Wastes, E.C., Roma.

(3) ZUCCONI, F., FORTE, and MONACO, A. Analisi della fitotossicità mediante bioassaggio Lepidium sativum. In Simposio "Recupero biologico ed utilizzazione agricola dei rifiuti urbani" Napoli ottobre 1983.

THE FATE AND BEHAVIOUR OF SELECTED HEAVY
METALS DURING PYROLYSIS OF SEWAGE SLUDGE

R.C. KISTLER, F. WIDMER and P.H. BRUNNER
Swiss Federal Institute of Technology Zurich

Summary

The distributions of chromium, nickel, copper, zinc, cadmium, mercury and lead between the solid residue, condensates and flue gas during pyrolysis of sewage sludge have been measured. Cr, Ni, Cu, Zn and Pb are retained in the solid residue up to 750°C. The volatilization of Cd started at 600°C. Hg was quantitatively volatilized at 350°C. The heavy metal balances were 100% ±15%.

1. INTRODUCTION

The use of sewage sludge on agricultural land is limited in densly populated areas or if the sludge contains high concentrations of toxic organic and/or inorganic constituents. In such cases the sludge has to be dewatered and disposed of in sanitary landfills or it has to be incinerated. Due to the high temperatures during combustion (800°-1000° C) the heavy metals with the highest volatility (Hg, Cd, As ...) may be released to the atmosphere in spite of air pollution control equipment.

It is expected that lowering the reaction temperature to 500°-600°C would prevent the metals (except Hg) from volatilization.

This study investigates the distribution of heavy metals between flue gas, condensates and solid residue during pyrolysis of dried sewage sludge.

2. EXPERIMENTAL

In an all-glass batch pyrolysis unit (fig. 1) (Duran, "hot" parts: quartz) dried sewage sludge was pyrolyzed at different temperatures. The vapour phase which contained condensable and non condensable gases was cooled in two stages to -75°C. The non condensable gases were filtered and scrubbed in two impingers with a saturated $KMnO_4$/10% H_2SO_4-solution.

Due to the low gas velocities in the reactor and due to the fact that the sewage sludge was kept in a sample boat, no particle "carry over" was observed in this experimental set-up.

The heavy metals Cr, Ni, Cu, Zn, Cd, Hg and Pb in the sewage sludge feed and in the pyrolysis products ash, condensates and scrubbing liquids were analized by Atomic Absorption Spectroscopy. Special emphasis was put on accuracy of sampling and analysis in order to calculate mass balances of the investigated metals. It was thus possible to keep the deviations between material input and outputs always below ±15%.

Figur 1:   Pyrolysis apparatus

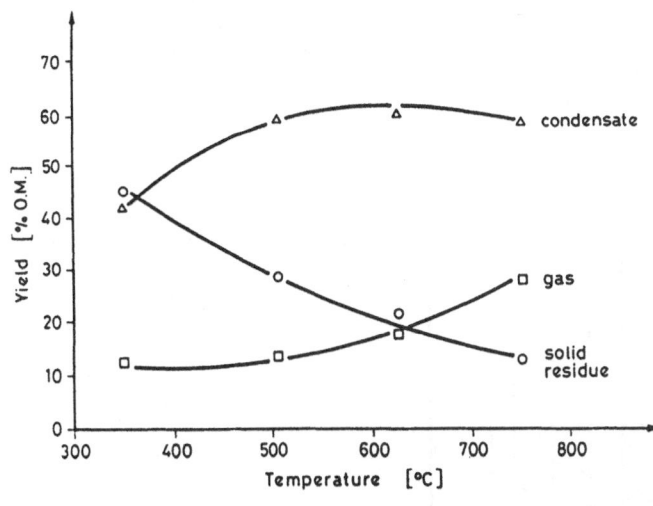

A - F Thermocouples

1. Tubular-furnace
2. Quartzreactor
3. Tar trap
4. Condenser T~20°C

5. Pressure gauge
6. Condenser T~-75°C
7. Quartzfibre filter
8. Scrubbers (KMnO₄/H₂SO₄)

9. Thermostat 25°C
10. Wet gas meter

3. RESULTS

3.1. Products of pyrolysis at different temperature

Figur 2 shows the temperature dependence of the product yields. The higher the temperature, the more gas and less solid residue is produced. The condensate yield has a maximum at $625^{\circ}C$.

In the investigated range (2-25 $^{\circ}C$/Min.) the yield of solid residue is not affected by the heating rate. At higher temperatures (>$625^{\circ}C$) and higher heating rates (>20 $^{\circ}C$/Min.) the yield of condensate decreases about 5-10% and the gas yield rises therefore 5-10%.

Figur 2:
Distribution of pyrolysis products at different temperatures

3.2. Distribution of Cr, Ni, Cu, Zn and Pb between gaseous and solid phase

In the temperature range between $350^{\circ}C$ and $750^{\circ}C$, the quantities of Cr, Ni, Cu, Zn and Pb in the condensate and scrubber liquids were neglectible. These metals are trapped in the solid residue and do not pose a thread to the atmospheric environment, unless they are transfered to the gas stream in particulates due to the high gas velocities in the reactor system (e.g. fluidized beds).

According to these results no volatile species of Cr, Ni, Cu, Zn and Pb are formed during pyrolysis. The residence time does not influence the distribution of these metals.

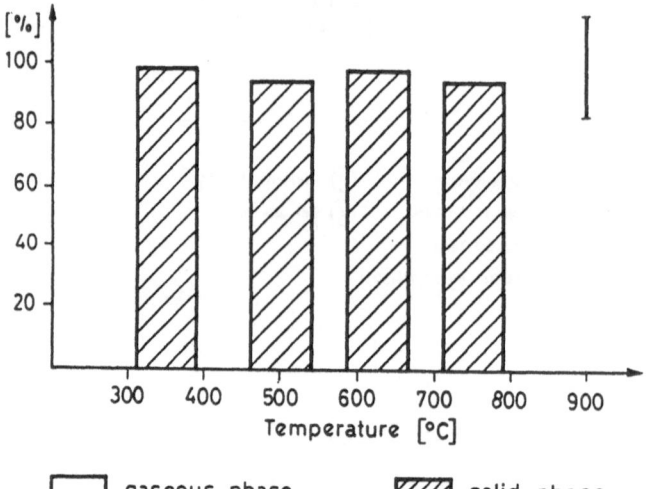

Figur 3:
Distribution of Cr,Ni, Cu, Zn and Pb between gaseous and solid phase

margin of error is shown by the bar on the right side

3.3. Volatilization of cadmium

The pyrolysis temperature has a marked influence on the volatilization of cadmium. At 500°C Cd is retained quantitatively in the solid pyrolysis residue, whereas more than 95% of Cd is found in the gasphase at 750°C. At 625°C the evaporation is only slow, and therefore the total amount of volatilized Cd depends on the residence time. At 750°C the rate of evaporation of Cd is at least as fast as the pyrolysis rate of sewage sludge. The evaporated Cd is easily trapped in a condenser at temperatures below 150°C.

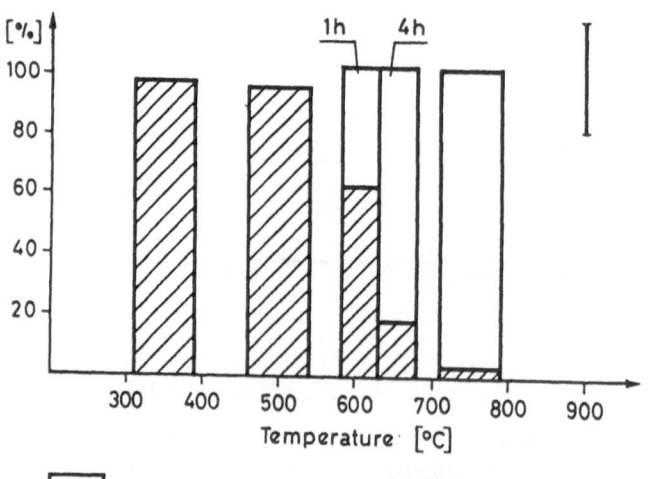

Figur 4:
Distribution of Cd between gaseous and solid phase

margin of error is shown by the bar on the right side

3.4. Emission of mercury

For all investigated temperatures (350°, 505°, 625°C) Hg is completely volatilized. The condensation of this metal was not achieved at
20°C (fig.1, no.4), but at -75°C (fig.1, no.6). A considerable amount
($\approx 10\%$) of Hg even passes this condenser and is captured by the first
scrubber (fig.1, no.8). Hg concentration in the second scrubber liquid
is always at or below detection limit. Since Hg is difficult to strip
from the flue gas, it should not be disposed of by the sewage system.

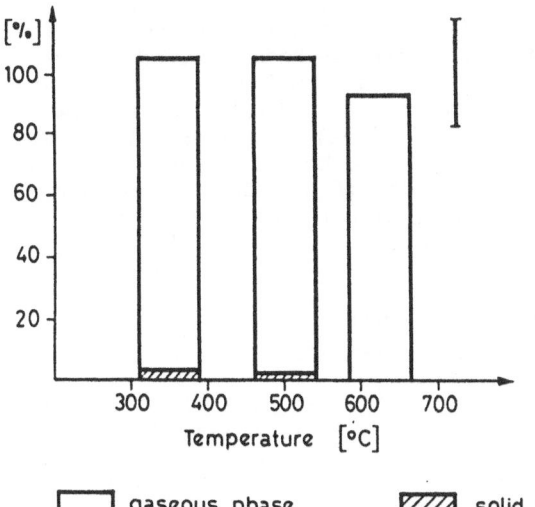

Figur 5:
Distribution of Hg
between gaseous and
solid phase

margin of error is
shown by the bar on
the right side

gaseous phase   solid phase

4. CONCLUSION

During the pyrolysis of sewage sludge in a laboratory reactor, most
metals were retained in the solid pyrolysis residue. Exception are
metals with a high volatility such as cadmium and mercury. While for
cadmium the shift to lower pyrolysis temperature ($<550^\circ$C) prevented the
emissions of this metal completely, mercury was quantitatively transfered to the gas phase at all investigated temperatures from 350° to
625°C. Since the elimination of mercury from the gas stream proved to be
difficult, this metal should be kept away from the sewage system.

LATEST TEST RESULTS IN SLUDGE DEWATERING WITH THE CHP-FILTER PRESS

U. Loll
Abwasser - Abfall - Aquatechnik
Heidelberger Landstr. 52
D-6100 Darmstadt

Summary

With the invention of the CHP-filter press the development of a continuous working high pressure press has been succeded by the first time at which the pressure on a suspension to be dewatered can be continuously increased and therefore dewatering results which were so far only able to be achieved with chamber-, plate- or membrane filter presses are obtained. The development characteristics, the constructive realization and the field of application as well as the efficiency of the equipment are described below.

1. CHARACTERISTICS OF THE CONVENTIONAL MACHINES FOR THE SOLID/ LIQUID SEPARATION OF SLUDGE-LIKE SUSPENSIONS

The use of the so far existing machinery for the solid/liquid separation, such as centrifuges, decanters, vacuum filters, screen- and belt-presses, as well as chamber-, plate-, and membrane filter presses include the following problems. In case of reasonable amount of machinery and aimed continuous operation in many cases of application the dewatering results are insufficient. In order to receive satisfactorily dewatering results, a considerable number of equipment and therefore high investment costs as well as a non-continuously working procedure had to be accepted.

The following characteristics are particularly relevant:

1.1. Centrifuges and decanters

Continuous working method, favourable relation between machinery and flow rate, high capacity, low dewatering results, bad separation cut. The separation cut and the dewatering results differ very high according to the specific weights in the sludge liquor. With increasing specific weight difference, the dewatering results improve.

1.2. Vacuum filters

Continuous working method, unfavourable relation between machinery and flow rate, high capacity, low dewatering results, good separation cut, partial vacuum up to approx. 0,6 bar.

Even the addition of press-belts does not show significant improvement in the dewatering results.

1.3. Screen- and belt filter presses

Continuous working method, reasonable relation between machinery and flow rate, small capacity, reasonable dewatering results, reasonable separation cut, short-time pressure of up to approx. 2,5 kp/cm².

Since the pressure is too low and the time of activity is too short, the real dewatering results can only be insignificantly improved.

1.4. Chamber-, plate-, and membrane filter presses

Non-continuous working method, very unfavourable relation between machinery dimensions and flow rate, high capacity, good dewatering result, good separation cut, long-term filtration pressure up to 20 bar, respectively 15 kp/cm² of membrane filter presses.

2. DEVELOPMENT CHARACTERISTICS AND DEWATERING PROCEDURE OF THE NEW CONTINUOUS WORKING HIGH PRESSURE FILTER PRESS

For the new development of the described continuous working high pressure filter press, the following requirements in technology, efficiency and optimal machinery were postulated:
- continuous sludge feeding
- continuous increase of pressure
- continuous cake outlet
- continuous filter cloth washing
- highest possible regulation of the flow rate
- good dewatering results
- high separation cut
- low amount of machinery (small dimensions)
- low energy demand
- highest possible pressure up to 20 kp/cm²
- pressure control in amount and activity time.

The dewatering procedure of the high pressure filter press, of which the most relevant elements are shown in figure 1, runs as follows:
- The suspension is fed continuously into the machine by a helical rotor pump via a static pre-dewatering-chamber stage V (which is not shown in fig. 1 due to lack of space).
- A closed sludge chamber (stage K, R, and H) changes its volume according to the dewatering progress).

The sludge chamber consists of two endless filter belts, two endless channel belts, two endless asided seals, and two endless armoured belts consisting of high pressure plate elements.

All sludge chamber elements are in operation synchronuously within the machine. The dewatering in the closed chamber is effected by successive following pressure phases, being individually controlable in amount and activity period. Amount and kind of reaction of the individually chosen pressure always correspond to the specific dewaterability of the suspension to be dewatered.

The following pressure phases were passed through:
- Increasing hydrostatic filling- and filtration pressure up to 0,25 bar (stage V) turning into an
- increasing static pressure op to 0,5 kp/cm² (stage K), turning into an
- increasing dynamic pressure up to 2,5 kp/cm² (stage R), turning into an
- increasing static pressure up to 20 kp/cm² (stage H).

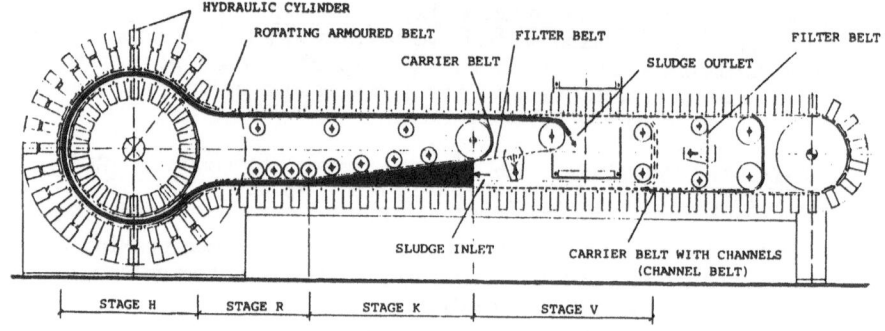

Fig. 1: Elements of the continous working high pressure filter press

Fig. 2: The CHP-filter press in operation on a sewage treatment plant

3. CONSTRUCTIVE REALIZATION OF THE REQUIRED PRESSURE REACTIONS

The hydrostatic filtration pressure up to 0,25 bar is produced by gravity of the filling hight of the suspension itself. Within a 0,6 m Ø filter chamber consisting of motionless filter cloth the filling hight up to 2,5 m can be controlled. The filter efficiency of the chamber is activated and increased by two speedcontrolled wipers.

By steadily reduction of the inflow and increasing solid matter concentration the hydrostatic filtration pressure turns into an increasing static pressure up to 0,5 kp/cm². This pressure is produced by the progressing sphenoidal reduction of the chamber channel belts, a number of pressure rolls, and the two side walls. Pressure rolls and side walls are built in, whilst the filter- and channel belts move synchronuously with the suspension, as controlled. The turning into the dynamic pressure is achieved infinitely.

The controlled dynamic pressure up to 2,5 kp/cm² is produced by in-creasing pressure of stationary built-in pressure rolls. All chamber ele-ments, like the two filter belts, the two channel belts, and the two asided seals are operating synchronuously. The chamber volume is automatically reduced with the progress of dewatering.

The increasingly controlled static pressure up to 20 kp/cm² is pro-duced by plate elements with built-in hydraulic cylinders. All chamber elements including the pressure producing plate elements with built-in hydraulic cylinders cooperate smoothely synchronous. The chamber volume is also reduced automatically with the progress of dewatering. According to building dimensions of the machine the total static pressure produced at the same time is up to 1500 Mp.

Since all pressure transmitting elements do not move relatively to each other, but only mutual and synchronously with the medium included, a smooth pressure transmission to the dewatering chamber and the suspension is possible. The pressure power only reacts within the pressure elements. Therefore outside effecting pressure and weight is not produced, and the foundation has to carry only the weight of the machine itself.

In figures 3 and 4 some important details of the construction are shown.

4. SPECIFIC MACHINE DATA AND EFFICIENCY RANGE OF THE VARIOUS MODELS

The new continuous working high pressure filter press is built in six different basic types according to various application. The most important technical data for the machine and the efficiency range of the various models are compiled in table 1 - 3. Special constructions for individual application where certain phases of dewatering have to be extended resp. shortened in order to receive an optimal dewatering result for problematic cases can be done.

All types can be equipped on request with a continuous operating sludge washer, which might be installed for the dewatering of centain sludges of mostly industrial sources.

Table 1: Technical machine data

type	filter cloth width	cylinder diameter	length	width		hight		total power capacity	total weight
	m	m	m	m	with stage V	m	with stage V	kW	Mp
H1/0,5/1,2	0,5	1,2	7,5	1,2	2,5	2,0	2,4	4,5	- 5
H2/1,0/1,2	1,0	1,2	7,5	1,7	3,0	2,0	2,4	6,8	-10
H3/1,5/1,2	1,5	1,2	7,5	2,2	3,5	2,0	2,4	7,8	-20
H4/1,5/1,5	1,5	1,5	8,5	2,2	3,5	2,3	2,9	7,8	-25
H5/1,5/1,8	1,5	1,8	9,5	2,2	3,5	2,6	3,4	9,2	-30
H6/1,5/2,1	1,5	2,1	10,5	2,2	3,5	2,9	3,9	9,2	-35

Fig. 3: Hydraulic system

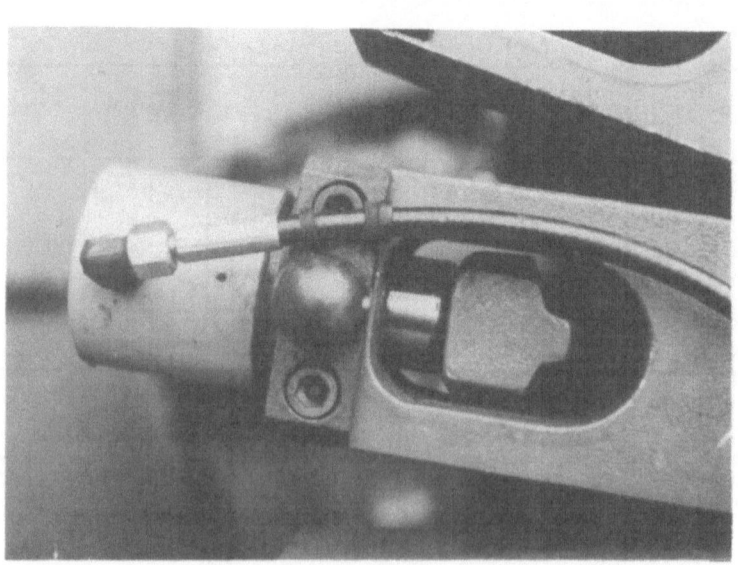

Fig. 4: Single hydraulic cylinder

Table 2: Pressure and time of active pressure

type	stage V pressure		stage K pressure		stage R pressure		stage H pressure		total time of active pressure min
	amount bar	time min	amount kp/m²	time min	amount kp/m²	time min	amount kp/m²	time min	
all types	0,05 to 0,25	1,5 to 15	up to 0,5	1 to 7	up to 2,5	1 to 7	up to 20	2,2 to 15	5,7 to 44

Table 3: Active static and dynamic filter cloth values – efficiency range of different machine types

Type	active static filter cloth values total-		active dynamic filter cloth value			effiency range of different types		
	-length	-area	length	area	supporting speed	cake thickness	cake production	sludge inflow
	m	m²	m/h	m²/h	m/sec	mm	m³/h	m³/h
H1/ 0,5/1,2	5,9 to 7,9	6,0 to 9,8	11 to 77	10 to 70	0,003 to 0,021	15 to 30	0,07 to 1,0	up to 10
H2/ 1,0/1,2	5,9 to 7,9	11,6 to 15,4	11 to 77	21 to 147	0,003 to 0,021	15 to 30	0,15 to 2,1	up to 20
H3/ 1,5/1,2	5,9 to 7,9	16,9 to 20,7	11 to 77	32 to 224	0,003 to 0,021	15 to 30	0,25 to 3,3	up to 40
H4/ 1,5/1,5	7,5 to 9,5	21,8 to 25,6	14 to 100	40 to 280	0,003 to 0,021	15 to 30	0,3 to 4,2	up to 60
H5/ 1,5/1,8	9,1 to 11,1	25,8 to 29,6	18 to 126	52 to 364	0,003 to 0,021	15 to 30	0,4 to 5,6	up to 80
H6/ 1,5/1,2	10,7 to 12,7	30,7 to 34,5	22 to 150	63 to 440	0,003 to 0,021	15 to 30	0,5 to 6,7	up to 100

5. FIELD OF APPLICATION AND CAPABILITY OF THE AGGREGATE

5.1. Field of application

The new continuous working high pressure filter press is suitable for the dewatering of sewage sludges, suspensions of industrial production and sediment sludges.

Since the amount as well as the time of pressure activity can be controlled in a wide range, the whole scale of aimed dewatering degrees can be covered. If requested, a conditioning of the material as well with organic as with anorganic coagulants resp. flocculants or thermal conditioning is possible. With the new kind of sewage sludge dewatering it is possible to produce a final product for land fill disposal or grain filter cake for agricultural use with one and the same machine.

The compact construction and the low weight of the machine make it suitable for many cases of implements, where conventional dewatering machines cannot meet any more the increased flow rate demand resp. the higher solid content of the filter cake.

5.2. Efficiency

Based on the high and long lasting active pressures, the new press achieved final solid matter qualities in the filter cake unter all so far tested applications, as they are only known from chamber-, plate-, and membrane filter presses. Especially in direct efficiency comparison with decanters and belt filter presses. The possible dewatering results in connection with the conditioning with organic flocculation aid are clearly exceeded. Since sewage sludge solid matter results ranged between 35 - 55 % dry matter content industrial suspensions partly were dewatered up to more than 75 % dry matter continuously.

6. PARALLEL TEST RESULTS WITH CENTRIFUGE, BELT FILTER PRESS, AND CHP-FILTER PRESS

In order to make the efficiency of the new CHP-filter press judgable, parallel test results under technical conditions with centrifuge, belt filter press and CHP-filter press are shown in the following:

Based on an anaerobic stabilized sewage sludge of a municipal water treatment plant with approx. 6,4 % dry solids content and after thickening by gravity a sludge conditioning with an organic polyelectrolyte (BASF CF 600) was performed before the dewatering.

The quantity of polymers used for all three dewatering processes was kept in the same range in order to have absolutely equal basic conditions for the three machines. The individual dewatering results for the dewatering presses are compiled in fig. 3.

During the test of the CHP-filter press two different final pressure stages in the system were applied. At an final pressure in the H-phase of 5 bar for about 4 minutes solid matter results between 34 and 38 % in the filter cake were reached.

At an increased final pressure of 15 bar for about 8 minutes a clear improvement of the results, namely between 40 and 44,5% solid matter was achieved. When comparing these results achieved by the technical tests as those achieved by Perchthaler and Stefou for the dewatering of sewage sludge with high pressure zones of a belt filter press (see fig. 4) a principle conformity in results is given.

REFERENCE

1. PERCHTHALER, H. and STEFOU, St. (1984). Filter belt presses with high-pressure zones. Recycling International, EF-Verlag Berlin, 416-421

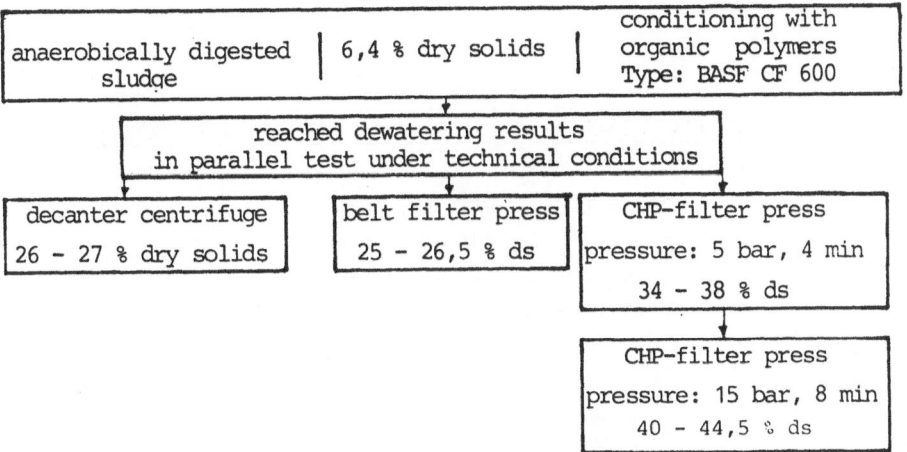

| anaerobically digested sludge | 6,4 % dry solids | conditioning with organic polymers Type: BASF CF 600 |

reached dewatering results
in parallel test under technical conditions

| decanter centrifuge 26 - 27 % dry solids | belt filter press 25 - 26,5 % ds | CHP-filter press pressure: 5 bar, 4 min 34 - 38 % ds |

CHP-filter press
pressure: 15 bar, 8 min
40 - 44,5 % ds

Fig. 5: Results of a parallel test with 3 different dewatering machines
with special reference to the CHP high pressure filter press

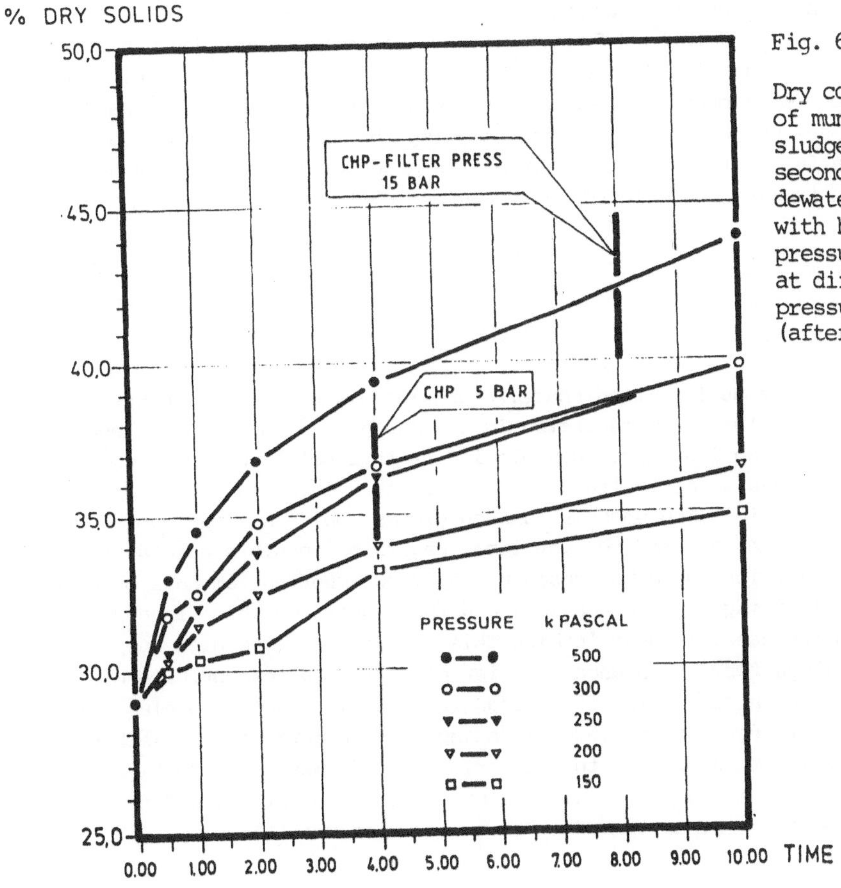

% DRY SOLIDS

Fig. 6:

Dry content
of municipal
sludge after
secondary
dewatering
with high-
pressure zones
at different
pressure
(after 1)

PRESSURE	k PASCAL
●—●	500
○—○	300
▼—▼	250
▽—▽	200
□—□	150

CHP-FILTER PRESS 15 BAR

CHP 5 BAR

A POT EXPERIENCE WITH A HIGH LEVEL COPPER SEWAGE SLUDGE

A. MANRIQUE, I.ARROYO, A.M.SANZ,
J.M.GARCIA DE BUSTOS,A.M.NEBREDA
Sección de Análisis Ambiental
Servicio de Investigaciones Agrarias
Junta de Castilla y León

Abstract

A pot experiment has been performed with alfalfa crop, under contro---
lled temperature and humidity conditions, in order to assess the res--
ponse of this crop when applying a sludge which is rich in copper. Different dosages of sludge have been used to obtain concentrations of -
50, 100, 200, 400 and 800 mg. of dry soil matter. For dosages greater/
than 300 mg. of CU/kg. we have obtained a diffusor effect which rea---
ched 20% of production loss when the dosage of copper added was 800 mg/
kg. dry weight of soil. When analysing the aerial parts and roots of -
alfalfa, we found that copper was retained by the roots, and only small
amounts were accumulated in the aerial parts.

1. INTRODUCTION

During the purification phase, approximately 90 % of the copper present
in the waste water is concentrated in the sludge (1). As a consequence of -
this, there in an important accumulation of this metal in sludge which may/
later on be used in agriculture. This fact may provoke soil contamination -
unless caution in taken when using such sludge. Quite a large number of - -
studies have been carried out until now regarding possible harmful effects/
(2, 3, 4, 5) which could be caused by applying important amounts of copper
to the soil, damaging both the soil and the vegetation which is developed -
on it. In our case, we have tackled this study due to the high copper con--
tent of sludge from the sewage purifier plant in the town of Palencia, which
comes from waste water tips of an industry that is devoted to electrolysis/
of this metal. Since this sludge is being used in agriculture, we have felt
it worth while to carry out this experiment in flower pots, to find the behaviour of this sludge in an alfalfa crop (Medicago sativa) under contro- -
lled temperature, humidity and illumination time conditions.

2. MATERIAL AND METHODS

As we have already said, for this experiment a sludge was used from — the sewage purifier plant of the town of Palencia, on a soil near our labo ratory and using an alfalfa crop (Medicago sativa). The characteristics of the sludge and soil were determined at the laboratory and were as follows:

Sludge:

pH - H$_2$O	7.1	K$_2$O %	0.24
N $_{total}$ %	0.91	M.O. %	27.8
P$_2$O$_5$ %	1.10	Humidity % 31.2	
		Conductivity 3,370 and mhos/cm.	

Metal content - aqua regia	Metal content NH$_4$ A$_c$/EDTA pH 4.65
Cd - 24.4 ppm	Cd - 12.6 ppm
Cr - 40 ppm	Cr - 1.4 ppm
Cu - 5.280 ppm	Cu - 2.500 ppm
Mn - 151 ppm	Mn - 74.1 ppm
Ni - 500 ppm	Ni - 150 ppm
Pb - 491 ppm	Pb - 31 ppm
Zn - 2.200 ppm	Zn - 356 ppm

Soil:

pH	7.8	Conductivity 882 & mhos/cm	
N$_T$ %	0.06 %	Sand 36.9 %	
P$_2$O$_5$ assimilable - 16.7 ppm		Lime 38.5 %	
K$_2$O assimilable - 316 ppm		Clay 24.6 %	
M.O. 0.78 %			

Heavy metal content Aqua regia	Metal content NH$_4$ A$_c$/EDTA pH 4.65
Cd - 0.06	Cd - 0.04
Cr - 18.5	Cr - 2.5
Cu - 24.5	Cu - 10.5
Mn - 196	Mn - 91.5
Ni - 18.7	Ni - 1.3
Pb - 0.02	Pb - 0.01
Zn - 72.6	Zn - 11.2

For this study, the plants were cultivated in 4 kg. plastic flower pots adding different dosages of sludge until copper concentrations of 0, 50, 100 200, 400 and 800 mg. of Cu/kg. dry soil matter were obtained. Four repeat — tests were made with each of the different dosages, and were placed in a --- growth chamber having the following temperature, humidity, and lighting time conditions:

	Temperature	Humidity	Time
Day	25º	80%	10 hours
Night	14º	80%	14 "

Under these conditions they were maintained for four months and three/ cuts were made to control the production in fresh and dry weight. The ti to lerance indexes were calculated and the ratios of concentration C, accumula tion A for copper, with the help of the repeat mediums, as other authors ha ve done (6).

The alfalfa samples (aerial parts and roots) were digested via humid – means with a mixture of $NO_3H - ClO_4H$ 2:1 v/v. As soon as the soil was dry,/ it was digested with boiling aqua regia for two hours to determine the to-- tal content and with Ac NH4 / EDTA at pH 4.65 for the assimilable ones. The copper was analysed in all the samples prepared in this way, using a PERKIN ELMER mod. 460 Atomic Absorption Spectrophotometer.

3. RESULTS AND DISCUSSION

The tolerance index ranges for each cut made and for the average of all the cuts are shown in table I. Here some minor differences are observed with the results of the three cuts that have been made, but generally speaking – it is found that for dosages of 50, 100, 200 ppm, the tolerance indexes are greater than 1.00, which means that by adding these dosages, there is a be- neficial effect on the alfalfa crop under review.

TABLE I

Tolerance indexes for each cut and average value

Dosage	1st cut	2nd cut	3rd cut	Average Value
Not treated	1.00	1.00	1.00	1.00
50 ppm	1.02	1.11	1.27	1.13
100 ppm	0.87	1.10	1.27	1.07
200 ppm	0.93	1.08	1.27	1.09
400 ppm	0.83	0.95	0.97	0.91
800 ppm	0.74	0.80	0.87	0.80

When we exceed the 200 ppm mark, there is an important jump in the to- lerance index, and moreover becomes less than 1.00, i.e. a prejudicial ef-- fect is detected. Although we do not have the intermediary dosages in the – experimental desingn, we calculate that the prejudicial effect of excess co pper is in the region of 300 p.p.m. It must furthermore be mentioned that – as the different crops are cut, the tolerance indexes increase and for dosa ges of 50, 100 and 200 are the same, such that the presence of copper at –– these dosages does not modify the behaviour of the crop.

The drop in production which we have observed is in consonance with fi gures obtained by other authors (7.8), although since the conditions are –– not the same, this should only be considered as orientative, reaching pro--

duction losses close on 20% for dosage of 800 p.p.m. of copper added in the form of sludge.

Really, we feel that the above mentioned value of 300 p.p.m. as production loss limit is quite high and we understand that this is because this is a soil with mildly alkaline pH, because as the pH increases, the copper/ is immobilized (9) in the soil.

Table II shows the copper concentrations in alfalfa, and soil after the experiment and Table III shows the concentration and accumulation in aerial parts and roots.

TABLE II

Average copper content in alfalfa and soil (expressed in mg/kg. dry matter)

	Aerial parts	Roots	Total Soil	EDTA soil
Not treated	11.80	21.86	20.6	10.5
50	17.56	32.1	78.1	31.2
100	14.32	44.93	124.93	52.6
200	14.05	54.68	203.40	98.7
400	16.15	92.83	433.7	187.2
800	15.84	154.34	738.6	376.4

TABLE III

Relation of concentration C and accumulation A in aerial parts and roots

	Aerial parts		Roots	
Dosage	C	A	C	A
50	1.48	0.28	1.46	0.49
100	1.11	0.05	2.05	0.56
200	1.19	0.05	2.50	0.37
400	1.36	0.05	4.24	0.40
800	1.34	0.03	7.06	0.36

In the above tables, we can observe that the copper concentration in aerial parts is very small, and does not undergo any increase when the added dosage is increased. When comparing ourresults with those of other authors/ (9.10), we observe that they are very similar to the ones which these authors found in soils with concentrations or normal ones. When observing the results of the roots however, a sharp increase is detected when the amount of copper which is added is increased, and we consequently understand that alfalfa retains the metals at root level. When comparing with the values of - phytotoxicity levels, (11) we find that these levels are not reached in our study; in other words, the negative effect of copper is not detected in the aerial parts of the plant. Furthermore, although there is a reduction in production at higher dosades, the quality of the alfalfa undergoes no variation which would render it unserviceable as cattle fodder.

The concentration and accumulation relations ratify all that has previosly been said regarding incorporation of copper in the roots, not in the aerial parts.

Making a correlation study between copper contents in roots and alfalfa production, we find that as the concentration increases in the roots, production drops (correlation index - 0.9804 with a significance level of 0.01) If, on the contrary we make the correlation between copper in roots and leaves, we find there is no correlation, in other words, the copper in the soil does not pass to the leaves through the roots. If however, when studying the correlation, we eliminate the results of dosages of 400 and 800 p.p.m. of - copper, we get a very higt correlation coefficient (0.9780) between the copper content in leaves and production, which means that in not very high dosages of sludge there is a beneficial effect on the alfalfa crop and that - it is with concentrations that are superior to 300 ppm when copper that is/ present in the sludge of the Palencia purifier plant begins to have negative effects on production.

REFERENCES

(1) LESTER,J.N., HARRISON,R.M., PERRY,R. (1979). The balance of heavy metals trough a sewage treatment works. Science of the Total Environ 12,13-23.

(2) BECKETT,P.L. (1980). Cooper in Sludge. Are the toxic effects of copper and other heavy metals additive?. EEC Workshop on Problems Encountered with Copper Bordeaux. Octubre 1980.

(3) DAVIS,R.D. (1980). Copper uptake from soil treated with sewage sludge/ and its implications for plant and animal health. EEC Workshop on Problems Encountered with copper Bordeaux. Octubre 1980.

(4) WEBBER,M.D.,SOON,Y.K.,BATES,T.E.,HAQ,A.U. (1980). Copper toxicity to -- crops resulting from land application of sewage sludge. EEC Workshop - on Problems Encountered with copper Bordeaux Octubre 1980.

(5) COTTENIE,A.,KIEKENS,L.,VAN LANDSCHOOT,G. (1983). Problems of the mobility and predictability of heavy metal uptake by plants. Processing -- and Use of Sewage Sludge Brighton. Setiembre 1983.

(6) COTTENIE,A. and CAMERLYINCK.R. (1981). Specific behaviour of somo mono cotyledon and dicotyledon species with regard to excess of trace metals In. Trace Metals in Agriculture and in Environment. Ed. A. Cottenie.

(7) WILLIAMS,J.H. (1975). Use of sewage sludge on agricultural land and -- the effects of metals on crops. Journal of the Institute of Water Pollution Control. 6 . 635-644

(8) MARKS,M.J., WILLIAMS,J.H.,CHUMBLEY,C.G.(1980). Field experiments tes-- ting the effects of metal contaminated sewage sludges on some vegeta-- ble crops. Inorganic Pollution and Agriculture MAFF 326/HMSO.

(9) HARTER,R.D. (1983). Effect of soil pH on adsorption of lead, copper, -- zinc and nickel. Soil Science Society of America Journal 47-1,47-51

(10) CAST Commiter (1976). Application of sewage sludge to cropland. Appraisal of potential hazards of the heavy metals to plants and Animals. EPA 430/9-76.013

(11) CHETELAT,A.A. (1978). Impact des elements trace sur la croissance des plants cultivées (une etude bibliographique). La Recherche Agronomique en Suisse. 17 3/4 211-227

CHEMICAL PROPERTIES OF SEWAGE SLUDGES
PRODUCED IN THE PIEDMONTESE AREA (ITALY)

G. PICCONE, C. SAPETTI, B.BIASIOL, G. DE LUCA and F. AJMONE MARSAN
Istituto di Chimica agraria. University of Turin (Italy)
A. TROMBETTA
Assessorato Ambiente ed Energia. Regione Piemonte

SUMMARY
Sewage sludges from representative treatment plants of the Piedmont region (Italy) were investigated to verify their suitability to agricultural uses.
A major limitation is the concentration of heavy metals from wastewaters discharged by the industry of the area.
The main aim of this research was therefore to set the conditions for obtaining a fertilizing effect without causing toxic accumulation in crops.

1. INTRODUCTION

The Istituto di Chimica Agraria and the Istituto di Microbiologia e Industrie Agrarie of the University of Turin in collaboration with the Istituto per le Piante da Legno e l'Ambiente (IPLA) and the Assessorato per l'Ambiente e l'Energia of the Regione Piemonte have been carrying out for three years a joint research project to examine the possibility of employing sewage sludges in agriculture.

The major subjects were the following:
- Preparation of disposal regulation according to the national laws (319/76 and 915/82).
- Testing of the analytical methods for sludges, soils and plant tissues to verify the effects of sludges.
- Characterization of sludges from representative treatment plants.
- Toxicity and accumulation of heavy metals in soils and crops and fertilizing effects.

A preliminary study on the influence of sewage sludges on wheat grain production and composition conducted in collaboration with the IPLA is also being presented to this Meeting.(2)

This study deals with the characterization of the sludges to assess their suitability to the use as fertilizer in relation to the variation of the composition with time, the nutrients content and the concentration of potentially toxic substances.

Only sludges from domestic or mixed (domestic plus industrial) wastewaters were examined.

2. TREATMENT PLANTS IN PIEDMONT

The number of ready to operate and operating plants is remarkably high. Of the 926 planned plants in the region, 191 are already regularly running. Furthermore, 64 associations of communes are provided with plants, 41 of which are working.

The great variety of plants and the consequent difficulty of classification is to be noted. If communal and association plants are considered separately a possible way of classifying them is to take in to account the number of served inhabitants thus excluding plants of small size (less than 5.000 inhabitants) whose production causes little disposal problems. The type of treatment is also to be considered, with special regard to sludge production (aerobic or anaerobic digestion, drying beds). The composition of wastewaters, roughly related to the ratio industrial domestic waters and to the type of industry, can be another criterion of classification.

3. CRITERIA FOR PLANTS SELECTION

On the basis of the above indicated parameters the plants can be classified as shown in tables I and II.

Table I. Communal treatment plants.

Province	Total n. of plants	Regularly working plants	Served pop. >5000	Aerobic working	Anaerobic working	drying beds only
Alessandria	179	37	3	2	-	1
Asti	91	22	3	1	1	i
Cuneo	184	59	10	1	2	7
Novara	127	22	3	1	1	1
Vercelli	132	8	1	1	-	-
Torino	213	43	8	2	1	5
TOTAL	926	191	28	8	5	15

Plants were selected that provided a regular production of sludge discarding plants under modernization; the obsolete ones were also excluded.

Table II. Association treatment plants

Province	Total n. of plants	Regularly working plants	Aerobic working	Anaerobic working
Alessandria	12	8	5	3
Asti	2	–	–	–
Cuneo	5	5	2	3
Novara	21	16	12	4
Vercelli	8	6	6	–
Torino	16	6	3	3
TOTAL	64	41	28	13

4. SELECTED TREATMENT PLANTS DESCRIPTION

The following plants were chosen as representative:

- Biella (province of Vercelli). The plant serves a population of 35.000 and 15.000 population equivalents (PE). Detergents and dyes are the most important polluting agents discharged by the textile industry. Primary and secondary treatment with anaerobic digestion, screen filtration and drying.
- Bra (province of Cuneo). Communal plant serving 22.000 inhabitants and 8.000 PE. Activated sludges and total oxidation, secondary aerobic digestion and screen filtration. Industrial waters of various type are treated.
- Collegno (province of Torino). Association plant in an industrial suburban area serving a population of 65.000 plus 65.000 PE. Metal and mechanical and related industries. Two settling tanks and anaerobic digestion. Sludge dried by screen filtration.
- Canelli (province of Asti). Communal plant; served population: 10.500. Activated sludges, aerobic digestion, centrifugation and drying beds.
- Cassano (province of Alessandria). Association plant; 3.000 residents and 60.000 PE served. Anaerobic digestion and drying by screen filtration.
- Pianezza (province of Torino). Association plant serving 25.000 residents and 9.000 PE. Technically similar to Collegno.
- Fossano (province of Cuneo). It serves a population of 12.000. Activated sludges with total oxidation, aerobic digestion, screen filtration. No industrial discharges are treated.
- Verzuolo (province of Cuneo). 5.000 people are served. Activated sludges with total oxidation, drying beds.

5. CHEMICAL COMPOSITION OF SLUDGES

The following analytical parameters were determined: pH, salinity, moisture content, ashes, C, N, P, K, Cu, Zn, Pb, Ni, Cr, Cd, As, Hg, B, anionic surfactants and volatile phenols.

Currently available analytical methods were used with modifications for the specific type of sludge when needed (1), (4), (6).

6. RESULTS AND DISCUSSION

The statistical analysis of the data showed significant changes in C, ashes and As contents with the season.

A significant correlation resulted between the served area and pH, P, Ni, Cu, Zn, Pb, Cr, salinity, ashes, C and surfactant.

The Pianezza and Biella plants (medium size, anaerobic) are different in pH values from the others; the same holds for Pianezza and Collegno surfactants load.

Plants serving industrial areas (Pianezza and Collegno) show a different concentration of heavy metals from the others (Tab.III).

Table III. Heavy metals significantly differing between plants.

	Bra	Collegno	Canelli	Cassano	Pianezza	Fossano	Verzuolo
Biella	Pb,Ni Cr	Cu,Zn	Pb,Cr	Pb,Ni Cr	Cu,Zn Cr	Cu	Pb,Cr
Bra		Cu,Zn Cr		Pb	Cu,Zn,Pb Ni,Cr	Pb	
Collegno			Cu,Zn Cr	Cu,Zn Pb,Cr	Cr	Cu,Zn	Cu,Zn Pb
Canelli					Cu,Zn,Pb Ni,Cr	Cu,Pb	
Cassano					Cu,Zn,Pb Ni,Cr	Zn,Pb	
Pianezza						Cu,Ni Cr	Cu,Zn,Pb Ni,Cr
Fossano							Cu,Pb

REFERENCES

(1) ANDERSON, A. (1976). On the determination of some heavy metals in organic material. Swedish J. Agric. Res. 6, 145-150.

(2) BARBERIS, R., CONSIGLIO, M., PICCONE, G., DE LUCA, G., TROMBETTA, A. (1985). Productivity and quality of cereal crops grown on sludge treated soils. Symposium on Processing and Use of Organic Sludge and Liquid Agricultural Wastes, Roma.

(3) GENEVINI, P.L., VISMARA, R., MEZZANOTTE, V. (1983). Utilizzo agricolo dei fanghi di depurazione. Ingegneria Ambientale, 12, 361-493

(4) MUNTAU, H., LESCHBER, R., BAUDO, R., L'HERMITE, P. (1983). Problemi inerenti l'analisi dei fanghi di depurazione destinati all'agricoltura. Acqua-Aria, 9 , 951-954.

(5) NAPPI, P., BARBERIS, R., CONSIGLIO, M., IODICE, R., TROMBETTA, A. (1985). Evaluation of sludge phytotoxicity by biological essais. Symposium on Processing and Use of Organic Sludge and Liquid Agricultural Wastes, Roma.

(6) SOMMERS, L.E., NELSON, D.W., YOST, K.I. (1976). Variable nature of chemical composition of sewage sludges. J. Environ Qual., 5, 303-306.

AUTOTHERMIC SLUDGE INCINERATION

P. OCKIER

W.Z.K.

Coastal Water Authority
Ostend, Belgium

Summary

The possiblities of autothermic sludge incineration in a fluidized
bed furnace are explained by means of thermal balances.
More stressed is the existing sludge treatment plant of Bruges.

1. INTRODUCTION
    Incineration is an expensive methode of sludge treatment.
Decisive are the fuel costs. A self sufficient incineration during continuous operation is becoming an absolute must.

2. SLUDGE INCINERATION IN A FLUIDIZED BED FURNACE
    The fluidized bed furnace is the most suitable furnace for sludge
incineration because of a good burn out, a low air factor and a concurrent
system which guarantees an odourless combustion.
The flue gas temperature amounts to 750-850° degree.
Different possibilities of energy recovery on flue gases are used in
practice :

             - preheating of combustion air
             - preheating of sludge in an indirect dryer by means of ther-
               mal oil (1) (2) or steam (3)
             - thermal conditioning of sludge (4) (5).
    The figures 1 & 2 give the results of the heat balance of a fluidized
bed furnace. The operation conditions are mentioned. The difference between those figures is a thermal sludge predrying of 0,2 kg water/kg wet
sludge. Table I is concluded from the curves.

TABLE I
Autothermic sludge incineration

Air factor : 1,2 Flue gas temperature exit furnace : 800°C	Centrifuge beltpress	Filterpress[*]
1. Moderate digested sludge (3.000 kcal/ 　　　　　　　　　　　　　kg dry solids) 1.1. without predrying 　- without air preheating 　- with air preheating to 600°C 1.2. with predrying 　- without air preheating 　- with air preheating to 600°C	 39 29 34 26	 46 35 40 31
2. Raw sludge (3.500 kcal/kg dry solids) 2.1. without predrying 　- without air preheating 　- with air preheating to 600°C 2.2. with predrying 　- without air preheating 　- with air preheating to 600°C	 35 26 31 23	 40 31 36 28

[*]Filterpress : 25% anorganic additional materials based on
　　　　　　　　sludge dry solids

It gives the dry solids content by which autothermic operation occurs under the operation conditions mentioned before.

When a centrifuge or beltpress is used an air preheater and a sludge dryer are advised.

In the case of a filterpress, air preheating is sufficient for raw sludge. Disadvantages are that, after the filterpress, a buffer has to be foreseen and the quantity of ashes quasi doubles. New on the market is the continuous working high pressure filterpress (6). The sludge is conditioned by polyelectrolytes. The results of dry solids content are comparable with those of a classical filterpress.

3. THE SLUDGE TREATMENT PLANT OF BRUGES (1) (2)

The wastewater treatment plant of Bruges is built for 305.000 population equivalents (1 p.e. = 54 g BOD), from which 116.000 from the industry.

The water treatment exists of an activated sludge process. The figure 3 gives a survey of the sludge treatment process. The mixed primary and activated sludge is thickened and dewatered in centrifuges. A part is thermally dried and mixed with the mechanically dewatered sludge to about 30 % dry solids content. That sludge is burned in a fluidized bed furnace. The air is preheated by the flue gases, which are further cooled to about 320°C by air mixing and finally dedusted in an electrostatic precipitator.

The sludge is indirectly dryed in a multiple hearth dryer by means of thermal oil, which is warmed up in a heat exchanger between flue gases and oil. In the multiple hearth dryer the sludge vapours are diluted by air to avoid condensation in the following pipes. The sludge vapours are conducted to the furnace to burn the volatile solids. After the dry sludge storage bin, sludge can be removed to agriculture.

The figure 4 gives an example of material balance for sludge dry solids. The multiple hearth dryer is designed on the one side to dry enough
sludge so that there is an autothermic incineration and on the other hand
it permits a sludge storage (for removal to agriculture or to mix it during low centrifugal efficiency and/or low heat value of the dry solids).

The figure 5 shows the thermal balance of sludge with 23 % dry solids
and a heat value of 3.500 kcal/kg dry solids. Without thermal drying they
need 150 kg/h supplemental fuel.

Two alternative draw ups are investigated for the heat exchanger between flue gases and thermal oil, namely before and after the elctrostatic
precipitator. This last solution was selected because it garantees a safer
operation - the flue gas temperature is low (limited to 320°C) so that
                fewer material problems occur and degradation does not exist
                for the oil (425°C) ;
             - the flue gases are already dedusted so that there exists
                slighter danger for erosion.
The disadvantages are :
             - lower beginning temperature and so the requirement for a big-
                ger heat exchanging surface ;
             - bigger cooling air required before the electrostatic precipi-
                tator.
Sludge incineration is started in 1985.

REFERENCES

(1)   P.OCKIER, Economische en energetische aspekten van de slibverwerking.
      Symposium IAWPRC Oostende, 1983, Becewa NR 70, 1983, 55-63.
(2)   E.PUTTAERT, Problematiek betreffende diverse verwerkingssystemen
      voor slib - autonome slibverbranding, 1984, K.V.I.V.-voordracht
      te Sint-Niklaas.
(3)   E. SCHURR and W.VATER, Energiesparendes Schlammverbrennungsofen
      im Stuttgarter Hauptklärwerk Mühlhausen in Betrieb genommen, Korres-
      pondenz Abwasser, 6/1982, 405-409.
(4)   J.P.LEGLISE and J.C.CORNIER, Heat recovery from sludge. Symposium
      EWPCA - ISWA Munich, 1984, 533-542.
(5)   J.C.CORNIER, Maximum use of sludge energy to achieve a complete
      treatment. Congress Berlin, 1984, 484-489.
(6)   U. LOLL, Continuous working high pressure filterpress, the new gene-
      ration of the chamber - and membrane filterpresses. Congress Berlin,
      1984, 422-427.
(7)   TH.A.A.VAN OVERBEEK, Slibverbranding, al dan niet gekombineerd met
      huisvuil, H_2O, 1982, nr 6, 112-117.
(8)   H.J.EGGINK, Verbrandingswaarde van zuiveringsslib en de vereiste
      ovenkapaciteit. Symposium IAWPRC Oostende, 1983, H_2O, 1983, nr 16,
      492-496.
(9)   F.A.FASTENAU, Thermische balans van slibverbranding in wervelbedovens
      en toepassingsmogelijkheden, H_2O, 1985, nr 6, 127-132.

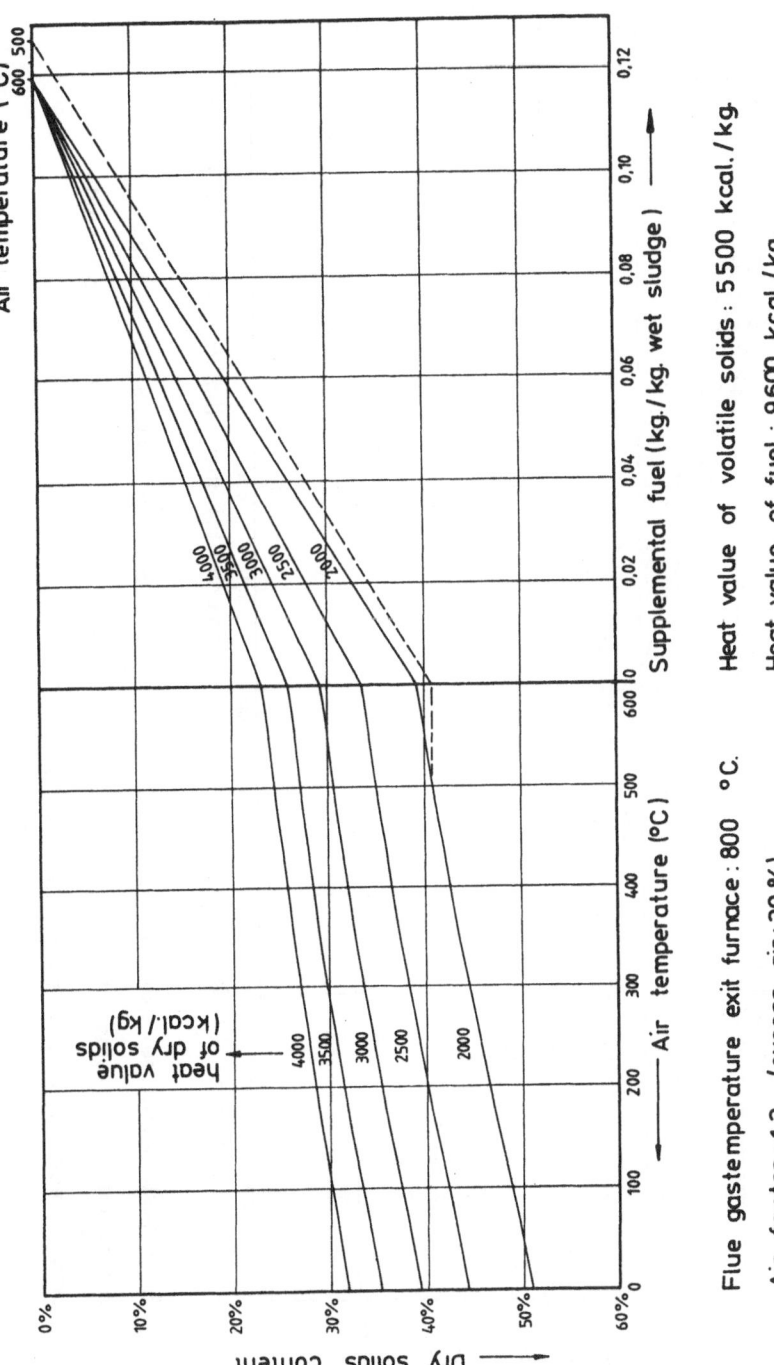

Flue gastemperature exit furnace : 800 °C.   Heat value of volatile solids : 5 500 kcal./kg.

Air factor : 1,2   (excess air : 20 %)   Heat value of fuel : 9600 kcal./kg.

Fig. 1 : Incineration curves of sludge without predrying

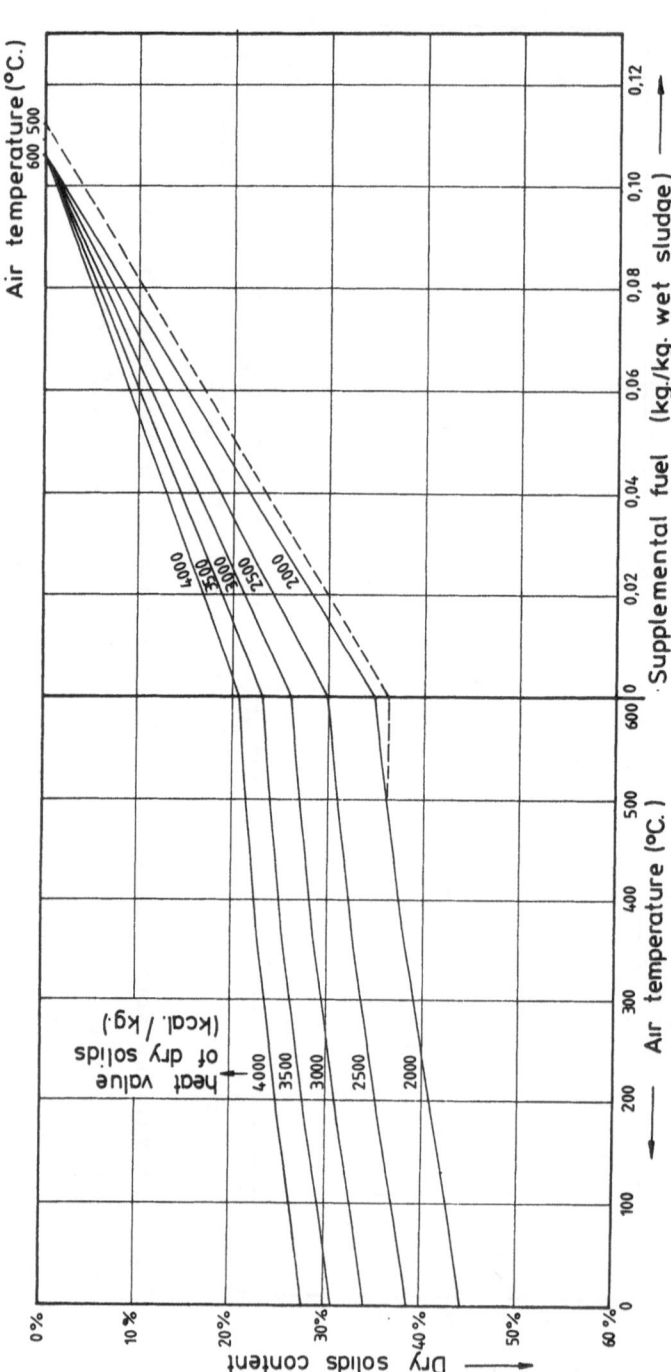

Air temperature (°C.)

Dry solids content

heat value
of dry solids
(kcal./kg.)

4000
3500
3000
2500
2000

Air temperature (°C.)

Supplemental fuel (kg./kg. wet sludge)

Flue gastemperature exit furnace : 800 °C.   Heat value of fuel : 9600 kcal./kg.

Air factor : 1,2 (excess air : 20 %)   Preheating : 0,2 kg. water / kg wet sludge

Heat value of volatile solids : 5500 kcal./kg.

Fig. 2 : Incineration curves of sludge with predrying

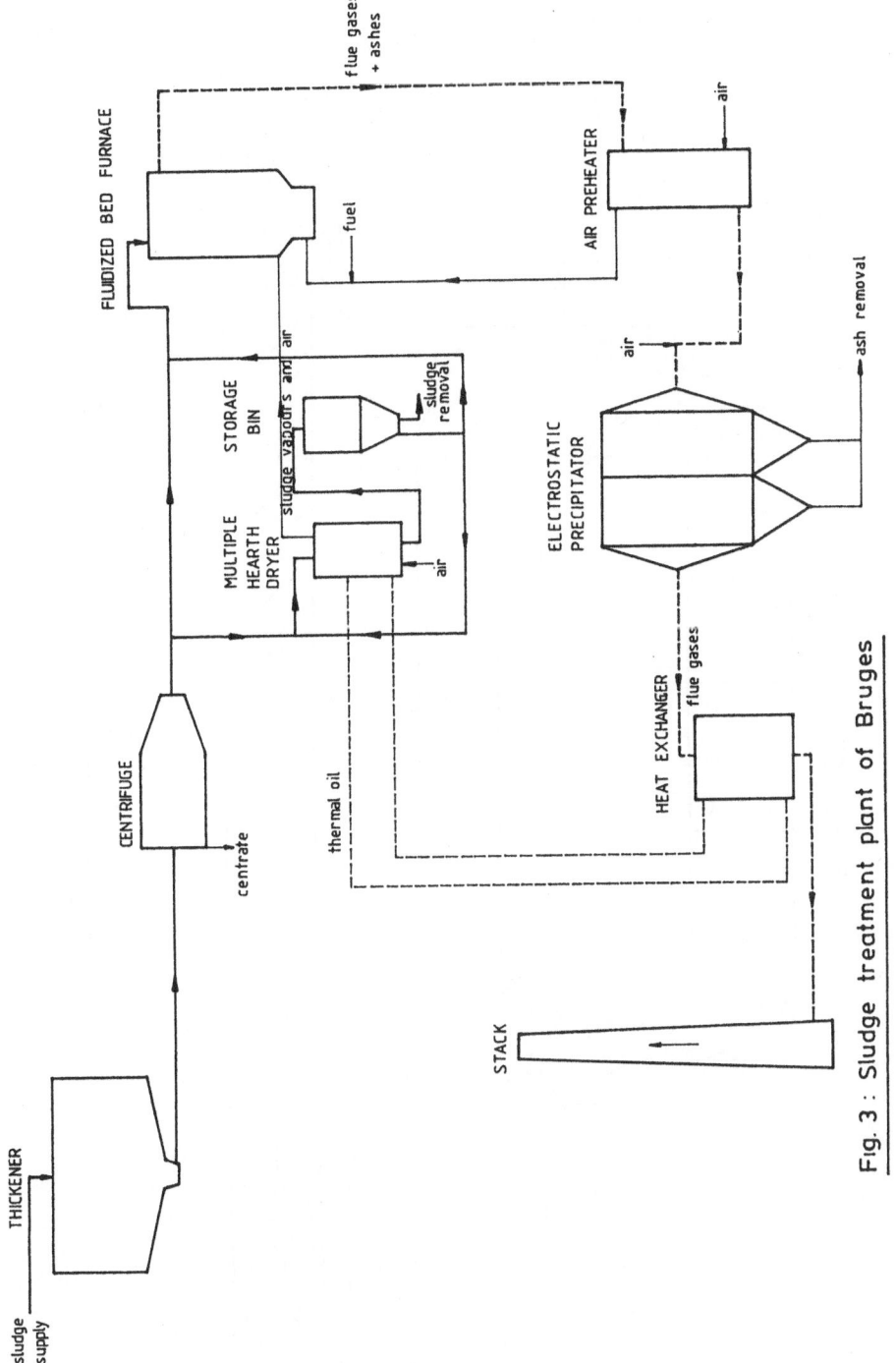

Fig. 3 : Sludge treatment plant of Bruges

- 475 -

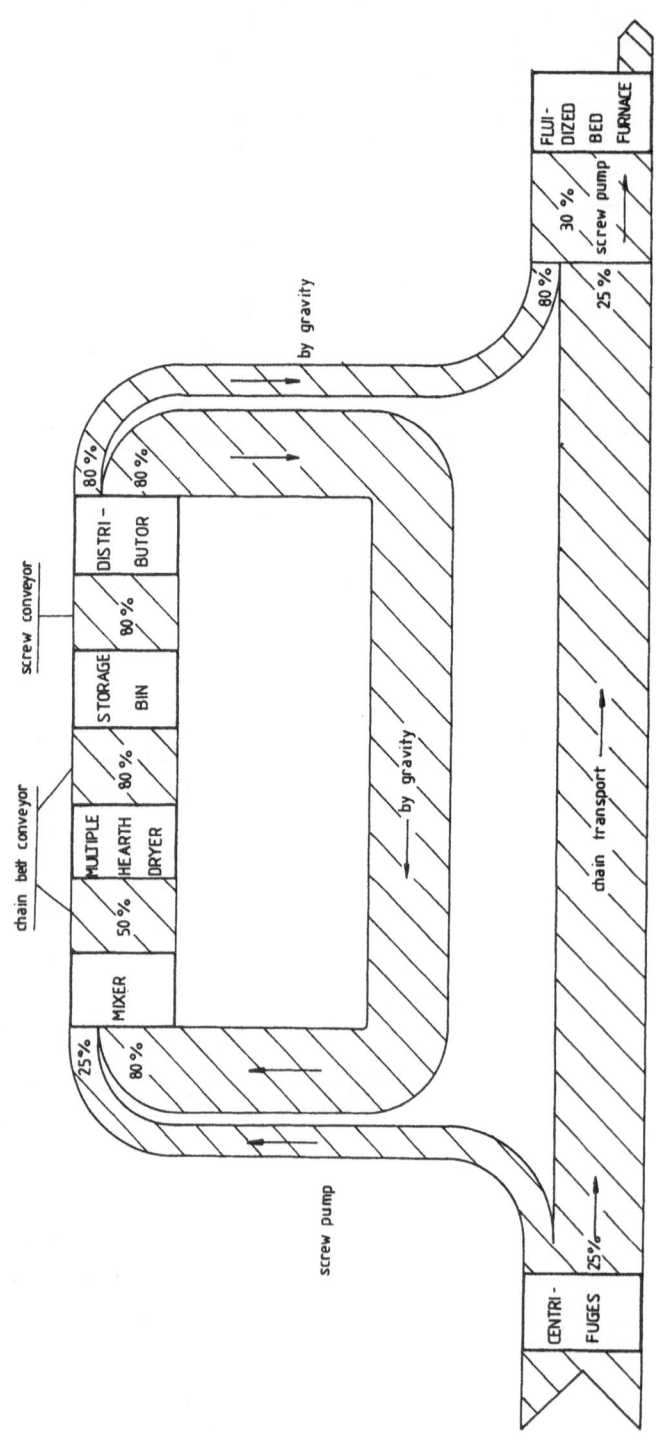

Fig. 4: Material balance of sludge dry solids

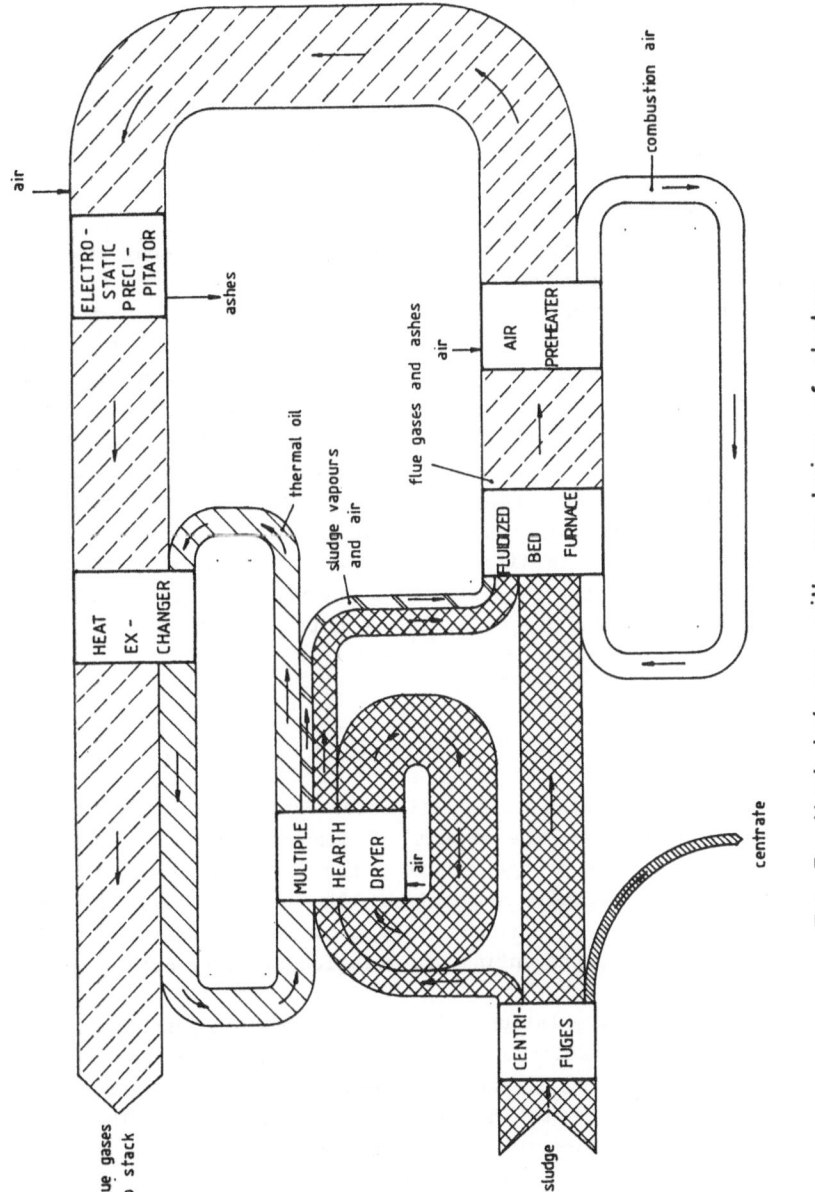

Fig. 5 : Heat balance with predrying of sludge

air

ELECTRO-
STATIC
PRECI-
PITATOR

ashes

thermal oil

sludge vapours
and air

combustion air

flue gases and ashes

air

AIR
PREHEATER

FLUIDIZED
BED
FURNACE

HEAT
EX-
CHANGER

MULTIPLE
HEARTH
DRYER

air

flue gases
to stack

CENTRI-
FUGES

sludge

centrate

MODIFICATION OF HEAVY METALS SOLUBILITY IN SOIL TREATED WITH SEWAGE SLUDGE

G. PETRUZZELLI, G. GUIDI and L. LUBRANO
Institute of Soil Chemistry C.N.R. Pisa Italy

Summary

Aerobic and anaerobic sewage sludges were applied to a sandy loam soil. Sludges were apllied at a rate corresponding to 30 t/ha/yr of organic matter and planted with corn. Soil samples were collected at the beginning of the experiment, every year, one month after sludge application and at harvest time. A sequential fractionation (H_2O, KNO_3, DTPA) was used to study the solubility of heavy metals in the soil. The exchangeable form of metals but Pb showed an increase after each addition of sludge and a decrease at harvest time. A similar trend was found for metals in complexed or adsorbed forms. Interactions among heavy metals and organic matter were also studied. In untreated soil metals were linked only by organic matter of a low molecular weight; in the sludge-treated soils metals distributed between fractions of humic materials of both high and low molecular weight.

1. INTRODUCTION

In Italy, where arable soils are intensively cultivated and where there is a deficiency in the amount of manure applied to soil. the agronomic utilization of sewage sludge for its high content of organic matter has been increasingly emphasized The use of sewage sludge or compost on soil must be in a manner that produces beneficial effects on soil productivity, while preventing accumulation of heavy metals in soils and crops.

Evidence that accumulation of heavy metals is associated with possible environmental deterioration stressed the need for a better knowledge of the chemistry of these elements in soil. Since the greatest amount of heavy metals in the human diet arises from soil. The present investigation was undertaken to evaluate the influence of repeated sewage sludge applications. on the solubility of heavy metals in soil, in a field experiment which lasted five years.

2. MATERIALS AND METHODS

Field trials were carried out on plots (500 m^2) of a sandy loam soil which contained on a dry weight basis 0.9% organic matter, 10.0% clay, 14.1% silt, 75.9% sand. The pH was 5.8, $CaCO_3$ was absent and the cation exchange capacity was 13.4 me/100 g.

Treatments included a control, aerobic (AES) and anaerobic (ANS) sludge, compost of the organic fraction of urban refuse with aerobic (CAS) and anaerobic (CANS) sludge. The yearly addition rates of sludges and composts were made equivalent 150 tons of manure/ha/yr on an organic carbon basis that corresponded to about 30 metrie trons of dry matter/ha/year. Organic materials were surface-applied and were plowed in before seed-bed preparation. All plots planted to corn were also treated with inorganic fertilizers (250, 120 and 120 kg of N, P_2O_5 and K_2O respectively/ha/year) because in the experiment sludges and composts were being mainly considered on the basis of their supply of organic matter to soil.

Soil samples (0-20 cm depth) were collected at the beginning of the experiment before any treatment (t_0) and every year one month after the sludge application (odd times) and at harvest time (even times). Solubility of heavy metals in the soil was investigated by a sequential extraction procedure with H_2O, 1M KNO_3, and DTPA (Diethylentriaminopenta-cetic acid). Heavy metals were determined by atomic adsorption spectro-scopy (Perkin Elmer mod 403, equipped with a background corrector). Linkages between heavy metals and organic matter were investigated by extracting the organic matter from the sludges and soils with 0.5 N NaOH. Samples of alkali-extracted organic matter were gel-filtred on Sephadex G-200 with 0.1 Borax as an eluant. In each chromatographic fraction heavy metals were determined by atomic adsorption.

3. RESULTS AND DISCUSSION

Among the various treatments, only the addition of aerobic sludge brought about a rise in soil pH. The pH level dropped in most of the sludge soils. However, the water soluble fraction in the investigated soils was generally negligible, nevertheless in the last year of the experiment at time t_7 the soluble Zn was 1.4 and 1.1 µg/g of soil for AES and ANS respectively.

Solubilization of heavy metals by KNO_3 and DTPA over time are reported in Tables 1-5. Certain trends are common to almost all investigated metals: samples collected one month after the incorporation of organic materials in the soil showed an increased extractability both in KNO_3 and DTPA while the opposite was found at harvest time. In the case of AES exchangeable metals (KNO_3 extraction) generally dropped to those found at t_0. On the contrary Pb increased with time showing a KNO_3

extractability more than doubled. Similar data were obtained for ANS where Pb showed a more anomalous trend, with great variations in the five years of the experiment: i.e. extractable amount at $t_4 = t_0$, while at t_8 it was three times t_0.

In soil treated with composted sludges KNO_3 extraction solubilized nearly the same quantities of heavy metals. A certain increase was found at the end of the experiment for Cd: from 0.7 ppm to 1.4 ppm and for Pb from: 3.0 ppm to 8.1 ppm. Extraction by DTPA from soil treated with composted sludges showed a remarkable increase in extractable Zn after five years of experiments for CAS treatment from 2.0 ppm (t_0) to 8.5 ppm (t_8) and for Pb from 6.7 ppm (t_0) to 17.1 ppm (t_8). Similar increases were found for Pb also in CANS treated soil, while solubilization of Ni and Cu by DTPA increased slightly after the repeated addition of CAS and CANS to soil.

Considering the not composted sludges, on average, with respect to the total content exchangeable metals ranged from 7% (Cu) to 28% (Ni) for AES and from 9% (Zn) to 18% (Pb) for ANS. Greater quantities were found in the complexed or adsorbed form, roughly from 9% (Ni) to 42% (Zn) for AES and from 8% (Ni) to 38% (Cu) for ANS. As a concluding remark it must be stressed that even if the total content of heavy metals in the soil remained nearly the same, the fractions extractable by KNO_3 and DTPA greatly changed from one sampling time to another; moreover the quantities in available forms largely depended on the composition of the utilized sludges which varied with time, as is apparent, for example, for the AES treatment, where at time t_6 the amounts of Zn, Cu, and Pb, remained at harvest time much higher than the other years. because of their anomalous content in the sludge of that year. It is noteworthy that these high quantities of the three metals were available the following year, as shown by the decrease to a much lower value of extractability after harvest time at t_8.

Heavy metals in soil have been considered divided into the following chemical pools:water soluble, easily exchangeable and adsorbed or complexed, which are in reversible equilibrium with one another. Further chemical pools have been postulated but they are not important as a source of heavy metals for plants. Each pool corresponding to a different chemical form of heavy metals in soils has been estimated by the previously described sequential extraction procedure. In the course of the whole experiment good correlations were found between plant uptake and DTPA extractability (data not reported here).

In this field study, over a five years period, it was observed an increase in soil organic matter content after AES treatment (1). The fact that a certain amount of the sludge organic matter is resistant to the decomposition agree with previous findings (2,3) and it is of particular importance, since humic substances could increase the adsorption capacity of a soil for heavy metals. Moreover even if a part of the native organic matter decomposes when sludges are added to soil the humification process produces new organic matter in the soil In this field experiment, in particular we have found a change in the quality of the soil organic matter, with the appearence of humic compounds of high molecular weight. Such compounds have been demonstrated to play an essential role in retaining heavy metals irreversibly (4).

Gel-filtration experiments showed that in untreated soil metals were linked only by organic matter of a low molecular weight on the contrary in the sludge-treated soils they were complexed by humic materials of both high and low molecular weight.

These data are of particular interest since interactions among heavy metals and organic matter may alter the solubility and the mobility of these elements in soil

REFERENCES
(1) LEVI-MINZI R., RIFFALDI R.. GUIDI G. and POGGIO G. Chemical characterization of soil organic matter in a field study with sewage sludges and composts. In: "Long term effects of sewage sludge and farm slurries application" D. Reidel Publ. Co. (In press).
(2) VARANKA M.W., ZABLOCKI Z.M. and HINESLY T.D. The effect of digestion sludge on soil biological activity. J. Water Pollut. Control Fed. 48, 1728-40 (1976).
(3) TERRY R.E., NELSON D.W. and SOMMERS L.E. Decomposition of anaerobically digested sewage sludge as affected by soil environmental conditions. J. Environ. Qual. 8, 342-47 (1979).
(4) PETRUZZELLI G., GUIDI G. and LUBRANO L. Cadmium occurrence in soil organic matter and its availability to wheat seedlings. Water Air Soil Pollut. 9. 263-69 (1978).

Table 1 - Zinc extractability by KNO_3 and DTPA. Data are expressed as µg/g soil.

Time	AES		ANS		CAS		CANS	
	KNO_3	DTPA	KNO_3	DTPA	KNO_3	DTPA	KNO_3	DTPA
t_0	2.6	2.0	2.5	1.9	2.6	2.0	2.4	1.9
t_1	7.1	20	8.3	8.4	5.2	10	5.2	4.8
t_2	5.5	10	4.1	5.0	3.2	2.2	3.8	0.5
t_3	5.8	16	4.5	5.9	5.7	3.9	4.5	2.6
t_4	5.1	11	4.4	3.6	3.9	3.4	3.8	2.4
t_5	5.6	45	5.9	7.5	3.7	6.2	4.3	4.2
t_6	5.5	38	3.5	6.2	3.5	3.3	3.1	3.5
t_7	6.6	39	3.7	7.5	3.9	6.1	3.6	6.9
t_8	3.3	14	3.4	5.3	2.6	8.5	3.2	4.2

Table 2 - Copper extractability by KNO_3 and DTPA. Data are expressed as µg/g soil.

Time	AES		ANS		CAS		CANS	
	KNO_3	DTPA	KNO_3	DTPA	KNO_3	DTPA	KNO_3	DTPA
t_0	3.1	17	3.2	17	3.2	16	3.1	17
t_1	6.6	31	14.3	22	7.0	22	7.7	21
t_2	3.2	23	3.2	8.9	3.6	17	4.0	13
t_3	4.4	27	8.9	12	4.9	21	4.5	14
t_4	4.6	17	2.5	11	3.1	18	3.1	12
t_5	5.1	33	4.7	20	5.6	25.2	4.7	20.5
t_6	3.3	30	3.6	19	3.3	19.9	3.3	15.2
t_7	3.6	31	4.3	22.4	3.5	23.1	4.6	22.7
t_8	2.2	18.3	2.6	19.1	3.1	20	3.2	20.6

Table 3 – Nickel extractability by KNO_3 and DTPA. Data are expressed as µg/g soil.

Time	AES		ANS		CAS		CANS	
	KNO_3	DTPA	KNO_3	DTPA	KNO_3	DTPA	KNO_3	DTPA
t_0	8.0	3.1	8.0	3.2	8.1	3.1	8.0	3.0
t_1	10	4.9	9.5	4.2	12	4.9	8.8	4.8
t_2	8.0	5.3	8.0	3.7	11	3.7	8.0	2.5
t_3	15	5.8	10	4.1	15	4.2	10	3.1
t_4	8.3	4.8	7.6	3.7	13	3.3	9.3	2.1
t_5	8.5	10.3	8.5	5.6	14	6.8	11	4.6
t_6	6.0	7.1	8.5	5.0	9	6.5	8.5	4.0
t_7	10.1	7.4	8.5	5.7	9.8	7.1	10.1	5.6
t_8	8.9	6.7	7.2	4.0	8.4	5.2	9.4	4.5

Table 4 – Lead extractability by KNO_3 and DTPA. Data are expressed as µg/g soil.

Time	AES		ANS		CAS		CANS	
	KNO_3	DTPA	KNO_3	DTPA	KNO_3	DTPA	KNO_3	DTPA
t_0	3.0	6.7	3.0	6.6	3.1	6.7	3.0	6.7
t_1	4.0	8.6	5.6	9.0	4.0	12	2.8	7.5
t_2	9.3	10	12	3.8	12	9.0	12	5.3
t_3	9.3	18	4.6	12	3.7	19	3.2	10
t_4	13	15	2.2	7.3	6.9	16	7.0	8.7
t_5	8.3	15.6	8.3	13.5	16.6	20.8	11.6	13.5
t_6	13.3	21.2	10.1	11.1	18.6	11.1	10.6	11.1
t_7	9.8	20.9	10.2	11.8	8.9	22.0	8.0	15.7
t_8	9.7	21	10.3	7.8	8.9	17.1	8.1	14.4

Table 5 - Cadmium extractability by KNO_3 and DTPA. Data are expressed as µg/g soil.

Time	AES		ANS		CAS		CANS	
	KNO_3	DTPA	KNO_3	DTPA	KNO_3	DTPA	KNO_3	DTPA
t_0	0.8	–	0.8	–	0.7	–	0.8	–
t_1	1.8	–	2.1	–	1.8	–	1.8	–
t_2	0.9	–	0.5	–	0.8	–	0.8	–
t_3	1.9	–	2.2	–	2.6	–	2.1	–
t_4	1.8	–	1.6	–	2.3	–	1.4	–
t_5	2.0	1.0	2.0	0.4	2.7	0.9	1.9	0.8
t_6	1.6	0.7	1.5	0.3	1.3	0.5	1.4	0.3
t_7	1.6	0.6	1.7	0.8	1.4	0.9	1.6	0.6
t_8	1.3	0.5	1.6	0.6	1.3	0.5	1.4	0.5

METHODS OF TREATING PIG SLURRY TO INCREASE THE VOLUMES WHICH CAN BE USED ON CROPS

Piccinini S., Cortellini L., Bonazzi G.
Centro Ricerche Produzioni Animali
Via Crispi, 3 - 42100 Reggio Emilia - Italy

Summary

In this paper the current situation regarding the treatment and agro nomic use of pig manure is analysed; the lack of success in the pol lution-treatment system applied to such a type of waste-water, the faults in legislation and the subsequent confusion on the part of those operating in the pig sector are brought to notice. Finally four possible solutions for the areas with high density pig breeding are given. The most favourable solution is that of agronomic sprea-ding which is the least costly while at the same time allowing a minimisation of risks to the environment.

1. INTRODUCTION

Emilia-Romagna, a region which represents a twentieth of the total land area of Italy, contains 2,300,000 head, a quarter of the national pig population. This population is not distributed uniformly throughout the region, but is concentrated in the area where Parmesan cheese is produced. Swine rearing in this area is an age-old tradition since it is linked to production of whey, which has always been used as a feed stuff.

The great increase in the number of swine reared in the area in the last 20 years is due to the increases in national pork consumption and to the presence, in this same area, of the country's largest butchering and pork-processing industries.

The average size of the farms is 600 head, 48% of the pig population being contained in farms with more than 1000 head, the vast size of which along with the dense concentration of these farms in the area heighten the problem of manure disposal.

In Italy regulations concerning manure disposal are contained in the national law n° 319/76, entitled "Water Pollution Control Act." This law was successively integrated by another law (n° 650/79), which allotted certain tasks to the various Regions for the upkeep of the law itself and has provided financial backing to contribute to pollution control work.

The law establishes two categories of farming enterprises:

a) INDUSTRIAL-TYPE farms: i.e. farms without agricultural land or with in sufficient agricultural land for manure disposal (> 4 tons of live weight/Ha);

b) AGRICULTURAL-TYPE farms: i.e. farms with sufficient agricultural land for manure disposal (≤ 4 tons of live weight/Ha.) which produce part of the feed needed for their stock (the quantity is unspecified) on this land.

Farms of the former type are sybject to the ruling of the National Law, while those of the latter type are subject to regulations enforced by each single Region.

It has been shown that on a territorial level, the contribution to pollution by pig manure represents a small part of the total amount which is mainly due to urban, industrial and widespread agricultural waste (fertilization with mineral fertilizers).

In the hydrographic Po basin, according to the theory formulated by the IRSA (1) on the distribution of the loads of phosphorous carried along to the Adriatic, pig waste along with cattle waste affect the total amount by 17%, not to be ignored but nonetheless lower than that recorded for municipal waste-water, and detergents which represent respectively 25% and 30,5%. However it is equally true that a single effluent contains a very high polluting rate. If we take BOD_5 as a parameter to indicate the level of organic load and nitrogen as a chemical species responsible, together with phosphorous, for the phenomenon of eutrophycation of river life, we have some idea of the impact of this kind of effluent on the environment.

Table 1 shows the figures for the concentration of manure from piggeries of the kind annexed to cheese dairies, in comparison to the figures shown for municipal waste.

Tabel 1: BOD_5, TKN and P_t in pig slurry (cleaning with water) and in municipal waste.

PARAMETERS	PIG SLURRY	MUNICIPAL WASTE
BOD_5 (mg/1)	3500-8000	200-400
Total Kjeldhal Nitrogen (mg/l N)	800-1800	20-85
Total Phosphorous (mg/l P)	300-1000	6-20

The figures involved are very high, in the case of Nitrogen even 20 times above those for municipal waste.

It is with these figures that we must begin if the reasons which have brought about the lack of success of the pollution treatment system applied to pig waste are to be understood. Law n°319/76 making legal limits the same for all wastes, did not take into acount the specific nature of zootechnic wastes, and did not consider that for several parameters such limits are practically impossible to reach consistently, despite the use of costly and complex plants. Hence the great confusion of those operating in a production sector in which, since the passing of law n°319/76, the following facts have been noted:
- insufficient indication, on the part of the organisations whose task is to oversee the application of the law, as to the technology to be used in order to adhere to the extremely restricting limits imposed on the unloading of effluent in ground surface water;
- insufficency and delays in the distribution of contributions to those equipped with a pollution treatment plant. Thus to the very high running costs of the plants are added the equally high interest rates;
- differences in treatment received among breeders: those who, at the time, were defined as having agricultural-type farm were able to put off the required adjustments as stipulated by the Regional Authorities in their regulations on waste for this type of farm, by benefitting from the delays which continue even now;

- the harmful effects on the environment brought about by this state of things are well known.

Agronomic spreading in the absence of any precise ruling or directives (the Emilia-Romagna Region is the only one to have issued a law specifically on waste for agricultural type farms) is lither not carried out, hence unloading into surface water is continued, or it is carried out irrationally, unloading excessive amounts over a few hectares of agricultural soil with the extremely grave risk of polluting the underground water and irreversibly degrading the soil itself.

Pollution treatment systems, where adopted, offer no better picture. From a study carried out by the C.R.P.A. (2), in the context of a research and experimentation project of the Emilia-Romagna Region on the performance of pollution treatment plants currently in existence, it applears that often these plants are lacking in components essential to their proper performance. The equaliser tank is hardly ever present, the primary settling tank is often missing and the sludge rotation system is non-functioning even its operation is carried out in a not altogether proper fashion. The farmer has received very sketchy training from the construction firms and he operates the plant in a very empiric way, considering it as separate from the farm's production scheme, without however the help of adequate technical assistance.

2. POSSIBLE SOLUTIONS

The points discussed lead to the conclusion that the problem of spreading pig slurry is far from being resolved. Not only that, but we should admit that the suggestions made as time goes by, even if valid in certain situations, for the most part rarely provide definitive solutions. Agronomic spreading is the least costly solution, capable at the same time of minimizing risks to the environment. Much work is still needed if the rules for proper use are to be laid down and respected.

Agronomic spreading is the solution which can and must be adopted wherever the density of pig-breeding is relatively low, with even distribution throughout the territory.

More complex are the solutions in areas with high density of breeding and heads. There is almost always found an excess of manure either due to the kind of cultivation unsuitable for its use or due to the existence of environmental limits.

The solutions in sight for these areas are:

a - moving piggeries to more suitable areas for those unsuitably housed, for example piggeries which are situated in built-up areas.

b - Connecting up the piggeries to sewerage collectors and unloading all the excess waste produced in the area or a part of it for pollution treatment, mixed in with municipal waste, in the treatment terminal of the right size. It is a solution which is becoming increasingly practical as the network of public pollution treatment systems extends.

c - Collecting excess wastes and subsequent agronomic use in areas picked out as particularly suitable (inter-farm solution).

d - agronomic spreading carried out within single farms.

It is to be presumed that, in most of the situations, the four solutions must be adopted together.

Solutions b, c and d presuppose almost always that in order to be put into effect, the wastes are subjected to treatments suitable for stabilizing and deodorizing them and for reducing the nitrogen and phospho rous load.

In figure 1 the most straight-forward and least costly lines of treatment are summarised graphically, both from an investment point of view and a management one.

As reagards solution b, simply separating solids from liquids (screening and/or sedimentation) may prove to be sufficient as a pre-treatment.

The simplest line is that of separating solids from liquids and the subsequent storage of the clarified part in a lagoon. The separated solids (2-4% of the volume of the wastes) must reach their destination outside the farm or inter-farm land if the nitrogen and phosporous content in the waste is to be considered effectively reduced.

As regards consortia, more efficient systems of mechanical separation can also be envisaged (eg. centrifuges) which however cannot be considered in the case of farm treatments, due to management complexity and high costs both in investment and management. Depending on the lengh of storage time (from a minimum of 3 months up to a 6 months maximum), the nitrogen will be reduced noticeably due to loss into the atmosphere.

The amount of nitrogen and phosphorous segregated in the sludge can not on the other hand be considered, since sooner or later this sludge will have to be taken back and spread on available farm land.

A period of storage lasting more than 6 months has not been conside red on purpose, as a choice of this kind, as well as being costly (also as invested farm surface area) can lead to a concentration of the sprea-ding of waste in short periods, in particular in the period summer-autunn when crop production drops. The reduced removal capacity of the crops can increase the risk of pollution in the winter period due to the run-off of the residual nitrogen.

In the second line of treatment a greater reduction is achieved by means of the addition of a settling tank between the screening and stora ge. The thickened sludge which is obtained (maximum concentration of dry matter: 5%) can be removed and spread on land outside the farm or treat ment centre.

In this way, having removed the nutrients contained in the scree-need solids and those contained in the thickened sludge, the percentage of reduction in the clarified liquid is raised considerably (roughly 50% for nitrogen, 55% for phosphorous).

Of course the right conditions are necessary for the operation to be successful. The volume of sludge to be transported is about a quarter of the volume of waste treated, consequently, sufficient land must be found, and the further the land, the higher the costs to be faced.

The operation has a chance of succeeding in the plains of the Vene to, Emilia and Lombardia, where the number of land farms without cattle are increasing in number, where the demand for organic fertilizers is on the increase.

The use of a settling tank is justified only in the case of highly diluted wastes. Above a level of 2.5% dry matter the operation appears to be of dubious use.

As regards treatment consortia, where qualified personnel are to hand, the settling tank can be replaced by more complex systems such as flotation units, which can improve the separation efficiency of the su-spended solid fraction.

In the third line of treatment the insertion between screening and storage of an aeration section is envisaged, with a retention time and

Fig. 1: Methods of treating pig slurry.

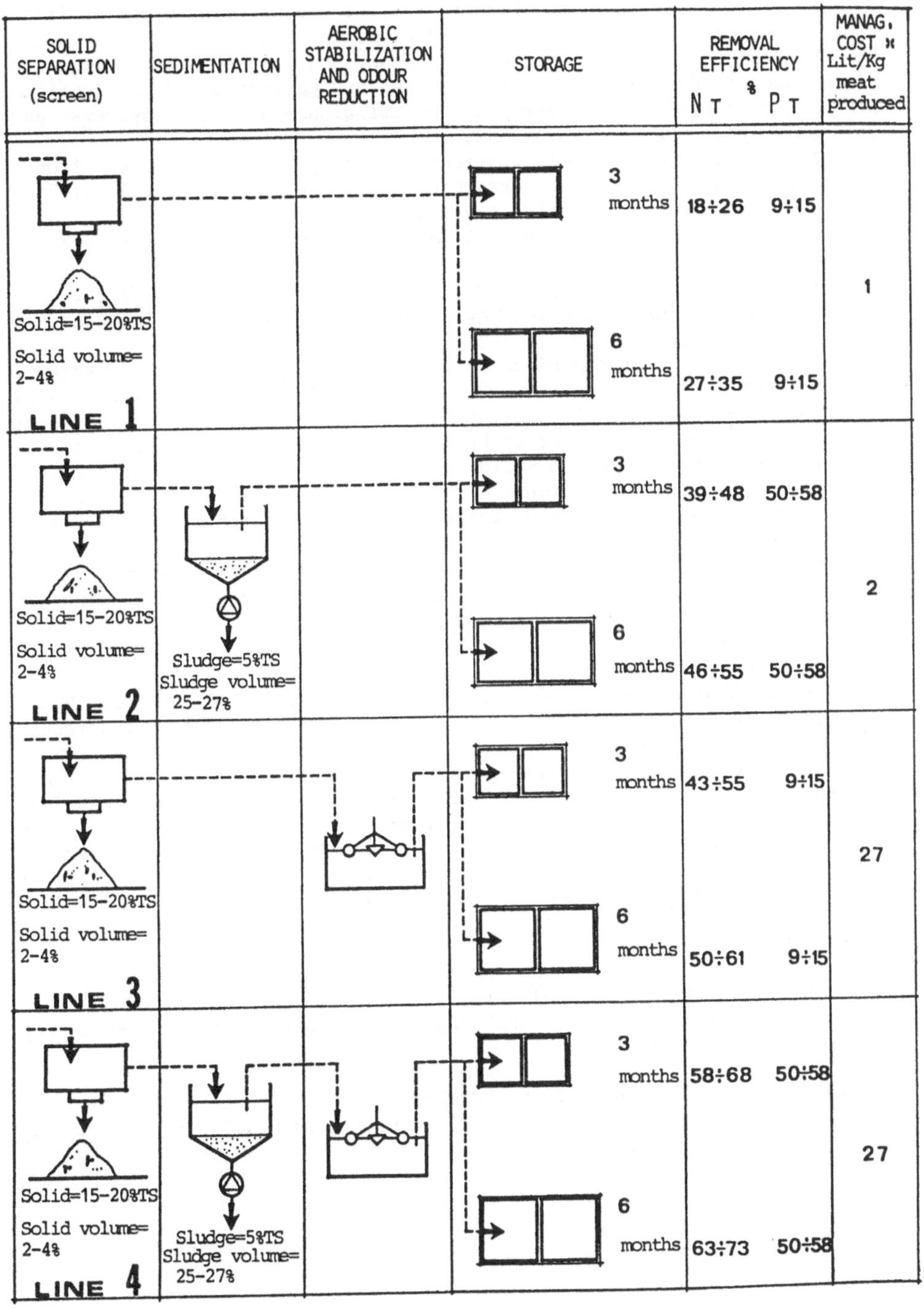

SOLID SEPARATION (screen)	SEDIMENTATION	AEROBIC STABILIZATION AND ODOUR REDUCTION	STORAGE	REMOVAL EFFICIENCY % N_T	P_T	MANAG. COST × Lit/Kg meat produced
Solid=15-20%TS; Solid volume=2-4% LINE 1			3 months	18÷26	9÷15	1
			6 months	27÷35	9÷15	
Solid=15-20%TS; Solid volume=2-4% LINE 2	Sludge=5%TS; Sludge volume=25-27%		3 months	39÷48	50÷58	2
			6 months	46÷55	50÷58	
Solid=15-20%TS; Solid volume=2-4% LINE 3			3 months	43÷55	9÷15	27
			6 months	50÷61	9÷15	
Solid=15-20%TS; Solid volume=2-4% LINE 4	Sludge=5%TS; Sludge volume=25-27%		3 months	58÷68	50÷58	27
			6 months	63÷73	50÷58	

* Management cost: transport costs for solid and thickened sludge are not considered, nor is depreciation.

installed power suitable for promoting good stabilisation and deodorization of the waste. A secondary effect, one that is nonetheless essential to the final result, is the high removal of the nitrogen by volatilization of ammonia and by nitrification-denitrification.

Management costs are relatively high, however the right definition of the amounts of oxygen to be transferred and perfection of the aeration systems can contribute to improving the efficiency of these systems and consequently to reducing such costs.

In the fourth line of treatment a more complete system is introduced, one which is capable of reaching high levels of removal without being as complexas a pollution treatment plant. ·

The investment costs are obviously the highest as are management costs.

It is in fact necessary to add on the costs of transporting the thickened sludge, for the spreading of which the same points as expressed for line 2 apply.

In the case of treatment consortia centres, instead of the transportation of thickened sludge as such outside the area, its stabilization in an anaerobic digestion plant may be of interest. The digested sludge, given the large quantities at stake, could be convenienctly subjected to the treatment of separating solids from liquids, for example with a filter press or a beltfilter press. A solid proportion (about 25% of dry matter) would thereby be obtained, for which transportation outside the area would certainly give fewer problems than the sludge as such.

The production of biogas would therefore consitute an advantage, capable by means of the plant's indipendent energy source of reducing remarkably the liabilities in the budget.

3. CONCLUSION

The technological systems have been fully tested and have the bonus of being flexible and capable of being adapted to specific situations. They can be adopted both in consortia solutions (agronomic spreading or pollution treatment mixed in with municipal waste) and because of their straighformardness, in farm solutions which foresee agronomic spreading. What is needed at the moment is the setting up of several projects which have the financial and technical support of the local authority because of the value which they have in demonstrating the practicability of staighforward and relatively low-cost solutions.

REFERENCES

1) R. MARCHETTI "Indagini sul problema dell'entrofizzazione delle acque costiere dell'Emilia-Romagna" Regione Emilia-Romagna, Assessorato ambiente e difesa del suolo.

2) G. BONAZZI, L. CORTELLINI, A. FERRARI "Indagine sulla diffusione dei depuratori su scarichi suinicoli in Emilia-Romagna" - Rivista di suinicoltura n. 9 Settembre 1984.

CHEMICAL COMPOSITION OF SLUDGES
FROM SEWAGE TREATMENT SYSTEMS IN THE EMILIA ROMAGNA REGION

N. ROSSI, R. RASTELLI and P.L. GRAZIANO
Istituto Chimica Agraria, Università Bologna (Italy)

N. DE MARTIN
Ente Regionale Sviluppo Agricolo per l'Emilia Romagna, Bologna (Italy)

Summary

Sludges from urban sewage treatment systems in the Emilia Romagna
Region of Italy were analyzed chemically for the following: dry
matter, pH, ash, total nitrogen, inorganic nitrogen, sulphur, phosphorous, potassium, calcium, magnesium, sodium, iron, manganese,
zinc, copper, lithium, cobalt, nickel, lead, chromium, cadmium,
arsenic, mercury, and molybdenum.   During the period 1980-1983,
samples of sludge were collected from sewage treatment systems whose
capacities exceeded 20,000 equivalent inhabitants, grouped as follows:
systems serving i) small-medium cities located along the Adriatic
Riviera, ii) small-medium cities located in the interior of the
Region and iii) medium-large cities located in the interior of the
Region.   The sludges examined can be considered as fertilizing materials of mediocre agronomical value.   They contain only modest
quantities of plant nutrients, but have a substantial quantity of
organic matter (around 50% dry weight basis) and generally the heavy
metal content is not too high.   With the exception of Zn for some
samples, the heavy metal contents are below those proposed by other
countries of the EEC as limiting values acceptable for agricultural
use of sludges.

1. INTRODUCTION

Various solutions have been proposed for the disposal of the large
quantities of sludge from urban sewage treatment systems.   Among these,
those solutions which take into consideration the possibility of recovering useful substances contained in the sludges are of particular interest.
Basically organic sludges containing substances which are useful and non—
toxic for agriculture can be disposed of by spreading them on the soil,
including soil destined for agricultural use (2,3,6).   In fact, sludges
coming from urban sewage treatment systems can have a fertilizing effect
since they do contain components which are among those elements essential

for plant nutrition (e.g., nitrogen, phosphorous, potassium and other macro- and microelements) as well as considerable amounts of organic matter, a useful soil conditioner.

Before applying sludges to agricultural land, however, it is indispensable that the sludge be analyzed chemically in order to determine the quantities of the various plant nutrients it contains so that the quantities of normal fertilizers which must be added to supply the soil with the correct amounts of nutrients can be calculated. In addition, chemical analyses of the sludges are doubly important because it also is necessary to quantify the concentrations of certain toxic elements contained in the sludge; of particular importance are the quantities of heavy metals. Excessive quantities of these elements are undesirable not only because they are toxic to the plant and also can accumulate in the soil to which the sludges are applied but also because they can be absorbed by plants, influencing their growth and entering into the food chain. For this reason, determination of the heavy metal content in sludges destined for disposal on agricultural land is of particular importance. Research in this respect must be carried out not only for those elements which are highly toxic such as Pb, Cd and Hg, but also for other metals such as Cu, Mn and Zn which beneath a certain limit are indispensable for plant growth, but which above certain limits hinder growth.

In view of this possible agricultural utilization of the sludges, at the beginning of the 1980's when a discrete number of urban sewage treatment systems were already in operation, the Regione Emilia Romagna felt it worthwhile to assign a research project to the E.R.S.A. (Ente Regionale Sviluppo Agricolo) directed towards determining the feasibility of an agricultural use for the sludges from urban sewage treatment systems operating in the Region. A systematic investigation of the chemical composition of the sludges coming from the various systems studied under the E.R.S.A. project was carried out at the Institute of Agricultural Chemistry (University of Bologna) and the analytical results obtained are reported and discussed in this paper.

2. MATERIALS AND METHODS

Samples of sludge were taken from all the urban sewage treatment systems having a capacity of at least 20,000 equivalent inhabitants and grouped into the following 3 categories: systems serving i) small-medium cities located along the Adriatic Riviera, ii) small-medium cities located in the interior of the Region and iii) medium-large cities located in the interior of the Region.

Samples of sludge from the sewage treatment systems in the cities studied located in the Adriatic Riviera were collected during the summer of 1980; those collected from the cities in the interior of the Emilia Romagna Region were collected mainly in the summer of 1981 and in part also during the summers of 1982 and 1983. A more thorough characterization of the sludges from the Bologna sewage treatment system was undertaken with samples collected monthly over a period of two years (April 1980 to March 1982).

The determinations of dry matter content, pH and nitrogen content were carried out on the as sampled material (i.e., not dried). Dry matter content was determined by drying the samples to constant weight in an oven at 75-80 C. The ash content of the dry matter was determined by heating the dried sludges in a muffle furnace at 450-500 C for 5 hours. For the determination of the various elements, sample dissolution was obtained by wet ashing with a nitric-perchloric acid mixture or by dry ashing in a muffle furnace followed by a hydrofluoric acid treatment to remove any silicon present. The quantitative determinations of the elements present in the dissolved materials were carried out by colorimetry (P), turbidometry (S), distillation and titration (N) and atomic absorption spectrophotometry (cations in general). Analyses for some of the elements present at very low concentrations (Mo, Co and Pb) were obtained by the atomic absorption technique using a graphite furnace rather than the flame, while for others (As and Hg) the hydride generation system was used.

3. RESULTS

Key characteristics of the sewage treatment systems from which the sludges were sampled for analysis are reported in Table I. Reported in Table II are the average values of the quantities of each of the elements analyzed for the sludges of the various small-medium cities of Groups i) and ii), located along the Adriatic Riviera and in the interior of the Emilia Romagna Region, respectively. Analogous data are given in Table III for the sludges from medium-large cities in the interior of the Region (Group iii)) and for the city of Bologna. Since the values for the individual elements in the sludges varied considerably from city to city, the standard deviations from the averages also are given in Tables II and III, as an indication of the extent of the variations observed. The data reported in Table III for the city of Bologna are the values obtained from the statistical analysis of the chemical data for the samples collected monthy over a 2-year period. For comparison, Tables II and III also contain the values of the upper limit of heavy metals acceptable in sludges destined for agricultural use, obtained from existing Norms in some EEC countries (1).

4. DISCUSSION AND CONCLUSIONS

An examination of the results obtained in this study shows that the sludges produced from the urban sewage treatment systems in the Emilia Romagna Region are materials with a neutral pH and widely variable dry matter contents. For this reason, in order to make meaningful comparisons, the analytical data must be presented on a dry weight basis. In general, the sludges had the following overall composition (dry weight basis): organic matter, 50%; total N, 3-4%, 10-15% of which in the form of soluble inorganic N; S, 1%; P_2O_5, 2-4% and K_2O, 0.5-1%.

As indicated by the high values of the standard deviations from the average (Tables II and III), the parameters which characterize the chemical composition of the sludges analyzed vary considerably among

Table I - Urban sewage treatment systems operating in the Emilia Romagna
Region at the beginning of the 1980's from which the samples
of sludge were taken to be analyzed

Locality	Capacity (equivalent inhabitants)	Type of sludge produced	Disposal Method at time of sampling*	Sampling date
Systems serving small-medium cities along the Adriatic Riviera				
Cattolica (FO)	50,000	aerobic	a,b	July 1980
Misano (FO)	35,000	aerobic	a,b	July 1980
Riccione (FO)	120,000	aerobic	a,b	July 1980
Rimini (FO)	160,000	aerobic	a,b	July 1980
Bellaria (FO)	46,000	aerobic & anaerobic	a,b	July 1980
Gatteo a Mare (FO)	34,000	aerobic	a,b	July 1980
Cesenatico (FO)	66,000	aerobic	a,b	July 1980
Cervia (RA)	200,000	anaerobic	a,b	July 1980
Lido Adriano (RA)	12,500	aerobic	a	July 1980
Systems serving small-medium cities in the interior of the Region				
Cesena I (FO)	60,000	aerobic	a	June 1981
Cesena II (FO)	60,000	aerobic	a	June 1981
Faenza (RA)	70,000	aerobic	a	June 1981
Lugo (RA)	30,000	aerobic	a	June 1981
S. Lazzaro (BO)	20,000	aerobic	a	June 1981
Calderara (BO)	12,000	aerobic	a	Oct. 1983
Fidenza (PR)	36,000	aerobic	a	June 1981
Systems serving medium-large cities in the interior of the Region				
Bologna	450,000	–	c	monthly from April 1980 to March 1982
Forlì	79,000	aerobic	a	July 1980
Ravenna	140,000	aerobic	a	Sept. 1983
Ferrara	45,000	anaerobic	a	Sept. 1983
Parma	230,000	anaerobic	a,b	Sept. 1983
Reggio Emilia	155,000	anaerobic	a	June 1981

* a = landfill
  b = spreading on agricultural land
  c = burning

Table II - Chemical data for sludges from small-medium cities of Group i)
(located along the Adriatic Riviera) and Group ii) (located in
the interior of the Emilia Romagna Region) and upper EEC limits
of heavy metals acceptable in sludges for agricultural use

		Group i)*		Group ii)**		EEC limits (1)
		Average	Std.Dev.	Average	Std.Dev.	
		(as sampled basis)				
pH		6.6	0.3	6.9	0.3	
Dry matter	(%)	39	27	63	34	
		(dry weight basis)				
Ash	(%)	37	7	50	15	
Total nitrogen	(N %)	4.43	0.64	4.04	0.92	
Inorganic nitrogen	(N %)	0.27	0.12	0.30	0.12	
Total sulphur	(S %)	0.93	0.15	0.80	0.15	
Phosphorous anhydride	(P_2O_5 %)	1.40	1.16	4.05	0.76	
Potassium oxide	(K_2O %)	0.39	0.17	0.57	0.19	
Calcium	(Ca %)	4.46	1.29	4.00	1.17	
Magnesium	(Mg %)	0.93	0.17	0.87	0.14	
Sodium	(Na %)	0.27	0.12	0.14	0.05	
Iron	(Fe %)	1.44	0.45	1.65	0.53	
Manganese	(Mn ppm)	1481	1124	1035	780	
Zinc	(Zn ppm)	2853	1059	3540[++]	1724	3000
Copper	(Cu ppm)	271	59	408	176	1500
Lithium	(Li ppm)	5.3	3.5	10.3	4.2	
Cobalt	(Co ppm)	7.2	2.1	13.6	2.9	20[a]
Nickel	(Ni ppm)	42	27	130[++]	151	400
Lead	(Pb ppm)	207	56	119	57	1000
Chromium	(Cr ppm)	45	77	453[++]	736	1000
Cadmium	(Cd ppm)	3.4	0.9	6.2	1.6	40
Arsenic	(As ppm)	5.6	3.8	2.2	0.4	
Mercury	(Hg ppm)	2.2	0.2	1.7	0.5	25[b]
Molybdenum	(Mo ppm)	2.4	0.9	2.3	0.6	25[b]

* Cities of Group i): Cattolica, Misano, Riccione, Rimini, Bellaria,
  Gatteo a Mare, Cesenatico, Cervia and Lido Adriano
** Cities of Group ii): Cesena I, Cesena II, Faenza, Lugo, S. Lazzaro,
  Calderara and Fidenza
++ Some of the individual values forming this average exceed those pro-
  posed as upper acceptable limits
a) French Norms; b) German Norms

Table III – Chemical data for sludges from medium-large cities of Group iii) (located in the interior of the Emilia Romagna Region) and from the city of Bologna (based on data for samples collected monthly over a 2-year period) along with the upper EEC limits of heavy metals acceptable in sludges for agricultural use

		Group iii)*		City of Bologna		EEC limits (1)
		Average	Std.Dev.	Average	Std.Dev.	
		(as sampled basis)				
pH		6.6	0.3	6.5	0.4	
Dry matter	(%)	59	32	24	4	
		(dry weight basis)				
Ash	(%)	60	12	43	9	
Total nitrogen	(N %)	3.00	1.38	2.94	0.71	
Inorganic nitrogen	(N %)	0.28	0.14	0.49	0.14	
Total sulphur	(S %)	1.07	0.52	0.60	0.16	
Phosphorous anhydride	(P_2O_5 %)	3.41	1.03	1.93	0.37	
Potassium oxide	(K_2O %)	0.91	0.91	0.49	0.18	
Calcium	(Ca %)	5.10	1.53	4.89	1.06	
Magnesium	(Mg %)	0.88	0.24	0.70	0.12	
Sodium	(Na %)	0.21	0.12	0.07	0.02	
Iron	(Fe %)	2.07	1.24	1.34	0.82	
Manganese	(Mn ppm)	725	298	386	69	
Zinc	(Zn ppm)	2012[++]	863	1655	294	3000
Copper	(Cu ppm)	323	130	288	83	1500
Lithium	(Li ppm)	10.0	4.4	15.4	6.0	
Cobalt	(Co ppm)	11.8	3.1	12.9	1.7	20[a]
Nickel	(Ni ppm)	92	26	95	15	400
Lead	(Pb ppm)	198	154	432	150	1000
Chromium	(Cr ppm)	135	112	138	22	1000
Cadmium	(Cd ppm)	7.5	5.7	18.5	6.13	40
Arsenic	(As ppm)	3.1	0.7	3.3	0.7	
Mercury	(Hg ppm)	1.6	0.5	1.6	0.3	25[b]
Molybdenum	(Mo ppm)	2.4	1.1	3.0	1.1	25[b]

* Cities of Group iii): Bologna, Forlì, Ravenna, Ferrara, Parma, and Reggio Emilia

++ Some of the individual values forming this average exceed those proposed as upper acceptable limits

a) French Norms; b) German Norms

samples coming from different sewage treatment plants. It was expected that the sludge samples from different sized cities or from cities in different geographical locations (seaside as compared to interior zones of the Region) would show some differences in their chemical compositions. This, however, was not the case. Indeed, no big differences were found in either the range of values or the average values of the various elements in the groups of sludges from different sized cities or different geographical location. Variations over time of the compositions of sludges taken from the same sewage treatment system are quite limited as illustrated by the data for the sewage treatment system of Bologna (Table III). During the two years of observations carried out monthly for the Bologna sewage treatment system, no particular variations from year to year were noted, nor were any seasonal variations observed. The constancy with time of the chemical composition of these materials is in agreement with that reported in previous studies (4,5). For longer periods of time, probably this constance of composition may no longer be valid because of modifications in sewage treatment processes made mandatory by new laws specifying various treatments to be used, dephosphatization, etc.

In regard to the heavy metal contents, in general the sludges examined can be considered suitable for disposal on agricultural soil. The exception to this general statement regards the values of Zn, which for some cities were greater than the acceptable limit of 3000 ppm. Cities for which the Zn values exceeded the acceptable limit were the following: Group i), small-medium cities along the Adriatic Riviera - Misano (3354 ppm), Rimini (5391 ppm), and Cesenatico (3147 ppm).
Group ii), small-medium cities in the interior of the Region - Cesena I (4810 ppm), Cesena II (4162 ppm), Faenza (5642 ppm), and San Lazzaro (4717 ppm).
Group iii), medium-large cities in the interior of the Region - Parma (3084 ppm).

Another case where some elements exceeded the acceptable limits was the sludge from Calderara where the value for Ni (471 ppm) exceeded the 400 ppm limit and that for Cr (2083 ppm) exceeded the 1000 ppm limit.

In brief, on the basis of the analytical data obtained in this study, it can be concluded that, with the previously mentioned limitations relative to the excessive Zn contents and excepting the sludges from the city of Calderara, there are no serious counterindications for the disposal of the sludges produced from urban sewage treatment systems in the Emilia Romagna Region by spreading them on agricultural land.

In most cases, even the high values of Zn probably should not be considered prohibitive to the application of these sludges on soils in the Region. It would be necessary in these cases, however, to apply smaller quantities of sludge than those recommended when all the heavy metal contents are within the prescribed limits. In addition, the major part of the soils under cultivation in the Emilia Romagna Region are calcareous soils with a relatively high C.E.C. as well as relatively good quantities of phosphates, all factors which, in general, reduce the availability of Zn to plants and thus making such soils better able to tolerate relatively high Zn contents in the sludge applied.   The best solution, however, to the problem of high Zn contents in the sludges would be to determine the source of the Zn in the sewage so that it can be excluded from entering the urban sewage systems.

Spreading the sludges from urban sewage treatment systems directly on the soil is one of the preferred disposal methods because it usually is less costly than other methods of destroying or disposing of these materials, and at the same time favors agriculture by reducing the necessary expenditures for traditional fertilizers and conditioners.   It must be kept in mind, however, that the suitability of such sludges for agricultural use should be verified periodically for the individual sewage treatment systems, especially for those where the heavy metals contents   are close to the limiting values.

Acknowledgement

This research was supported by a grant from the Regione Emilia Romagna.

REFERENCES

(1) Commissione delle Comunità Europee (1982).   Proposta di direttiva del Consiglio concernente l'utillizzazione in agricoltura dei fanghi residuati dai processi di depurazione.   Gazzetta Ufficiale delle Comunità Europee. 256/3-264/7.

(2) E.R.S.A. (Ente Regionale di Sviluppo Agricolo per l'Emilia Romagna) (1981). Utilizzazione agricola dei fanghi prodotti dai depuratori urbani. p. 128, Bologna.

(3) GENEVINI, P.L., VISMARA, R., MEZZANOTTE, V. (1983). Utilizzo agricolo dei fanghi di depurazione. Ingegneria Ambientale, 12, 9, p. 133.

(4) LEVI-MINZI, R., RIFFALDI, R., SOLDATINI, G.F., PINI, R. (1981). Variazione nel tempo della composizione chimica dei fanghi di depurazione. Agrochimica, 25, 2, 168-176.

(5) RIFFALDI, R., LEVI-MINZI, R., SARTORI, F., CONTI, G. (1979). Elementi di fertilità e metalli pesanti nei fanghi provenienti da impianti di depurazione. Agric. Ital. 108, 61-71.

(6) SANTORI, M. (1980). La utilizzazione agricola dei fanghi in agricoltura. IRSA-CNR, Roma.

APPLICATION OF CACL$_2$-EXTRACTION FOR ASSESSMENT OF CADMIUM AND ZINC MOBILITY IN A WASTEWATER-POLLUTED SOIL

C. SALT and A. KLOKE[*]

Institute for Ecology, Botanical Section, Technical University, Rothenburgstr.12, D-1000 Berlin 41

[*]Federal Biological Research Centre for Agriculture and Forestry, Königin-Luisestr. 19, D-1000 Berlin 33
Federal Republic of Germany

SUMMARY

During a field study of heavy metal pollution of plants and soil on a sewage farm in West-Berlin, soil extraction with calciumchloride was employed as a method for assessing the plant available fractions of cadmium and zinc (watersoluble ions and complexes, readily exchangeable cations). Total and extractable soil metal contents were set into relationship with pH, cation exchange capacity and organic matter content. - As far as is known, usually less than 5 per cent of the heavy metals in naturally enriched soils are plant available. This investigation shows, that in anthropogenically contaminated soils, especially those having received sewage sludge or wastewater, up to 30 per cent of the total content of cadmium and zinc can be plant available. In accordance with other studies it is confirmed that plants growing on such soils contain higher amounts of heavy metals.

1. INTRODUCTION

The West-Berlin sewage farm of Karolinenhöhe has been used for wastewater disposal since 1900, which has led to accumulation of heavy metals in the top layer of the soil, a sandy cambisol (brown-earth) (ref.1). Until 1984 a wide range of crops were cultivated on small sunken fields, covering an area of 84 ha, which are periodically flooded with wastewater. Because of the high cadmium content of

some plants, detected during this investigation, food production for human consumption has now completely ceased.

2. SAMPLING AND ANALYSIS

During the summer and autumn of 1984 a total of 48 soil
samples were taken out of the root space of 61 plants (radish, carrot, potatoe, rye, cabbage, parsley, stinging
nettle/Urtica dioica, scentless mayweed/Matricaria inodora,
meadow grass/Poa trivialis+annua).

Soils were analysed for aqua-regia as well as 0.05 M
$CaCl_2$-extractable cadmium and zinc by ICP and AAS (graphite
furnace) respectively. Potential cation exchange capacity
and exchangeable cations (ref.2), pH(CaCl) and organic
matter content(Carlo Erba N- and C-Analyser) were determined. The plants, after washing with deionized water and
drying at 85°C, were digested with nitric acid in a Büchi-
Digestor prior to ICP-analysis.

The heterogenous distribution of heavy metals in the
soil and lack of uniform plant growth made it difficult
to collect representative samples.

	$\overline{X}$(mean)	R(range)
pH		4.4-5.9
Organic matter(%)	2.1	0.8-4.6
pCEC(mval/100g)	8.7	4.9-16
EC(mval/100g)	6.2	3.1-11
Fe_t(mg/kg)	5000	3400-7400
Mn_t(mg/kg)	96	27-180
Zn_t(mg/kg)	240	72-640
Zn_{ex}(mg/kg)	30	1.4-80
Zn_{ex}(%)	14	1.7-30
Cd_t(mg/kg)	3.3	1.1-8.6
Cd_{ex}(mg/kg)	0.44	0.11-1.3
Cd_{ex}(%)	14	4-39

Table I. - Characteristics and heavy metal contents of sewage-
treated sandy cambisol(brown-earth)

3. RESULTS AND DISCUSSION

A summary of analysed soil properties and soil metal concentrations is given in table I. Both cadmium and zinc are extracted by $CaCl_2$ at an average of 14 per cent, which points to the possibility that these metals are bound by the same mechanisms in the soil. The strong positive relationship between total and extractable amounts up to a high level indicates that binding could mainly be determined by adsorptive processes (ref.3).

Cd_t versus Cd_{ex} - R^2=0.9146***   n=48

Zn_t versus Zn_{ex} - R^2=0.8185***   n=48

Due to the fact that soil samples were taken from only one soil type, it is not surprising to find that potential and effective cation exchange capacity are positively correlated with the organic matter content (see Fig.1). The variation of CEC, not explained by organic substance, could either derive from iron and manganese oxides or clay minerals.

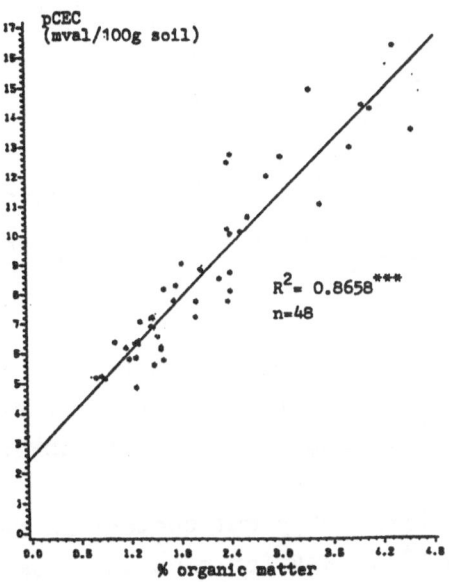

Fig.1 - Potential cation exchange capacity versus % organic matter

Total zinc shows a stronger linear relationship to
pCEC (R^2=0.8000***) and organic matter (R^2=0.7727***) than
total cadmium (R^2=0.5762*** and R^2=0.4945***).

Over the investigated pH-range of 4.4 to 5.9 the zinc
extractability linearly decreases with increasing pH, while
the cadmium extractability does not appear to be as pH-
dependent (see Fig.2+3). This is confirmed by other authors
(ref.4+5) and might be explained by the fact that cadmium
is stronger adsorbed by organic substances than zinc and
consequently not as easily released when the pH decreases.
Rapid mobilization of cadmium can be expected when the pH
of the investigated cambisol drops to 4 and lower.

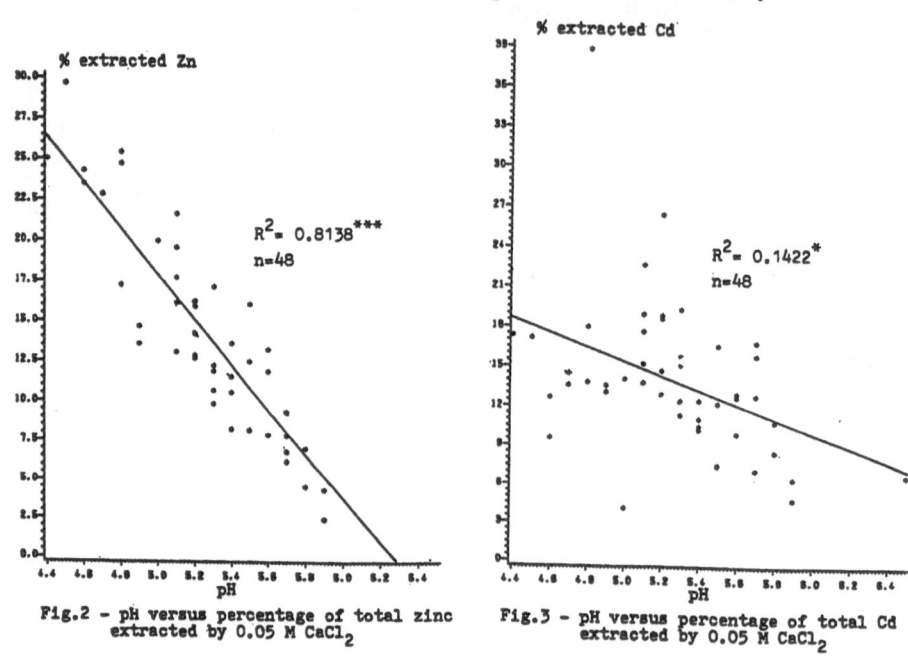

Fig.2 – pH versus percentage of total zinc
extracted by 0.05 M CaCl$_2$

Fig.3 – pH versus percentage of total Cd
extracted by 0.05 M CaCl$_2$

The average clay content of all samples is estimated
to be around 10 per cent. The negative correlation of pH and
the quotient pCEC/EC points to a similar clay composition of
all samples (ref.6) (see Fig.4).

Regression analysis of plant content in relation to soil
content gives high R^2-values for zinc and cadmium in almost
all cases, regardless of whether extractable or total metal

is employed as independent variable (see table II). The difference between regression with total or extractable metal is much smaller than reported in other studies (ref.7), probably becaus only one single type of soil was investigated.

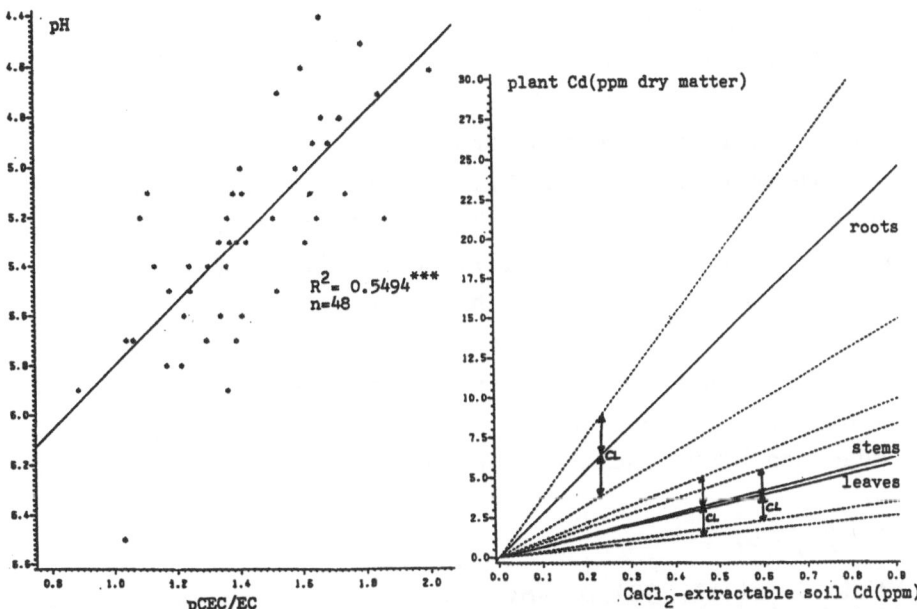

Fig.4 - pH versus pCEC/EC
pCEC=potential cation exchange capacity
EC=exchangeable cations(effective CEC)

Fig.5 - Cd-content of different fractions of parsley versus CaCl$_2$-extractable soil Cd, n=6, CL=confidence limits for mean values (95%)
For R^2-values see Table II

plant	Cd$_t$	Cd$_{ex}$	Zn$_t$	Zn$_{ex}$
parsley - leaves	0.8306	0.9205	0.7832	0.9280
- stems	0.7705	0.8538	0.8067	0.8625
- roots	0.8589	0.9260	0.8066	0.9779
radish roots	0.8145	0.8798	0.6985	0.9650
carrot roots	0.8993	0.9373	0.9231	0.9674
rye grains	0.9454	0.9269	0.8466	0.7966
potato tubers	0.8602	0.7827	0.7894	0.8368

Table II- R^2-values for regression of plant metal content (ppm dry matter) onto soil content(ppm), significant at least at the 5% level, n=6

The fact that $CaCl_2$-soluble cadmium and zinc gave slightly higher R^2-values for most plants (tested for significance by Wilcoxon's Signed Ranks Test) has to be judged carefully since sample sizes for each plant species are small (n=6) and the coefficients of determination consequently have very wide confidence limits (see example in Fig.5). A more detailed study of 2 species, with 50 samples of each, is now in preparation.

The results show that $CaCl_2$-extraction, a method proposed by the EC for routine analysis in assessing plant available cadmium and zinc, gives a better prediction of plant contents than total soil metal in most cases. Soil properties, influencing mobility of cadmium and zinc in a wastewater-treated cambisol, have similar effects to those described by other authors for different soil types. It would appear that in the investigated case the bulk of cadmium and zinc present in the soil is either in the soluble or adsorbed state and therefore highly plant available.

4. REFERENCES

1.Milde G. et al.1981.Dynamics of soil and grounwater pollution by irrigation of sewage.Proc.Int.Symp.Quality of groundwater.Stud.Env.Sc.Vol17.297-303.Elsevier.
2.Bower C.A. et al.1952.Exchangeable cation analysis of saline and alkaline soils.Soil Sc.73.251-261.
3.Herms U.,Tent L.1982.Cadmiumgehalte in Spülfeldern aus Hafenschlick und in darauf angebauten Kulturpflanzen-eine Felderhebungsuntersuchung.Landw.Forsch.SH 33.251-261.
4.Merkel D., Köster W.1976.Der Nachweis einer Zinktoxizität bei Kulturpflanzen durch die Bodenuntersuchung mit Hilfe der $CaCl_2$-Methode.Landw.Forsch.SH 33.274-281.
5.Häni H.,Gupta S.1985.Reasons to use neutral salt solutions to assess the metal impact on plants and soils.in: R.Leschber,R.D.Davis and P.l'Hermite.Chemical methods for assessing bio-available metals in sludges and soils. Elsevier Appl.Publ.London-New York.42-47.
6.Bundesanstalt für Geowissenschaften und Rohstoffe. Projektbericht:Vergleich von Untersuchungsmethoden zur Bestimmung der KAK und der austauschbaren Kationen tropischer und subtropischer Böden.1978.Berichterstatter:Grüneberg et al.
7.Sauerbeck D.R.,Styperek P.1985.Evaluation of chemical methods for assessing the Cd and Zn availability from different soils and sources. in ref.5.49-66.

ANAEROBIC CONTACT DIGESTION OF BIOCHEMICAL SLUDGE
(FROM SIMULTANEOUS PRECIPITATION) -
RESULTS OF A SEMI FULL-SCALE STUDY

UNTO TANTTU
M.Sc. (Civ. Eng.)
Plancenter Ltd, Finland

Summary

For more than one year a part of the mixed primary-secondary
sludge from an activated sludge/simultaneous precipitation plant
was treated in a two phase anaerobic contact reactor, volume 35
m^3. The raw sludge characteristics were as follows: TS 6,3 %, VS
3,6 %. In mesophilic temperature area the achieved results were:
hydraulic retention time 9 - 12 d, digester efficiency of
organic compounds about 50 %, biogas yield 0,9 - 1,1 m^3/kg
$VS_{destroyed}$. Also thermophilic temperature was tested during 12
weeks. No significant difference compared with the mesophilic
process was noticed except the energy balance. In the mesophilic
process about 48 % of the total biogas amount was used for
heating the equipment, while extra heating energy was needed to
keep the right temperature in the thermophilic process.

1. INTRODUCTION

Sludge digestion is used in Finland only in large sewage
treatment plants because of its numerous advantages. Conventional
anaerobic sludge treatment requires high investment costs and for
that reason digestion has not been used in small and medium size
plants.

The aim of the study described in this paper was to clear up the
suitability of a prefabricated reactor for municipal sludge digestion
and energy recovery. The results can be utilized in hundreds of small
and medium size sewage treatment plants in Finland.

The startup of the digestion process began on March 1983 and the
study took 16 months for experiments and data collection.

2. RAW SLUDGE CHARACTERISTICS

The sludge to be treated is mixed primary-secondary sludge from
an activated sludge process. Phosphorus is precipitated by
ferrosulphate ($FeSO_4$ x $7H_2O$) simultaneously with biological process.
The whole process scheme is illustrated in Fig. 1.

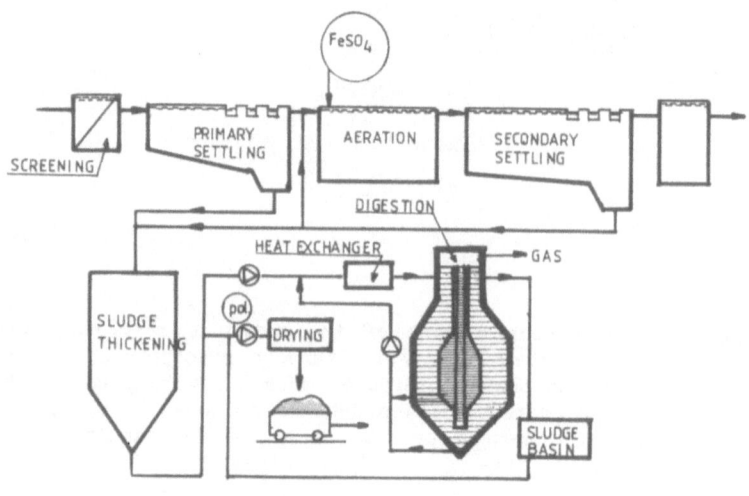

Fig. 1. Arrangement of test equipment at sewage treatment plant

During test periods the characteristics of raw sludge were as follows:

	Min.	Max.	Average
Total solids, % TS	4,5	9,0	6,3
Volatile solids, % VS	2,6	5,5	3,6

3. TEST EQUIPMENT

The reactor used is prefabricated so called ACP-reactor manufactured by Finnish contractor YIT Ltd. Fig. 2. illustrates the main features of the reactor and auxiliary facilities. The reactor consists of two steel made cylinders lying one inside the other. The outermost part is working as the first phase of the process while the inner one is the second phase. The circulation from the inner part to the outermost is arranged by both pumping and propeller situated in the central pipe.

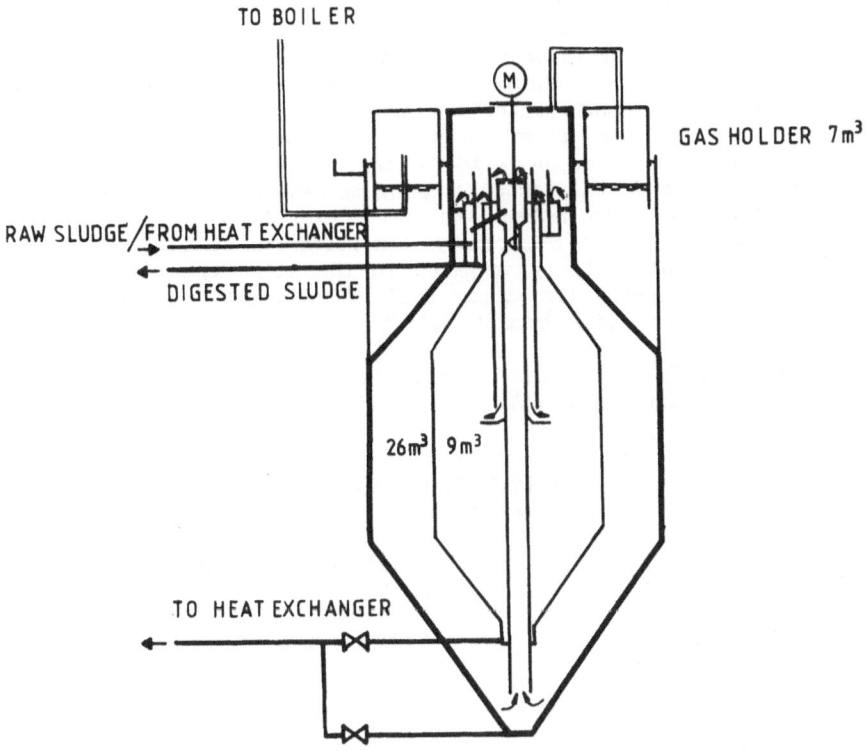

TO BOILER

GAS HOLDER 7m³

RAW SLUDGE/FROM HEAT EXCHANGER

DIGESTED SLUDGE

26m³ 9m³

TO HEAT EXCHANGER

Fig. 2. Test reactor

4. LOADING

The raw sludge was pumped automatically at intervals of one hour for a period of four minutes, which means 3 - 4 m³ raw sludge per day depending on the concentration of the sludge. Hydraulic retention time was correspondingly 9 - 12 days. The following parameters can be stated as results:

Volatile solids loading	3 - 4	kgVS/m³ d
Gas production	0,9 - 1,1	m³/kg VS
Digester efficiency	50	%

The "In-Out" test using volatile solids indicates the digester efficiency.

5. DIGESTION PROCESS CONTROL

Five factors must be in balance to accomplish the digestion: bacteria, sludge (food), loading, mixing and environment. Controls are needed to confirm satisfactory operation and to indicate an action that would bring about change (1).

Decomposition of organic material in digester is a continuous two-step process. In the first step, the organic material is converted into organic acids by acid-forming bacteria. The organic acids are used as food in the second step by methane-forming bacteria which convert the acids into methane and carbon dioxide gases.

The methane-formers are extremely sensitive to environmental changes. They grow quite slowly compared to the acid formers since they get very little energy from their food. This causes the methane formers to be very sensitive even to slight changes in loading, pH and temperature.

Internal controls show what is happening inside the digester. Four tests were run frequently: temperature, volatile acids, alkalinity and pH.

The best control is obtained by monitoring the ratio between volatile acids and alkalinity. The volatile acid/alkalinity relationship can vary from less than 0,1 to about 0,35 without significant changes in digestion.

The following VA/Alk ratios were listed as conclusion from the test run:

	VA/Alk-ratio
Mesophilic digestion	
Startup	0,4 - 0,6
Run	0,05 - 0,2
Thermophilic digestion	
Startup	0,6 - 0,8
Run	0,4 - 0,6

It took at least 1 month to establish a stable mesophilic digestion. The stabilitation of the thermophilic digestion takes rather long time, up to one year. In the test run a stable state was not achieved, as it can be seen from VA/Alk-ratio.

Because of the high alkalinity in the digester, the pH remained at a constant level. The alkalinity ranged from 4 500 - 6 000 $mgCaCO_3/l$, sometimes as high as 7 000 mg/l. pH values were as follows:

Mesophilic digestion	7,0 - 7,3
Thermophilic digestion	7,4 - 7,5

One of the most important environmental requirements was the proper temperature. The process temperature for normal digestion was $35^{\circ}C$. The temperature should not wary more than 1 degree $^{\circ}C$ per day after the operating temperature has been reached. The methane-formers are affected by changes in temperature, but the acid formers are not as sensitive to temperature changes.

Some difficulties, especially in the sludge recirculation system, changed the temperature more than 1°C per day. When the temperature dropped, the following changes occurred.

1. The volatile acids increased

2. The CO_2 content of the gas increased

3. The gas production rate decreased.

6. HEAT BALANCE
Heat balance of the test equipment were evaluated on the basis of the test period from April to June, 1985. The produced gas was utilized mainly for heating of test equipment and raw sludge. Gas was burned daily according to process temperature trying to keep the temperature near 35°C. Gas production results are shown in Table 1.

Table 1. Gas production results on test period

	Burned gas m^3/d	Excess gas m^3/d	Total gas production m^3/d
min.	10	19	41
average	29,5	32,3	61,8
max.	51	46	77

The efficiency of the boiler plant was approximately 50 %. Methane content ranged from 65 to 71 %.

During 12 weeks test period of the thermophilic temperature in the autumn 1984 it was observed that the biogas yield was not enough for keeping the right temperature in the process.

REFERENCES

(1) EPA, Operation manual anaerobic sludge digestion, February 1976.

THE ALTERNATIVE "EARTHWORM" IN THE ORGANIC WASTES RECYCLE

U. TOMATI, A. GRAPPELLI, E. GALLI
Institute of Radiobiochemistry and Plant Ecophysiology
National Research Council

Summary

The use of earthworm in organic residue management has been proposed as a practical process which could be applied on a limitate scale and under favorable conditions.
Earthworms consume all organic wastes reducing their volume of about 50 %. As a result, a casting with a good fertilizing value and a protein biomass which could be used as a foodstuff are produced.
Earthworms stimulate microbial activities which the composting processes are dependent by.
Conditions of culture, casting analysis and earthworm biomass composition are reported. Feasibility of vermiculture as an economically option in organic waste recovery is proposed on the basis of the recent literature about this topic.

INTRODUCTION

The appropriate disposal of organic residues is one of the most critical tasks now facing all advanced societies. Recycling technologies applied to organic residues management are more and more object of research directed toward seeking better ways in resources conservation and environment protection.
The use of earthworm in organic residues management has been proposed as a practical process, at least on a limitate scale and under certain favorable conditions (1, 2, 3). Earthworm ability to consume organic wastes and conditions for vermicompost has definitively established by many AA and discussed both by scientists and practical workers (4-13).
Vermicomposting is a result of a combined action of the earthworms and microflora living in their intestine and in growth medium. Earthworms stimulate the composting process by mechanical and biochemical actions. The mixing and aereating of the substrate, the breaking up of waste as it passes through the intestine, are purely mechanical processes. The biochemical effects are mainly dependent on microbial activities.

As a consequence of a such digestion, sludge is stabilized and the large particles and irregular shapes of solid waste are reduced to much smaller particles of relatively uniform shape and size. Earthworms consume organic waste, strongly reducing its volume (about 50%) and expelling the digested materials as casting, which is a useful soil amendment, odourless, with attractive aesthetic and handling properties, and may be easily stored for a long time waiting for its agricultural use.

The processing of large amounts of organic waste may eventually produce large quantities of excess earthworms which can be recycled into new vermicomposting beds or utilized as a biomass as an animal foodstuff.

CONDITIONS OF CULTURE

The optimum conditions for supporting a worm population has been studied by researches, worm breeders and worm farmers (10, 11, 14, 15).

Temperature
Earthworms exhibit a fairly complex response to changes in temperature. In general the optimal range for the most rapid feeding and waste conversion is between 13° to 22°C.

pH
Earthworms generally find their optimum environment at pH 6.0 to 8.0.

Moisture content
The optimum range of moisture content is between 50 to 80 percent.

Aeration
Earthworms are sensitive to anaerobic conditions, so in the vermicomposting operations it is essential to assure aeration requirements.

Substrate
Earthworms thrive in a medium of 9 to 15 percent protein. A moderate C:N ratio (15:1 to 35:1) is considered desirable for a maximal weight gain of the biomass. As an example, fresh bovine feces contain about 14 to 15 percent protein, sludge vary from 12 to 38 percent protein, mixed municipal refuse contains only about 4 percent protein (obviously the biodegradable fraction only is suitable to be worked out by earthworms).

****

Successful techniques of vermicomposting produce castings and earthworms.

CASTING
Earthworm egesta are usually called "casting". It is a good soil amendment and it can be considered as an attractive fertilizer, because its content in available nutrients. Castings are less rich in comparison to the starting materials, especially

in regard to nitrogen and organic substances content. However, they are richer in nutrients more readily available to plants and,because of the microbial biodegradation they have undergone, they are rich in microbial metabolites, particularly active on plant metabolism.
The content in available nutrients, the presence of a large microbial population and biologically active metabolites, are among the factors from which the fertilizing value of casting in dependent on. Generally casting could be considered as a good organic soil amendment, with attractive structural properties (tab.1).

EARTHWORM BIOMASS AS AN ANIMAL FOODSTUFF
The earthworm body is mostly muscle. Dried earthworm contains 60-65% protein, 7-10% fat, 2-20% carbohydrate, 2-3% mineral, a wide range of vitamins and of essential aminoacids (tab.2)(16). Because its composition, vermimeal could be regarded as a potentially valuable protein feed additive for domestic animals.

****

Experiments using earthworms in organic waste management has been carried out in a number of countries. Experimental works are now investigating the use of earthworms in waste management by looking at the engineering parameters of the system. The results show that vermicomposting may be used as a feasible alternative in current systems of waste management.
Because the castings are derived from wastes, there exist a potential hazard for contamination of the product by heavy metals or organic polluting substances. Worms apparently are quite capable of concentrating some heavy metals to levels of toxicity for some little animals. It is not clear whether earthworm consumption changes the availability of metals to plants.

ENVIRONMENTAL AND PUBLIC HEALTH ASPECT OF VERMICOMPOSTING
Potential risks are dependent on toxic substances and heavy metals present in the starting materials.
Another point is the survival of pathogens in earthworm worked materials. Although data seem to confirm a reduction in some pathogens species, especially Salmonella, the relative hazard to public health needs to be documented (10, 11, 15).

REFERENCES

(1) COLLIER, J. (1978).Use of earthworms in sludge lagoons.
In: Utilization of Soil Organisms in sludge management,
Proceedings. R.Hartenstein, (ed.) SUNY College of Environ-
mental Science & Forestry, Syracuse, N.Y. pp.6-7.

(2) GREEN, Ed and PENTON, S. (1980). Full scale vermicomposting at the Lufkin water pollution control plant. In: Workshop on the role of earthworms in the stabilization of
organic residues, Proceedings. M.Appelhof, comp. pp.229-31.

(3) PINCENCE, A., DONOVAL, J., BATES J. (1980). Vermicomposting of municipal solid wastes and municipal wastewater
sludges. In: Workshop on the role of earthworms in the
stabilization of organic residues, Proceedings. M.Appelhof,
comp. pp. 241-254.

(4) HARTENSTEIN, R. and MITCHELL, M.J. (1978). Utilization of
earthworms and micro-organisms in stabilization, deconta-
mination and detoxification of residual sludges from treatment of wastewater. Final report to the National Science
Foundation, SUNY College of Environmental Science and Forestry.

(5) GRAFF, O. (1980). Preliminar experiments of vermicomposting of different waste materials using Eudrilus eugenie
Kinberg. In: Workshop on the role of earthworms in the
stabilization of organic residues, Proceedings. M.Appelhof
comp. pp. 178-191.

(6) MITCHELL, M.J. (1980). Effects of Eisenia foetida on decomposition process in laboratory microcosms, sewage sludge drying beds and sludge amended soils. In:Workshop on
the role of earthworms in the stabilization of organic residues, Proceedings. M.Appelhof, comp. pp. 192-204.

(7) HUHTA, V. (1980). Results of preliminary experiments cul-
turing Eisenia foetida on Different Types of Sewage Sludge
Animal and Human Excreta mixed with Low Nitrogen Organic
Materials. In: Workshop on the role of earthworms in the
stabilization of organic residues, Proceedings. M. Appelhof, comp. pp. 220-228.

(8) SABINE, J. (1980). Vermiculture as an option for resource
recovery in the intensive animal industries. In: Workshop
on the role of earthworms in the stabilization of organic
residues, Proceedings. M. Appelhof, comp. pp. 207-219.

(9) TOMATI, U., GRAPPELLI, A., GALLI, E., ROSSI, W. (1983).
Fertilizers from vermiculture as an option for organic
wastes recovery. Agrochimica XXVII (2-3), pp. 244-251.

(10) CAMP DRESSER and MCKEE, Inc. (1980). Compendium on solid
waste management by vermicomposting. Environmental Protection Agency, 600/8-80-033, Cincinnati, OH.

(11) CAMP DRESSER and MCKEE, Inc. (1981). Engeneering assess-

ment of vermicomposting municipal wastewater sludge.
Environmental Protection Agency, 600/2-81-075, Cincinnati
OH.

(12) GRAPPELLI, A., TOMATI, U., GALLI, E. (1983). Il vermicom-
postaggio nello smaltimento congiunto di fanghi di risulta
e rifiuti solidi urbani. International symposium on agricultural and environmental prospects in earthworm farming.
Proceedings Rome.pp.73-80.

(13) GRAPPELLI, A., GALLI, E., TOMATI, U. (1985). Olive oil
wastewaters recycle as fertilizer via wormcomposting.
International symposium on earthworms, Proceedings. Carpi
Bologna.

(14) APPELHOF,M. Ed(1980). Workshop on the role of earthworms
in the stabilization of organic residues. Kalamazoo, Michigan.

(15) TOMATI, U., GRAPPELLI, A., Ed(1983). International symposium
on agricultural and environmental prospects in earthworm
farming. Rome Italy.

(16) AMERIO, M. BERTOLINELLI, M. (1983) Caratteristiche chimi-
co-nutritive del lombrico: applicazioni nel settore zoo-
tecnico. International symposium on agricultural and environmental prospects in earthworm farming, Proceedings,
Rome. pp. 141-153.

Table 1. Chemical analysis, microbial population and growth regulators content of earthworm casting.

Chemical analysis		
Water content	51.60 %	
pH	6.50	
C (total)	16.78 %	d.w.
C (inorganic)	1.37 %	"
N (total)	1.63 %	"
$N-NO_3^-$	0.40 %	"
P (total)	0.92 %	"
$P-PO_4^{-3}$	0.14 %	"
K_2O	1.61 %	"
Ca (total)	8.60 %	"
Ca (available)	0.14 %	"
Mg (total)	2.51 %	"
Mg (available)	0.45 %	"
Na	3.03 %	"
Fe	910.10 ppm	d.w.
Mn	218.40 ppm	"
Cu	7.20 ppm	"
B	0.35 ppm	"
Zn	68.30 ppm	"

Microbial population (Number of cells/g d.w.)	
Bacteria	1.8×10^8
Actinomycetes	2.8×10^6
Fungi	2.0×10^5

Growth Regulators (μg equiv./g d.w.)	
Gibberellins (GA_3)	2.75
Cytokinins (IPA)	1.05
Auxins (IAA)	3.80

Table 2. Amino acids composition of worm meal (percentage on crude protein).
from: M.Amerio and M.Bertolinelli; International symposium on agricultural and environmental prospects in earthworm farming. Rome, Italy, 1983.

	Total a.a.	Free a.a.
Lysine	5.93	0.47
Histidine	1.97	0.09
Arginine	6.22	0.42
Aspartic acid	9.16	0.17
Threonine	3.86	0.19
Serine	4.48	0.20
Glutamic acid	12.68	0.33
Proline	3.88	0.09
Glycine	4.60	0.14
Alanine	5.43	0.30
Valine	4.71	0.33
Methionine	1.65	0.14
Isoleucine	4.22	0.31
Leucine	7.25	0.64
Tyrosine	3.65	0.22
Phenylalanine	3.99	0.25
Tryptophan	0.92	--
Cystine	1.00	--
Hydroxyproline	0.62	--
Taurine	--	0.17
Asparagine	--	0.46

SLURRY METER INSTRUCTIONS

H. TUNNEY
The Agricultural Institute, Ireland

1. Collect representative sample of animal manure slurry.

2. Put slurry in a plastic bucket and stir well.

3. Place slurry meter in slurry and read dry matter immediately.

4. Read corresponding N, P and K content from Table 1 below.

5. Cattle slurry higher than 5% dry matter (d.m.) and pig slurry higher than 8% is normally too viscous for accurate measurement. In such cases, mix slurry with an equal volume of water, stir well and take d.m. reading with Slurry Meter. Double the reading obtained to get the correct d.m. of the undiluted slurry. If necessary make further dilutions.

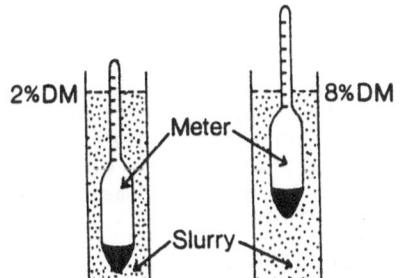

% Dry Matter	Concentration
0 – 5	Low
5 – 10	Medium
10 – 15	High

Note: depth of slurry in bucket should be 25 to 30 cm.

Table 1. Total N, P and K content of cattle and pig slurry in kg/tonne (kg/m³).

Cattle Slurry				Pig Slurry		
N	P	K	% Dry Matter	N	P	K
1.5	0.2	1.6	– 2 –	2.5	0.5	1.2
2.5	0.4	2.2	– 4 –	4.5	1.0	1.7
3.5	0.6	2.7	– 6 –	5.5	1.5	2.1
4.2	0.8	3.1	– 8 –	6.0	2.0	2.3
5.0	1.0	3.5	– 10 –	6.5	2.5	2.5
5.5	1.2	3.7	– 12 –	7.0	3.0	2.6
6.0	1.4	3.9	– 14 –	7.2	3.5	2.7

The higher the dry the higher the fertiliser value. If a slurry sample containing mostly urine is collected the N and K values will be higher and P values lower than shown.

It can be assumed that the N value shown in Table 1 will be half as effective and the P and K will be as effective as their chemical fertiliser equivalents.

Consult your agricultural advisor for information on the correct rates to apply and how you can reduce fertiliser costs by recycling animal manures.

COMPOSTING OF SEWAGE SLUDGE CONTAINING POLYELECTROLYTES

J.J. van den Berg
Grontmij n.v., De Bilt, the Netherlands

Summary

In the Netherlands research is being performed on composting of sewage sludge containing polyelectrolytes, using the
aerated static pile method. After a literature study experiments were carried out. The results showed that progress
of the composting process is enhanced by applying moisture-regulating and porosity-increasing additives. Any influence of polyelectrolytes on the composting process has
not been proved. Additives were mainly used to obtain sufficient porosity and moisture-regulation, and not to influence the C/N-ratio of the mixture to be composted.

1   INTRODUCTION

In the Netherlands, agriculture is the main market for
useful disposal of sewage sludge. Today, however, there is an
increasing interest in alternative ways of processing or utilization. Therefore the Dutch Foundation for Applied Wastewater
Research (STORA) has prepared an investigation on:
- composting of sewage sludge;
- preparation of black earth from sewage sludge.

The investigation consisted of three parts. In the first part
a literature study on the main technical and financial aspects
of these processing options was carried out. The literature
study did not supply the answers to a number of questions,
although many experiments were examined. The results were not
or hardly comparable because the composting experiments of
other investigations, were not performed under similar circumstances. Literature on black earth preparation is relatively
scarce.

In the second part of the investigation a number of experiment
were performed to gain information on:
- treatment of materials (sludge and additives);
- equipment required;
- process management (total process time, process monitor-
  ing/control and side effects such as the influence of processes on the soil);
- post-processing;
- the degree of disinfection;
- approximate running costs.

These experiments were focussed on the practical aspects of
the processing methods. Different basic materials and additives were treated in the same way, so that the experimental
results were comparable.

In the third part of the investigation the results of the experiments were evaluated and compared with theoretical models (1).

This paper presents an outline of the set-up of the experiments, a summary of the results and the main conclusions.

2 BASIC MATERIALS

Sludge types

The following types of sludge were used:
- aerobically stabilized sludge, conditioned with polymers and dewatered with a belt screen press to a dry matter content of 15-20%;
- anaerobically stabilized sludge, conditioned with polymers and dewatered with a belt screen press to a dry matter content of 20-25%.

Additives

Function of the additives is primarily to increase the porosity of the pile and the dry matter content of the mass to be composted. The following additives were used:
- wood chips;
- wood shavings;
- rubber scrap;
- wood blocks;
- straw.

3 COMPOSTING METHOD APPLIED

The aerated static pile method, with aeration by pressure or suction, was applied for the composting process.

The required quantity of air was calculated by the Haug method (1).

4 RESULTS AND CONCLUSIONS

Both types of sludge used in the experiments can be composted; the basic material is stabilized by decomposition of organic matter and moisture is removed by evaporation. No indications have been found that the polymers have any influence on the composting process.

The results of the experiments confirm the assumption that additives should have porosity-increasing and moisture-regulating properties. Effective moisture absorption was achieved when wood shavings with a dry matter content of at least 60% were used as additive.

The application of wood chips for moisture regulation did not provide sufficient permanent porosity, even with a sludge/additive ratio up to 1:3 by volume. To improve the porosity of the pile, secondary additives were used, such as wood blocks or rubber scrap.

The application of porosity increasing additives combined with moisture-regulating additives proved to have a positive influence on the composting process.

According to the literature, additives with a high carbon content would have an influence on the C/N ratio. This influence was not demonstrated by the experiments. Such influence is considered unlikely.

Pelleted compost is proved to be unsuitable as additive for composting of sewage sludge. The pellets have insufficient moisture-regulating capacity and are liable to be damaged by mechanical forces during mixing.

After completion of the composting process the porosity-increasing additive can be recovered completely; also 50-60% of the wood shavings used as moisture-regulating additive can be recovered. Straw used as moisture-regulating additive could not be separated from the end-product.

The following method of supplying and mixing basic materials and additives was found to be satisfactory. After the basic materials and moisture-regulating additives had been mixed thoroughly with a manure spreader, the porosity-increasing additive was mixed through with a bucket crane. Using a hydraulic excavator the mixture was piled on the aeration tubes, which had been laid on a bed of wood shavings or chips.

The supply of air needed for the process and the removal of decomposition products was accomplished by forced pressure or suction aeration. During suction aeration, the amount of air required for the process was generally more than calculated. In the experiments with pressure aeration the required amount of air was equal to or less than calculated. During experiments under a roof, the required amount of air was less than calculated both with suction and pressure aeration.

The temperature was used as process control parameter, being the most practical choice because of its direct measurability. The temperature observations gave a reasonable indication of the temperature in the piles, on the basis of which the forced aeration could be regulated. Assessing the progress of the composting process by determining the carbon and nitrogen contents and the ratio of these elements proved to be impossible.

Moisture content has been proved to be an unsuitable process parameter, because of the problems with sampling and the influence of precipitation. A high moisture content has a negative influence on the composting process. A minimum dry-matter content of 40% in the mixture to be composted, as mentioned in literature, seems to be correct.

During composting, CO_2 and NH_3 were removed via the ventilation system and the surface of the piles. The concentrations of the components varied strongly. The nuisance caused by the odour of the sludge, ammonia and amines, was strongly dependent on the degree and method of stabilisation of the sludge.

For determining the extent to which the sludge had been stabilized during composting, the chemical oxygen demand was measured. In the experiments the average decrease was 63%. The COD of the end-product was influenced by remainders of additives, so the value of these results should not be overestimated.

The content of micro-elements in the end-product is strongly

influenced by the presence of covering compost. It was impossible to take samples representative for the composted materials. However, it has become clear that the following effects may occur:
- increase of the concentrations due to decomposition of organic matter and the covering compost;
- decrease of the concentrations due to an increase of the amount of dry matter from additives and covering compost.

The C/N ratio of the end-product is of the same order of magnitude as that of the top soil of clay lands and grasslands, i.e. 10 to 15.

The results of microbiological analyses indicated that the hygienic reliability of the end-products was increased. Remarkably, the concentration of heat-resistant f-specific phages, which were designated as indicator organisms, declined more sharply than the concentration of micro-organisms with a lower heat resistance (2).

A cost price calculation has been made for composting a sludge production of 6000 m^3/year with a dry matter content of 20% (corresponding with a sewage treatment plant of 75,000 p.e.). The cost of open air composting is roughly DFL 440 per tonne of dry matter, which corresponds with DFL 88 per m^3 of dewatered sludge.

When a simple roof is applied, the cost price will increase to about DFL 495 per tonne of dry matter (DFL 99 per m^3 dewatered sludge).

5 RECOMMENDATIONS

For composting of sewage sludge it is recommended that two types of additives be applied:
- moisture-regulating additive; volume ratio of sludge: additive = 1:2 to 2.5. The dry matter content of the additive should be at least 60%;
- porosity-increasing additive; volume ratio of sludge: additive = 1:1 to 1.5, such as wood blocks or rubber scrap.

Sludge and moisture-regulating additives should be mixed thoroughly.

Next porosity-increasing material can be added, after which the mixture can be piled on the aeration tubes.

In occasional composting, double walled HDPE tubes can be used for the air supply. This material has a reasonable resistance against the prevailing temperatures and radial loading; the tubes can be used several times. When composting is more or less continuous, application of a more durable construction could be considered. When selecting the tube material, account should be taken of the fact that the air sucked in will contain ammonia. To ensure that the air is distributed uniformly, it is recommended that the tubes be placed on a bed of fine additive. Each pile may contain more than one aeration tube. The distance between the surface of the pile and the aeration tubes should be less than 1.50 m, to avoid short circuiting. With a pile length of more than 10 m, it is advised to install the ventilator in the middle of a pile or to install ventilators at both ends of the aeration tubes.

With composting in the open air, the shape of the pile should promote the runoff of precipitation; preferably the piles should be covered with compost, to reduce heat losses, and plastic sheet, to promote runoff of precipitation.

It is recommended that aeration can be regulated, to maintain an average temperature of 50-55°C in the pile. Both pressure and suction ventilation should be possible. Pressure ventilation is especially applied for a rapid removal of moisture. If possible, aeration should be done in the daytime; supply of relatively cold air by night should be limited.

REFERENCES
(1) Haug, R.I., Compost engineering. Ann Arbor Science Publishers, Inc (1980)
(2) Burge, W. D., Colacicco, D.S., Cramer, W. N., Criteria for achieving pathogen destruction during composting. JWPCF 53,12: 1683 e.v. (1981).

AN ALTERNATIVE APPLICATION FOR SEWAGE SLUDGE: BLACK EARTH

J.J. van den Berg
Grontmij n.v., De Bilt, the Netherlands

Summary
Preparation of black earth is receiving more and more interest. As an alternative processing option for sewage sludge. Black earth is a product that can be applied for greenery provisions, for forestry applications, as covering soil for waste tips, and for the slopes and shoulders of roads. It is produced by mixing sludge with a mineral additive, after which oxidation and maturing processes take place in a natural way. These processes are stimulated by turning the mixture regularly with a rotary cultivator. The moisture-absorbing capacity of the end-product is comparable with that of soils of natural origin, with the same chemical composition.

1. INTRODUCTION

In soil science terms, black earth is defined as a soil mixture, which serves to improve the growing environment in soils that are not or hardly suitable for vegetable growth. To achieve this goal, black earth should be capable of holding sufficient moisture, have a sufficiently high basic level of nutrients, and possesses soil physics properties adequate for vegetable growth.

The basic material for black earth preparation is partially dewatered sludge, which is insufficiently matured. The production process is determined by the quality of the dewatered sludge, the physical and chemical maturing of the sludge,and the nature and application method of the additives (usually sand). The quality of the dewatered sludge depends on the treatment process and the dewatering method. In general, naturally dewatered sludge -especially anaerobically digested sludgeis best suitable for black earth preparation. Mechanicaly dewatered sludge which has been conditioned chemically can be processed to black earth only with great difficulty and many extra operations. Mechanically dewatered and chemically conditioned sludge matures poorly and is difficult to dewater further; only direct mixing with additives and regular turning of the mixture can give some improvement.

The manural value of the black earth produced from sludge is dependent of the nature and the quality of the sludge and the additives and of the mixing ratio of the various components. In the most common sludge types and mixing ratios all nutrients are present in sufficient quantities in the first instance, while even an ample stock of phosphate and magnesium is present. The pathogenic germs are usually strongly reduced to numbers under the M.I.D.-50 value, while the content of heavy metals and toxic substances is dependent of the basic materials and the mixing ratios so that no general statement can be made about this aspect.

The process of preparing black earth from sewage sludge can be classified according to the dewatering condition of the sludge (spadeable or liquid) at the moment it is mixed with the additives. In the processing of naturally dewatered sludge the best results are obtained if the sludge is already in an advanced stage of maturition before it is mixed with sand and any other additives. In this case it is relatively simple to prepare black earth with favourable physical characteristics. If the sludge is dewatered mechanically, after conditioning with, for instance, polymers or lime and iron chloride, it should be mixed immediately with additives. Only in this way is this sludge capable of further maturing and dewatering.

On a practical scale, good results have been obtained in the Netherlands with black earth preparation from naturally dewatered, spadeable sludge.

2. POSSIBLE APPLICATIONS OF BLACK EARTH

New residential areas

A great demand for black earth is often felt in extension plans of municipalities. Especially in areas where inadequate drainage or bearing capacity requires that the original ground level be raised, application of a layer of black earth as a growing medium for the vegetation is necessary. This is because the fill used for raising the ground level is generally sand, which contains hardly any fertilizing substances or organic matter, and is therefore lacking in nutritive and moisture-absorbing elements. The thickness of the layer to be applied depends on the underlying soil profile, the groundwater regime and the vegetation that is to be planted.

Waste tips

In setting up and finishing off waste tips, a substantial amount of covering soil is required. The various layers of waste as well as -eventually- the entire tip have to be covered with soil. On the one hand this prevents nuisance for the surroundings (odour, vermin, birds), while on the other hand the finished tip can be planted. The required covering soil can be supplied partially be excavating soil at the tipping site in advance and putting this soil in store. To what extent this is possible depends on the soil profile and the prevailing groundwater levels. The rest of the required soil has to be hauled from elsewhere; black earth prepared from sewage sludge can be a suitable product for this application.

Recreational objects

Black earth can also be an important component in the realisation of recreational objects, such as sports and playing grounds and grassed areas around outdoor swimming-pools

If these objects are located on soils that are poor and sensitive to drought by nature, enriching with black earth is necessary, for which preferable humic, sandy soil mixtures are applied, so as to increase the moisture-absorbing capacity. Black earth produced from sludge can be used for this purpose.

Public gardens and parks

To improve the growing conditions for the vegetation in public gardens and parks in existing residential areas, there may be a need for an appropriate substratum. Exchanging the existing topsoil with black earth can supply this need so that the urban greenery has a reasonable chance of developing propperly.

Tree soil

A special application with regard to urban greenery is the use of black earth as tree soil. Also in this case poor soil is exchanged for a product with a reasonable content of manuring substances and organic matter. This exchange takes place with existing plantation as well as with relocation of (large) trees. By striving for optimum growing conditions in the plant hole, the chance of a successful relocation of the tree is increased. The composition of the black earth can be adjusted to the specific demands of a certain tree species. A striking example of such a specific composition is the use of black earth from sewage sludge for the Floriade ornamental park in Amsterdam. Because of the widely differing types of vegetation, soils of various compositions have been used in this one object, in order to create optimum growing conditions for every species.

Forestry

Also in forestry, where maximum wood production is strived for, application of black earth may have possibilities, especially for deciduous trees. Black earth with specific composition as regards fertilization and structure may be used to achieve optimum growth rates of the plantations. This can improve efficiency in forestry.

Nurseries

In addition to its use in more or less permanent cultures, black earth can be applied in nurseries, notably in the cultivation of ornamental plants.

Road construction

Finally, one of the applications of black earth prepared from sludge could be road construction. When new roads or motorways are constructed, it is common to apply a sand bed over the entire width of the pavement and the berms. For the berms, measures should then be taken to prevent the sand from being blown away, for instance by planting vegetation, usually grass. The sand from the cunette offers insufficient guarantee for a rapid emergence and growth of the grass, so it is desirable to apply a layer of black earth. This layer can then also serve as a substratum for any higher vegetation.

3. PRACTICAL TRIALS ON PREPARATION OF BLACK EARTH

Practical trials were carried out to establish whether it was possible with simple means to prepare black earth from mechanically dewatered sludge, using the 'rotary cultivator process'.

Basic materials

Sludge types

The following sludge types were involved in the trials:
- anaerobically stabilized sludge, conditioned with lime and iron chloride and dewatered with filter presses to a dry matter content of about 35% (FS);
- a mixture of digested and aerobically mineralised sludge, conditioned with poly-electrolytes and dewatered with centrifuges to 18 to 20% dry matter (CS)

Additives

In black earth preparation, in contrast with composting, additives are not recycled. Material that is mixed with the sewage sludge will remain in this mixture definitively. Therefore, in black earth preparation the additives are chosen such that only minimal quantities are needed, while low price and ample availability are also of great importance. In most cases sand with low humus and loam contents is used.

Mixing ratios

The mixing ratio of sewage sludge and additives is determined on the basis of factors which are related to the nature and composition of the components to be mixed, as well as to the destination of the end-product. These factors are the chemical composition and the physical qualities, and the manuring value and the pollution level, respectively.

Processing method

The rotary cultivator process consists of applying a layer of a few centimetres of dewatered sewage sludge on a substratum consisting of additive. The sludge is applied with a manure spreader, after which a rotary cultivator is used to mix the sludge layer through a 0.2 m thick layer of the substratum. To homogenize the mixture and to promote the maturing processes, the cultivator operation is repeated once or several times. The desired final mixing ratio is achieved by repeating the above process of applying and mixing as often as required.

Process monitoring

The process of preparing black earth from sewage sludge is basically a dewatering and maturing process. Factors which influence this process are:
- nature and composition of the sludge and additives;
- weather conditions;
- thickness of the mixture layer and the consequent aeration of the mixture;
- homogeneity of the mixture, dimensions of the sludge lumps and sludge cakes;
- frequency of turning or other operations.

The progress of the process was judged with visual and o-dour observations. During these observations attention was paid to aspects such as:
- changes in structure, such as disintegration of the lumps and occurrence of a pore structure;
- the occurrence of (micro)biological activity, such as bacteria colonies and mycelium from fungi;
- colour differences in the material, which could indicate differences in drying and maturing;
- the occurrence of vegetation.

The basic materials (sludge and additive) and the end-product (black earth) were analysed with regard to the following parameters:
- dry matter   - copper
- raw ash/organic matter - chromium
- nitrogen   - zinc
- phosphate   - lead
- potassium   - mercury
- calcium   - cadmium
- magnesium   - arsenic
- pH   - nickel

Furthermore, from all samples the grain size distribution was determined. In a limited number of samples of the end-products, the relationship between moisture content and moisture tension was determined, with which pF curves were constructed to represent this relationship. Examples are given in Figures 1 and 2.

Between the initial and final sampling, intermediate samples were taken to gain insight into the decomposition of organic matter and possible pH fluctuations in time. Furthermore, the oxygen content of the pore air in the sludge mixtures was measured.

4. CONCLUSIONS AND RECOMMENDATIONS ON BLACK EARTH PREPARATION
The trials have shown that both filter press cakes and centrifuge or band screen press sludge conditioned with poly-

mers can be processed into black earth. A precondition for obtaining a good end-product is that the process should take place in a layer of limited thickness (0.20 to 0.25 m). When a thicker layer is applied, anaerobic conditions are likely to occur, which halts the maturing process. Furthermore, in this situation no moisture can be removed by evaporation.

Assuming that a limited thickness of the layer is applied, processing in the open air is possible. Mixing of sewage sludge with the additive (sand) can be done effectively with a tractor-mounted rotary cultivator. First the sludge has to be deposited in the appropriate quantity on the sand layer to be mixed through, after which the cultivator operation can take place. This operation is only successful if the sludge is deposited in thin layers (not more than 0.04 m at a time).

This depositing requires a system in which the application equipment does not have to travel on the layers to be mixed; a manure spreader with sideward delivery meets this condition. When hard filter press cakes are being processed, it is advisable to crumble the cakes prior to their deposition.

The number of layers of 0.04 m each is determined by the composition of the products to be mixed, their dry matter contents, and the demands that are made on the end-product. If the basic materials are sewage sludge with 20% dry matter, of which 60% is organic matter, and sand with low humus and loam contents, then 5 applications are possible to produce an end-product with ample possibilities for use. This product then contains 5 to 7.5% organic matter. Depending on the weather conditions, this process will take 4 to 6 months.

Another possibility, using the same materials, is to strive for a maximum mixing ratio which still results in a product that is usable on a waste tip. It was found that up to 8 or 9 applications in a period of about 9 months give a product that is still manageable.

The rotary cultivator method of sludge processing in the open air thus offers the possibility of preparing a good vegetable earth, while it also offers perspective for processing mechanically dewatered sewage sludge into a product suitable for tipping. In the preparation of black earth, the material appears to have the proper chemical and granular composition after the production period of 4 to 6 months; the product then requires further maturing and structuring before it can be regarded as actual black earth. The latter process, in which soil life and root development play an important role, can be accelerated by providing the black earth with vegetation as soon as possible.

Both the production of a useful product, black earth, and the conversion of sewage sludge into a product suitable for tipping can be carried out at relatively low cost (respectively DFL 180 and DFL 145 per tonne of dry matter). The cost price calculation shows that black earth preparations is already possible on a small scale.

Moisture absorbing capacity of black earth

Importance of the moisture curve

The moisture curve of a soil is a piece of information from which an important part of the physical growing environment of the plant can be deduced. The curve represents the relationship between the moisture tension and the moisture content, from which, for instance, the amount of moisture available for a plant can be calculated.

Measuring only the moisture content of the soil does not provide all the information required to judge how wet or dry the soil is, at least not if different soils are to be compared with each other. A soil with a high clay content, for instance, may feel dry, while sandy soil with the same moisture content feels clearly moist. On the clay soil with that moisture content a plant could dry out, while on the sandy soil vegetable growth is still well possible.

A better indication of the moistness of the soil is given by the moisture tension. This represents the force by which the water is retained in the soil and thus gives important information on the potential for uptake of soil moisture by plants.

The moisture curve is determined by taking undisturbed samples of the soil in rings of 100 or 50 cm^3. In the laboratory, the moisture content is determined at different tension values (cm water column). The range of moisture tension values is very large, namely from 1 cm water column to nearly 16,000 cm. To make graphical representation possible, the moisture tension is expressed as the logarithm (base 10) of the moisture tension, which is known as pF. The moisture curve is the curve in which the pF is plotted against the moisture content: the pF curve.

Every specific soil type has its own pF curve, which depends mainly on the clay or loam content, the organic matter content, and the density. Important points in the pF curve are the moisture content at pF=2 and pF=4.2. The moisture content at pF=2 is the moisture condition at field capacity. It is the moisture content of a soil which, after having been saturated with water, has been drained freely for some time. It is usually a good indication of the moisture situation in spring in the topsoil of high-lying soils. (In lower soils the moisture content remains higher because of capillary rise from the groundwater). pF=4.2 is the wilting point for plant growth, which means that water that is bound more strongly to the soil than corresponds with pF=4.2 cannot be released by the suction from the plant.

The field capacity (moisture content at pF=2) is a good approximation of the upper limit of the amount of moisture that is stored in the soil and can be used by the plant. The wilting point (pF=4.2) indicates the lower limit of this amount. The difference between the moisture contents at pF=2 and pF=4.2 indicates the amount of moisture available for the plant which is present in the root zone of the profile. This amount of available moisture depends, just like the shape of the pF curve, on the granular composition of the soil type. A comparison between natural soil types and the black earth produced from filter press sludge or centrifuge sludge is presented in Figures 1 and 2.

In figure 1A it is shown that the moisture volume at pF=2 is about 44% and at pF=4.2 about 13%. The moisture content available in the soil volume is 44-13= 31%. In a well-rooted soil profile with a thickness of 500 mm, 0.31 x 500 = 153 mm moisture is available.

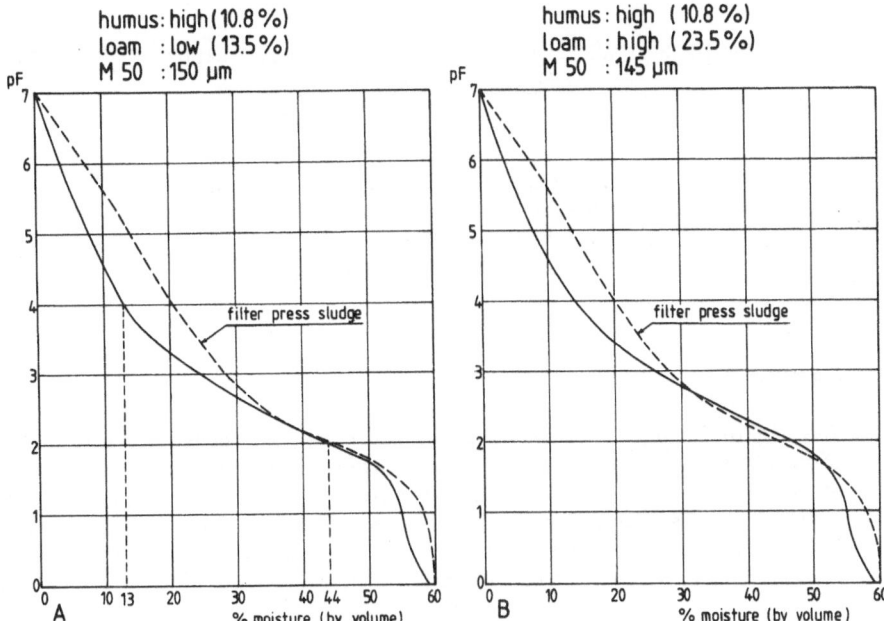

Figure 1: Comparison of pF curve of filter press sludge with that of very humic sand with low or high loam content

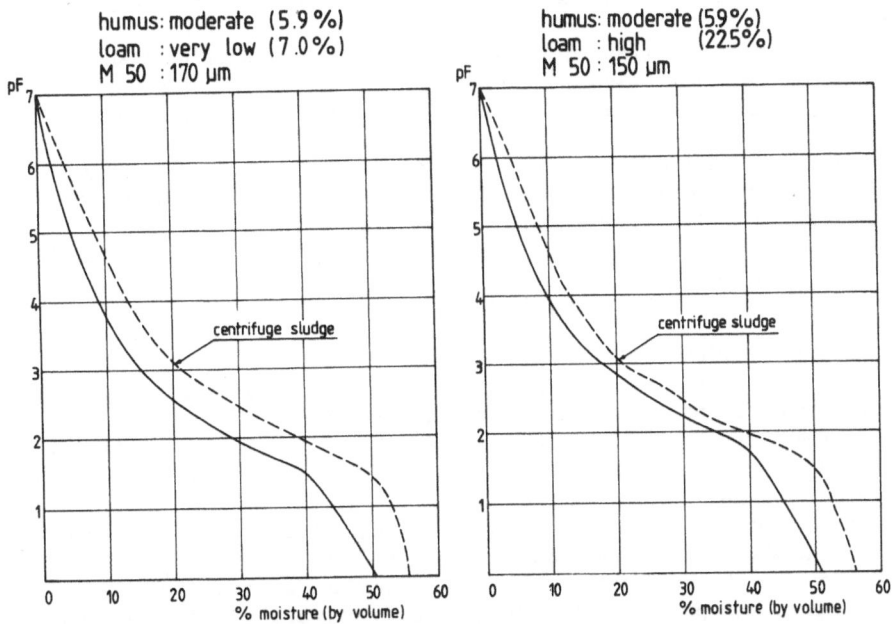

Figure 2: Comparison of pF curve of centrifuge sludge with that of humic sand with very low or high loam content

TOXIC ORGANIC SUBSTANCES IN SEWAGE SLUDGES : CASE STUDY OF TRANSFER BETWEEN SOIL AND PLANT

M.A. Wegmann, R.Ch. Daniel, H. Häni, A.Iannone
Swiss federal research station for agricultural chemistry
and hygiene of environment
CH-3097 Liebefeld-Bern (Switzerland)

Summary

The use of sewage sludges in agriculture has increased organic pollution of soils. A study of these pollutant transfers has been started at our research centre. The penetration in the plant of organic pollutants was particularly studied. Does it take place at root level or by air transport? To answer this question, we carried out 2 trials on contaminated sand with pure chemicals (Aroclor 1254).

1. INTRODUCTION

The use of sewage sludges in agriculture leads to organic pollution of soil.

A study of the transfer of these products from soil to plant has been started. We are now able to present a first trial model of english ray-grass (lolium perenne) grown on sand contaminated with polychlorobiphenyls (PCB).

We were interested in the penetration mode of pollutants from soil to plant. One can think that the transfer occurs through the roots or, on the contrary, under the effect of the ambiant temperature; that is to say that the pollutants near the surface of the soil evaporate to deposit themselves on the plant.

To try to answer the question, we have carried out 2 separated trials.

2. MATERIAL AND METHODS

In pots containing 2 kg of sand contaminated with Aroclor 1254, english ray-grass was grown using a nutrient solution
a) in the first trial, "surface contamination", the top 300 grams layer of sand (2-3 cm thick) was contaminated with 3 different concentrations of Aroclor 1254: 10; 100 and 1000 µg/pot
b) in the second trial, "total contamination", the whole of the sand was contaminated by the same amounts of PCB.

To test the eventuality of aerial transfer, we used the following disposition of the pots (see figure 1) in order to avoid a possible contamination between pots.

The ray-grass analysis was divided into two groups: the analysis of the green parts and of the roots.

The trials sand was analysed in the following manner: in the "surface contaminated" trial, it was divided in 3 horizontal layers so as to display any washing of PCB, whereas only a portion of "totally contaminated" sand was taken.

Both the plant material and the sand were analysed according to the annexed method.

3. RESULTS AND DISCUSSION

Sand: in "surface contamination", 90% of the PCB were left in the superior layer (see table I). These products were therefore not washed away by watering, but stayed adsorbed on the sand.

In both trials, no contamination of the blanks was observed.

Ray-grass (green part):
"Surface contamination": in both cuts we observed a sudden increase in contamination when the levels of 100 µg and 1000 µg/pot were attained.

"Total contamination": the observed previous phenomenon occured only for a 1000 µg/pot load, and, the obtained values were also weaker.

We can deduce from these results that a higher surface contamination seems to lead to aerial contamination.

In contrast to the contaminated sands, both trials showed a slight contamination of the blanks. This can be explained if we admit aerial transfer: figure 2 shows that it is easier for the PCB from one pot to deposit on the ray-grass of another in the neighbourhood than on its sand.

Ray-grass (roots): the contamination of the blanks is similar in both trials (see table I).

But the similitude stops here!

In the "total contamination" trial, the roots are in close contact with contaminated sand and this leads to high contamination. As an example, a PCB concentration in the sand of 500 ppb (1000 µg/pot) produces a contamination of almost 14000 ppb (14 ppm!) at the root level.

4. CONCLUSION

The results show that the soil pollution by PCB leads to a more or less important plant contamination depending on the analysed part.

Moreover, these results speak in favour of aerial transport, although internal transport cannot been ruled out.

Whether the recovered PCB are adsorbed on the surface of the plant has not, up to now been established.

Moreover, if we consider the recovery, that is to say the ratio of recovered PCB to the total added PCB, we observe good recoveries in the case of "total contamination" (83-96%), but poor recoveries in the case of "surface contamination" (28-58%). Further work is needed in the search of a satisfactory explanation of this phenomenon.

5. ANNEX

Analysis: the method allows the determination of PCB in plant material, sand, soils and sewage sludges.

Extraction: - Place the sample in a conical flask containing the following mixture:

$$
\begin{array}{ll}
\text{Methanol} & (90 \text{ ml}) \\
\text{Water} & (10 \text{ ml}) \\
\text{NaOH} & (8 \text{ g})
\end{array}
$$

- The mixture is placed for 2 hours in a ultrasonic bath at 55°C.

Table I: the table is a synopsis of the analytical results for ray-grass (green part and roots) and for the sand. The results are stated in ng/g of fresh plant.

	added amount of PCB (µg)	0	10	100	1000
Surface contamination:	green part cut I	6	2	20	123
	green part cut II	11	11	17	330
	roots	12	76	322	1352
	sand layer I	nd	5	22	333
	sand layer II	nd	nd	2	16
	sand layer III	nd	nd	nd	nd
Total contamination:	green part cut I	13	18	15	59
	green part cut II	19	18	22	50
	roots	15	143	1101	13871
	sand	nd	1	12	155

- nd: not detected
- the letters I, II and III are the superior, middle and bottom layers of the sand

                          - Filter.
Clean-up: - Extract the filtrate with 3 x 30 ml of cyclohexan.
                 - Wash the cyclohexan phase with 30 ml of water.
                 - Treat it with 5 ml of H_2SO_4 conc. (A large part of the
                   co-extracted matter is thus destroyed).
                 - Wash the organic phase twice with 30 ml of water.
                 - Dry on Na_2SO_4.
                 - Concentrate to about 1 ml with a rotavapor.
                 - Prepare the following chromatography column:
                   - length: 200 mm; i.d.: 16 mm
                   - fill half of the column with pentan
                   - Add 15 g of deactivated alumina (2% water)
                   - Rinse with 30 ml of pentan
                   - Add the concentrated extract
                   - eluate as follows:
                     - Pentan (20 ml): disregarded
                     - Hexan (60 ml): this fraction contains the PCB
                   - This fraction is then concentrated to exactly 1 ml.
     The quantitative analysis of the PCB ist carried out by gas chromato-
graphy equipped with a electron capture detector
                     - capillary column: 50 m
                     - cross-bonded phase: SE 54

REFERENCES

(1) TARRADELLAS, J. and DIERCXSENS, P., Schweiz. Arch. Tierheilk., <u>125</u>,
     589-605, 1983.
(2) FRIES, G.F. and MARROW, G.S., J. Agric. Food Chem., <u>29</u>, 757-759, 1981.

Figure 1: Disposition of the pots

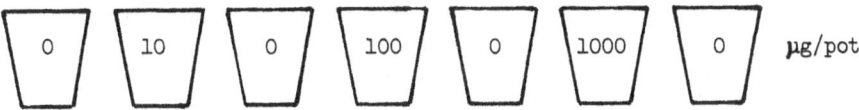

Figure 2: Growth of the ray-grass

VARIETAL TOLERANCE IN CEREALS TO METAL CONTAMINATION IN A SEWAGE TREATED SOIL

J H WILLIAMS, K A SMITH AND J R JONES

SOIL SCIENCE DEPARTMENT, MAFF, WOLVERHAMPTON, WEST MIDLANDS, UK

Summary

Advisory experience and some preliminary pot-work had shown differences in metal uptake between roots and shoots of cereal cultivars. The purpose of the present study was to compare the tolerance of winter barley and wheat cultivars, grown under field conditions, to high soil concentrations of zinc, copper and cadmium resulting from past applications of sewage over many years.

In 1983, three winter wheat varieties, viz. Avalon, Aquilla and Rapier and three winter barley varieties, viz. Igri, Sonja and Pirate were compared in the field in a randomised block design (4 replicates). In 1984, the field experiment was repeated on an adjacent area of the same field with the addition of a fourth winter wheat variety, viz. Longbow. Visual assessments of growth were made and samples of roots and shoots taken at the end of the tillering stage, for metal analysis. Yields were taken at harvest and samples of grain and straw analysed for their metal content.

The soil was a fine sandy loam with a pH of 7.1. "Total" soil metal concentrations were of the order of 700, 100 and 5 mg kg^{-1} for Zn, Cu and Cd respectively with EDTA extractable values of 280, 70 and 4 mg kg^{-1}. Visual differences in growth and colour between the wheat varieties were marginal whereas severe chlorosis was apparent in all plots of Igri barley from an early growth stage (December) in both years – Pirate was intermediate and Sonja showed no visual toxicity symptoms. Differences in metal uptake between years were large. In 1983 the shoot levels of Zn, Cu and Cd were more than double those in 1984. Shoot uptake of all 3 metals in Igri barley were consistently higher than in the other varieties.

The variety Sonja out-yielded the other two barley varieties in both years. Whereas Pirate out-yielded Igri in 1983, the reverse was true in 1984. Yield differences between wheat varieties were much less and only Longbow in 1984 yielded significantly better than the other two varieties and this may well be due to the improved genetic potential of this new cultivar. Visual differences between the barley varieties early in the season were reflected in shoot metal uptake which influenced final yield.

1. INTRODUCTION

Advisory experience and some preliminary pot trials had shown differences in metal uptake between roots and shoots of different cereal cultivars (1). There are large areas of land around industrial conurbations in the UK which have built up high levels of metals as a result of continuous applications of sewage over many years. Much of this land is of a loamy sand to sandy loam texture and favourable for continuous arable cropping with cereals, sugar beet and potatoes. It is of importance to the farmers concerned to know which particular cultivars are more tolerant than others to the high metal levels present in these soils. It is with this in mind that the present study was initiated and designed to compare the tolerance of winter barley and winter wheat cultivars to these contaminants.

2. TREATMENTS AND METHODS

Three cultivars of winter wheat (Avalon, Aquilla and Rapier) and 3 cultivars of winter barley (Igri, Sonja and Pirate) were compared in the 1983 field trial in a randomised block design with 4 replicates.

In 1984, the experiment was repeated on an adjacent area in the same field with the addition of a 4th winter wheat cultivar (Longbow).

The soil was a pebbly fine sandy loam overlying a sandy clay loam at 60 cm and was moderately well drained.

Visual assessments of growth were made and samples of roots and shoots taken at growth stage 30/31, near the end of tillering for metal content. Yields were taken at harvest and samples of grain and straw analysed for their metal content.

3. RESULTS

Soil samples were taken in the spring of each year and analysed for their 'total' and EDTA extractable metal contents. The 'total' was determined by refluxing with a mixture of nitric and perchloric acids. The results are shown in Table 1.

TABLE 1 - Soil Metal Contents.

Total (mg/kg)	1983	1984
Zinc	670-770	700-830
Copper	90-105	85-115
Cadmium	4.5-5.5	4.8-7.6
EDTA extractable (mg/l)		
Zinc	235-300	250-375
Copper	63-73	65-82
Cadmium	3.6-4.5	4.1-5.1

The levels of zinc present in these soils are more than twice the maximum recommended in the EEC Draft Directive and in the present MAFF UK Guidelines. Copper levels are just at the maximum recommended in the Draft Directive and MAFF UK Guidelines whilst cadmium levels are up to twice the recommended maxima.

3.1 Yield Data

Yields in t/ha at 85% dry matter are shown in Table 2 below:-

TABLE II - Grain Yields in t/ha (85% DM).

Year	Wheat Variety				Barley Variety			SE Diff. (17 d.f)
	Avalon	Aquilla	Rapier	Longbow	Igri	Sonja	Pirate	
1983	6.61	5.52	5.89	-	4.64	6.46	5.71	+ 0.72
1984	6.18	5.98	6.18	7.23	5.78	6.30	5.06	+ 0.28

Visual symptoms of metal toxicity were clearly visible in the barley varieties by early December and persisted throughout the winter. The variety Igri was most severely affected with bright chlorosis of the whole plant. Sonja barley was virtually unaffected and Pirate appeared to be intermediate in 1983 and 1984. Winter wheat varieties showed no discoloration but growth and vigour was poor, more particularly in the varieties Rapier and Longbow by the end of May.

Sonja barley outyielded the other varieties giving significantly better yields than Igri in 1983 and both Igri and Pirate in 1984. In normal uncontaminated situations, Igri and Pirate would be expected to outyield Sonja.

Avalon was the highest yielding wheat in 1983 but not significantly better than the other two varieties. In 1984, the variety Longbow was included and this significantly outyielded all the other varieties between which there was virtually no yield difference. Genetically the wheat variety Longbow should normally be expected to outyield Avalon, Aquilla and Rapier.

3.2 Metal Uptake

Metal concentrations in the grain and straw, and in the shoots and roots at late tillering stage are shown in Tables 3 and 4 for 1983 and 1984 respectively.

At the tillering stage, metal concentrations were much more concentrated in the roots than in the shoots and very much higher zinc and cadmium levels in 1983 than in 1984. Zinc, copper and cadmium levels were appreciably higher in shoots of Igri barley compared with Sonja or Pirate especially in 1983. There were no significant differences in shoot metal concentrations of the wheat varieties. Although there were differences in root metal concentrations, they were not significant.

At harvest, Sonja showed higher concentrations of zinc in both straw and grain compared to Igri or Pirate barley in both years. Copper levels were higher in grain than straw with an appreciable difference in grain levels between Pirate and the other two varieties in 1984. Cadmium levels, on the other hand, were consistently higher in the straw than in the grain. Differences between varieties were small in 1984 but, in 1983, Sonja and Igri showed significantly higher cadmium concentrations in straw than did Pirate. Differences between the wheat varieties were much smaller but with a tendency for Avalon to show higher straw cadmium concentrations.

TABLE III - Crop metal content (mg/kg of dry matter) - 1983

	W. Wheats			W. Barleys			SED(+) (15 d.f.)	% COV
	Avalon	Aquilla	Rapier	Igri	Sonja	Pirate		
1. Grain								
(a) Zinc	84.3	86.5	84.5	104.0	111.0+	94.0	6.59	9.9
(b) Copper	5.2	5.7	5.5	8.2	7.8	9.2	0.89	18.2
(c) Cadmium	0.48	0.48	0.51	0.31	0.50	0.31	0.11	37.7
2. Straw								
(a) Zinc	188.5	174.5	191.5	130.8	142.8+	100.0	19.90	18.2
(b) Copper	2.0	2.2	2.8	5.6++	4.8	3.9	0.54	21.4
(c) Cadmium	1.2	1.0	1.1	1.3+++	1.25++	0.6	0.12	16.3
3. Shoots (G.S. 25-27)								
(a) Zinc	700	694	669	781+++	519	488	48.70	10.7
(b) Copper	24.5	27.3	23.0	48.3+++	39.8+	32.5	2.96	12.9
(c) Cadmium	7.1+	6.8	5.9	7.7+++	3.9	2.7	0.63	15.8
4. Roots (G.S. 25-27)								
(a) Zinc	1350	1350	1587	938	1006	850	145.60	17.4
(b) Copper	116	129	126	133	142	143	16.10	17.3
(c) Cadmium	31.3	30.3	35.8	27.3++	22.5	17.3	3.21	16.6

Comparisons with lowest value for wheat and barley taken as standards.
+ = 5% Sig.Diff. above standard; ++ 1%; +++ 0.1%.

TABLE IV - Crop metal content (mg/kg of dry matter) - 1984

	W. Wheats				W. Barleys			SED (+/-) (17 d.f.)	% COV
	Avalon	Aquilla	Rapier	Longbow	Igri	Sonja	Pirate		
1. Grain									
(a) Zinc	74	74	77	76	94+	107+++	87	3.4	5.8
(b) Copper	5.7	4.5	3.8	4.7	17.1+	17.6+	11.3	2.42	37.1
(c) Cadmium	0.58	0.68	0.74+	0.61	0.53	0.65	0.53	0.075	17.1
2. Straw									
(a) Zinc	166+	131	137	128	107	119+	85	13.6	15.4
(b) Copper	2.1	2.4	2.7	1.4	4.7	3.9	3.7	0.49	23.4
(c) Cadmium	1.5++	1.1	1.3	1.2	0.9	0.9	0.7	0.12	15.4
3. Shoots (G.S. 31)									
(a) Zinc	245	272	229	254	262+	221	178	26.1	15.5
(b) Copper	9.8	13.1	11.7	12.0	21.1++	19.2+	15.7	1.53	14.8
(c) Cadmium	4.2+	4.1+	3.4	3.9	3.1+++	2.2	1.6	0.35	15.4
4. Roots (G.S. 31)									
(a) Zinc	565	563	588	631	488	475	513	49.1	12.7
(b) Copper	84.6	87.9	91.6	87.4	93.5	87.3	99.9	7.61	11.9
(c) Cadmium	18.3+	15.4	15.8	15.3	12.3	10.1	10.6	1.22	12.3

Comparisons with lowest value for wheat and barley taken as standards.
+ = 5% Sig.Diff. above standard; ++ 1%; +++ 0.1%.

4.0 DISCUSSION

During the course of advisory work it had been noted that certain barley varieties were obviously more susceptible to high soil metal concentrations of zinc and copper than others. In particular, the highly recommended barley variety, Igri, did not appear to be particularly tolerant to the conditions and it was felt necessary to establish which varieties, if any, were tolerant of the high soil metal levels in terms of yield and metal content.

Of the barleys, the variety Igri, which normally possesses the highest yielding potential of those tested, was obviously susceptible and was outyielded by Sonja. The depressed yield of Igri was associated with the highest shoot metal concentrations of zinc and copper. In the case of the wheats, yield differences were smaller and could be ascribed, at least in part, to the differences in normal genetic potential of these varieties with Avalon outyielding Aquilla and Rapier in 1983 and Longbow outyielding Avalon, Aquilla and Rapier in 1984. Whereas there were appreciable differences in shoot concentrations of zinc, copper and cadmium varietal differences in concentrations of these metals in grain and straw were small.

REFERENCES

(1) WILLIAMS, J.H. AND JONES, J.R. (1981). Metal uptake – tolerance in wheat and barley. ADAS Research and Development Report (Soil Science, Midlands and Western Region).

EFFECT OF SAMPLE STORAGE ON THE EXTRACTION OF METALS FROM RAW, ACTIVATED AND DIGESTED SLUDGES

J.V. TOWNER, J.A. CAMPBELL and R.D DAVIS
Environmental Impact Group, Water Research Centre, Henley Road
Medmenham, P O Box 16, Marlow, Buckinghamshire SL7 2HD, UK

Summary

Liquid storage of sludge at 4 $^{\circ}$C was examined as a method of sample preservation. Raw, activated and digested sludges from Beckton STW were stored at 4 $^{\circ}$C, and periodically sampled (after 0, 29, 79, 107 and 134 days) for analysis of the speciation of Zn, Cu, Pb, Ni, Cr and Cd using a sequential extraction procedure. The procedure employs 5 reagents: KNO_3, KF, $Na_4P_2O_7$, EDTA and HNO_3. Statistical evaluation of the extraction data indicates that significant time-dependent variations in the fractionation of all six metals occurred over the storage periods tested. It is concluded that liquid storage of sludge at 4 $^{\circ}$C is inadequate to preserve the integrity of sewage sludge samples for the investigation of trace metal speciation.

1. INTRODUCTION

The potential environmental consequences of sludge disposal depend not only on the gross concentration of metals present in the sludge, but also on their speciation. Sequential extraction schemes are frequently used for evaluating the speciation of trace metals in sewage sludges. In practice, sequential extraction schemes categorise the speciation of metals in terms of operationally defined fractions; i.e. the binding forms of trace metals are characterised by the susceptibility of fractions of trace metals to attach by a succession of chemical extractants. Nevertheless, such procedures do provide insight into the affinity of metals for sludges from which it may be possible to predict their behaviour following disposal. As part of a programme looking at speciation of metals in sludge before and after disposal it was thought necessary to examine changes in speciation during storage, with a view to finding a way of storing batches of sludge with minimum effect on metal speciation.

Liquid sludges are not stable materials, and a variety of chemical, micro-biological and physical changes take place continuously. These processes may profoundly affect the speciation of metals in sludge; Fig.1 summarises a number of the processes which might redistribute metals between different binding sites. Thus, it is probable that the fractionation of metals in sludge is influenced by both the mode and duration of storage between sampling and extraction. Oake et al (1) have demonstrated that the drying of sludge may alter the distribution of metals relative to that found for fresh wet sludge. Furthermore, Thomson et al (2) have reported that the freezing of estuarine sediments causes the restructuring of amorphous colloids and the lysis of bacterial cells, which results in the subsequent release of metals upon thawing. The wet storage of sludge under refrigeration might be considered as an alternative sample preservation method. The present paper reports an investigation into the effects of liquid storage at 4 °C upon the fractionation of metals in sludge.

2. METHODS

Samples of raw, activated and digested sludges were obtained from a large sewage treatment works near the London conurbation. On receipt, the sludges were homogenised and three subsamples of each were analysed for the speciation of Zn, Cu, Pb, Ni, Cr and Cd using a modification (1) of the sequential extraction procedure described by Stover et al (3). The distribution of metals in the sludges on receipt were chosen as controls against which fractionation changes on storage were assessed, since prompt extraction is the method that is most likely to reflect the integrity of the sludges. The remainder of the liquid sludges were then stored at 4 °C, and sampled as before for sequential extraction analysis after periods of 29, 79, 107 and 134 days; the activated sludge was not extracted on the 134th day since insufficient sample was available.

3. EXPERIMENTAL

In order to fractionate the liquid sludges an appropriate volume of thoroughly mixed sludge, sufficient to contain 1 g dry solids, was centrifuged at 2000xg for 20 minutes. The supernatant was aspirated and discarded; previous analysis has shown that less than 0.5% of the total metal resides in the supernatant (Booker, unpubl.). The resultant pellet was then fractionated as follows: One grammme of sludge was extracted with the following sequence of reagents: (1) 50 ml of 1M KNO_3; (2) 80 ml of 0.5M KF; (3) 80 ml of 0.1M $Na_4P_2O_7$; (4) 80 ml of 0.1M Na_2EDTA; (5) 50 ml of 6M HNO_3. All the extractions were of 16 hours duration and conducted with mechanical agitation. The supernatant extract fluids were separated from the sludge residues by centrifugation at 2000xg for 20 minutes, followed by filtration through 0.45 um polycarbonate filters. The supernatants were analysed simultaneously by inductively-coupled plasma emission spectrophotometry for the six metals of interest.

4. RESULTS AND DISCUSSION

The mean concentrations of the extractable metal associated with each fraction of the control samples are summarised in Table 1. Friedman's non-parametric analysis of variance (4) was used to test if the duration of storage had a significant effect (less than 0.05) on the concentrations of trace metals in each fraction of the stored sludges relative to those in the control samples. The sludge storage periods after which significant changes in the concentrations of metals in individual fractions were detected are also indicated in Table 1. The statistical analysis suggests that although the concentrations of metals in some of the individual fractions of sludge did not vary significantly over the entirety of the storage period tested, significant changes in the overall distribution of metals between the sum of the constituent fractions generally occurred within 29 days of the start of the experiment. Only in the case of the distribution of Ni in the activated sludge were no changes detected after 29 days storage. However, after 79 days storage signficant changes in the distribution of this element were confirmed in the activated sludge.

No recognisable trends in metal redistribution between the various fractions with time were apparent. The data shown in Fig.2 for the time dependent variations in the concentrations of Zn in the individual sludge fractions are exemplary of the fluctuations found for all of the metals.

Fig.1. Summary of processes which might redistribute metals between binding sites in sludge.

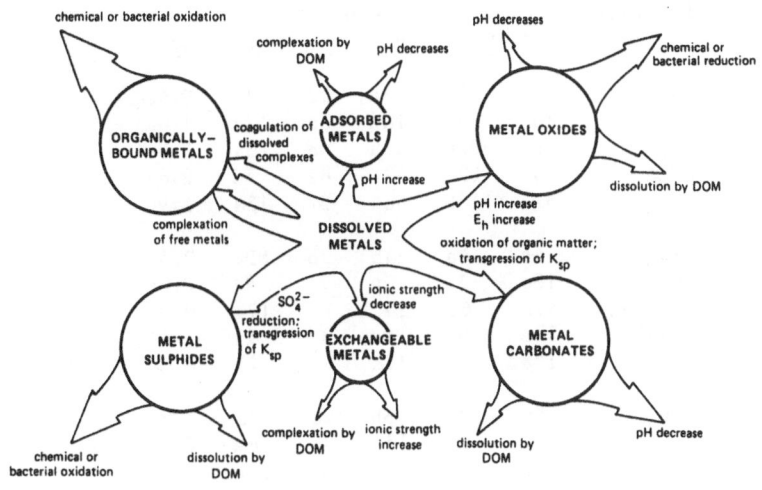

Fig.2 Changes in mean extractable Zn concentrations (mg/kg dry solids) in relation to time of storage at 4°C of three types of sewage sludge (bars indicate standard error of the mean).

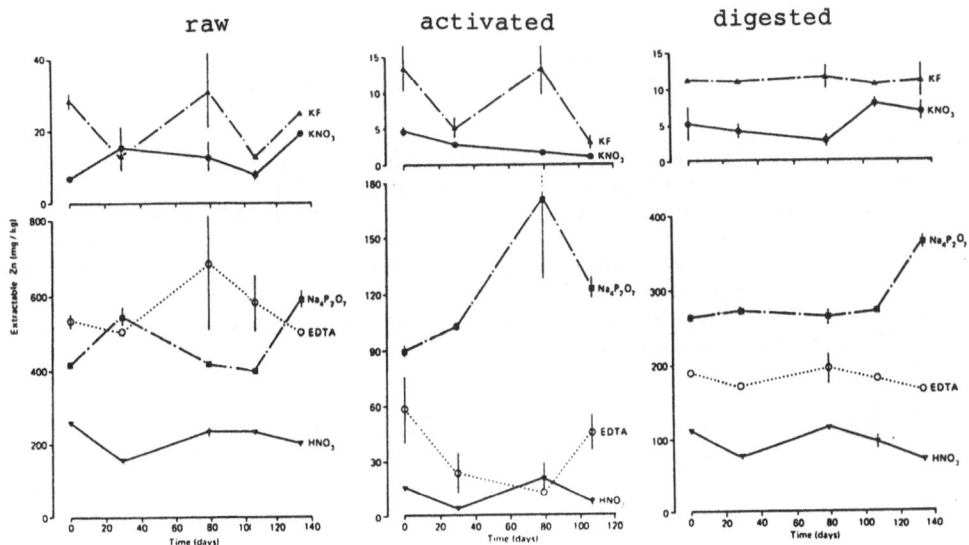

The lack of discernible patterns of metal redistribution suggests that microbial activitiy and physico-chemical conditions fluctuated considerably over the period of storage.

TABLE 1 Mean concentrations (mg/kg) of extractable metal associated with each fraction in the control (Day 0).

Sludge	Fraction	Element					
		Zn	Cu	Pb	Ni	Cr	Cd
Raw	KNO_3	7a	2b	9a	5a	1	0.4a
	KF	28a	6c	18a	8a	13a	1.0a
	$Na_4P_2O_7$	418a	32b	147a	4a	43a	2.5a
	EDTA	533	15a	142	8a	7d	2.6
	HNO_3	257a	115b	66a	3	13a	1.9
Activated	KNO_3	4a	4a	11a	3b	NDb	0.4
	KF	13a	3a	9a	5	4a	0.9a
	$Na_4P_2O_7$	89a	6b	24	1	17a	0.9
	EDTA	58a	8	19	3	1	1.1
	HNO_3	15a	13a	11a	1	2b	0.3a
Digested	KNO_3	5a	2	5b	2a	1a	NDb
	KF	11	8a	15a	7	8a	0.3a
	$Na_4P_2O_7$	263d	21a	66a	2	36a	1.6a
	EDTA	190a	11c	55a	6a	5	2.0b
	HNO_3	112a	68b	26a	2a	10c	1.3a

a – value after 29 days storage differed significantly from control (P less than 0.05)
b – value after 79 days storage differed significantly from control (P less than 0.05)
c – value after 107 days storage differed significantly from control (P less than 0.05)
d – value after 134 days storage differed significantly from control (P less than 0.05)
Values without superscripts did not vary significantly from the control over the entire storage duration.
ND – denotes that the metal was not detected in the extract.

CONCLUSIONS
 Statistically significant changes in the distribution of metals between sludge fractions were generally evident. Only the distribution of Ni in activated sludge remained unchanged after 29 days, however, significant changes in the distribution of this element were evident after 79 days of liquid storage. It is therefore concluded that the liquid storage of sludge at 4 °C is an inadequate method for preserving the integrity of sewage sludge samples for use in experiments where trace metal speciation is a critical factor.

REFERENCES
(1) OAKE, R.J., BOOKER, C.S. and DAVIS, R.D. (1984). Fractionation of heavy metals in sewage sludges. Wat.Sci.Technol., 17. 587-598

(2) THOMSON, E.A., LUOMA, S.N., CAIN, D.J. and JOHANSSON, C. (1980). The effect of sample storage on the extraction of Cu, Zn, Fe, Mn and organic matter from oxidised sediments. Water, Air, Soil Pollut., 14, 215-233
(3) STOVER, R.E., SOMMERS, L.E. and SILVIERA, D.J. (1976). Evaluation of metals in wastewater sludge. Journal WPCF, 48, 2165-2175.
(4) SIEGEL, S. (1976). Non-parametric statistics for the behavioural sciences. McGraw-Hill, New York.

WORKING PARTY 1

re. Poster No 29, "Composting of sewage sludge containing
polyelectrolytes" - J.J. van den Berg

A. FARNETI

You described the composting of sewage sludge that had been treated with
polyelectrolytes to aid dewatering. What effect does the polyelectrolyte
have on the composting process and on the final composted material? Also
what is the fate of the polyelectrolyte?

J.J. van den BERG

We thought that the polyelectrolyte could have an effect on the
decomposition of organic matter and so we composted, using the aerated
static pile method, some aerobically stabilized sludge that had been
conditioned with polyelectrolytes. We found that the polyelectrolyte
seemed to have no influence on the composting process or on the quality
of the finished product. I am not able to comment on the fate of the
polyelectrolyte - it was not something we examined.

re. Poster No 17, " Latest test result in sludge dewatering with the CHP
- Filter press" - U. Loll

L. SPINOSA

Is the CHP Filter press commercially available?

U. LOLL

Yes it is in some countries, for example, in America. However, in a
number of others, eg UK, the licencing agreement has not yet been
completed and it will be perhaps a year or so before the CHP Filter press
is commercially available there.

WORKING PARTY 2

re. Poster No 25, "Application of $CaCl_2$ - extraction for assessment of
Cadmium and Zinc mobility in a waste water polluted soil" - C. Salt

R.L. LESCHLER

I would like to contrast Mrs Salt's poster with the paper given during
Session 2 by Haeni and Gupta on "Chemical methods for the biological
characterisation of metal in sludge and soil". $CaCl_2$ extraction has been
shown to be a valuable tool for determining extractable heavy metals.
This was demonstrated by Sauerbeck and his co-workers and confirmed by
Mrs Salt's work. There was an agreement made at the Munster Seminar
involving both Working Parties 2 and 5 of COST 681 that $CaCl_2$ extraction
should form the basis for a European standard. I am therefore most
surprised to learn that our Swiss colleagues have not referred to this

"European agreement": and apparently have not even mentioned it to their own government when specifically requested. I am concerned about what will happen in Switzerland following the setting up of an internal standard based on a different extractant when all other countries will be using standards based on $CaCl_2$ extraction; and that chosen because it gave the greatest accuracy. Surely the inter-comparison of results from the two standards will be almost impossible.

H. HAENI

I agree with Dr Leschber that it is preferable to have a common standard. I am not opposed to $CaCl_2$ as an extractant but we need much more experience of its use. For instance, I have no data using $CaCl_2$ but a lot using $NaNO_3$.

C. SALT

I do accept that even using $CaCl_2$ there can be difficulties. For example, analysis of contaminated soil using $CaCl_2$ extraction and spectrophotometric analysis. Here, there can be problems with reproducibility. However, I think it unlikely that the extraction and analysis could be achieved at all using $NaNO_3$.

SALMON

I agree with some of the points made by Dr Haeni. Probably a neutral salt is best for extraction without the need for changing the pH value. However I prefer $CaCl_2$ as it extracts more of the heavy metals and in a ratio similar to plant uptake. I know some UK experts do not like $CaCl_2$ as they think it makes the heavy metals appear to be too available. In general, though, I think it is satisfactory especially as the higher levels of heavy metals, extracted make the analytical work that much easier. I do believe, however, that at the moment there is not sufficient data to set guide-values.

J.C. TJELL

I am concerned that any extraction methods will be included in the proposed "Sludge to Land" Directive - there just is not enough data available at the moment.

WORKING PARTY 4

re: Poster No 27 - "The alternative "Earthworm" in the organic wastes recycle". - U. Tomati, A. Grappelli and E. Galli

A.M. BRUCE

Can earthworms survive low temperatures?

U. TOMATI

The worms are very hardy and can live under extreme conditions. However, the optimum is between 24° and $32\ ^{\circ}C$. Certainly, if the temperature falls below about 10 $^{\circ}$C the worms stop transforming the organic matter.

It is interesting to note that the worms and the soil can have quite different temperatures.

re: Poster No 28 - Slurry meter - H. Tunney

A.M. BRUCE

Do the gas bubbles in the sludge affect the reading of specific gravity and indirectly the measured solids content?

H. TUNNEY

The gas bubbles are boiled off by applying a high vacuum to the sludge, and so do not affect the reading greatly. I believe that the instrument has good advantages for the Water Industry where a reasonable accuracy is required.

re: Poster 13 - Sludge composting and utilisation - G. Hubert

J.E. HALL

I am unsure whether composting falls withing the intersts of Working Party 1 or 4. What is the rational for composting? Ease of operation, cost and a product to sell must be important criteria. There are other factors to take into account though; for example, relationship between price of composted material and the supply and demand ratio. Storage costs are also an important factor. I would also like to see an indication of the maturity of the composted material established and believe that there should be a standardised method to assess this. I would like to ask our Swiss colleague why he chose to compost already digested sludge?

G. HUBERT

We used digested sludge because that was our locally available material.

A.M. BRUCE

The potential odour problem needs to be added to the list of criteria for composting referred to by Mr Hall. I note that at the Rutte Recycling plant in The Netherlands, they have had to fit a roof over the plant to control the odours.

J.J. VAN DEN BERG

That is not strictly true. We could control the odours with an extraction system. The roof is not really necessary for this purpose. What it will do though is to help us to control the plant in the most efficient way in all weathers.

SESSION REPORTS

Processing and costs

Characterization of sludge

Management considerations in sludge and effluent disposal

PROCESSING AND COSTS

A.M. BRUCE
Water Research Centre, Stevenage, United Kingdom

It is just about two years since the previous COST 68 Symposium was held, in Brighton. During this interval, research and development on the many aspects of sewage sludge and animal manure processing have continued at a steady pace and progress has been made. But we must not expect too much in the way of dramatic new discoveries, new developments or marked reductions in costs in the relatively short space of two years. It has been said that distinct developments and trends in the field of sludge and slurry processing are really only discernible over ten year periods or longer. Thus, it is not surprising that we do not have much really 'new' to discuss in this session but, rather, some useful reviews, some important surveys and some interesting pointers for future research. All this should help to concentrate our research resources, and our minds, on the most important areas for future research and the most likely fruitful areas for costs reductions and improved plant performance.

Dr Moller's paper gives us a mature, perceptive view of what the future is likely to hold, in Europe at least, in regard to permitted sludge disposal practices and the consequent changes that will be necessary in processing requirements. He predicts confidently that disposal of sludge to the sea will have to be curtailed, that disposal to landfill will require sludge to be dewatered to a high solids content to ensure physical stability, and that positive disinfection as well as stabilisation will become a fairly general requirement for sludges utilised in agriculture. All these trends will necessitate a greater degree of sophistication in sludge processing techniques. For example, high pressure dewatering, and combined or sequential disinfection and stabilisation stages. Much of the new technology to achieve these ends is already available, or is under active development, at least on the pilot scale. What is needed now is the appropriate research and development to permit the introduction of the more sophisicated technology without significantly increasing processing costs.

The contribution by Nielsen is a concise review of the current 'state of the art' of biological treatment for liquid animal manures. The objectives of farm waste treatment are seen to be both control of environmental pollution and the achievement of an improved fertiliser value of the treated manure. In terms of design constraints, treatment systems for farms are considered to be much more demanding than those for sludges at municipal sewage works. In particular, it is absolutely vital that that the plant should be easy to operate with minimal labour requirements and that it should be very robust and easily repaired. In addition to all this, the overall operation has to be economic for the farmer. Odour control is a crucial factor in the economics of treatment. With anaerobic treatment, the utilisation of digester gas can also affect the economic balance favourably but it appears that, so far, efficient and economic use is difficult to achieve. Much development work has been done on the small and pilot scale in both anaerobic and aerobic systems. The way forward now is to implement many of the ideas arising from research and make available to the farmer reliable and economic full-scale plants for his use. He will then be in a position to respond if the increasing

pressure of environmental legislation demands that treatment systems
be installed on a wider scale than at present.

In his overview of anaerobic digestion, Bruce highlights the
importance of the mesophilic process for stabilisation and pathogen
reduction of sewage sludges. Over 50 per cent of all the sewage sludge
produced in the European Economic Community countries is subjected to
mesophilic digestion and there is the prospect of many more plants in
the future. The message is clear, therefore, and that is that further
research to improve the efficiency and economics of the process could
yield large benefits in terms of reduced costs. Many improvements have
been made over the years in terms of both design and operational
aspects of digestion but there is still room for improvement.
Operational problems related to foaming, inadequate mixing and
inhibition still occur and some research input in these areas might be
appropriate. The energy contribution of anaerobic digestion on a
national scale has sometimes been over-rated in the past but on a
local scale there is no doubt of its importance and ways of improving
the economics of gas utilisation for energy production should still be
sought. The only serious rival to anaerobic digestion is mesophilic
(or thermophilic) aerobic digestion and much of the early development
of this process was done by Loll in FR Germany. It is of interest to
note that Dr Loll is an active member of COST 681 Working Party 1. The
costs of the two processes are certainly very comparable and it is a
matter of what design assumptions are used as to whether or not one
process or the other appears the more economic in particular
circumstances. As far as the use of anaerobic digestion for farm
wastes is concerned, it is still of much less importance than it is
for sewage sludges but its role in this area will probably grow in the
next ten years. The basic microbiology of the process is the same for
both types of waste. Great strides had been made in elucidating the
complex chemical pathways involved in both hydrolysis /acidogenisis
and in methanogenesis. The question is whether or not further research
in the field of fundamental biochemistry and microbiology is likely to
yield significant advances in process design or control systems for
anaerobic digestion. The principle of hydrogen monitoring already
seems to be one worthwhile benefit from this type of research and it
seeems essential that further advances in basic knowledge should be
pursued with the hope that it will reap more practical benefits in due
course.

Composting of sewage sludge has been acquiring a more important
role in recent years though its role is still insignificant in most
areas compared to that of anaerobic digestion. The 'state of the art
review' by Duvoort van Engers and Coppola reveals considerable
differences in the popularity of the method among the various
countries surveyed. Scandanavian countries seem to have the greatest
preference for composting but the reason for this is not clear. It
may be associated with low population density or, possibly, the ready
availability of bulkiing materials such as wood chips. Both
'in-vessel' and windrow composting seem equally used but the
advantages and disadvantages of the respective systems need to be more
clearly defined. Whatever the reasons for employing a composting
system, there is little evidence that the economics of the process
weigh heavily in the considerations. We need to acquire much more
accurate costings for the various systems in use so that they can be
compared with other stabilisation methods. More information is also
required on the availability of reliable commercial outlets for

composted sewage sludge.

The wider question of how to determine the optimum combination of sludge treatment processes for particular circumstances is tackled in the paper by Spinosa and Lotito. They stress the importance of identifying the disposal outlet first and then deciding on the treatment processes to be used. For optimisation, it is essential to be able to model the various unit processes (ie thickening, stabilisation, conditioning, dewatering) and some examples are given in the paper. Most of these are based on American experience and it is necessary to validate them for European conditions. Sewage sludge is an extremely variable material and it is not appropriate to assume the same behaviour in every case. Models must allow fo the in-putting of specific values for a given sludge and these values should preferably be determined by experimental tests.

In their contribution on the economics of the various treatment and disposal options, Di Pinto and Mininni conclude that the disposal of liquid sludge to farm land is generally not economic unless the disposal site is very close to the treatment works. The use of pipelines for transportation is by no means a new idea but it is one that should always be considered in appropriate situations. For dewatered sludge, farm utilisation is often the most economic option and, of the other options, landfill is usually more attractive than incineration. But the long term future of landfill sites is uncertain and the incineration route may well become of much more importance in the future. Research to improve the costs of incineration seems therefore to be worthwhile.

Perhaps the most interesting 'new' ideas in this symposium flowed from the paper on 'alternative uses' by Frost and Campbell. Sewage sludge is a renewable resource but how best to exploit it? The authors showed that the possibilities are numerous but the most promising prospect seems to be fuel production by low-temperature pyrolysis. Its future prospects will depend much on the future cost of energy and on the reliability and costs of the pyrolysis process. It is interesting to note that the process was originally devised by Bayer in FR Germany but further development work has been largely done in Canada and the UK mainly as a result of the rapid communication of new ideas between countries which has been effected through membership of COST 681 Working Party 1.

Finally, it is appropriate to make reference to the paper, in another Session, on olfactometric methods of odour measurement (Voorburg et al). This is a 'landmark' paper because it represents a distillation of knowledge on a very specialised scientific topic from a group comprising the leading experts on odour measurement in Europe. This group was formed specially as a sub-group of COST 681 Working Party 1 and was given very specific terms of reference and a short reporting timescale. The results, seen in the paper, are likely to be of great value for both researchers and scientists concerned with assessing odours in relation to sludge treatment and disposal operations. The principle of commissioning groups of specially selected experts to report on ways of dealing with very specific problems in a short timescale would seem to have a lot to recommend it. It is probable that the COST 681 Working Parties will employ this principle again in the future to good effect. Certainly, if this leads to the more rapid implementation of research to the direct benefit of those responsible for treating and disposal of our sewage sludge then it is a path that should be followed.

DISCUSSION

re: <u>Session 1</u>

J. TARADELLAS

We seem to have reached a log-jam with our thinking on the treatment and disposal of sewage sludge. There seems to be no research on the underlying mechanisms which are so important. We need a fresh look at these basics and consider them in parallel with the overall global approach.

P. BALMER

We certainly need to carry out some more work on lime treatment of sewage sludge which is a good low capital cost option. Furthermore, landfill disposal is important but we know little about the management. Two additional areas of work that I would like to suggest is chemical treatment of sludge and improved guidelines for the disposal. Thus we can reduce the levels of cadmium in sludge to about 3 ppm by chemical treatment. What we need to know is where the cadmium is coming from in the first place. As regards guidelines, we need to take account of the fact that people do not eat only sludge-grown crops.

G. FLEMING

We seem to have only highlighted the negtive aspects of sewage sludge. It does have its advantages though. For example, spreading sewage sludge on agricultural land can provide small amounts of trace elements such as selenium and cobalt to deficient soils.

G. MININNI

I can't agree with Mr Taradella's point that little is known about the mechanism of sludge treatment and disposal. For example, there is now so much more known about composting. What we do need to look at though is the disposal of industrial wastes. One way is to seal them into solid matrices from which they can be released very slowly or even not at all into the environment.

CHARACTERIZATION OF SLUDGE

A.H. HAVELAAR
National Institute of Public Health
Bilthoven, The Netherlands

E.B. PIKE and R.D. DAVIS
Water Research Centre, Environmental Directorate
Medmenham, UK

I. Physico-Chemical Characterization

Characterization of sludge depends in the first place on obtaining a representative sample on which to carry out determinations of chemical and biological quality. Sampling is therefore of fundamental importance but is all too frequently taken for granted and inadequate. The paper by Gomez, Leschber and Colin addressed the problem of how to sample sludge and emphasised the importance of establishing a statistically-based sampling regime derived from detailed assessment of how quality varies with time. There are obvious difficulties in producing standard guidelines for sampling a material such as sewage sludge where conditions at any two given wastewater plants are likely to be quite different. Nevertheless this problem must be tackled if sludge quality is to be characterised accurately which is an essential requirement for decision-making at both the experimental and operational level. The problem of sampling does not end with the sludge since soil and plant analysis will also be needed where sludge is utilized on agricultural land. Procedures for soil sampling are better defined than for sludge sampling but conventional methods for soil developed to assess texture and fertility may not always be appropriate for sludge-treated soils where the aim is to assess levels of contaminants such as metals. From the discussion to this paper it was made clear that sampling regimes should be practical and feasible and not based only on statistical theory.

Taking samples for the determination of organic contaminants poses particular problems in handling, storage and preparation for analysis. Interest in organic contaminants in relation to environmental aspects of sludge disposal is growing but the analytical problems are formidable. Tarradellas, Muntau and Beck described the results of an interlaboratory comparison exercise on the analysis of polychlorinated biphenyls (PCBs) in which 39 laboratories submitted results. Agreement between the results obtained by different laboratories needed to improve. To this end it was thought that more strictly defined procedures would help especially as regards extraction techniques used in sample preparation.

The state-of-the-art in metal analysis has now advanced to focus on chemical methods for assessing bioavailable metals in soils, in addition to total concentrations. Following on from the COST 681 Seminar on this subject in Munster in 1984, Häni and Gupta reported their latest findings using a neutral salt solution as the extractant. They had extended their work with sodium nitrate and were now able to provide critical soil concentrations for sodium nitrate extractable levels of lead, cadmium, copper, nickel and zinc in soils. These critical values would probably be incorporated in new Swiss guidelines for sludge utilization on land to be published early in 1986.

Apart from the chemical characterization of sludge, the paper by

<u>Loll</u> drew attention to the importance of physical characterization as a criterion in sludge treatment and disposal. Presumably this would be particularly important where large quantities of dewatered sludge were to be put in sanitary landfills and is also relevant to sludge dewaterability and pumpability. The physical parameters mentioned were bearing capacity, density, shearing strength, conditionability, specific resistance to filtration and compressability coefficient.

II. <u>Biological Characterization</u>

The increasing installation of municipal sewage treatment plants increases pressures for disposal, and land use is relied upon heavily in most European countries. Ideally, reduction of risks from pathogens in sewage sludge should be based upon the following sequence of steps:
- presence and numbers of pathogens in raw sludge,
- dose-response data,
- definition of acceptable risk,
- survival of pathogens during sludge processing and application,
- designing of treatment processes and development of guidelines,
- monitoring of treatment processes for indicator organisms.
The following is a brief summary of the 'state of the art' and in particular of the progress made during the Symposium.

ad a. Major concern still is with salmonellae, viruses and parasites, the latter with growing emphasis on coccidia and protozoans. There are no systematic surveys of pathogens in raw sludge. Relatively many data are available for salmonellae in certain regions (Switzerland, Germany, United Kingdom, the Netherlands). Data on parasites are scarce, and no methods are yet available for the epidemiologically most significant viruses. Monitoring projects should be developed and these should use standardized methodology.

ad b. Dose-response data can be defined for certain cases, such as infection of cattle with <u>Taenia</u> <u>saginata</u> eggs. The relationship between cysts produced and eggs swallowed is not proportional: there seems to be a lower limit of infection ($\sim$ 6 eggs) and generalized infection seems only possible in the case of swallowing of whole proglottides ($\sim 10^5$ egs). Viability is not related to infectivity. However, it is uncommon to find straight forward infection routes. In practice, complex infection networks exist, and multiplication of pathogens may take place en route (e.g. salmonellae in the intestines of food animals). Because of this and the factors which determine innate and acquired immunity, there is no simple relationship between numbers of pathogens in sludge and prevalence of human or animal infection. Further progress will require the development of mathematical models.

ad c. No attempts at definition of acceptable risk have been made and this approach does not seem to be feasible in the near future. Further developments will rely heavily on surveillance data and systems of surveillance and reporting of data should be improved. This might stimulate the conduct of well-designed specific epidemiological studies.

ad d. As treatment criteria cannot be defined from risk assessment at present, arbitrary standards will have to be set. These should define the required degree of inactivation during processing, which in turn should be related to subsequent sludge use and constraints made. Both Working Party 3 and the U.S. EPA have divided treatment processes into two categories of efficiency.

"Processes to significantly reduce pathogens" will inactivate at least 90% of all relevant pathogens and sludge treated in this way can be brought on land with further restrictions after application.

"Processes to further reduce pathogens" will inactivate pathogens to below detection level, this needs to be quantified. Sludge treated in this way could be used without further constraint.

A basis for a kinetical model of pathogen die-off, employing death rate constants, was presented and also specific data on parasites were given. It seems that <u>Taenia saginata</u> eggs are most resistant at temperatures above $50^{o}C$, but that <u>Ascaris</u> eggs are most resistant to lime (i.e. high pH) and storage. Points to be clarified are the use of seed organisms (lab strains) versus indigenous organisms, the documentation of experimental conditions and standardization of study design and reporting.

ad e. Development of guidelines and rating of treatment processes has been possible for several years on the basis of existing data and common sense. New developments and more conceptual approaches may – and should – lead to re-evaluation and updating of old recommendations and new data were reported. Importantly, real world situations offen diverge from labscale models, in particular because of absence of batch or plugs flow characteristics. Process studies should preferably be carried out with pathogens themselves.

With regard to guidelines, it was again emphasized that they should take local situations into account. Also, the need for a code of good hygienic practice for sludge handlers was stressed. Transport of pathogens from sludge application sites – in particular viral pollution of groundwater – also is an aspect not covered in present guidelines but could be important in some areas. New experience with regard to soil injection of raw sludge – both favourbale because of development of reliable injection equipment, and unfavourable because of earthworm activity and persistence of pathogens in the soil – was reported in the discussions and it will be interesting to see the detailed reports in the near future.

ad f. The use of indicator organisms should be restricted to monitoring of existing treatment processes, that were previously accepted on the basis of pathogen analysis. The discussions on choice of organisms and numerical values are just emerging and will require more international study. Nevertheless, consensus seems to appear on faecal streptococci, eventually supplemented with enterobacteriaceae as suitable models. Interest is also growing for the heat resistant f2 phage. Differences in opinion still do exist with regard to numerical values and this subject will be important for future work.

Contributions were made to the discussion of the papers by E.Lund (DK), J.E.Hall (UK), J.Tarradellas (CH), P.J.Mathnews (UK) and A.M.Bruce (UK). Their comments have been incorporated in this report.

MANAGEMENT CONSIDERATIONS IN SLUDGE AND EFFLUENT DISPOSAL

J.E. HALL and R.D. DAVIS
Water Research Centre, Environment Directorate
Medmenham, UK

The theme for this session was management considerations and is an important area for this conference to review as it is the point where scientific research and the realities of sludge disposal meet. Sewage sludge and livestock slurries are a resource that if used with due care, will benefit agriculture and it is clear that if the efficiency of use of these materials can be improved, many of the problems associated with their use would be reduced. A particular problem readily apparent to the public at large is the odour produced by sludges and slurries. Voorburg described in his paper a series of recommendations on olfactometric measurements to standardise methods of odour assessment which try to make odour testing as objective as possible. The measurement of odour intensity or offensiveness is much more difficult and subjective than odour strength and further work is required on this.

A difficulty in improving the nutrient efficiency of organic manures is knowing what are the concentrations of the nitrogen, phosphorus etc present as this may vary tenfold from average analyses. The rapid methods described by Tunney are an important development with the Slurry Meter and the Agros Meter together having significant potential for both sludges as well as slurries, in measuring dry solids, total nutrients and available nitrogen contents. The paper by Williams and Hall described recent developments in improving the efficiency of nitrogen utilisation. Increasing N efficiency is important not only from the fertiliser point of view but the amount of nitrogen that potentially could be lost to the environment is also reduced. Correct timing and rate of applications are perhaps the most important with the late winter-early spring period being preferred but efficiency can also be improved by split applications, applying to growing crops and soil injection; the latter has the greatest promise as it has a number of other agronomic and operational advantages, principally that it can also eliminate odour. Nitrification inhibitors and anaerobic digestion of slurry (unlike sewage sludge) do not appear to be very effective at improving N efficiency. Satisfactory organic matter levels are important for long term structure and fertility of soils but are often too low generally due to current farming practices. The longer term organic matter value of sludge and manure was considered by the work of Lineres, Juste, Tauzin and Gomez which indicated that farmyard manure had a longer lasting effect on soil than digested sludge.

The paper by Fleming and Davis looked at contamination problems in relation to land use drawing together results presented by contributors to COST 681 meetings over several years. The criteria identified as a basis for assessing hazard were risks to the food chain or of phytotoxicity, zootoxicity, or contamination of surface or groundwater. On this basis it was possible to assign a hazard rating to each outlet for sludge and it was concluded that horticulture, gardens and allotments represented high-risk outlets; arable land, grassland and forestry represented medium-risk whilst use of sludge for land reclamation, on green areas or in viniculture was associated with comparatively low risk. It was emphasised that potential contamination problems could be minimized by control of sludge application, good land management, monitoring, and control of industrial discharges to the sewer.

Long-term effects were dealt with by Sauerbeck on the basis of data from a large number of field trials mainly in Germany. It was concluded that lead and chromium were unlikely to present a problem unless soil concentrations were extraordinarily high. The present German threshold values of 100 ppm in soil for these elements were extremely cautious. For copper the 100 ppm limit value in soil was about right but perhaps too high for some crops on light, sandy soils. The nickel limit of 50 ppm in soil guaranteed no phytotoxicity problems and needed to be taken seriously only where most of the nickel was pollution derived and not of geological origin. The limit for zinc in soil of 300 ppm appeared to be about right on balance. The present limit of 3 ppm for cadmium in soil was shown to result in excessive cadmium concentrations in some staple food crops and needed to be reduced to 2 ppm in soils used for food production. The criteria of main concern about cadmium was the soil concentration likely to produce about 1 ppm in cereal grains. On the basis of the German results it seemed that a soil concentration of cadmium of 2 ppm would be needed to ensure this. However, looking back to Fig 1 in the paper by Fleming and Davis it is clear that for 3 contrasting soils typical of UK agriculture the concentration of 1 ppm of cadmium in wheat grain is not reached until the soil concentration of cadmium is at least 8 ppm. This would presumably be due to the different soil conditions in most of the UK compared with Germany where the soils are predominantly light and sandy.

In general terms Sauerbeck thought that grasses were less sensitive than broad leaf crops to metals in the soil. Concerning changes in availability of metals with time it was very difficult to judge what would happen on the basis of a few years or decades. Any incorporation of persistent contaminants would result in a gradual deterioration of soils, Inputs of contaminants to soils should be reduced by all possible means by tracing and closing sources without compromise.

Juste and Solda described results of an experiment with maize started in 1976. Juste drew attention to the importance of metal anatagonisms and soil temperature in affecting metal uptake by crops. The metal contents (nickel and cadmium) of the sludge used in the experiment were about one hundred times higher than normal for sewage sludge and an industrial sludge of this kind would never be used in agriculture. Consequently the results of the trial may have little application. Sauerbeck however incorporated some of Juste and Solda's data into his review.

The session concluded with a review of metal limits for sludge utilization in agriculture by Tjell and Webber. This paper identified large variations in limit values set by different countries and attempted to identify the reasons for these differences according to the approach adopted in setting limits which, it was argued, could be historical, pragmatic or scientific. Whether the sludge was used on land as a source of nitrogen or phosphorus affected the amount applied and hence the quantity of metals added. Prevailing soil conditions, especially pH value, were of fundamental importance also. The scientific approach to guidelines was dealt with in detail especially with reference to cadmium. Three options were identified for unifying sludge utilization regulations: 1. Elaborate rules taking account of different conditions (e.g. soil pH value) in each country. 2. Strict uniform rules. 3. Individual rules for each country. However, it semed difficult to argue that sludge regulations ought to be internationally standardized.

DISCUSSION of paper by Fleming and Davis
It was thought that salinity could be included in the assessment of phytotoxicity especially where sludge was used in horticulture as germination of some vegetable seeds was very sensitive to high salinity levels in soil solution. Liming was not a practical way of reducing metal uptake on some soils such as peats. Perhaps more attention should be paid to iron content and form which was another way of controlling metal availability in acid soils.
(Contributors to discussion: Sbaraglia, Kuntze)

DISCUSSION of papers by Sauerbeck; Juste and Solda; and Tjell and Webber
It was suggested that the speciation of metals in soil was important in the understanding of availability. The use of selective extractants gave an indirect indication of this but more data on the performance of extractants was needed from field experiments. It was likely that the relationship between extractability by particular chemicals and availability to plants was essentially empirical for the extractants now in use. The use of soil solution analysis was of interest in this context. Ultimately it would be of interest to know how forms of metals changed after entry into the plant or animal and the specific enzymes or enzyme systems they affected.

Professor Sauerbeck had used 1 ppm cadmium in wheat grain as a criterion for reducing the soil limit for cadmium in Germany from 3 ppm to 2 ppm. However data in the paper by Fleming and Davis (Fig.1) showing cadmium concentrations in wheat grain in relation to cadmium concentrations in three contrasting soils broadly representative of UK agriculture showed that 1 ppm cadmium in wheat grain will not be reached until the soil concentration is about 8 ppm.

It was thought that hexavalent chromium (the more toxic form) was unlikely to occur in sludge and would not persist in soil.

CONCLUSIONS AND RECOMMENDATIONS

H.M.J. SCHELTINGA
Staatstoezicht op de Volksgezondheid
Hoofdinspectie voor de Hygiene van het Milieu
Pels Rijckenstraat 1
NL-6800 DR ARNHEM

This 4th Symposium reflected in a very condensed way the work, that within the framework of the Concerted Action COST 681, has been done by the working parties.

The original objectives of the Symposium were to give a comprehensive view on completed research in Europe in general and more specifically of the results obtained from COST 681 "Treatment and Use of Organic Sludge and Liquid Agricultural Wastes" in the last two years.

Aims of special interest were:

- to evaluate the technical aspects of the processing of sludge and slurries,

- to assess the constraints raised by the production of a safe end-product, i.e. to identify problems of heavy metals and organic compounds, and to evaluate the hygienic constraints to the use of sludge and slurries in agriculture due to pathogens,

- to assess the possible impact on the environment of sludge treatment and disposal,

- to contribute knowledge necessary for optimum use of slurries and sludge in agriculture with benefit for the farmer,

- to study the research needs and priorities required over the next years for an efficient and safe management of sludge and effluent disposal,

- to reinforce the contacts between the scientists of the laboratories participating in the research work in view of improving the coordination between the different activities related to the implementation of the research programme.

Many thanks are due to all the scientist from Member States as well as from Non-Member States, who made it possible to solve many problems in this field of research and practice during these years.

This feeling was also expressed last year by the evaluation panel of the Environmental Protection R&D Programme at a meeting with COST 681 representatives in Brussels.The value of the easy communication between scientists from 15 countries cannot be overestimated.It was reflected in the fact that many papers in this Symposium were prepared by more than one author from different countries. The long list of publications of proceedings from symposia and workshops organised by the working parties was highly appreciated as a means to disseminate knowledge beyond the circle of those directly involved in the Action.

Professionally and academic outstanding results have been achieved and published.

The research coordinated by COST 681 is the only Concerted Action from which the whole set of results has been used in order to establish a proposal for a Community Directive on the application of sludge in agriculture.

Improvements were proposed by the panel too. Certain specific suggestions may be very useful to bring us to some thoughts on our future work.

Recently a proposal for a Council Decision was made, adopting three multi-annual research and development programmes in the field of environment (1986-1990):

1. Environment Protection
2. Climatology
3. Major technical hazards

For the future of COST 681, the first programme, environment protection, is most interesting. The proposal covers contract research and Concerted Actions.

In the scientific content of the programme subjects are listed that fully fit into the future work of our COST Action, for example:

-   eutrophication
-   effects of agricultural practice
-   organic wastes, such as sewage sludge and animal manure
-   analytical methods

Furthermore to the subject of groundwater pollution, chemical and biological,increased attention should be paid. The evaluation panel, referred to earlier, asked for more coordination in this area, mainly because of influences from agriculture in many European countries.

The research and development programme will also be implemented, as said, by concerted actions. A renewal of the 5 existing actions, among them COST 681, is announced.

The Community COST Concertation Committee 681 will now and in the beginning of 1986 discuss intensively about this renewal. In my view, more attention than so far, has to be given to the many problems in the field of manure disposal in agriculture. On the other hand certain needs for research in the field of sewage sludge are indicated during this Symposium.

We will invite the national representatives to consult their experts involved in COST activities and others from outside our action to give their view on the proposal that the Committee will prepare early in spring 1986. It should in our view reflect the expectations that the many countries participating will formulate.

LIST OF PARTICIPANTS

AHTIANEN, M.
National Board of Waters
Water District Office
of North Karelia
Torikatu 36A
PL 69
SF - 80101 JOENSUU

AICHBERGER, K.
Landw.-chem. Bundesanstalt
Wieningerstrasse 8
A - 4025 LINZ

AMMANN, P.
EPFL - Génie Biologique
Ecublens
CH - 1015 LAUSANNE

ANASTASI, P.
A.C.E.R.
Ple Ostiense, 2
I - 00154 ROME

BACCHIN, P.
ENIRICERCHE
I - 00015 MONTEROTONDO (ROMA)

BALMER, P.
Chalmers University of
Technology
Fach
S - 41 296 GOTEBORG

BARIDEAU, L.
Groupe Valorisation des Boues
Faculté des Sciences Agronomiques
B - 5800 GEMBLOUX

BISSOLOTTI,
Politecnico di Milano
Istituto di Ingeniera Sanitaria
Via Fratelli Gorlini, 1
I - 20151 MILANO

BOARI, G.
Istituto di Ricerca sulle Acque
Via F. De Blasio 5
I - 70123 BARI

BON, P.H.
Hoogheem Raadschap West-Brabant
Bouvignelaan 5
NL - 4036 AA BREDA

BONAZZI, G.
Centro Richerche Produzione
Animali
Via Crispi, 3
I - 42100 REGGIO EMILIA

BRAMRYD, T.
Department of Plant Ecology
University of Lund
Helgonavägen 5
S - 223 62 LUND

BRUCE, A.M.
Water Research Centre
Elder Way
UK - SG1 1TH STEVENAGE HERTS

BRUNNER, P.M.
EAWAG
CH - 8600 DUBENDORF

CALATRAVA, R.
Empresa Aguas De Cordoba SA
Cronista Rey Diaz, 2
E - 14006 CORDOBA

CALCIMAI, M.
C.N.R. Ist. Chimica del Terreno
Via Corridoni, 78
I - 56100 PISA

CANDINAS, T.
Forschungsanstalt für Agricultur-
chemie und Umwelthygiene
Schwarzenburgstr. 155
CH - 3097 LIEBEFELD (BE)

DI PINTO, A.C.
Istituto di Ricerca sulle Acque
C.N.R.
Via Reno 1
I - 00198 ROMA

DUVOORT-VAN-ENGERS, L.E.
Rijksinstituut voor Volksgezondheid
en Milieuhygiene
Anthonie van Leeuwenhoeklaan, 9
Postbus 1
NL - 3720 BA BILTHOVEN

EGGINK, H.J.
Waterboard "de Dommel"
Eikenlaan 6
NL - 5271 RS SINT-MICHIELS GESTEL

FARNETI, A.
Snam Progetti/Ecol.
Via Papiria
I - 61032 FANO (PESARO)

FLEMING, G.A.
The Agricultural Institute
Johnstown Castle Research Centre
IRL - WEXFORD

FROST, R.C.
Water Research Centre
Elder Way
UK - SG1 1TH STEVENAGE HERTS

GALLI, I.
I.R.E.V. - C.N.R.
Via Salaria km 29.300
Area della Ricerche di Roma
I - MONTEROTONDO SCALO (Roma)

GENEVINI, P.
Istituto di Chimica Agraria
Università Milano
Via Celoria 2
I - 20133 MILANO

GOMEZ, A.
I.N.R.A.
Station d'Agronomie
Centre de Recherches de Bordeaux
Domaine de la Grande Ferrade
F - 33140 PONT DE LA MAYE

GRANA CASTAGNETTI, M.
Provincia di Modena
Vl. Martiri Liberta, 34
I - 41100 MODENA

GRECO, P.R.
Regione Lazio-Ass. Agricultura
Via Pisana, 1301
I - CAP 00163 ROMA

GUETTIER, Ph.
Secrétariat d'Etat
à l'Environnement
14, Boulevard du Général Leclerc
F - 92524 NEUILLY

GUIDI, G.
Laboratoria per la Chimica
del Terreno
C.N.R.
Via Corridoni, 78
I - 56100 PISA

GUPTA, S.K.
Eidgenössische Forschungsanstalt
für Agrikulturchemie und
Umwelthygiene
Schwarzenburgstr. 155
CH - 3097 LIEBEFELD (BE)

HAENI, H.
Forschungsanstalt für
Agrikulturchemie
und Umwelthygiene
Schwarzenburgstr. 155
CH - 3097 LIEBEFELD (BE)

HALL, J.E.
Water Research Centre
Medmenham Laboratory
Henley Road
Medmenham
P.O. Box 16
UK - SL7 2 HD MARLOW, BUCKS

HAVELAAR, A.H.
National Institute of
Public Health
Anthonie van Leeuwenhoeklaan 9
NL - 3720 BA BILTHOVEN

HERAS, R.J.
Ayuntamiento Madrid
Barcelo, 6
E - 28004 MADRID

HUBERT, G.
IGR - EPFL
Hydrologie et Aménagements
Ecublens
CH - 1015 LAUSANNE

INDELICATO, S.
C.S.E.I.
Via Cifoli, 27
I - 95123 CATANIA

JOHANSSON, J.
Natioanal Swedish Board for
Technical Development
Box 43200
S - 100 72 STOCKHOLM

JUSTE, M.
I.N.R.A. Station d'Agronomie
Centre de Recherches de Bordeaux
Domaine de la Grande Ferrade
F - 33140 PONT DE LA MAYE

KISTLER, R.
Swiss Federal Institute of Technology
IVUK
ETH-Zentrum
CH - 8092 ZURICH

KRATZER, D.
S.I.E.C. Vevey-Montreux
Quai Maria Belgia, 18
CH - 1800 VEVEY

KUNTZE, R.
Niedersächsisches Landesamt
für Bodenforschung
Bodentechnologisches Institut
Friedrich-Missler-Str. 46-50
D - 2800 BREMEN

LANG, J.
M.A.F.F.
Room 378
Great Westminster House
UK - SWI P2AE LONDON

LATOSTENMAA, H.
National Board of Waters
P.B. 250
SF - 00101 HELSINKI

LENER, M.
ENIRICHERCHE
I -00015 MONTEROTONDO (ROMA)

LESCHBER, R.
Institute for Water-, Soil and
Air-Hygiene
Federal Health Office
Corrensplatz 1
D - 1000 BERLIN 33

LEVI MINZI, R.
Istituto di Chimica Agraria
Università degli Studi di Pisa
Via S. Michele degli Scalzi, 2
I - 56100 PISA

L'HERMITE, P.
Commission of the European
Communities
DG Science, Research &
Development
200, rue de la Loi
B - 1049 BRUSSELS

LIBERTI, L.
Istituto di Ricerca sulle Acque
Via F. De Blasio 5
I - 70123 BARI

LINERES,
I.N.R.A. Station d'Agronomie
Centre de Recherches de Bordeaux
Domaine de la Grance Ferrade
F - 33140 PONT DE LA MAYE

LOLL, U.
Abwasser-Abfall-Aquatechnik
Heidelbergerlandstrasse 52
D - 6100 DARMSTADT

LORE, F.
Istituto di Ricerca sulle Acque
Via F. De Blasio 5
I - 70123 BARI

LOTITO, V.
Istituto di Ricerca sulle Acque
Via F. De Blasio 5
I - 70123 BARI

LUBRANO, L.
Laboratoria per le Chimica del Terreno
C.N.R.
Via Çorridoni, 78
I - 56100 PISA

LUND, E.
Royal Veterinary &
Agricultural University
Bülowsvej 13
DK - 1870 KØBENHAVN

MANRIQUE, A.
Delegación Territorial de Agricultura
Sección de Análisis Ambiental
Servicio Investigación Agraria
Avd. General Vigon 23
E - 09006 BURGOS

MANSTRETTA, M.
Enichem Agricoltura
RISP
Via Medici del Vascello, 26
I -20138 MILANO

MARAZITI, A.
Am.m.ne Prov.le
V. Telesio, 21
I - PERUGIA

MARTINI, P.
A.C.E.A.
Ple Ostiense, 2
I - 00154 ROMA

MASCETTI, A.
COMODEPUR Spa
Viale Innocenzo XI, 50
I - 22100 COMO

MATTHEWS, P.J.
Anglian Water Authority
Ambury Road
Huntingdon
UK - PE18 6NZ CAMBS

MEIJER, H.A.
Water Authority Hollandse Eilanden
en Waarden
P.O. Box 469
NL - 3300 AL DORDRECHT

MEIJER, T.
Waterschap Zuiveringschap Limburg
P.O. Box 314
NL -6040 AH ROERMOND

MEZZANOTTE, V.
Università degli Studi Milano
Istituto di Chimica Agraria
Via Celoria 2
F - 20100 MILANO

MININNI, G.
Istituto di Ricerca sulle Acque
C.N.R.
Via Reno 1
I - 00198 ROMA

MOLINAS, G.
A.C.E.A.
Ple Ostiense, 2
I - 00154 ROMA

MUNTAU,
Commission of the European
Communities
Joint Research Centre
Ispra Establishment
I - ISPRA (VARESE)

NEGRI, M.C.
Ecooleco SpA
Cassinazzo di Baselico
I - 27010 GIUSSAGO (PV)

NEWMAN, P.J.
WRC Environment
Henley Road
Medmenham
P.O. Box 16
UK - SL7 2HD MARLOW BUCKS

NIELSEN, V.C.
Ministry of Agriculture, Fisheries
and Food
ADAS Farm Waste Unit
78-81 Basingstoke Road
UK - RG2 OEF READING - BERKSHIRE

OCKIER, P.
Waterzuiveringsmaatschappij van
het Kustbekken
Zandvoordestraat 375
B - 8400 OOSTENDE

OTT, H.
Commission of the European
Communities
DG Science, Research and
Development
200, rue de la Loi
B - 1049 BRUSSELS

PAGLIAI, M.
Laboratoria per la Chimica
del Terreno
C.N.R.
Via Corridoni, 78
I - 56100 PISA

PARAINI, G.
Ist. Agronomica - Univ.
Via S. Michele degli Scalzi
I - 56100 PISA

PASSINO, R.
Istituto di Ricerco sulle Acque
C.N.R.
Via Reno 1
I - 00198 ROMA

PESCHECK, E.
Bundesministerium für Umweltschutz
Stubenring 1
A - 1010 WIEN

PETRUZZELLI, G.
Istituto Chimical del Terreno
C.N.R.
Via Corridoni 78
I - 56100 PISA

PICCININI, S.
Centro Richerche Produzioni
Animali
Crispi 3
I - 42100 REGGIO EMILIA

PICCONE, G.
Istituto di Chimica Agraria -
Università
Via P. Giuria 15
I - 10126 TORINO

PIKE, E.B.
Water Research Centre
Elder Way
UK - SG1 1TH STEVENAGE HERTS

POGGIO, G.
C.N.R. - Istituto Chimical
del Terreno
Via Corridoni 78
I - 56100 PISA

PUOLANNE, Y.J.
National Board of Waters
Pohjoinen Rautatiekatu 21B
P.O. Box 250
SF - 00101 HELSINKI 10

RAMSAY, R.C.
Department of the Environment
Room B4/46
Romney House
43, Marsham Street
UK - SWI 3PY LONDON

RINGELE, A.
Instituut voor Hygiëne en
Epidemiologie
Juliette Wytsmanstraat 4
B - 1050 BRUSSELS

ROSSI, N.
Istituto di Chimica Agraria
Università di Bologna
S. Giacomo 7
I - 40126 BOLOGNA

TERZANO, C.
A.C.E.A.
Ple Ostiense, 2
I - 00154 ROMA

TJELL, J.Ch.
Technical University of Denmark
Dept. of Environmental Engineering
Building 115
DK - 2800 LYNGBY

TOMATI, U.
National Council of Research
Via Salaria Km 29,300
I - 00016 MONTEROTONDO SCALO (ROMA)

TOMEI, M.C.
Istituto di Ricerca sulle Acque
C.N.R.
Via Reno 1
I - 00198 ROMA

TROMBETTA, A.
Regione Piemonte
Via Principe Amedeo, 17
I - 10123 TORINO

TUNNEY, H.
The Agricultural Institute
Johnstown Castle Research Centre
IRL - WEXFORD

ULMGREN, L.
Scandiaconsult AB
Box 4560
S - 102 65 STOCKHOLM

UNMARINO, G.
Stazione Sperimentale
Via N. Poggioreale, 39
I - 80100 NAPOLI

VAN DEN BERG, J.J.
Grontmij N.V.
Postbox 203
NL - 3730 AE DE BILT

VEDY, J.-C.
Ecole Polytechnique Fédérale
Ecublens
CH - 1015 LAUSANNE

VERHAAGEN, J.
Zuiveringschap West Overijssel
Klein Doesburg, 4
NL - HEERDE

VILLALOBOS, J.L.
Aguas Filtradas, S.A.
Jacometrezo, 4
E - 28013 MADRID

VISCONTI, A.
Cassa Mezzogiorno
16, Via Valnerina
I - 00100 ROMA

VOORBURG, J.H.
Rijks Agrarische Afwalwater Dienst
Kemperbergerweg 67
NL - 6816 RM ARNHEM

WEGMANN, M.-A.
Station Fédérale de Recherches en
Chimie Agricole et sur l'Hygiène
de l'Environnement
CH - 3097 LIEBEFELD

WILLIAMS, J.H.
Ministry of Agriculture
Woodthorne
Wolverhampton
UK - WV6 8TQ STAFFS

ZINGALES, F.
Ist. di Chimica Industriale
Università di Padova
Via Marzolo, 9
I - 35125 PADOVA

ZUCCHI, A.
Amn.ne Prov.le di Cremona
C.So V. Emanuele II, 17
I - 26100 CREMONA

INDEX OF AUTHORS

AHTIAINEN, M., 366
AJMONE MARSAN, F., 465
AMMANN, P., 393
ARROYO, I., 460

BARBERIS, R., 436, 441
BARIDEAU, L., 374
BECK, H., 124
BIASIOL, B., 465
BLOCK, J.C., 210
BONAZZI, G., 485
BRAMRYD, T., 381
BRUCE, A.M., 39, 552
BRUNNER, P.H., 448

CAMPBELL, H.W., 94
CAMPBELL, J.A., 543
CARRINGTON, E.G., 198
CHEN, Y., 389
COLIN, F., 112
CONSIGLIO, M., 436, 441
COPPOLA, S., 59, 407
CORTELLINI, L., 485
COTTON, A., 393, 402

DANIEL, R.CH., 532
DAVIS, R.D., 304, 543, 556, 559
DE BERTOLDI, M., 178
DE FRAJA FRANGIPANE, E., 414
DE LUCA, G., 436, 465
DE MARTIN, N., 491
DI PINTO, A.C., 76
DUVOORT-VAN ENGERS, L.E., 59

FLEMING, G.A., 304
FROST, R.C., 94

GALLI, E., 510
GARCIA DE BUSTOS, J.M., 460
GENEVINI, P.L., 427
GOMEZ, A., 112, 290

GRAPPELLI, A., 510
GRAZIANO, P.L., 491
GUIDI, G., 478
GUPTA, S., 157

HADAR, Y., 389
HÄNI, H., 157, 532
HALL, J.E., 258, 559
HANGARTNER, M., 236
HARTUNG, J., 236
HAVELAAR, A.H., 210, 556
HUBERT, G., 431

IANNONE, A., 532

JAAKKOLA, A., 366
JODICE, R., 441
JØRGENSEN, P.H., 215
JONES, J.R., 537
JUSTE, C., 290, 336

KISTLER, R.C., 448
KLOKE, A., 499

LESCHBER, R., 112, 225
LINERES, M., 290
LOLL, U., 168, 452
LOTITO, V., 76
LUBRANO, L., 478
LUND, E., 215, 225

MAIRE, N., 393
MANRIQUE, A., 460
MATTHEWS, P.J., 225
MELANEN, M., 366
MELKAS, M., 366
MEZZANOTTE, V., 414
MININNI, G., 76
MOLLER, U., 2
MUNTAU, H., 124

NAPPI, P., 441
NEBREDA, A.M., 460

NEGRI, M.C., 427
NIELSEN, V.C., 25

OCKIER, P., 470

PETRUZZELLI, G., 478
PICCININI, S., 485
PICCONE, G., 436, 465
PIKE, E.B., 198, 556

RASTELLI, R., 491
REGAMEY, P., 431
ROMANO, F., 407
ROSSI, N., 491

SALT, C., 499
SANZ, A.M., 460
SAPETTI, C., 465
SAUERBECK, D.R., 318
SCHELTINGA, H.M.J., 563
SMITH, K.A., 537
SOLDA, P., 336
SPINOSA, L., 76
STRAUCH, D., 178
STYPEREK, P., 318

TANTTU, U., 505
TARRADELLAS, J., 124
TAUZIN, J., 290
TJELL, J.C., 348
TOMATI, U., 510
TOWNER, J.V., 543
TROMBETTA, A., 436, 441, 465
TUNNEY, H., 243, 517

VAN DEN BERG, J.J., 518, 523
VILLANI, F., 407
VISMARA, R., 414
VOORBURG, J.H., 236

WEGMANN, M.A., 532
WIDMER, F., 448
WILLIAMS, J.H., 258, 537